Computational and Physical Processes in Mechanics and Thermal Sciences

COMPUTER METHODS FOR ENGINEERING WITH MATLAB® APPLICATIONS

SECOND EDITION

Series in Computational and Physical Processes in Mechanics and Thermal Sciences

A Series of Reference Books and Textbooks

Series Editors

W. J. Minkowycz

Mechanical and Industrial Engineering
University of Illinois at Chicago
Chicago, Illinois

E. M. Sparrow

Mechanical Engineering
University of Minnesota, Twin Cities
Minneapolis, Minnesota

Computer Methods for Engineering with MATLAB® Applications, Second Edition, *Yogesh Jaluria*

Numerical Heat Transfer and Fluid Flow, *Suhas V. Patankar*

Heat Conduction Using Green's Functions, Second Edition, *Kevin D. Cole, James V. Beck, A. Haji-Sheikh, and Bahman Litkouhi*

Numerical Heat Transfer, *T.M. Shih*

Finite Element Analysis in Heat Transfer, *Gianni Comini, Stefano Del Guidice, and Carlo Nonino*

Computer Methods for Engineers, *Yogesh Jaluria*

Computational Fluid Mechanics and Heat Transfer, Second Edition, *John C. Tannehill, Dale A. Anderson, and Richard H. Pletcher*

Computational Grids, *Graham F. Casey*

Modern Computational Methods, *Herbert A. Konig*

The Intermediate Finite Element Method: Fluid Flow and Heat Transfer Applications, *Juan C. Henrich and Darrell W. Pepper*

Modeling and Dynamics of Regenerative Heat Transfer, *A. John Willmott*

Computational Heat Transfer, Second Edition, *Yogesh Jaluria and Kenneth Torrance*

The Finite Element Method: Basic Concepts and Applications, Second Edition, *Darrell W. Pepper and Juan C. Heinrich*

Computational Methods in Heat and Mass Transfer, *Pradip Majumdar*

Computational and Physical Processes in Mechanics and Thermal Sciences

COMPUTER METHODS FOR ENGINEERING WITH MATLAB® APPLICATIONS

SECOND EDITION

YOGESH JALURIA

CRC Press is an imprint of the
Taylor & Francis Group, an **informa** business

CRC Press
Taylor & Francis Group
6000 Broken Sound Parkway NW, Suite 300
Boca Raton, FL 33487-2742

Printed in the United States of America on acid-free paper
Version Date: 20110427

International Standard Book Number: 978-1-59169-036-8 (Hardback)

Library of Congress Cataloging-in-Publication Data

Jaluria, Yogesh.
Computer methods for engineering with MATLAB applications / Yogesh Jaluria.
p. cm. -- (Series in computational and physical processes in mechanics and thermal sciences)
Includes bibliographical references and index.
ISBN 978-1-59169-036-8 (hardback)
1. Engineering mathematics--Data processing. 2. Numerical analysis--Data processing. 3. Engineering--Data processing. 4. MATLAB. I. Title.

TA345.J29 2011
620.00285'53--dc22 2011014687

Visit the Taylor & Francis Web site at
http://www.taylorandfrancis.com

and the CRC Press Web site at
http://www.crcpress.com

Contents

Preface to the Second Edition

Computer methods continue to be critical in the analysis, simulation, design and optimization of engineering processes and systems. Computational approaches are needed to solve the complex mathematical equations that typically arise in engineering problems, for correlating experimental data, and for obtaining numerical results that are used for improving existing processes and developing new ones. The second edition follows the basic ideas, discussions, approaches, and presentation employed in the first edition. The focus is clearly on engineering processes and systems and on the equations that characterize and describe these. Computer methods that are employed to solve these equations and the nature and validity of the numerical results obtained are discussed for a variety of problems. The main thrust is on the discussion of the various numerical methods that are available for a given problem, on the presentation of the basic aspects of the methods, discussing their applicability, efficiency and behavior, and then applying these to typical problems chosen from various engineering disciplines.

Besides discussing the solution of different types of mathematical equations, a large number of engineering examples and problems were chosen to present the choice of the method, development of the numerical algorithm and use of the computer to solve the problem. A systematic approach is followed to obtain physically realistic, valid and accurate results through numerical modeling. Examples from many different engineering areas are employed to explain the various elements involved in the numerical solution and to make the presentation relevant and interesting. Similarly, a large number of solved examples and exercises are included to supplement the discussion and to illustrate the ideas and methods presented in the text. The book continues the thinking that the basic purpose of the computational approach is to provide physical insight and to obtain inputs for analysis and design of practical systems. Thus, the solution methodology is linked to both the computer and to the fundamental nature of the problem to allow the student to appreciate the basic aspects of the numerical approach.

The book is appropriate as a textbook for engineering undergraduate courses on computer methods at the sophomore or junior levels. Because the background of students at the sophomore level may not be sufficient for some of the topics covered, such as partial differential equations, a few such topics may be avoided for sophomore students and may be included in the junior or senior courses. The book is also appropriate as a reference on computational methods for various other basic and applied undergraduate courses in mechanical engineering and in other engineering disciplines. The book will also be useful as a reference for engineers who are interested in using computer methods for analysis, simulation, design, or data analysis.

The second edition is a substantially revised and updated version of the earlier book. Recent advances in available computational facilities, both in software and in hardware, are included. In several places, the presentation has been simplified and clarified to make it easier to follow. Certainly, the main difference from the first edition is the extensive use of MATLAB®, instead of a high-level programming language like Fortran, for numerical modeling. This is done in view of the current trend in engineering education where MATLAB has emerged as the dominant environment for the numerical solution of basic mathematical equations. Much of the discussion on computer solution is thus directed at MATLAB and a large number of MATLAB commands and programs are given in the text, as well as in the Appendix, in order to facilitate the presentation as well as to provide ready access to MATLAB programs for solving exercises given in the text and other similar problems. In many cases, the programs are focused on the example or problem being considered, in order to encourage the readers to develop their own computer programs for specific problems. However, the programs can be easily modified for different circumstances and parameters. Available MATLAB functions and commands are frequently employed to generate results that can be used for comparisons with the results obtained from more detailed and versatile programs. Fortran has not been abandoned because of its continued importance in engineering and the existence of substantial software in Fortran for many complex problems. Several important Fortran programs are included in the Appendix to illustrate the ease with which one could go from one computational environment or language to another and to allow those interested in Fortran to use these for their specific problems. Additional exercises and examples are included in all the chapters. References have been added on new topics included in the book and references in the first edition have been updated.

The methods, discussions, and computer programs presented in this textbook are the result of many years of teaching computer methods to engineering undergraduate students, in required as well as elective courses. The inputs from many colleagues and graduate students, as well as undergraduate students, who took the courses from me, have been valuable in selecting the topics, the depth of coverage, the computer programs presented here and many other aspects related to computer methods for solving engineering problems. Inputs from those who have used the first edition in their courses, particularly from Professor Wally Minkowycz, have been particularly valuable. The support and assistance provided by the editorial staff of Taylor & Francis, particularly by Jessica Vakili and Jonathan Plant, have been valuable in the development of the second edition.

The book would never have been completed without the strong support and encouragement of my wife, Anuradha. Our children, Ankur, Aseem, and Pratik, as well as Pratik's wife Leslie and son Vyan, have also been sources of inspiration and encouragement for me and have contributed in their own way to my efforts over the years. I greatly appreciate the patience and understanding of my family that made it possible for me to spend extensive periods of time on the book.

MATLAB® is a registered trademark of The MathWorks, Inc. For product information, please contact:

The MathWorks, Inc.
3 Apple Hill Drive
Natick, MA 01760-2098 USA
Tel: 508 647 7000
Fax: 508-647-7001
E-mail: info@mathworks.com
Web: www.mathworks.com

Preface to the First Edition

The use of computational methods in the analysis and simulation of engineering processes and systems has grown tremendously over recent years. Increasing national and international competition has made it imperative to improve existing facilities and to develop new ones for a wide variety of applications. Because of the constraints imposed on detailed experimentation needed for design and optimization of systems, due to excessive time, manpower, and financial requirements, computer simulation is extensively employed to obtain the desired information. Analytical methods are generally very restrictive in their applicability to practical problems, and numerical methods are usually necessary. In addition to the growing need for numerical solutions to engineering problems, we have also seen substantial improvements in the computational facilities available, both in software and in hardware, over the last decade. All of these changes have made it more important than ever for engineers and engineering students to develop expertise in numerical methods and to use them for solving problems of practical interest.

In recognition of the growing importance of computer methods in engineering, many courses in engineering curricula now include the numerical solution of engineering problems on the basis of numerical analysis taught earlier at the sophomore or junior level. Generally, engineering students are first exposed to the computational procedure through a course on programming, frequently employing Fortran as the programming language. Numerical methods are then taught at a later stage to introduce the basic concepts of numerical analysis and to allow the students to numerically solve important mathematical problems such as integration, matrix inversion, root solving, and solution of differential equations. However, since the basic purpose of the computational approach is to provide physical insight and to obtain valuable information for the analysis and design of practical systems, such courses have been integrated into the engineering curricula at most universities. This implies that the solution methodology is coupled with the computer on one hand and with the physical or chemical nature of the problem on the other. The numerical procedure, as well as the results, are considered in terms of actual problems to permit the student to develop a physical feel for the numerical approach to engineering problems.

Traditionally, numerical analysis courses have been mathematically oriented. Although this orientation brings in some very important and fundamental aspects of numerical analysis, it lacks in the application of the methodology to actual problems. It is extremely important to integrate the basic understanding of the methods with their actual use on the computer. Unless the students learn to choose and implement a computational scheme on the computers available, they will not develop a satisfactory appreciation or understanding of the numerical technique. In addition, recent advances in computational facilities, such as structured programming, interactive computer usage, and graphics output, must be introduced so that the most efficient procedure is adopted for a given problem. The incorporation of problems derived from various engineering disciplines aids in this learning process and also makes it interesting and enjoyable. In addition, it reinforces the important point that the physical or chemical background of the given problem forms an important element in the selection of the method and in the evaluation of the accuracy of the results obtained.

This book, directed at computer methods for engineering, integrates the treatment of numerical analysis with the physical background of the problems being solved and with the implementation of the methods on available computers, employing several recent advances in this field. Although a large number of books are available on numerical analysis, not many satisfactorily discuss the implementation of the methodology on the computer, and even fewer discuss the implications of the physical nature of the problem in the numerical solution. This book recognizes the need for a satisfactory incorporation of these concepts into the mathematical treatment of numerical analysis. It couples numerical methods for a variety of mathematical problems with the use of these methods for the solution of engineering problems on the computer.

Numerical methods for important mathematical operations, such as integration, differentiation, root solving, and solution of algebraic systems, are discussed in detail. The solution of differential equations, both ordinary and partial, is presented. Curve fitting, which is an important consideration in engineering problems, is also discussed. A large number of problems from basic sciences and various engineering disciplines are chosen to illustrate the use of these methods. The problems chosen are relatively simple so that they can easily be understood by students at the sophomore/junior level. However, in several cases, the basic background of the problem is outlined so as to bring the important points into proper focus. The importance of the physical or chemical background of the problem in the selection of the method, the choice of numerical parameters, the estimation of the accuracy of the results, and the overall validity of the results is discussed. The book mainly uses Fortran 77 to demonstrate the implementation of the numerical methods on the computer, because of the overwhelming importance of this language in engineering applications. However, a few programs in *Basic* are also given to bring out the similarities between the two languages and the ease with which one may switch from one to the other. A discussion of other languages and important aspects in computational procedure is included. A large number of examples, with the corresponding programs, are given. The programs are written specifically for these examples, so that the students must develop their own programs for the large number of problems given at the end of the chapters. Several important features that are currently employed in computational procedure are demonstrated in these programs. Recent trends in this area are outlined, and their significance for engineering applications is discussed. The students are strongly encouraged in every way to develop their own computer programs, since this is an essential ingredient for learning computer methods.

Most of the material covered in this book has been employed by the author for courses at the sophomore and junior levels. Since the background of students at the sophomore level may not be sufficient for some of the topics covered, such as partial differential equations, this particular topic and a few sections marked with an asterisk may be avoided by sophomore students. The book can also be used at the senior level, if such a course is included in the curriculum at this level. The material included is quite adequate for a one-semester course. However, the best time to teach this course is probably at the junior level, so that the students can fully understand the material and then use it in courses taught at higher levels. The book is also appropriate for professional engineers in various disciplines and as a reference for courses that employ computational methods as an important element in the presentation. The book considers problems from diverse engineering applications, and the treatment is at a level appropriate for engineering students of all disciplines.

I owe tremendous gratitude to several colleagues and students who have contributed to my understanding and enjoyment of computational methods for engineering applications. First, I would like to thank Dr. Frank Kreith, who suggested that I write this book and contributed several very valuable suggestions on the presentation. I would also like to acknowledge several stimulating and interesting discussions on the subject with Professors Dave Briggs and Abdel Zebib. Professor Samuel Temkin provided me with tremendous support and encouragement. Dr. M. V. Karwe helped with the numerical solution of some problems. Also of considerable value was the support provided by the staff of Allyn and Bacon, Inc., particularly by Ray Short. The manuscript and its several versions were typed with great patience and competence by Diane Belford and Lynn Ruggiero.

I would like to dedicate this book to my parents, who have always encouraged, supported, and inspired me to strive for the best I could achieve. The greatest contributions to this effort have been the encouragement and support of my wife, Anuradha, and of our children, Pratik, Aseem, and Ankur, who had to bear long hours that kept me away, working on this book, with patience and understanding.

The author extends special thanks to the following reviewers whose contributions have enriched the text: Professor Clayton Crowe, Washington State University; Professor Rodney W. Douglass, University of Nebraska; Professor S. V. Patankar, University of Minnesota; Dr. James F. Welty, U.S. Department of Energy.

Author

Yogesh Jaluria is currently a board of governors professor at Rutgers, the State University of New Jersey, New Brunswick, New Jersey, and the chairman of the Mechanical and Aerospace Engineering Department. He received his BS from the Indian Institute of Technology (IIT), Delhi, India, and his MS and PhD in mechanical engineering from Cornell University. He worked at Bell Labs and at IIT, Kanpur, before joining Rutgers University in 1980.

Professor Jaluria has contributed more than 450 technical articles, including over 170 in archival journals and 16 chapters in books. He has two patents in materials processing and is the author/coauthor of seven books. He is also editor/coeditor of 13 conference proceedings, two books, and three special issues of archival journals. Professor Jaluria received the prestigious 2007 Kern Award from the American Society of Chemical Engineers (AIChE), the 2003 Robert Henry Thurston Lecture Award from the American Society of Mechanical Engineers (ASME), and the 2002 Max Jakob Memorial Award, the highest international recognition for eminent achievement in the field of heat transfer, from ASME and the AIChE.

In 2001, he was named a board of governors professor of mechanical and aerospace engineering at Rutgers University. He received the 2000 Freeman Scholar Award for work on fluid flow in materials processing, the 1999 Worcester Reed Warner Medal for extensive contributions to the engineering literature, and the 1995 Heat Transfer Memorial Award for significant research contributions to the science of heat transfer, all from ASME. He served as the chair of the Heat Transfer Division of ASME during 2002–2003. He was the editor of the ASME *Journal of Heat Transfer*, the preeminent publication in this field, during 2005–2010, and is on the editorial boards of several international journals.

1 Introduction

1.1 INTRODUCTORY REMARKS

Over the past three decades, there has been a tremendous increase in the use of computers for engineering problems. This increase has been mainly due to the growing need to optimize systems and processes in order to raise productivity and reduce costs. With increasing worldwide competition, it has become necessary to modernize existing engineering facilities and develop new ones through analysis and design. Consequently, we have seen a considerable improvement in engineering systems, particularly those related to electronic circuitry, materials processing, biotechnology, transportation, and energy generation. The concern with safety, including homeland security, and with our environment has also led to detailed investigations of existing engineering processes and to substantial improvements in many of these to reduce the impact on our environment and to make their use safer.

Because of the complexities involved in most engineering applications, analytical methods based on mathematical techniques are usually unable to provide a solution to the equations that characterize their behavior, and computational methods are needed to obtain quantitative information on physical quantities of interest. Even though analytical solutions are obtained in a few simple cases, the form of the solution itself may be quite involved, since the results are frequently expressed as a series or in terms of integrals and complex functions. In such cases, the computer is needed to extract the desired information from the analytical solution. Also, the problem may have to be solved several times with different sets of data, making it advantageous to use the computer rather than analytical methods.

There has also been a phenomenal increase in the availability of computers over the recent years. With the advent of microcomputers, such as personal computers (PCs), computational facilities have become widely available. The computational power available has also increased dramatically in individual, single-processor, machines, or *serial* computers, as well as in linked multiple machines or processors that result in a *parallel* computing cluster. There is every indication that these trends will continue, making computers even more accessible and powerful. Although most practical engineering problems still require larger and faster computers (such as supercomputers, minicomputers, or parallel computing systems), microcomputers do allow the solution of many common problems and are also useful in testing numerical procedures that may subsequently be employed on larger or parallel machines. The availability of a wide variety of microprocessors has also substantially affected the control and operation of systems through automation and expanded the reach of computational software.

Along with the revolution in computer hardware, there has inevitably been one in the available software as well, making the use of computers for scientific and engineering problems easier than ever. Thus, for a wide range of problems, the programs available in the computer library, commercially available software, or user-friendly computational environment may be used effectively. However, it is generally necessary to understand the basic techniques involved in order to modify the program for satisfactory application to a given problem. In industrial systems, the use of commercially available programs is particularly important, since the processes are often quite involved and interest lies in obtaining the needed information as rapidly as possible. For simpler problems, such as those related to individual physical and chemical processes that constitute the overall system, it is often easier and more desirable to personally write the computer program or use an appropriate computational environment, rather than use a commercially available code written specifically for a given problem. Therefore, it is important to understand computational methods relevant to engineering applications and to use them in physical problems that are of interest to various disciplines.

Computer-aided design, simulation-based design and optimization, and computer-aided manufacturing are important areas that have grown substantially in the very recent past. These areas have arisen from the need to optimize on the one hand and the growing availability of the computers on the other. They are interdisciplinary in nature, particularly simulation-based design, which is of interest in such diverse fields as electronic systems and structural design. The basic approach in this case is to numerically solve the governing equations, choose physical parameters to simulate existing processes and systems, and finally vary these parameters to optimize the design for existing and future systems. Several other similar applications of computer methods have arisen in recent years, making it imperative to link the computational approach to the physical or chemical aspects of the problem under consideration.

In view of the growth of computer usage and availability in the recent years, it is surprising that much of the mathematical background underlying numerical analysis and computer logic has been available for several centuries. Binary logic operations, which use 2 as the base, instead of 10 employed in the decimal system, and which form the basis for most present digital computing, have been known and used for quite some time. Francis Bacon used binary codes in the early seventeenth century to transmit secret messages. In 1804, Joseph Marie Jacquard used punched cards with binary codes and logic to operate looms. A mathematical theory for binary logic was developed by George Boole during the nineteenth century. Similarly, adding machines and mechanical calculators were developed centuries ago, such as the one developed by Blaise Pascal in the seventeenth century. Charles Babbage designed the first automatic digital computer in 1833, with several features similar to those of modern computers. However, this machine was never constructed.

Modern digital computers were developed largely after World War II. A high-speed electronic digital computer was developed during the period from 1945 to 1952 under the direction of John von Neumann at the Institute for Advanced Study in Princeton, New Jersey. Binary digits, which can be represented by the opening or closing of a switch, were stored electrostatically in cathode-ray tubes. Several thousand vacuum tubes were used for computer memory, which had to be again stored about a thousand times per second due to the decay of electrostatic charge. Much of the logic behind this machine has persisted in modern computers. The major advancement has been in electronic hardware, particularly in the development of transistors, integrated circuits, microelectronics, and now nanoscale devices and systems. As a result, there has been a considerable reduction in size and cost of electronic digital computers and also a substantial increase in their capability, speed, and reliability. The availability of PCs has brought computational techniques within easy access for a wide variety of problems, both for students and for professional engineers. Therefore, the coming years may be expected to improve the available computational facilities even further through the advancement in both computer software and hardware. It is also evident that PCs, with an interface with larger machines or with other machines in a parallel computing environment for more complicated problems, will continue to grow in availability and usage. Thus, it is important to learn the computational techniques relevant to engineering problems on the basis of the currently available computational facilities, while considering expected future trends as well.

Several important and useful features have been incorporated in the modern computer systems. Among the most important of these is an *interactive* use of the computer, rather than the previously common *batch* operation mode. Frequently, an interpretive compiler is used so that each program statement entered into the computer is screened for syntax errors and a message issued if any error has been committed. The interactive mode allows one to enter variables and make changes in the program, as the need arises after each run of the program. The execution may also be stopped to make modifications and then continued. Therefore, the interactive mode is very well suited for the initial stages of program development, when the testing and debugging of the program is being done, and for obtaining the trends for a wide range of input parameters. For instance, if the values of x at which a nonlinear equation $f(x) = 0$ is satisfied (known as roots of the equation) are to be determined, the interactive mode may be used very effectively to obtain the general behavior of $f(x)$ over the range of interest in x. Various values of x may be entered and the corresponding values of $f(x)$ obtained. A graph of $f(x)$ versus x may easily be plotted using available software. The information obtained may then be used to select the method for finding the roots and also to obtain suitable initial guesses for the roots. Figure 1.1 shows a few examples where the plot of $f(x)$ versus x would be particularly useful in root finding.

The batch operation mode involves feeding the complete job into the computer and then running it with no interaction with the operator until the job is executed. This mode is appropriate for obtaining the numerical results for different parametric values after the program has been developed and debugged, particularly for large programs. Other important features available with present computer systems are graphics facilities, which plot the computed results, and interfacing between various computers, which allows program development to be carried out on small computers in the interactive mode. Once the program has been completed, debugged, and tested, the numerical code may be transmitted to a larger computer or to a parallel computer system, which would generally be more efficient for computing and will have greater storage capability, and run in the batch mode to obtain the desired computed results. Of course, with the increasing computational power and storage capacity of individual machines and workstations, code developments, as well as extensive computational runs, are often carried out on the same unit.

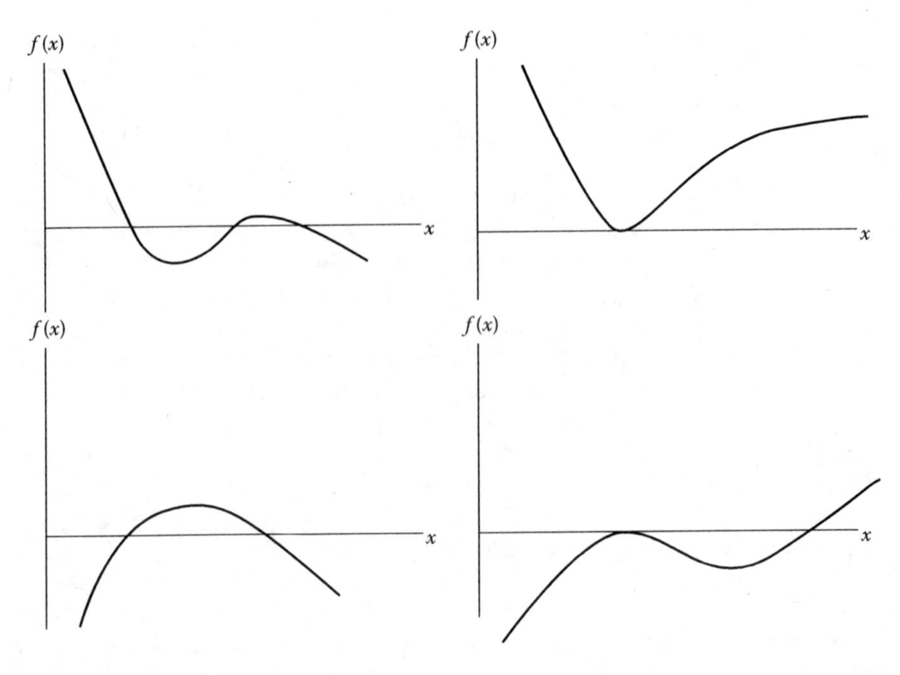

FIGURE 1.1 Some examples of the plotting of the function $f(x)$ versus x to determine the approximate values of the roots of the equation $f(x) = 0$.

1.2 NUMERICAL SOLUTION

The development of a computational procedure, or *algorithm*, to solve a given problem requires knowledge of both the available numerical methods and the methodology to interface with the computer. Since several methods are generally available for a given application, it is important to understand the applicability and advantages of each method compared to those of the other methods. For instance, a system of linear equations may be solved by a wide variety of methods, including direct methods, which give a solution in a definite number of steps, and iterative methods, which involve a repeated solution of the equations until a chosen convergence criterion is satisfied. The choice of the method for a given problem depends mainly on the nature and number of the equations. Direct methods are suitable for smaller systems and iterative methods for large sets of equations. Also, if the same system of equations must be solved several times with different constants on the right-hand side of the equality sign, methods based on matrix inversion are often preferable since the different solutions may be obtained easily once the coefficient matrix has been inverted. Similarly, in curve fitting, the method to be adopted is strongly dependent on the nature and form of the given data. If the data have been provided at uniform intervals of the independent variable, certain specialized methods may be used, taking advantage of the uniform distribution of data.

Sometimes, several methods are applicable for a given problem, and the selection of the method becomes a matter of personal choice. The previous experience with a particular method may be an important consideration in its selection. Also, the availability of certain programs in the computer library may make it advantageous to choose a given method. Many specialized methods have been developed for specific applications. Such methods are often limited in their applicability, although they may be the most efficient ones when applied to the problem for which they are particularly suited. For instance, certain methods for finding the roots of an algebraic equation are applicable only to polynomial equations and are popular choices for this application. They cannot be used for other types of algebraic equations, say, transcendental equations that involve transcendental functions such as exponential, logarithm, and trigonometric functions. Similarly, direct methods for solving systems of equations apply only for linear equations. Iterative methods are generally necessary for a system of nonlinear equations.

It is evident from the preceding discussion that the selection of the most appropriate numerical method for a given problem is an important consideration and is generally based on the nature of the problem. Once the method has been selected, one proceeds to implement it on the computer. The program is written in a programming language or in the computational environment available on the computer system to be employed. Although Fortran, with its many versions like Fortran 77, Fortran 90, Fortran 2003, and Fortran 2008, has been used extensively in engineering applications on most minicomputers and mainframe systems, *Basic*, *C*, *C++*, and other languages developed in recent years have often been used on PCs. MATLAB® is probably the most commonly used computational software being used today on both PCs and servers to solve mathematical problems that arise in engineering and scientific applications. Most of the numerical solutions discussed in this book, therefore, employ MATLAB.

The computer program written in the chosen programming language is converted into machine language by the computer. This process, known as compilation of the program, is achieved by using the relevant software, termed the *compiler*, available on the computer. An *operating system* is used for the control of the program and the computer resources. The editing of the program, for making changes and corrections, is done with the help of the editing system available on the given computer. The compilation, editing, and execution of the program are governed by the operating system of the computer and therefore vary with the machine. Similarly, the *job control language*, which interfaces the programmer with the computer, depends on the computer system. For those who may not be familiar with the terms mentioned here, Chapter 2 outlines the basic features of a computer system.

The interpretation of the numerical results obtained is also an important consideration, since it relates to the accuracy and the correctness of the numerical solution. The computational scheme may be employed to yield results for a wide range of input variables, so that the results may be considered in terms of the physical or chemical nature of the problem being investigated. If possible, a comparison is made with available analytical results in order to determine the accuracy of the computed results. The *verification* and *validation* of the numerical scheme involve ensuring that the results obtained are accurate and valid. These are particularly important if a commercially available computer program or one available in the public domain is being employed to solve a given problem. It is also important to determine the range of governing parameters over which the scheme can be used to yield accurate numerical results. These considerations are discussed in the following sections. Once the accuracy and validity of the results have been verified, the desired results may be obtained in a tabulated or graphical form.

1.3 IMPORTANCE OF ANALYTICAL RESULTS

As mentioned earlier, the equations that arise in most engineering problems are too complicated to be solved analytically, and computational techniques must be used to obtain the numerical values needed. Analytical solutions are often obtained only in very simplified circumstances. Also, as indicated before, analytical results are frequently given in terms of convergent series, integrals, and complicated functions, such as transcendental functions, Bessel functions, and so on. In engineering, we are largely interested in numerical values corresponding to given input data, and the computer is frequently needed to obtain the desired numerical information from a given analytical solution. However, analytical results, whenever available, are extremely important in evaluating the accuracy of the numerical scheme and in validating the model. Similarly, analytical results may be used to study the convergence characteristics of the numerical method and to decide if the correct solution has been obtained.

As an example, let us consider the solution of the differential equation that governs the variation, with time t, of the charge q of a capacitor in an electrical circuit that also contains a voltage source and a resistance. If the initial charge in the capacitor is Q and the voltage input, resistance, and capacitance are denoted by E, R, and C, respectively, the governing equation is obtained as follows (Young et al., 2000):

$$R\frac{\mathrm{d}q}{\mathrm{d}t} + \frac{q}{C} = E \tag{1.1}$$

If R, C, and E are constants, the preceding equation may be solved mathematically to obtain

$$q = Q\mathrm{e}^{-t/RC} + EC(1-\mathrm{e}^{-t/RC}) = EC + (Q - EC)\mathrm{e}^{-t/RC} \tag{1.2}$$

The physical problem and the analytical solution are sketched in Figure 1.2. The charge q decreases from the initial value of Q to a steady-state value of EC, if $EC < Q$. Similarly, q increases to a steady charge of EC, if $EC > Q$.

Several other physical problems are governed by equations similar to Equation 1.1. The temperature T(t) of a small, heated metal block being cooled by a stream of air, the moisture content of a wet body drying in air, and the pressure of gas in a container with an opening are often governed by equations of the same form as Equation 1.1. However, in actual practice, the parameters, such as R, C, and E, may be the nonlinear functions of the charge or voltage and may, in some cases, also vary independently with time. For instance, nonlinear conductors, such as vacuum tubes, do not obey Ohm's law, and heat and mass transfer processes operating at the surface of a given

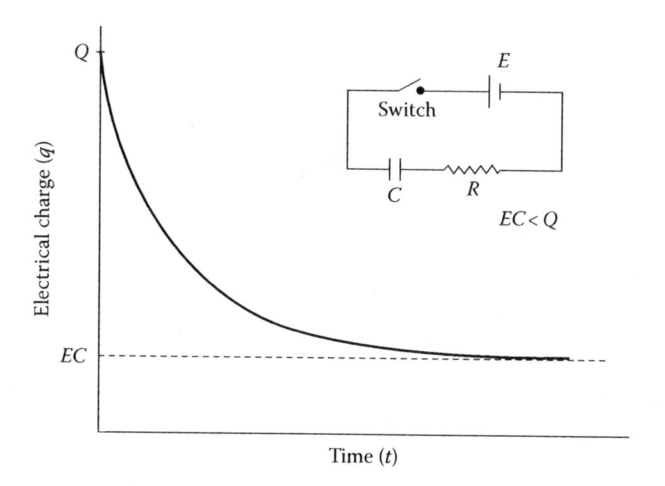

FIGURE 1.2 Variation with time t of the charge q in a capacitor, which is originally at charge Q, due to the closing of the switch in the electrical circuit shown.

object generally depend on the temperature, concentration, and pressure, making the differential equation nonlinear. The governing equation may, in general, be written as $d\phi/dt = -H(\phi, t)\phi + B$, where ϕ is the dependent variable, $H(\phi, t)$ is a functional parameter, and B is a constant. If q is replaced by ϕ in Equation 1.1, then $H(\phi, t) = 1/RC$ and $B = E/R$. This equation is linear in ϕ, or q, since H and B are constants, resulting in only the first power of ϕ to appear in the equation.

If H is not a constant but a function of ϕ as $H(\phi, t)$, an analytical solution is often not obtained because of the nonlinear expression $-H(\phi, t)\ \phi$ that arises on the right-hand side of the differential equation. In such circumstances, a numerical solution of the differential equation may be obtained by choosing a time step Δt and advancing time to compute ϕ as a function of time, starting with the given initial condition. This computation is done until an insignificant change is observed in $\phi(t)$ from one time level to the next, thereby indicating that the temperature has reached steady state, given by $d\phi/dt = 0$. However, since an analytical solution is available for the simplified circumstance of Equation 1.1, the numerical scheme should first be used to solve the problem with H taken as a constant and the computed results compared with the analytical solution. This comparison will allow determination of the anticipated accuracy of the numerical results and will also check the correctness of the procedure. Such a comparison is particularly valuable in complicated problems where an error in the numerical scheme may go undetected. Fortunately, many physical and chemical problems can be formulated in terms of idealized circumstances, which lead to simplified equations that can be solved analytically. Chapter 8 discusses several methods for solving ordinary differential equations (ODEs) and demonstrates again the importance of available analytical results.

Similarly, in numerical differentiation and integration, the computational scheme may be tested by employing simple functions whose derivatives and integrals can be obtained analytically. In radiative heat transfer, for instance, integration over the wavelength λ of the radiation is frequently needed to determine the total energy lost or gained, Q, per unit area, at a surface. The expression for Q is

$$Q = \int_{0}^{\infty} f(\lambda)\,d\lambda \tag{1.3}$$

where $f(\lambda)$ is known as the monochromatic emissive power and is often a fairly complicated function of the wavelength λ, generally obtained from a curve fit of experimental measurements. However, the radiation from a blackbody, which is an idealized circumstance, is well known in physics and is given by Planck's law, which expresses $f(\lambda)$ as

$$f(\lambda) = \frac{c_1}{\lambda^5\left[\exp(c_2/\lambda T) - 1\right]} \tag{1.4}$$

where T is the surface temperature on the Kelvin scale and c_1, c_2 are the known constants. Figure 1.3 shows the variation of $f(\lambda)$ with λ for the ideal surface of a

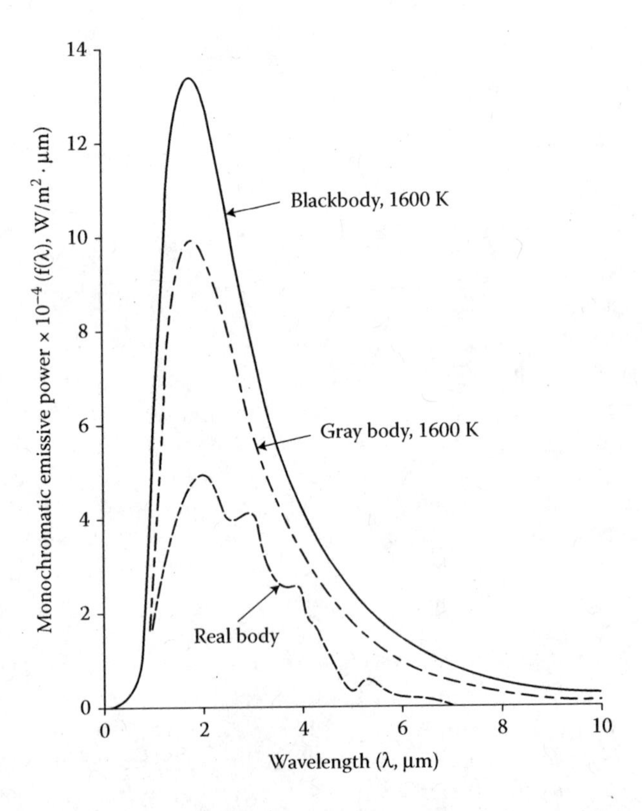

FIGURE 1.3 Variation of the emissive power $f(\lambda)$ with the wavelength λ for thermal radiation by a blackbody, a gray body, and a real surface.

blackbody, for a real or practical surface, and for a gray body for which $f(\lambda)$ is a constant fraction of that for a blackbody at all λ.

For a blackbody, the integral in Equation 1.3 has been evaluated analytically and is given by

$$Q = \sigma T^4 \tag{1.5}$$

where σ is known as the Stefan–Boltzmann constant and whose numerical value is given in the literature as 5.67×10^{-8} W/m^2 K^4. Therefore, the computational scheme developed for numerically determining Q for a wide variety of engineering surfaces, and thus different $f(\lambda)$, may first be applied to blackbody radiation and the results compared with the analytical solution given by Equation 1.5 to determine the accuracy and validity of the numerical method.

The numerical solution of large systems of linear or nonlinear equations is often needed in engineering problems. Since small sets of equations, typically three or four equations, can be solved analytically, the numerical procedure for solving systems of simultaneous algebraic equations may be employed for a small number of equations and the numerical results compared with the analytical values, to determine the accuracy and correctness of the numerical solution.

In numerical methods based on iteration, a convergence criterion ε is employed to decide when to terminate the iteration. Generally, the convergence criterion is applied to a physical variable in the problem, and computation is stopped when the change from one iteration to the next is less than the chosen value of ε. A relationship between ε and the accuracy of the numerical results may be obtained by a comparison of the computed values with the analytical solution that may be available for a simplified circumstance. This information can then be employed in the choice of the convergence criterion. If analytical results are not available, an extensive testing of the numerical procedure, over wide ranges of the initial guess, convergence criterion, and time step Δt, for example, in the problem given by Equation 1.1, must be carried out to ensure that the numerical results are essentially independent of the values chosen and that the desired accuracy level has been achieved. Figure 1.4 sketches typical computed iterative and converged solutions to the ODEs that govern a particular flow circumstance. The questions related to iterative convergence and to the choice of the numerical parameters, such as ε and Δt, are extremely important and are discussed in detail in Chapter 2.

1.4 PHYSICAL CONSIDERATIONS

The physical or basic considerations that give rise to a given mathematical expression or equation can often be used very effectively in selecting the numerical method, in choosing an acceptable solution from the several that may be obtained, and in testing the method for accuracy and correctness. In most engineering problems, the basic nature of the desired solution is known, along with the range in which it lies. Let us consider, for example, the free fall of a body of arbitrary shape in air. A terminal velocity is attained due to the balancing of the gravitational force by the frictional drag force (Halliday et al., 2004). Depending on the size and shape of the body, an

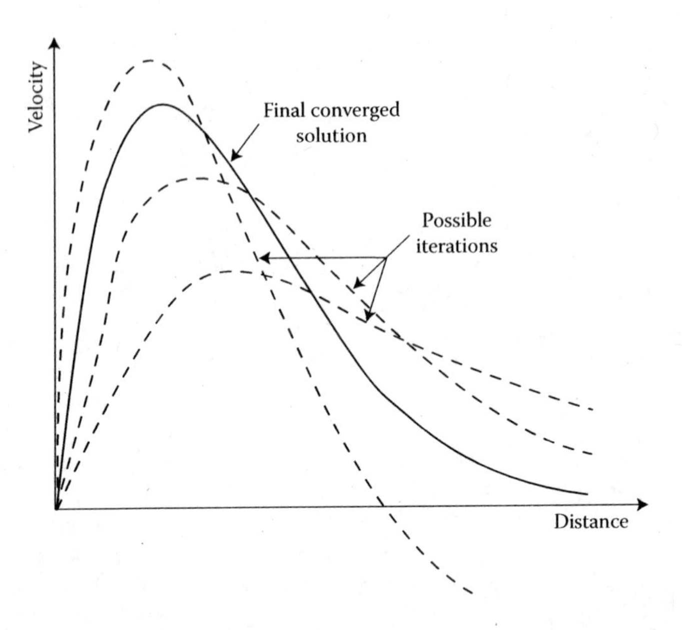

FIGURE 1.4 Typical iterations, leading to a converged result, in the numerical solution of ODEs that determine the velocity profile in a flow.

expression for drag may be obtained from considerations of air flow around the body. For a flat plate, a commonly employed expression for the frictional force is $(AV^{13/7} - BV)$, where V is the speed at which the plate is moving in stationary air and A and B are constants that depend on the length of the plate and the properties of air at the given temperature. Then, if m is the mass of the plate and g the magnitude of gravitational acceleration, the terminal velocity is the root of the equation

$$AV^{13/7} - BV = mg \tag{1.6}$$

From a physical consideration of the problem, we know that the terminal velocity must have a unique, positive value. The range in which the value lies may also be estimated from the available results for other bodies, for example, the sphere. A similar equation is obtained for bodies of other shapes and sizes. In many cases, the expression for drag is obtained from a curve fit of experimental results and is given as a fairly complicated function of the velocity V. A solution of the resulting force balance equation will then yield the terminal velocity for the given body. The method for solving the above equation may be selected knowing that the root is real, distinct, and positive. As discussed in Chapter 5, the secant method and the Newton–Raphson method are two efficient computation schemes that may be employed for this problem. If a method that determines all possible roots of the equation is used, the physical considerations are employed in choosing the correct solution. Since the solution is expected to be unique, the other roots must be complex numbers, negative or beyond the expected range of values.

The physical background of the mathematical problem being solved numerically is particularly important in the solution of nonlinear equations, such as the polynomial equation, Equation 1.6, or transcendental equations. Some examples of the latter are as follows:

$$\tan x = \frac{B}{x} \tag{1.7}$$

$$\log x + 2x^2 = 4 \tag{1.8}$$

$$e^x + x^2 - 2x = 2 \tag{1.9}$$

Nonlinear equations arise very frequently in engineering problems, such as those related to fluid flow, heat transfer, chemical reactions, and dynamics of bodies. The problems encountered may involve finding the roots of a given nonlinear equation or solving a system of nonlinear equations. Since the characteristics of nonlinear equations are generally much more complicated than those of linear equations and since several solutions are feasible, the physical aspects of the problem are used in the development of the computational procedure and in deciding which solutions are acceptable. Even for solving a system of linear equations by iterative methods, physical considerations are often important in obtaining the starting values. Linear and nonlinear equations are also frequently obtained in the numerical solution of partial differential equations (PDEs). The physical nature of the quantities to be computed is usually employed in the choice of the method, the initial guess, the grid to *discretize* a computational region, the desired accuracy level, and the convergence criterion for the termination of the numerical scheme. Since analytical solutions are rarely available, the numerical results obtained are generally considered in terms of the fundamental nature of the problem in order to determine the validity of the numerical scheme.

Curve fitting is another area in which the physical or basic considerations underlying the given problem are of particular importance in developing the computational scheme. Numerical methods are generally used to obtain the best fit to a given set of data. In such cases, it is important to know the expected trends on the basis of the physical aspects of the problem, so that the best fit obtained is a true representative of the process involved.

Consider, for example, the mean daily ambient air temperature at a given location. We wish to obtain a mathematical expression from the 365 data points that represent the measurements of the average daily temperature over a year. We could obtain a 364th-order polynomial from the given data. However, to do so would involve a substantial computational effort, both in obtaining the polynomial and in the subsequent usage of the polynomial in relevant problems. Moreover, the air temperatures fluctuate due to environmental disturbances. Consequently, we are interested in obtaining an expression that represents a best fit to the data and also characterizes the variation over the year. Since we know that the variation is periodic, with a time period of 365 days, we may try to fit the measurements to a sinusoidal variation. Examples of some of the distributions that may be employed are as follows:

$$T_a = A \sin [\omega(t - a)] \tag{1.10}$$

$$T_a = A \sin \omega t + B \cos \omega t \tag{1.11}$$

$$T_a = A \sin \omega t + b \sin 2\omega t \tag{1.12}$$

where T_a is the ambient temperature; ω is the frequency, given as $2\pi/365$; t is the time in days; and A, B, and a are constants to be determined numerically from a best fit. The first equation is frequently used, with fairly satisfactory results. Figure 1.5 shows the resulting curve fit qualitatively. Similar considerations are employed in obtaining empirical correlations from experimental data and for representing material property data, such as those of interest in thermodynamics, by a best fit.

Numerical simulation of engineering systems is important in design and optimization. It involves the mathematical modeling of components and physical or chemical processes that comprise the given problem to simplify the problem, followed by a numerical solution of the governing equations obtained. The input parameters, initially chosen on the basis of available data, are varied until a close agreement between the physical system and the numerical model is obtained. Once an existing system or process has been numerically simulated, the effects of variations in design on the performance of the system may be studied numerically, leading to optimization. At various stages in such a study, the physical or chemical aspects of the problem are employed. In fact, the comparison between the numerical model and

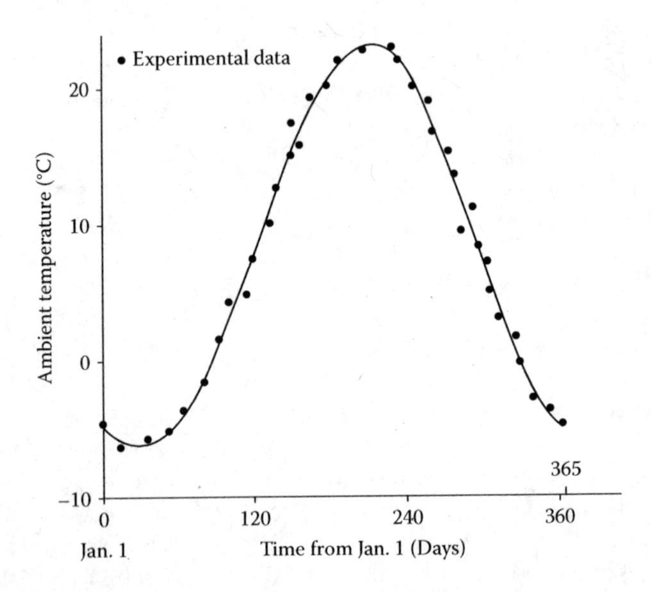

FIGURE 1.5 Sketch of the best-fit curve to the experimental data on the ambient temperature variation over the year at a given location.

the actual system forms the basis for the development of the numerical scheme and for the study of the numerical results obtained.

Therefore, in the presentation of numerical methods for engineering problems, actual problems need to be considered, in order to demonstrate the importance of the physical background of the problem in the selection of the method and in determining if the numerical results are accurate and valid. The general features of the various methods are important and must also be studied in detail. However, some of the important aspects can be best understood in terms of the underlying physical considerations. Therefore, simple examples from several areas of engineering interest are employed in this book.

1.5 APPLICATION OF COMPUTER METHODS TO ENGINEERING PROBLEMS

Computational techniques are used in engineering for a wide variety of applications. Several examples of problems that are generally solved on the computer have been given in the preceding discussion. Numerical methods for engineering application may best be considered in terms of the various mathematical problems that commonly arise in engineering. Computer methods for the solution of these problems may then be considered, using examples of mathematical expressions and equations from various engineering disciplines. This approach would allow a consideration of the various methods that may be employed for obtaining the numerical solution of a particular mathematical problem, say, integration, while employing examples from engineering to bring out the importance of physical considerations in obtaining accurate and valid results. This book employs this approach to present and discuss computer methods for engineering.

Various types of mathematical equations are encountered in engineering applications, such as linear and nonlinear algebraic equations and ordinary and PDEs. Frequently, systems of equations, which are linked with each other through the unknown variables, are obtained. PDEs arise in areas such as heat transfer, fluid mechanics, elasticity, electrostatics, and combustion. These equations are usually solved by *finite-difference* or *finite-element* methods, which convert the problem into a system of algebraic equations by applying the PDEs at a finite number of grid points or integrating them over finite regions. ODEs are also sometimes solved by these methods. Therefore, the solution of a system of algebraic equations is very important in engineering applications, and many methods have been developed to solve the different types of equations that are frequently encountered. Sets of algebraic equations are also directly obtained in many physical problems, such as those of interest in thermodynamics, economics, vibrations, structural analysis, and electrical networks. Although linear systems are particularly important, many engineering problems result in systems of nonlinear equations, which must be solved iteratively to obtain the solution. However, in most cases, nonlinear systems are formulated so that the methods for linear equations may be employed iteratively to converge to the desired solution.

In many engineering problems, the roots of a nonlinear algebraic equation, transcendental or polynomial, are to be determined. Such problems arise, for instance, in the determination of the temperature of a body from an energy balance, the terminal velocity of a body falling under gravity, the density of a gas from its equation of state, and vibration frequencies from the characteristic equation of a given system. Again, various methods are available, some of which are applicable only to polynomial equations, while others may be used for finding the real or complex roots of other types of equations. Depending on the nature of the problem, the appropriate method may be selected. If not much prior information is available on the nature and approximate magnitude of the roots, the general behavior of the function $f(x)$ that constitutes the given equation, $f(x) = 0$, where x is the unknown, may be investigated numerically. The numerical method for the solution may then be chosen on the basis of the information obtained on the variation of $f(x)$ with x.

ODEs are important in several areas of engineering interest, such as heat and mass transfer, dynamics, fluid flow, chemical reactions, electrical circuit analysis, and elasticity. In some cases, PDEs can be transformed into ODEs. Frequently, several ODEs that are coupled through the unknowns are to be solved simultaneously. The solution procedure depends on the nature of the problem, particularly on the order of the equation, that is, the highest-order derivative in the equation, and the boundary conditions. For instance, the following second-order ordinary differential is obtained for a resonant electrical circuit:

$$A\frac{d^2V}{dt^2} + B\frac{dV}{dt} + V = 0 \tag{1.13}$$

where V is the voltage across a capacitor, A and B are constants that depend on the resistance, inductance, and capacitance in the circuit, and t is time. If the initial conditions are given as

$$V = V_0 \quad \text{and} \quad \frac{dV}{dt} = 0 \ \text{ at } t = 0 \tag{1.14}$$

we have an *initial-value* problem, in which the integration of the equation may be started at the given time $t = 0$ and incremented to larger time to obtain the solution. If one of the conditions is given at a different time, a *boundary-value* problem is obtained, in which a correction scheme is needed to satisfy the given conditions. Similarly, the boundary conditions may be given at two different spatial locations, or two different values of the independent variable. Then, iteration is generally employed to converge to the solution.

Besides algebraic and differential equations, several other mathematical problems arise in engineering. Numerical differentiation and integration are needed in many cases, often as part of a more complicated problem. Numerical integration over time is needed, for instance, in determining the total energy lost or gained by an object, such as at the surface of a lake. Similarly, integration of velocity across a cross section of a channel gives the total volume flow rate in the channel. Numerical differentiation is needed, for example, in the determination of the acceleration of a particle from the measured variation of its velocity with time. Rate processes are important in engineering, and numerical differentiation is frequently employed for obtaining the rates of change of various physical quantities. Numerical techniques are also needed in interpolation and extrapolation, employing curve fitting of given data. In some cases, an exact fit which yields the exact value at the given data points is appropriate. However, more frequently, a best fit of the data is employed so that the general features of the results may be represented by a correlating equation, without forcing the curve to pass through each data point, as seen earlier in Figure 1.5. Software for graphics can be employed advantageously with the computer solution of engineering problems to present the numerical results.

In summary, a consideration of numerical methods for engineering application involves a wide variety of mathematical problems, as outlined here. It is important to understand the advantages and limitations of a particular method for solving a given problem. The numerical procedure and the results obtained must also be related to the physical or basic background of the problem in order to ensure the validity of the computational scheme and to choose an acceptable solution. Similarly, a comparison between the numerical and analytical results must be made, whenever possible, to check the accuracy of the results obtained. The development of the numerical scheme for a given problem may be discussed in several ways. A practical approach is to take the mathematical problem arising from the actual circumstance, present the computer program, and discuss the numerical results in terms of the physical aspects of the problem and available analytical results. It is this approach that is followed here. The computational software chosen is MATLAB, which is presently the most widely used computational environment for the application of computer methods to engineering problems. However, other languages and software may also be employed by suitably modifying the given programs, as discussed in Chapter 2. Of particular importance in the use of numerical techniques for solving engineering problems is the need to check the computational scheme for accuracy and to correctly interpret the numerical results obtained. In this book, these and other aspects mentioned earlier will be considered in terms of various examples taken from several engineering disciplines, including aeronautical, chemical, civil, electrical, industrial, and mechanical engineering.

1.6 OUTLINE AND SCOPE OF THE BOOK

1.6.1 Basic Features

This book presents the mathematical background as well as the application of computational techniques to problems of engineering interest. The material is developed by the derivation of the formulas for each method, followed by a discussion of the accuracy, computational effort, storage requirements, and range of applicability of the method. For each problem area considered, for example, root solving, several methods are discussed, emphasizing the ones that are most extensively employed. A comparison between various methods applicable for a particular type of mathematical problem is made, in order to indicate the advantages and disadvantages of a given method. Of particular interest in such a comparison are the associated errors, ease in programming, computing time and storage needed, and flexibility in the application to a wide variety of problems. The circumstances under which a given method would be the preferred one are outlined. This consideration is an important one, since several methods are frequently available for problems that arise in engineering applications and the choice of the most appropriate method is highly desirable, in order to minimize the computing resources needed and to obtain the required accuracy level.

Following a detailed discussion of the mathematical background and the derivation of the relevant formulas for each numerical method, the computational procedure for applying the technique is discussed. The important considerations underlying the development of the numerical scheme are discussed, along with the difficulties that may be encountered. Appropriate MATLAB commands and schemes are outlined, whenever appropriate, or reference is made to programs in Appendix B to illustrate the numerical solution. Finally, examples based on actual engineering or mathematical problems are given, for most of the methods considered, and the computer program is outlined. Again, the important features of the program are discussed and the numerical results obtained are presented and discussed. The emphasis is on presenting the basic algorithm of the method in terms of its application to an actual physical, chemical, or mathematical problem. Although the program is discussed as part of the example and is, therefore, geared to the solution of the specific problem considered, a few modifications in the program can easily be made to use it for the solution of other problems of similar nature. This approach of writing a problem-oriented computer program presents the program simply as a sample and encourages the reader to write his or her own program on the basis of the information given, making the program as efficient as possible and employing ongoing improvements in available computational facilities. General programs that can be used for a wide range of problems are also presented in many cases.

1.6.2 Computer Programs

Many useful features are incorporated in the computer programs given in the book. Both interactive and batch operation modes are utilized. In the former case, the input data are fed and the results are obtained interactively by the operator. This makes an interactive use of the computer preferable for short computer runs and for program development. The batch mode, in which the entire program is entered with the input data and the computer gives the results after the complete run, is preferred for large runs and complicated programs, after the program has been developed, tested, and debugged. Although most programs are written for the MATLAB environment, several programs are also given in Fortran, in order to indicate the similarities and differences between these and to demonstrate the ease with which the basic logic of the program can be employed in a different language or environment. Also, Fortran continues to be an important programming language for engineering problems. Subroutines or function files are useful in developing complicated programs and are employed wherever appropriate. In some cases, the outputs are stored in data files for future analysis or plotting and, in others, these are printed or plotted as soon as the computational runs are completed.

1.6.3 Examples and Problems

The examples and problems considered in this book are derived from topics of interest in the major engineering disciplines and in the basic sciences. The physical or basic background of the problem is outlined in order to enable the reader to follow the relevance of these considerations in the choice and testing of a particular numerical technique. Also, a selection of problems that arise in practical circumstances makes the discussion interesting and relevant to engineering applications. As discussed earlier in this chapter, numerical solutions must be considered not only in terms of the basic nature of the given problem but also in terms of any analytical solutions available, even if these are for very simple situations. These aspects are stressed in evaluating the numerical results for accuracy and validity. In solving problems of engineering interest, the available information on the given system or process must form the basis for the development of the numerical scheme and for the verification of the results obtained.

Both the problems and the examples tend to expand on the material covered, so that they contribute to an increased understanding of the discussion given in the text. Several new physical phenomena are also introduced in the problems to indicate the application of the methods presented to a much wider spectrum of engineering processes. Although the emphasis is, obviously, on the numerical solution, several problems are also directed at the mathematical background, particularly at the errors involved and the mathematical formulation for a numerical solution. In addition, many problems can be solved on a calculator in order to study a given numerical scheme.

Much of the material presented in this book has been used in courses taught at the sophomore and junior levels in engineering. A few of the topics covered may be somewhat advanced for sophomore students. Similarly, the physical background of the problems may not be familiar to some of the readers. Consequently, a brief discussion of the important aspects of the problem or example under consideration is included. In some cases, reference is also made to books that can be consulted for a more detailed coverage of the topic. A background in programming, such as a freshman-level, one-semester course, is assumed, although some of the important aspects are covered in Chapters 2 and 3 for completeness.

1.6.4 A Preview

The presentation of the numerical techniques for engineering application starts with Chapter 2 on the basic considerations in computer methods. This chapter outlines the important elements in computational procedure, including program development, numerical errors, accuracy, convergence, and other basic aspects. Although some of the discussion will be quite familiar to those experienced in computer programming, many of the aspects considered in this chapter are important in obtaining an accurate and valid solution to a problem of engineering interest. This chapter also outlines the current trends in computational methods and facilities, with respect to both the software development and the growing capability of computer systems.

A brief review of MATLAB is presented in Chapter 3 in order to discuss the main features of this computational environment. Commonly used commands and the basic procedures to develop a program in MATLAB are outlined. Standard software that can be used advantageously to solve mathematical problems, such as matrix inversion, root solving for polynomial equations, solution of a system of linear equations, and obtaining a best fit from given data, is presented and discussed. Since plotting of data is easily done in MATLAB, some simple plotting methods are presented. This chapter serves to give a brief discussion of programming in MATLAB, while referring to more extensive presentations in other books, and also outlines the terminology and nomenclature to be used in later chapters

The *Taylor* series, which forms an important element in the estimation of numerical truncation errors (TEs), is presented in Chapter 4, along with the numerical approximation of derivatives. Several methods for differentiation are presented, and many of the results presented here are employed in later chapters. Methods for finding the roots of nonlinear algebraic equations are discussed in Chapter 5. Several methods, which are based on the sign change, at the root, of the function $f(x)$ in the given equation $f(x) = 0$, are first considered. Efficient methods such as the secant and Newton's methods, which converge very rapidly, although they may also diverge in certain cases, are discussed in detail. Specialized methods for equations in which $f(x)$ is a polynomial are also discussed. Finally, a comparison between the various available methods is made.

The solution of simultaneous linear or nonlinear algebraic equations is an important problem in engineering applications and forms the subject of Chapter 6. Direct as well as iterative numerical methods are discussed, the latter being the inevitable approach for most nonlinear equations. Eigenvalue problems are also considered and the available methods outlined. Numerical methods for curve fitting of data are presented in Chapter 7, considering both the exact fit as well as the best-fit approach. Various techniques for interpolation are discussed, emphasizing popular methods such as Lagrange and Newton's interpolating polynomials. The least-squares method for a best fit is discussed in detail, and various forms of the function for curve fitting are considered.

Numerical integration forms the subject of Chapter 8, and several important methods, such as the trapezoidal and Simpson's rules, Romberg integration, and Gaussian quadrature, are discussed. The advantages of each method, its limitations, and the conditions under which it is preferred are considered in some detail. The associated errors and the resulting accuracy are also discussed. The numerical integration of improper integrals, whose limits of integration may be infinite or the integrand may become singular over the range of integration, is also presented.

The solution of differential equations is an important subject in engineering. Because of the complexity of typical engineering problems, numerical methods are generally needed. ODEs are considered in Chapter 9 and PDEs in Chapter 10. Both self-starting methods, such as Euler's and Runge–Kutta methods, and multistep methods, such as predictor–corrector methods, are considered for ODEs. Also, the associated errors, accuracy, stability, and convergence of these methods are considered, along with their efficiency in terms of the computational effort required. Several types of equations, including initial-value, boundary-value, and systems of equations, are considered and the relevant numerical techniques are presented. Again, a critical comparison between the various methods is made in order to guide the choice of the most suitable scheme for a given problem. Finite-difference methods, derived from the numerical approximation of derivatives given in Chapter 4, are also outlined for ODEs.

PDEs are included in this book largely for junior- and senior-level students and also for professional engineers. With the introductory background presented, the material could also be used for less advanced students. The material covered in Chapter 10 considers mainly linear equations of parabolic, elliptic, and hyperbolic type. The basic nature of the equations is discussed in detail, and important numerical methods for their solution are presented. The questions of accuracy, convergence, and stability are again considered. Finite difference methods are largely considered, with a brief introduction to finite element methods, since the former is easier to understand and can be developed on the basis of the material presented in Chapter 4. The methods for treating different types of boundary conditions are also outlined.

In all the topics considered here, a large number of examples and problems are given, so as to provide a strong physical and numerical base for the computational study of engineering problems. Since the best way to learn numerical methods is by applying the techniques available to different problems and developing one's own computer code, almost all the examples and many of the exercises demand the development of the relevant program and its use for obtaining the desired numerical results. Although a calculator may be used in several cases to study the computational steps in a given method, the readers are strongly encouraged to write computer programs for the problems given, using the discussions, formulas, and examples given in the text.

As mentioned earlier, this book is largely directed at the use of the MATLAB computing environment for solving engineering problems. However, many Fortran programs are also included in deference to the continued importance of this programming language in engineering. Extensive expertise and software exist in Fortran and it continues to be widely used, particularly for complex problems. However, the student or the reader can easily focus entirely on MATLAB, if desired, or a chosen mixture of the two computing software may be employed for instruction.

2 Basic Considerations in Computer Methods

2.1 INTRODUCTION

In the numerical solution of engineering problems, there are several important aspects that need to be considered in order to ensure the validity of the chosen approach for a given problem and the accuracy of the results obtained. The computational procedure involves a consideration of the methods available for solving the given problem, the appropriate programming language, the computational environment and software being employed, the computer and its operating system, and so forth, before proceeding to the development of the numerical scheme, or algorithm, and the corresponding program. Since these considerations are fundamental to most computer methods, this chapter discusses the general approach to the development of the computational scheme. Also considered are the interfacing with available computer software and the verification and validation of the numerical results by a comparison with available analytical and experimental results, as discussed in Chapter 1.

The consideration of numerical errors and the accuracy of the results is important in the numerical solution of any given problem. The various types of errors that arise in the computational approach are discussed, along with methods that may be employed for reducing the error. The accuracy of the solution may often be estimated by comparing the numerical results with those from the analytical solution for simpler problems, since the analytical solution of the given problem is presumably not available. Frequently, satisfactory analytical results are not available for comparison. In such cases, the numerical scheme itself is first employed to check the accuracy of the numerical results by ensuring that numerical parameters, such as the chosen time step and grid size, do not significantly affect the results. This process is often known as *verification* of the numerical method. Also, the basic nature of the problem being solved can often be employed as a check on the validity of the numerical scheme and the correctness of the results obtained. The accuracy of the numerical results can frequently be evaluated by substituting the solution obtained back into the algebraic or PDE being solved to determine how closely it satisfies the equation. Several other similar procedures are generally employed to check the accuracy of the numerical solution.

Consider, for example, the dynamics of a moving body whose displacement x is governed by the ODE $dx/dt = F(x, t)$, where t is time and $F(x, t)$ is a given function. We may assume that the analytical solution is not available, since if it were, there would be no need to solve the problem numerically. However, the numerical scheme may be employed to solve a simpler equation, say, $dx/dt = -ax + b$, where a and b are constants. The mathematical solution to this equation can be obtained as $x = ce^{-at} + b/a$, where c is a constant to be determined by applying the initial condition, that is, by using the given value x_0 of the displacement at time $t = 0$ or at any other specified time; see Figure 2.1. The accuracy of the numerical method may be estimated by comparing the numerical solution for this simple problem with the analytical solution. For a more complicated function $F(x, t)$, the following considerations may be used. The physical nature of the problem demands that the displacement be real and positive. Also, it would often be known whether it is periodic or whether it must increase, or decrease, with time. This information may be employed to select the correct solution in case multiple solutions arise and also to check the validity of the numerical scheme. Once the numerical solution $x(t)$ is obtained, numerical differentiation may be used to determine dx/dt for a few selected values of t. These may then be employed to check if the numerical values of x do indeed satisfy the equation $dx/dt = F(x, t)$ to the desired accuracy level. Finally, the step size Δt employed in the numerical scheme must be reduced until a further reduction in Δt does not significantly affect the numerical results. Of course, if any experimental results are available on the given problem, these may be effectively used for evaluating the accuracy of the numerical results.

The numerical methods for the solution of several problems are based on an iterative approach, in which the solution is gradually improved, starting with an initial, guessed value until the change in the solution from one step to the next becomes less than a chosen small quantity, known as the *convergence criterion* or parameter. In such cases, the convergence of the iterative procedure is an important consideration, and it is necessary to determine the conditions under which the scheme may diverge. If a particular method diverges for a given problem, the problem can sometimes be reformulated

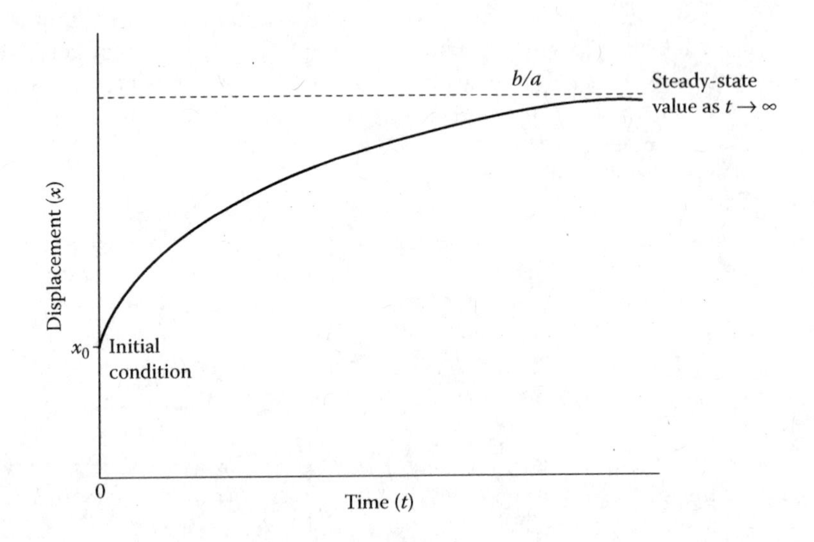

FIGURE 2.1 Sketch of the analytical solution of the differential equation $dx/dt = -ax + b$, where a and b are constants and $x = x_0$ at $t = 0$.

so that the scheme converges. Otherwise, a different method must be employed. *Numerical stability* is another important consideration that guides the selection of the method and of the grid, or step, size in the numerical scheme. Again, it is necessary to determine when numerical instability might arise and to take steps to avoid it.

This chapter discusses many of these considerations which are basic to most numerical methods. The general approach to the development of a numerical scheme is outlined, indicating various important aspects that need to be taken into account. The concepts of error, accuracy, iteration, convergence, and stability are discussed in general terms, by taking examples from various topics, such as root solving, numerical differentiation and integration, curve fitting, and solution of algebraic and differential equations, considered in greater detail in later chapters. The discussion in this chapter forms the basis for the development, application, verification, and validation of the numerical procedures for these and other topics of interest in engineering applications.

2.2 COMPUTATIONAL PROCEDURE

The general approach to the development and application of the computational procedure for solving a given problem is discussed in this section, indicating the important aspects that generally need to be considered for an efficient and accurate scheme. Although some of the considerations outlined here may not be applicable to a particular circumstance, it is important to recognize the important steps that lead to a successful numerical method. Most of the items included here are fairly straightforward and are quite familiar to those who have done a significant amount of numerical work. However, the systematic approach given here is helpful, particularly for those who are relatively less experienced in computer methods, in investigating the relevant aspects that determine the efficiency, accuracy, and validity of the numerical procedure. It is assumed that the mathematical formulation of the given physical or engineering problem has been completed and that an analytical solution is not easily obtainable, so that it has been decided to solve the problem numerically.

2.2.1 METHOD SELECTION

Frequently, several methods are available for the numerical solution of a given mathematical problem. The selection of the method to be employed, from among the several applicable methods, is an important consideration and is generally based on many relevant criteria, such as the following:

1. Accuracy
2. Efficiency
3. Numerical stability
4. Programming simplicity
5. Versatility
6. Computer storage requirements
7. Interfacing with available software
8. Previous experience with a given method

The accuracy of a given method is an important consideration in its selection for solving a particular problem. The evaluation of the accuracy of a method may be based on a comparison of the numerical results with available analytical results, as outlined in the preceding section, on an estimation of the associated numerical errors, or on various methods for checking the correctness of the numerical solution, such as substitution of the numerical results back into the equation being solved to determine the accuracy to which the numerical solution satisfies it. All these aspects, particularly the numerical errors that arise in computational methods, are discussed in detail later in this chapter.

The *efficiency* of a given method is generally based on the total number of arithmetic operations needed for solving the given problem. This is reduced to the number of arithmetic operations needed per computational step if the number of steps is fixed. One could also solve a given problem with different methods and determine the computational or central processing unit (CPU) time needed in each case, as obtained from the computer. However, the number of arithmetic operations, which include addition, subtraction, multiplication, and division, can often be determined by noting down the various mathematical manipulations performed, per step, in a given numerical scheme. If a particular method involves a smaller number of total arithmetic operations needed to solve the given problem, than another method, then it is more efficient. A higher efficiency of the method also implies shorter computer time and, thus, lower computational cost. For instance, matrix inversion methods for solving systems of linear equations, though convenient and widely used, are generally less efficient than other direct methods, as seen in a later chapter.

Numerical instability refers to the unbounded growth of numerical errors as computation proceeds. It is of particular concern in the solution of differential equations and, if present, can lead to an erroneous and unacceptable numerical solution. Therefore, it is important to determine the stability characteristics of the various methods that are applicable to a given problem. Frequently, the numerical scheme may be conditionally stable; that is, it may be stable within certain constraints that often limit the grid or step size. In the solution of parabolic PDEs, for instance, the explicit schemes, which are generally simpler to use, often restrict the step size to small values, making these schemes inefficient. Then the implicit methods, which usually do not have such constraints resulting from stability considerations, are preferred. Thus, the numerical instability of the method is an important consideration in its selection.

As listed before, several other considerations also play an important role in the selection of the method. These include simplicity in programming, versatility of the method, computer storage needed, and interfacing with available software. In engineering applications, the simplicity and versatility of the method are very important, since interest often lies in solving a wide variety of problems with the least amount of effort. This is particularly true for the design and optimization of systems that often involved a diversity of components and equations. Frequently, some sacrifice is made with respect to accuracy and efficiency in order to select a simpler and more versatile method. An example of this is the *Runge–Kutta* method, for solving ODEs. This method is often chosen over *predictor–corrector* methods, which are more efficient than the former but are also more complicated to program.

The computer storage requirements of the method are generally important in the simulation of large systems that are of interest in engineering applications. For example, the *Jacobi* method for solving a system of linear algebraic equations involves the storage of the matrices of the unknowns at two iterative steps, the present and the previous one, whereas the *Gauss–Seidel* method requires the storage of only the latest values. Thus, the latter method requires only about half the storage needed by the first method. It is also more efficient on conventional single-processor computers and is preferred.

The interfacing of the numerical method with the computer software is particularly important when available programs are being employed. For instance, if a matrix inversion program is available, methods based on the inverse of the matrix for solving a system of linear equations may be chosen. This is particularly true for MATLAB®, which has excellent matrix inversion software built into the system. Similarly, prior personal experience with a given method would be an important consideration in its selection.

2.2.2 Programming Language

After the numerical method for the solution of the given problem has been selected, the next step is the development of the computer program or code that allows one to interface with the computer system. However, before proceeding with the code development, one must select the programming language and the computer system to be used and become fully conversant with the selections made. The programming languages, often termed *high-level languages*, allow one to write the step-by-step instructions, or algorithm, for the computer in a form that is quite similar to ordinary English and algebra. The computer itself interprets and executes statements only in the machine language, and a compiler is employed by the computer to achieve the translation from the programming language to the machine language. The machine language program is then stored, providing direct access for immediate or later execution.

Several high-level programming languages have been developed over the years. In the past, the most widely used among these, for engineering and science, was Fortran, which stands for *formula translation*. It was originally developed by IBM in the 1950s for scientific and engineering applications and is now available in many versions, such as Fortran 77, Fortran 90, Fortran 95, and Fortran 2003. It is still commonly used and remains one of the important languages for high-performance scientific computing and for benchmarking and ranking the world's fastest supercomputers, partly because of extensive existing programs for a wide array of engineering problems. Fortran 90 and beyond are also well suited for use on parallel machines. Most Fortran programs are structured so that control flows from top to bottom, rather than one in which control is transferred from one point in the program to another in a seemingly random fashion. The structured system makes development as well as debugging relatively easy. Similarly, other important features, such as *object-oriented programming* that uses *objects*, which include information on the relevant data, methods, and their use to design the computer programs, have also been incorporated in recent versions. Several Fortran programs are given in this book to present the algorithm and the logic of the method, as well as to show the similarities with and differences from the MATLAB environment and to provide information for those who are well versed in this programming language. Many books are available on programming in Fortran and may be referenced for details on the language. See, for instance, the books by Metcalf, Reid, and Cohen (2004), Chapman (2007), and Chivers and Sleightholme (2009).

There are several other programming languages that have been employed for solving problems in science and engineering. These include *Basic*, *Pascal*, *C*, *Lisp*, and others. Among these, *Basic*, which stands for *beginner's all-purpose symbolic instruction code*, was also a widely used language, particularly on PCs, since it is generally simpler to use than Fortran and is well suited for small programs. However, it is not as versatile as Fortran and is often inconvenient for large, complex programs. Many improved versions of *Basic* have been developed in recent years, and many of the constraints that existed in the earlier versions have been eliminated. A useful version is *Visual Basic*, which is a relatively easy to learn and use programming language, because of its graphical development features and derivation from *Basic*.

Similarly, other programming languages have their special advantages and limitations. An important language is *C*, which is a general-purpose programming language developed in the last two decades. It is a relatively low-level language, implying that it is closer to assembly language than high-level languages such as Fortran. As a result, it is more difficult to move the program from one computer system to a different one. However, the language was designed to encourage machine-independent programming, allowing *C* programs to be compiled for a very wide variety of computer platforms and operating systems with little or no change to its source code. The language has several advantageous features in control flow and data structures because of which it is one of the most popular programming languages and is widely used on many different software platforms. *C* has greatly influenced many other popular programming languages, most notably *C++*, which originally began as an extension to *C*. For details on the *C* and *C++* languages, the books by Kernighan and Ritchie (1988), Kochan (2004), Prata (2005), King (2008), and Stroustrup (2000, 2009), among many other available books, may be consulted.

Several other programming languages have gained considerable importance in the last few years. Among these are languages that allow symbolic manipulation, that is, languages in which words, sentences, and expressions can be employed for programming. *Lisp*, which takes its name from *list programming*, is one such language that is important in the development of intelligence in computers. Similarly, *Prolog* and *Smalltalk* are languages used in generating artificial intelligence in engineering systems. For details on these languages, several references are available. See, for instance, the books by Winston and Horn (1989), Clocksin (2003), Clocksin and Mellish (2004), and Lalonde (2008).

Recent years have seen a tremendous growth in computational software, including programming languages and computational environments, making it convenient and efficient to carry out the numerical solution of the wide range of problems encountered in engineering applications. Some of these that may be mentioned are MATLAB, *Mathematica*, *SciLab*, *Maple*, GNU *Octave*, *R programming language*, and *Perl Data Language*. The more computationally intensive aspects in the software are often based on some variation of Fortran or *C*. The main computational environment used in this book is MATLAB and Chapter 3 is devoted to a brief discussion on the programming and implementation in this environment.

The computer program, written in a high-level language such as Fortran or C++, is implemented on the computer by means of an interpreter or a compiler. An interpreter examines each line of the program and checks it for the rules of the language before it is executed. The interpreted approach is very valuable during program development, since error messages are given as soon as a statement is entered. However, it is very slow in the execution of the program. A compiler, on the other hand, organizes the entire program into a set of machine instructions and locations, and several compilers are available. The compiler is often written for a given computer system and is generally a completely separate process undertaken before the program is run. Once the machine code has been produced by the compiler, the compiled program is stored and the program may be executed with a separate command. A single command that compiles and executes the program may also be used. The use of a compiler thus reduces the computer time for a given problem. Various compilers have their particular advantages and characteristics. For instance, Unix and Linux are particularly good at providing diagnostic error messages and are widely used.

From the above brief discussion of the various programming languages widely employed for engineering problems, it is obvious that the trend has been toward structured programming and interactive use of the computer, through an interpreter, which responds almost immediately, or an interactive compiler. Substantial improvements and modifications continue to be made in the available languages to simplify programming and to increase the versatility and capability of the language. Although it is difficult to keep up with all the advancements in the high-level languages, available interpreters and compilers and computational software, it is important to determine what is available on a given computer.

In general, an interactive use of the computer is preferable during program development, since the parameters of the problem may be entered by the operator at the terminal. The program may be compiled and executed to obtain the output as the program continues to execute. If the results are unacceptable, the execution may be stopped at any stage, and the input parameters varied and execution resumed. In the batch operation mode, the input parameters are part of the program, and the execution of the program must be completed before any changes can be made. Thus, at the initial stages of program development, interactive computer usage is particularly valuable. Once the program has been satisfactorily developed, detailed numerical results are best obtained by the batch operation mode on the computer.

Example 2.1

Compute the sum S of the series

$$S = 1 + x + x^2 + x^3 + \cdots + x^n + \cdots \tag{2.1}$$

where x is a variable whose value is to be entered into the program interactively. In order for the series to be convergent, $|x| < 1$. This series represents the binomial expansion of $1/(1 - x)$, which therefore gives the exact value SX of the series. Compare the exact and computed values of S to determine the numerical error. Discuss the dependence of the sum S on the number of terms n taken in the series.

SOLUTION

The value of x is to be entered and terms in Equation 2.1 are to be added sequentially. The basic considerations relevant to convergence are discussed in detail later in this chapter. However, it will suffice to mention here that each term in the series, given in Equation 2.1, is larger than the next term, for $|x| < 1$. Thus, the contribution of each additional term to the sum decreases as n is increased. This relationship is used as a check on the convergence, since it is not possible to take an infinite number of terms and since it is desirable to have the least number of terms that give S within an acceptable error. If SN represents the nth term and S the sum of the series up to and including this term, then the condition $SN/S < \varepsilon$, where ε is a chosen small quantity, such as 10^{-6}, which implies that the contribution of the nth term to the sum S is less than 10^{-4}%, can be employed to check the convergence and to terminate the computation if this condition is satisfied. The percentage error E is then given by $E = 100\ [(SX - S)/SX]$.

The preceding description of the procedure to solve the problem may be written in terms of the following steps:

1. Set the initial value of the sum S as zero.
2. Set the initial value of the term n as zero.
3. Enter the value of x.
4. Add the next term $SN = x^n$ to the sum S.
5. Check if the convergence criterion $SN/S < \varepsilon$ is satisfied.
6. If the convergence criterion is satisfied, stop and print the results on n, S, and E.
7. If the convergence criterion is not satisfied, advance n by 1 and go back to step 4.
8. Continue till convergence criterion is satisfied or a given maximum value of n is reached.

A fairly simple computer program can be written to follow these steps, as discussed below and shown in Figure 2.2 in Fortran 77. This program is presented to show the logic and the various steps involved and for those who are familiar with the language.

The program would then yield the number of terms needed for the preceding convergence criterion to be satisfied, the computed sum S of the series, and the percentage error E. Figure 2.3 presents the typical results obtained from this program. Here E is given in a format of the form 0.1E–04, or 0.1×10^{-4}, in order to check against the convergence criterion of $SX/S < 10^{-6}$. Clearly, the error is a function of ε, which may be chosen to keep the error within an acceptable value. Also, note that the number of terms needed increases with the value of x. This result is expected, since convergence is slower at the larger value of x, as discussed in most textbooks on advanced calculus; see, for instance, Larson et al. (2005) and Stewart (2007).

This is an interesting problem, which shows the effect of truncating a series after a certain number of terms and the use of a convergence criterion. The analytical result of the summation of the infinite series is known and can be used as a check on accuracy.

```
C     PROGRAM SERIES SUMMATION
C
C     HERE S IS THE SUM OF THE SERIES UP TO AND INCLUDING THE NTH
C     TERM, SN IS THE NTH TERM, SX IS THE EXACT VALUE OF THE
C     FUNCTION F(X)=1.0/(1.0-X), WHICH IS REPRESENTED BY THE
C     SERIES, AND ER IS THE ERROR.
C
C
C     ENTER INPUT QUANTITIES
C
          IMPLICIT REAL (A-H,O-Z)
          DO 5 I=1,5
          PRINT *, 'ENTER THE VALUE OF X'
          READ *, X
          N=0
          S=0.0
C
C     SUM THE SERIES
C
   1      SN=X**N
          S=S+SN
C
C     CONVERGENCE CHECK
C
          IF ((SN/S) .GT. 1E-06)THEN
          N=N+1
          GO TO 1
          ELSE
   6      WRITE (1,2)X
   2      FORMAT(2X, 'X=', F6.3)
          WRITE(1,7)N
   7      FORMAT(2X, 'THE REQUIRED NUMBER OF TERMS=',I5)
          WRITE(1,3)S
   3      FORMAT(2X, 'THE SUM OF THE SERIES=', F12.6)
C
C     COMPUTE THE ANALYTICAL VALUE OF THE SUM AND THE ERROR
C
          SX=1.0/(1.0-X)
          ER=((SX-S)/SX)*100.0
          WRITE(1,4)ER
   4      FORMAT(2X, 'THE ERROR=', E10.5,'PERCENT' /)
          END IF
   5      CONTINUE
          STOP
          END
```

FIGURE 2.2 Computer program in Fortran for the summation of the series given in Example 2.1.

```
 ENTER THE VALUE OF X
0.1
  X=0.100
  THE REQUIRED NUMBER OF TERMS=7
  THE SUM OF THE SERIES=1.111111
  THE ERROR=.10729E-04PERCENT

 ENTER THE VALUE OF X
0.3
  X=0.300
  THE REQUIRED NUMBER OF TERMS=13
  THE SUM OF THE SERIES=1.428571
  THE ERROR= .25034E-04PERCENT

 ENTER THE VALUE OF X
0.5
  X=0.500
  THE REQUIRED NUMBER OF TERMS=20
  THE SUM OF THE SERIES=1.999998
  THE ERROR= .95367E-04PERCENT

 ENTER THE VALUE OF X
0.7
  X=0.700
  THE REQUIRED NUMBER OF TERMS=37
  THE SUM OF THE SERIES=3.333328
  THE ERROR= .17166E-03PERCENT

 ENTER THE VALUE OF X
0.9
  X=0.900
  THE REQUIRED NUMBER OF TERMS=111
  THE SUM OF THE SERIES=9.999912
  THE ERROR= .85831E-03PERCENT
```

FIGURE 2.3 Results from the program in Fortran for Example 2.1.

2.2.3 Computer System

The next consideration in the numerical solution of a given problem pertains to the computer system. Frequently, several systems, ranging from PCs or workstations to minicomputers and mainframe computers, are available to engineers. Supercomputers may also be accessible for large-scale simulations of engineering systems. If several computers are available, the selection of the most appropriate one for a given problem is important. Once this selection has been made, or if only one computer system is available, one proceeds to obtain detailed information on the various elements of the system, such as the languages available, the operating system, the software available on the system, the input/output facilities, the memory/storage constraints, and the job control language, so as to implement the computer program being developed on the system.

As mentioned earlier, there are two main steps in the numerical solution of an engineering problem. The first involves the development of the computer code, and the second involves repeated execution of the program for a wide variety of input conditions and governing parameters to generate the numerical data needed for, say, the design and analysis of a given engineering system such as a furnace, a boiler, electronic equipment, a robot, a mechanical structure, or a chemical reactor. The computer requirements are usually quite different for these two steps. Code development involves frequent changes in the program and is thus best suited to an interactive use of the computer, preferably with an interpreter. The operating system, examples of which are Microsoft Windows, UNIX, and LINUX, controls the interaction with the computer, particularly the editor, and is an important component in the process. A screen editor, such as word processing programs and EMACS, which is available on many personal and minicomputers, allows one to make changes in the program very rapidly by moving the cursor to the desired location and making the needed modification. A line editor, on the other hand, allows changes to be made line by line, or in a collection of lines, and is much slower. The speed of the CPU, which finally runs the program, is not a very important consideration during code development. Similarly, the output facilities are not as important as at the second stage when computational results are being obtained, in tabular or graphical form.

Thus, during the development of the computer program, a good screen editor, which allows frequent changes and corrections in the program, is desirable. Also, the interpreter or compiler should provide adequate error diagnostics. PCs, workstations, and several minicomputers are particularly suited to code development because of the availability of most of the desirable features mentioned above.

Once the computer program has been developed, the desired numerical results for wide ranges of the governing parameters are obtained by repeatedly running the program with minor changes to enter the appropriate parametric values. Clearly, a rapid execution, with good output facilities, particularly graphics, is desirable at this stage. The editor and error diagnostics are not important. Also, an interactive use of the computer is not necessary. Thus, a batch execution of the developed program on a mainframe computer, or on a supercomputer, is the best method, particularly for large, computationally intensive programs. The program is loaded, compiled, and linked with computer memory before execution, which then proceeds rapidly.

2.2.4 Program Development

2.2.4.1 Algorithm

After the selection and the consideration of the important aspects of the method of solution, the programming language, and the computer system, one proceeds to the development of the computer program. However, before the program can be written, a step-by-step procedure, known as an *algorithm*, must be developed.

STEP 1. Start the calculation.
2. Input the limits x_1 and x_2 on x and the definition of the function $f(x)$.
3. Select the numerical parameters: Step size Δx and the convergence parameter ε.
4. Initialize: Take $x_i = x_1$.
5. Calculate the first derivative $f'(x_i)$
6. Check whether the magnitude of the derivative is within ε.
7. If $|f'(x_i)|>\varepsilon$, then advance x_i by Δx and check whether $x_i < x_2$. If $|f'(x_i)|<\varepsilon$, then go to Step 10.
8. Stop the calculation if $x_i > x_2$.
9. Calculate $f'(x_i)$ and again compare its magnitude with ε. Continue with Step 7 if $|f'(x_i)|>\varepsilon$.
10. If $|f'(x_i)|<\varepsilon$ then calculate the second derivative $f''(x_i)$.
11. If $f''(x_i)$ is positive or zero, advance x_i by Δx. Go to Step 8.
12. If $f''(x_i)$ is negative, a maximum is indicated.
13. Print the required results: x_i and $f(x_i)$.
14. Stop the calculation.

FIGURE 2.4 Representation of the algorithm for determining the value and location of the maximum of a given function $f(x)$ as a sequence of steps to be followed by the computer.

The method of solution is generally expressed in terms of the mathematical formulas involved in the computation. However, the computer must be programmed to follow a definite, logical, step-by-step procedure to perform the desired computation. The algorithm may be written as a sequence of steps to be followed. More frequently, the algorithm is represented graphically by means of a *flow chart*, which shows the steps in the form of a block diagram. Generally, a flow chart is used to outline the computational procedure, without giving the details of the actual computational steps, which are eventually entered into the actual program. Thus, a flow chart serves to indicate the logical sequence of programming steps and is frequently drawn before the program is developed.

The flow chart follows an accepted collection of symbols to represent input/output, decision, terminal, and computation. For example, let us consider the determination of the maximum of a function $f(x)$. In the optimization of engineering systems, one is frequently concerned with maximization or minimization of functions, under specified constraints. Let us assume that it is known that the given function $f(x)$ has a maximum in the range $x_1 < x < x_2$, where x is the independent variable. We know from mathematics that at the maximum, df/dx is zero and d^2f/dx^2 must be negative. Employing these characteristics of a maximum, one may write the algorithm as a sequence of steps, shown in Figure 2.4, or represented by a flow chart, shown in Figure 2.5.

For this problem, the computational procedure involves entering x_1 and x_2, advancing x with a chosen step size Δx, and computing the derivative df/dx. If the derivative is close enough to zero, as indicated by a chosen small quantity ε, a maximum or a minimum is obtained. Then the second derivative d^2f/dx^2 is computed. A maximum is obtained if d^2f/dx^2 is negative. In this case, the computation is

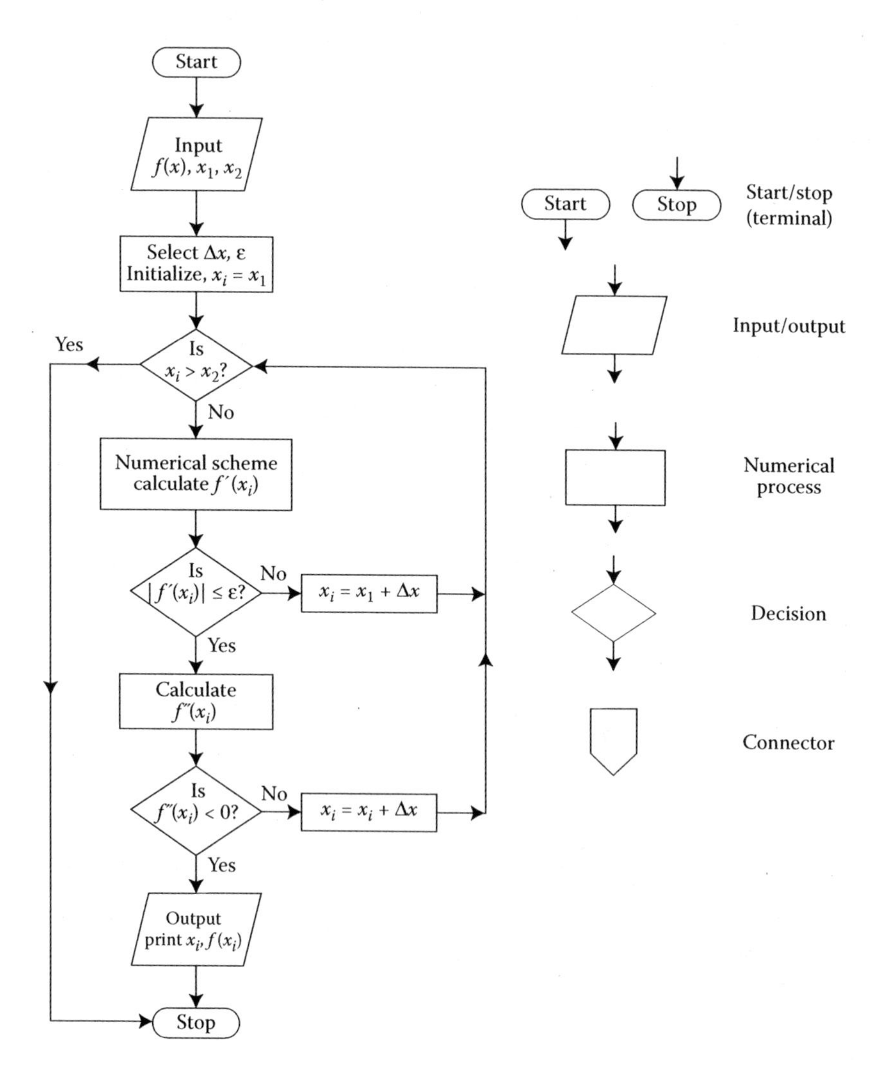

FIGURE 2.5 Flow chart representation of the algorithm outlined in Figure 2.4.

terminated and the output printed. However, if d^2f/dx^2 is positive, a minimum is indicated. A value of zero indicates a saddle or inflexion point. Then, the computation of df/dx is again carried out by advancing x until a maximum is obtained or until the upper limit on x (i.e., $x = x_2$) is attained. If a maximum is not obtained in the given domain and if $f(x)$ is known to have a maximum in the region, a larger value of ε may be selected and the procedure repeated. In fact, both ε and Δx must be varied to ensure that the location of the maximum is essentially independent of the values chosen.

As shown in Figures 2.4 and 2.5, a flow chart is a more convenient representation of an algorithm. The various symbols used for indicating the type or nature of a given step are also shown in Figure 2.5. The flow chart is a useful tool as long as it is used to give an outline of the overall process and not the detailed representations of individual steps. The numbered sequence of steps, given in Figure 2.4, can also be used instead, depending on the personal preference of the programmer. However, with experience, one could form a mental picture of the various steps in the algorithm, particularly for relatively simple problems, and proceed directly to computer programming.

2.2.4.2 Available Programs

Along with improvements in computer systems in recent years, with respect to memory and computational speed, there has been an explosive growth in software as well. A question, which is frequently asked these days, is if there is a need to develop numerical codes when many general purpose and specialized codes are easily available in the public and commercial domains. General-purpose programs such as *Fidap*, *Fluent*, *Phoenics*, *Nekton*, and *Ansys* are commercially available and can easily be used to study a wide variety of engineering problems. Software such as *Maple*, *MathCAD*, and MATLAB can be used for obtaining analytical and numerical solutions to a variety of mathematical problems such as differential equations, integration, root solving, and algebraic equations. Similarly, specialized codes such as *Polyflow* for polymer processing can be employed for specific problems and applications. In the public domain, several codes are available free of cost. These include programs for solving systems of linear equations, for solving ODEs, for inverting matrices, for curve fitting, and for providing graphical outputs of the computational results.

Commercially available software is generally expensive and usually does not provide the source code so that it is difficult to make changes in the code for a specific problem. In many cases, information on the algorithm, accuracy, discretization, convergence characteristics, range of applicability, and other important aspects associated with the software is not available in adequate detail. Despite the claims made with respect to the wide variety of problems a given software is capable of solving, one must judge each program very carefully and choose the one most suitable for a given application, keeping its cost, versatility, accuracy, and other features in mind. However, the general-purpose programs are finding wide use in industry, usually with specific changes made in the software to address the requirements of the given industry.

Computer programs in the public domain do not have many of these concerns and can often be adapted to a given computer system and linked with other software to solve a given problem. Thus, a program for solving a system of linear equations by cyclic reduction, fast Fourier transforms, or matrix decomposition may be used as part of the overall computer code to simulate an engineering problem. Certainly, software packages for producing graphical outputs are extensively used with the computational scheme generating the results. This approach of developing the core software and linking it with codes available in the public domain is a particularly attractive approach and is widely used.

Besides the easy availability of a wide variety of computer codes in the public and commercial domains that have led to considerable improvements and simplifications in numerical model development for engineering processes, several other advancements have occurred in recent years. These are expected to continue to have a significant impact on computational methods. Certainly, the most important development is that of parallel machines which employ several processors, instead of the single CPU used in traditional serial or sequential computing machines. As outlined in Section 2.2.5, multiple processors allow concurrent calculations to be carried out, resulting in a considerable speed up of the process. Similarly, considerable progress has been made in graphical representations of the results, employing color plots, contour plots, particle trajectories, two- and three-dimensional graphs, and vector field graphs, among other useful and interesting features.

The need to use supercomputers to solve complicated problems, such as those involving three-dimensional transport and turbulent flow, has led to improvements in computational techniques through vectorization of the variables, so that rather than treating each quantity in an array as a scalar the whole array is treated as a vector. Improvements in the user–computer interface, using languages such as *Visual Basic*, have also resulted in considerable ease in entering the relevant data such as geometry, operating conditions, and material characteristics. Information storage and retrieval, linking with the knowledge base on a given process or material, often using artificial intelligence techniques, and other new features in computer systems and software have had a considerable impact on traditional programming. It is expected that such advancement will continue in the future, resulting in valuable and desirable changes in the field of computational methods as well.

2.2.4.3 Validation

The final stage in the development of the computer program for solving a given problem is verification or validation of the numerical scheme. As discussed in Section 1.3, validation is done by a comparison of the numerical results with available analytical solutions and experimental results. However, the analytical solution of the problem being solved numerically is obviously not available, at least in a convenient form, making a numerical solution necessary. Therefore, the numerical scheme is generally validated by a comparison with the analytical solution available for simpler problems. For example, the algorithm shown in Figure 2.4 may be used with a simple analytic function whose maximum can easily be determined mathematically. Thus, a function such as $f(x) = 5 + 4x - 3x^3$, which can easily be shown to have a maximum at $x = 2/3$, may be chosen for the testing of the numerical scheme. The numerically obtained value may be compared with the analytical one to verify that the scheme is performing satisfactorily. Other, more complicated expressions may also be employed, if the corresponding analytical results are known, for the validation of the computer program. Similarly, experimental results are generally not available on the problem being solved. However, experimental data on similar systems or problems may be available. These data can then be used to validate the numerical solution.

2.2.5 Serial versus Parallel Computing

In this book, it is generally assumed that at a given instant only one computational step is being carried out on the computer. This assumption applies to most commonly used computers, such as PCs and minicomputers, for engineering calculations. The computational procedure in which the required calculations are performed sequentially, with each step being undertaken by the machine after the previous one is over, is known as *serial* or *sequential* computing. Thus, a single CPU is involved in the computation. However, in recent years, computers with multiple processors that allow concurrent calculations have been developed. Generally termed *parallel computers*, these machines represent the new generation of computing and have become important in the numerical simulation of complicated processes and systems.

In order to fully utilize these machines with multiple processing units, one must write the algorithm so as to employ the feature of parallel computing. Thus, statements must be given to direct various calculation steps to different units. Algorithms in which different steps are independent of each other are ideally suited for parallel computing, since each calculation step can easily be assigned to a given processor. Algorithms that involve strongly coupled steps cannot be solved very efficiently with parallel computing. Besides the calculation for each step, the processors need to communicate with each other at various stages in order to solve the overall problem. Thus, parallel computing involves developing algorithms that allow concurrent calculations and message passing between processors for greater efficiency. Depending on the problem and the algorithm, a considerable speed up of the computation can be obtained for a system consisting of n processors, a value approaching n indicating an excellent utilization of the parallel computing environment. Even though the assumption here is serial or sequential computing, the implications for parallel computing will be given at many places in the book. For details on parallel computing, see Grama et al. (2003) and Scott et al. (2005).

Example 2.2

A firm needs to borrow $50,000 to undertake improvements in its existing facilities. For the repayment of the loan, the firm wishes to pay only $1000 each month, beginning at the end of the first month after taking the loan, toward the principal and the interest. Considering possible interest rates as 8%, 10%, and 12%, determine the time required to pay off the loan for these three cases. Calculate the time required and the *future worth* (FW), or the value on the day the repayment is completed, of the money paid toward the loan. Also, determine the amount by which the final payment must be reduced to pay off the loan exactly.

SOLUTION

Let x denote the percent interest rate, so that an annual compounding yields an interest of x on $100. Then the annual interest on each dollar is $x/100$, denoted by x_1. Therefore, the FW of an amount P after n years is $P(1 + x_1)^n$, due to this interest which is compounded annually. Similarly, the *present worth* (PW), or the value today, of an amount R paid at the end of n years is $R/(1 + x_1)^n$. The concepts of PW and FW are very important in economic analysis; see, for instance, Stoecker (1989). First, we need to consider the PW of a series of uniform annual amounts R, paid at the end of each year starting at the end of the first year. If n is the total number of years, the PW of such a series of amounts is

$$\mathrm{PW} = \frac{R}{(1+x_1)^1} + \frac{R}{(1+x_1)^2} + \frac{R}{(1+x_1)^3} + \cdots + \frac{R}{(1+x_1)^n} \tag{2.2}$$

The series can be summed up to give

$$\text{PW} = R\frac{(1+x_1)^n - 1}{x_1(1+x_1)^n} \tag{2.3}$$

where $x_1 = x/100$ (since x is given as a percent).

Equation 2.2 follows from the fact that the PW of an amount P paid at the end of n years is given by PW $= P/(1 + x_1)^n$ and from the consideration of each lump-sum annual payment to yield the given series. Now, if we consider monthly payments, the total number of payments become m, where $m = 12n$, and the interest rate becomes x_m, where $x_m = x/(12 \times 100)$. Thus,

$$\text{PW} = R\frac{(1+x_m)^m - 1}{x_m(1+x_m)^m} \tag{2.4}$$

The FW of this series of amounts is obtained by simply multiplying the PW by $(1 + x_m)^m$. Therefore,

$$\text{FW} = R\frac{(1+x_m)^m - 1}{x_m} \tag{2.5}$$

Now, R is given as \$1000 and x as 8%, 10%, or 12%. We wish to compute the time, in months m, needed to repay the loan, and the FW of the total payment. The PW is \$50,000. Thus, m is to be computed from Equation 2.4, and the FW may then be obtained from Equation 2.5. The determination of m from Equation 2.4 is a root-solving problem, which will be presented in Chapter 5. Here, we shall use a very simple approach, since root-solving methods have not been discussed yet. For a given value of x_m, the value of m may be increased in steps of 1, starting with $m = 1$, and the PW computed from Equation 2.4, until the value of \$50,000 is reached. The computation stops when PW exceeds this amount, since a fixed payment of \$1000 is made each month. In practice, the monthly payment is adjusted to an appropriate value close to \$1000, so that the loan is paid off exactly.

Figure 2.6 shows the algorithm to be employed, in terms of a flow chart. The computational scheme is very simple for this problem and is based on a comparison between the PW of \$50,000 and the sum of the series in Equation 2.4, employing an increasing number of terms m. Once the latter exceeds the PW, the loan is paid off and the number of months needed is printed. Also, the FW, on the date when the loan is paid off, of the total payment made is computed from Equation 2.5. The PW of the total payment exceeds \$50,000, and the last payment may be reduced to avoid this excess payment or the monthly payments may be adjusted, as mentioned above. The FW of the loan is \$50,000 $(1 + x_m)^m$, and if this amount is subtracted from the computed FW of the payments, we obtain the amount by which the final payment may be reduced to pay off the loan exactly.

A computer program may easily be developed on the basis of this algorithm. Figure 2.7 presents a Fortran 77 program to give the logic and the various steps indicated in the algorithm.

Figure 2.8 presents the numerical results obtained from such a program. The inputs are entered and the print out gives the results, along with the input parameters to ensure that the correct values are being employed in the calculations. As seen here, the number of months needed to repay the loan increases with

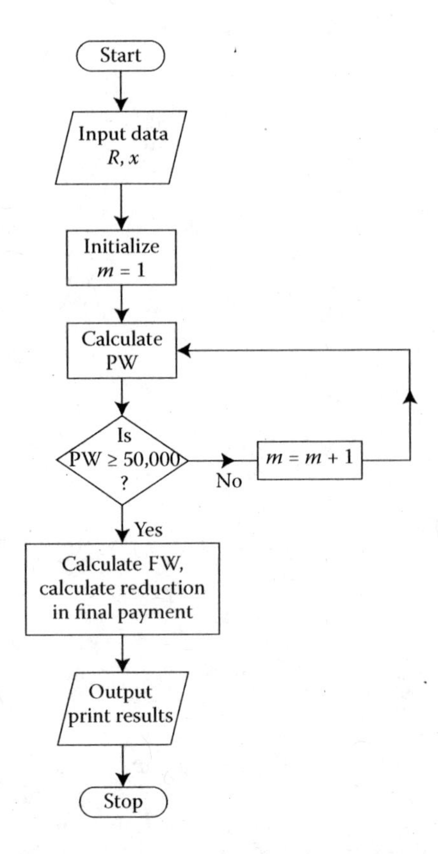

FIGURE 2.6 Flow chart for the problem in economics considered in Example 2.2.

the interest rate, as expected. Also, the FW increases. Note also that, since the monthly payment is kept constant, the total payment is more than the loan. To determine the amount needed to repay the loan exactly, subtract the FW of the loan from the FW of the total payment. This amount is the overpayment and is subtracted from the last month's payment of $1000 to obtain the reduction in the final payment if the loan is to be paid off exactly.

2.3 NUMERICAL ERRORS AND ACCURACY

A very important consideration in the solution of a given mathematical, chemical, physical, or engineering problem by computational methods is the accuracy of the numerical results obtained. The true measure of inaccuracy, or error, in the numerical solution is the difference between the numerical and the exact, or analytical, results. However, the analytical solution of the given problem is presumably not available, making it necessary to solve it numerically. Thus, alternative methods for estimating the errors involved and the accuracy of the numerical solution are needed. The dependence of the errors on the various parameters associated with the numerical procedure must also be determined, so that the accuracy of the solution may be improved by varying these parameters.

```
C                           PROGRAM ECONOMICS
C     R IS THE MONTHLY PAYMENT, X THE ANNUAL PERCENT INTEREST,
C     XM THE MONTHLY INTEREST PER DOLLAR, M THE NUMBER OF MONTHS,
C     PW THE PRESENT WORTH OF THE PAYMENTS, FW THE FUTURE WORTH
C     OF THE PAYMENTS, F THE ACTUAL FUTURE WORTH OF THE LOAN AND
C     RL THE REDUCTION IN THE FINAL PAYMENT IN ORDER TO PAY OFF
C     THE LOAN EXACTLY
C
C
C     ENTER INPUT VARIABLES
C
          IMPLICIT REAL (A-H,O-Z)
          DO 5, I=1,3
          PRINT *, 'ENTER MONTHLY DEPOSIT'
          READ (1,*)R
          PRINT *, 'ENTER INTEREST RATE'
          READ (1,*)X
          XM= X/(12.0*100.0)
          M=0
    1     M=M+1
C
C     COMPUTE PRESENT WORTH AND CHECK IF LOAN IS PAID OFF
C
          PW=R*((1.0+XM)**M-1.0)/(XM*(1.0+XM)**M)
          IF(PW.LT.50000.0)THEN
          GO TO 1
          ELSE
          WRITE(1,2)R,X
    2     FORMAT(/2X, 'MONTHLY DEPOSIT=',F9.4, 4X,' INTEREST
          RATE=',F6.3)
          WRITE(1,3)PW,M
    3     FORMAT(2X, 'PRESENT WORTH=', F12.3,4X, 'NUMBER OF
          MONTHS=',I5) C
C
C     COMPUTE THE FUTURE WORTH AND REDUCTION IN FINAL PAYMENT
C
          FW=PW*(1.0+XM)**M
          F=50000*(1.0+XM)**M
          RL=FW-F
          WRITE(1,4)FW
    4     FORMAT(2X, 'FUTURE WORTH=',F12.3)
          WRITE(1,9)RL
    9     FORMAT(2X,'REDUCTION IN FINAL PAYMENT=',F9.4//)
          END IF
    5     CONTINUE
          STOP
          END
```

FIGURE 2.7 Computer program in Fortran for the problem in Example 2.2.

```
        ENTER MONTHLY DEPOSIT
1000.0
        ENTER INTEREST RATE
8.0
        MONTHLY DEPOSIT=1000.0000    INTEREST RATE= 8.000
        PRESENT WORTH: 50647.547         NUMBER OF MONTHS= 62
        FUTURE WORTH= 76466.453
        REDUCTION IN FINAL PAYMENT= 977.6354

        ENTER MONTHLY DEPOSIT
1000.0
        ENTER INTEREST RATE
10.0
        MONTHLY DEPOSIT= 1000.0000    INTEREST RATE= 10.000
        PRESENT WORTH= 50029.789         NUMBER OF MONTHS= 65
        FURUTE WORTH= 85801.844
        REDUCTION IN FINAL PAYMENT= 51.0732

        ENTER MONTHLY DEPOSIT
1000.0
        ENTER INTEREST RATE
12.0
        MONTHLY DEPOSIT= 1000.0000    INTEREST RATE= 12.000
        PRESENT WORTH= 50168.523         NUMBER OF MONTHS= 70
        FURUTE WORTH= 100676.328
        REDUCTION IN FINAL PAYMENT= 338.1790
```

FIGURE 2.8 Numerical results obtained for Example 2.2.

There are several types of errors that arise in a computational solution. The two most important are the *round-off* (RO) and the *truncation* errors (TE). The former is related to the computer system used and to the number of significant figures retained in mathematical operations. An error is introduced in essentially every calculation since a finite number of significant figures or decimal places are retained and all real numbers are rounded off by the computer. In single precision, the number of significant figures retained ranges from 7 to about 14, depending on the computer system. The TE results from the replacement of an exact mathematical expression or equation by a numerical approximation. It refers to the difference between an exact expression and the corresponding truncated form, employed in the numerical solution. The resulting error in the solution, assuming the round-off error to be negligible, is known as *discretization* error. Of course, the discretization error is an idealization since all computational schemes would generally involve some round-off error.

2.3.1 Round-Off Error

The round-off error introduced in a given computation depends on the computer system used. The number of significant figures, and thus the number of decimal places retained, varies with the computer. In most cases, the last digit is rounded off to take into account the value of the digit after it. For example, the last retained digit is usually rounded up if the first discarded digit is 5 or larger. Otherwise, it is unchanged. Thus, if only four significant figures are to be retained, 4.3757 is rounded off to 4.376, and 4.3752 to 4.375. However, on some machines, the digits, beyond the ones that are to be retained, are simply chopped off. For many calculations, the round-off error is relatively unimportant, being much smaller than the TE, discussed in Section 2.3.2. However, it can affect the accuracy of the numerical solution and can be extremely important in certain problems.

The round-off error is fairly random in nature. If the last retained digit is rounded up, the error, obtained by subtracting the approximate value from the true value, is negative. If digits are discarded, the error is positive. Because of this random nature of the error, it does not cancel out in a given computation but rather tends to accumulate if later calculations are based on earlier ones. Thus, if a particular numerical scheme requires a large number of arithmetic operations, the cumulative effect of the round-off error can be quite significant.

It is difficult to determine the round-off error in a given numerical method. However, the error increases with the total number of arithmetic operations. Frequently, a count of the arithmetic operations in a computational step, or procedure, may be made. If a problem can be numerically solved by two methods, the one that requires a smaller number of arithmetic manipulations will have a smaller round-off error. It will also be more efficient, since the computational effort required is less. An example of such a consideration is the solution of a system of n linear algebraic equations by Gaussian or Gauss–Jordan elimination methods, discussed in Chapter 6. By counting the arithmetic steps involved in the solution, it can be shown that the former requires total arithmetic operations on the order of $n^3/3$, which is written as $O(n^3/3)$, and the latter $O(n^3/2)$. Thus, Gaussian elimination is more efficient and has smaller round-off error. Similarly, the multiplication of two $n \times n$ matrices can be shown to involve arithmetic operations on the order of n^3, or $O(n^3)$, implying greater round-off error and greater CPU time than the solution of n linear algebraic equations by the preceding methods.

Frequently, the numerical scheme involves dividing a given computational region into a finite number of subdivisions. For example, the length L of a rod may be subdivided into n divisions, where $n = L/\Delta x$ and Δx is termed the *step*, or *grid*, size along the x-direction, which coincides with the rod axis in this case; see Figure 2.9a. Thus, the total number of finite regions, or steps, is inversely proportional to Δx, implying that the number of arithmetic operations varies as $1/\Delta x$. Therefore, as Δx is reduced, the round-off error is expected to increase. This consideration is important since it indicates that the grid size may not be reduced indefinitely. In mathematical analysis, such as differentiation and integration, the desired results are obtained by taking the limiting condition of $\Delta x \to 0$. In numerical methods, however, an extremely small Δx would lead to an extremely large number of arithmetic operations and to an unacceptably high round-off error, as shown qualitatively in Figure 2.9b.

There are several circumstances for which the round-off error can be particularly important. For instance, in ill-conditioned matrices, discussed in Chapter 6, a small error in the computation due to round-off can lead to a large error in the solution. Similarly, in the solution of ODEs, considered in Chapter 9, round-off error can

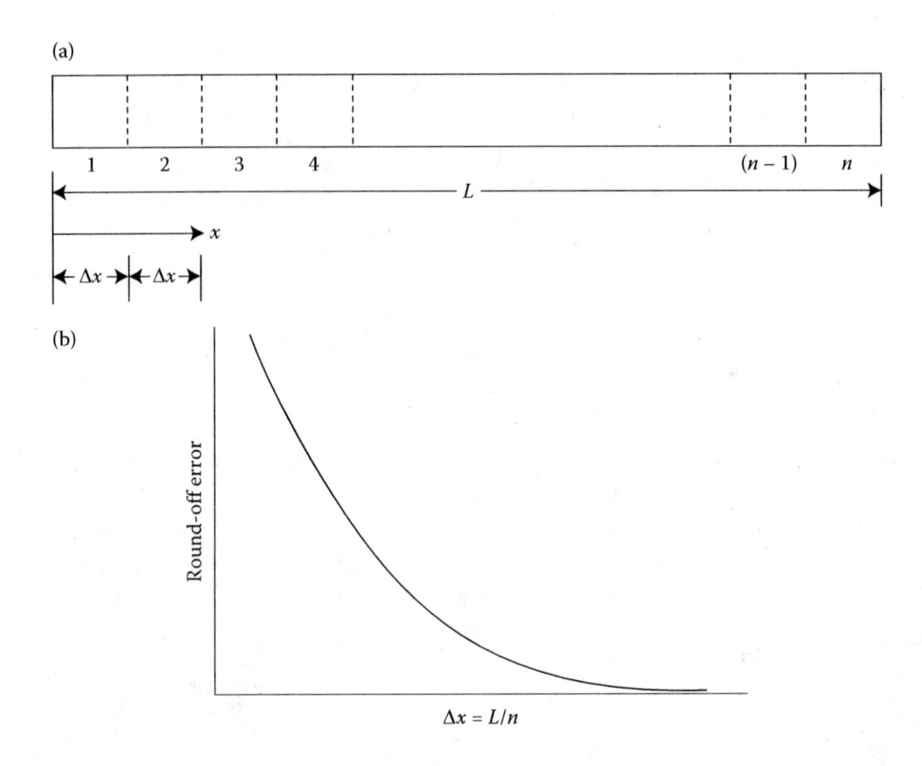

FIGURE 2.9 (a) Subdivision of a rod of length L into n intervals, each of length Δx, for a numerical scheme based on discretization of the length. (b) Qualitative representation of the variation of round-off error with the step size Δx.

accumulate and lead to erroneous results. If numerical instability is present in the scheme, the solution may be completely disrupted, as outlined later in this section. Consider, for example, the ODE $dy/dx = -1/x^2$, whose solution is $y = 1/x$, or $dy/dx = -1/(2x^2y)$, whose solution is $y = 1/\sqrt{x}$, if y is given as 1.0 at $x = 1.0$. In both cases, the solution decreases as x is increased and approaches zero as $x \to \infty$. Thus, at large x, y is small and the round-off error can affect the solution very substantially. Depending on the step size Δx, the value of x to which the solution is obtained, and the numerical scheme, the numerical solution may deviate significantly from the expected variation at large x, as shown in Figure 2.10. The accumulated round-off error is large, compared to the true solution, at these values of x. Thus, extending the computation to large x must be avoided in such cases. If numerical instability exists, the error could increase at a very rapid rate, often resulting in overflow and disruption of the solution. In many of these cases, double precision may be used to avoid the problems arising due to round-off error.

2.3.2 Truncation Error

TE is a function of the approximations used in the numerical scheme and is independent of the computer system. It arises because a function, which may be

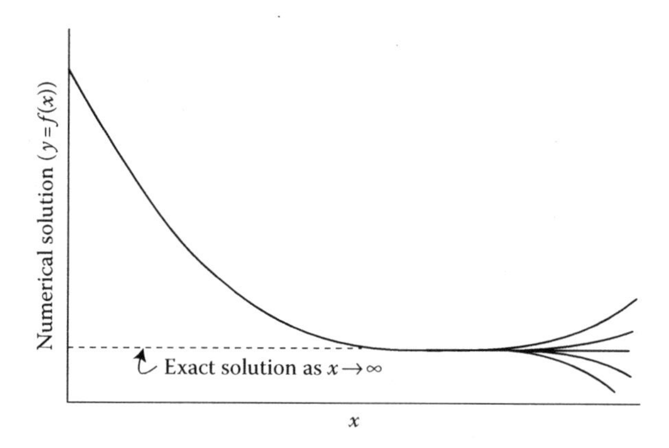

FIGURE 2.10 Possible effect of the round-off error, at large x, on the numerical solution of a differential equation, whose exact solution decays with increasing x to attain a constant value as $x \to \infty$.

represented by an infinite series, is truncated after a finite number of terms for approximating it numerically on the computer. The nature of such an approximation and the resulting error are discussed in greater detail in Chapter 4, on the basis of the Taylor series expansion of analytic functions. However, some of the important considerations are outlined here, in order to discuss the effect on accuracy and the methods to reduce the total error.

Consider, as an example, the binomial expansion of $1/(1 - x)$, as given by Equation 2.1. Then, for $|x| < 1$,

$$\frac{1}{1-x} = 1 + x + x^2 + x^3 + \cdots + x^n + \cdots \tag{2.6}$$

The variation of the function $f(x) = 1/(1 - x)$ versus x for $0 \le x \le 0.9$ is sketched in Figure 2.11. Now, the function $f(x)$ is also represented by the above infinite series. However, if the series is to be entered on a computer for representing the function, only a finite number of terms can be retained. The discarded terms, thus, give rise to the TE, which is the difference between the exact value of the function and its approximate value, obtained after truncation. Figure 2.11 shows the approximations if one, two, three, or four terms in the series are retained. Clearly, as expected, the approximation improves as a larger number of terms are retained.

Similarly, as given in most books on calculus, the function $f(x) = e^x$ may be represented by the following infinite series, which is known as the Taylor series expansion for the function about $x = 0$:

$$e^x = 1 + x + \frac{x^2}{2!} + \frac{x^3}{3!} + \frac{x^4}{4!} + \cdots \tag{2.7}$$

Again, a TE arises if a finite number of terms are used to represent the function on the computer. In numerical analysis, the computational region is often divided into a

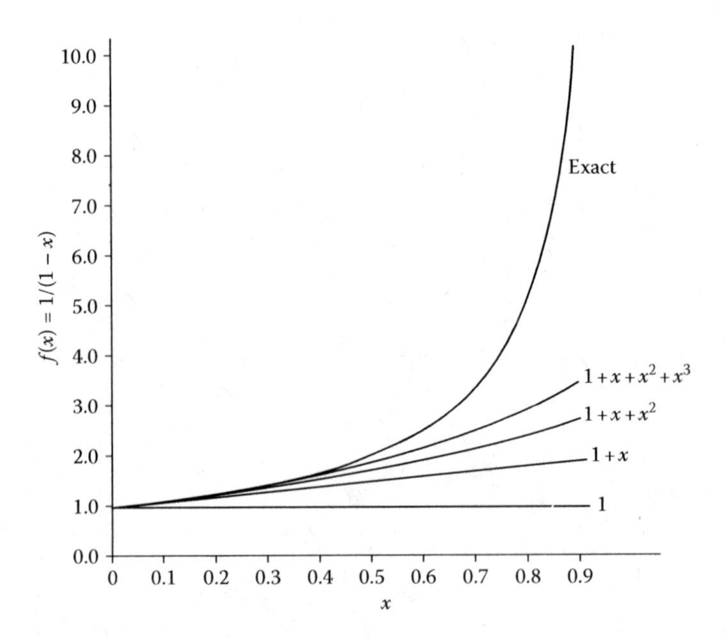

FIGURE 2.11 Approximations to the function $1/(1-x)$, represented by the series given in Example 2.1, when only one, two, three, or four terms in the series are retained. Also shown is the variation of the exact function with x.

finite number of subdivisions, as shown in Figure 2.9a. Then the numerical scheme is based on the values of a given function $f(x)$ at the finite number of grid points, and the resulting TE in the formulation depends on the grid size Δx. For example, if the series in Equation 2.7 is written for $x = \Delta x$, then

$$e^{\Delta x} = 1 + \Delta x + \frac{(\Delta x)^2}{2!} + \frac{(\Delta x)^3}{3!} + \frac{(\Delta x)^4}{4!} = \cdots \tag{2.8}$$

The TE resulting from the retention of a finite number of terms to represent $f(\Delta x)$, where $f(x) = e^{\Delta x}$, may be estimated from the above series. The error is generally written on the basis of the magnitude of the first discarded term. Thus, if only the first term is retained, the error is said to be on the order of Δx, that is, $O(\Delta x)$. Similarly, retaining two terms gives an error of $O[(\Delta x)^2]$, retaining three terms results in an error of $O[(\Delta x)^3]$, and so on. If the TE is $O(\Delta x)$, the scheme is said to be *first-order* accurate; if the error is $O[(\Delta x)^2]$, it is said to be *second-order* accurate; and so on. Since a higher-order error term indicates the retention of a larger number of terms, $O[(\Delta x)^p]$ represents a smaller TE than $O[(\Delta x)^q]$, where $p > q$. Thus, the error may be reduced by reducing Δx or by retaining more terms. The latter approach is generally known as *higher-order approximation*. A similar approach is used to derive the TE that is associated with a particular numerical scheme. Such derivations are particularly important for schemes employed in numerical integration and differentiation, and in the solution of ODEs and PDEs.

The preceding brief discussion indicates the importance of TE in characterizing the accuracy of a given numerical scheme. However, TE indicates only the error in the formulation of the numerical scheme. The resulting error in the numerical solution, neglecting round-off error, is the discretization error, as mentioned earlier. However, discretization error is much more difficult to determine than TE, since there is always some round-off error present and since the exact, analytical solution is generally not available. Consequently, the TE is generally taken as the most important measure of accuracy of a given numerical scheme.

2.3.3 Accuracy of Numerical Results

The round-off and TEs are the two main sources of inaccuracy in a numerical solution. However, several other errors may be present. An important one among these is the error due to incomplete convergence of an *iteration*, which is a frequently used approach to obtain a numerical solution. The criterion used for indicating convergence must be varied to ensure that the iterative scheme has indeed converged. Inaccurate results may also be due to errors in the input data for a given problem, in the computer program itself, or in the mathematical formulation of the physical or chemical problem. Although all these errors are important, numerical methods may be studied independently, assuming that adequate care has been taken to eliminate such errors. Thus, we shall be concerned largely with the round-off and TEs and with the resulting total error.

As discussed in the preceding sections, a decrease in the step size Δx leads to an increase in the number of computations and, thus, to an increase in the round-off error. On the other hand, the TE is reduced as the step size is reduced. The total error, resulting from the summation of these two errors, will therefore initially decrease as the step size is decreased, reach an optimum, and then increase again. Figure 2.12 shows, qualitatively, the variation of these errors with step size. Clearly,

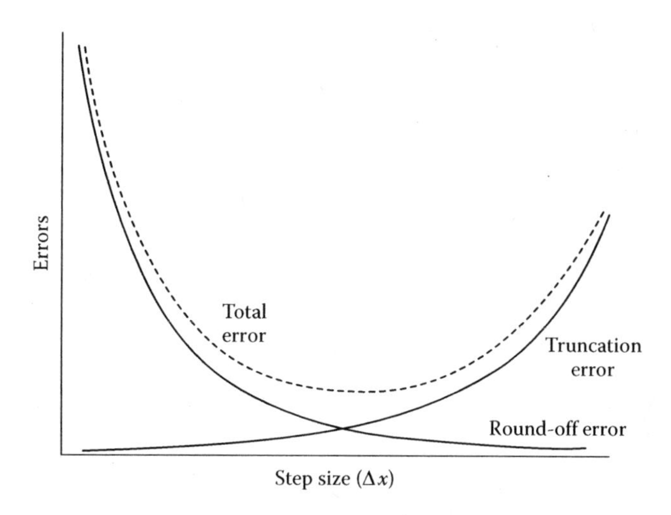

FIGURE 2.12 Sketch of the variation of the round-off, truncation, and total errors with step, or grid, size Δx.

a reduction in step size, or grid refinement, helps in error reduction to a point, beyond which the round-off error predominates. Thus, it is important to choose a step size that results in small TEs, without the associated penalty of large round-off errors.

The accuracy of a numerical solution can be best determined by a comparison between the numerical results and the analytical solution or experimental data, if available. However, the analytical solution or experimental result is generally not available for the given problem. Then such a comparison may be made by employing a problem that is simpler than the one being solved numerically and for which an analytical solution is available. For example, the numerical scheme for integrating an arbitrary function $h(x)$ may be used to integrate a simpler function, such as a polynomial, which can be integrated analytically. The numerical results can then be compared with analytical ones to quantify the accuracy of the method.

Various other methods are also employed to check the accuracy of the numerical results. One method is to put the obtained solution back into the equation being solved and check if the equation is satisfied. For example, after solving the matrix equation $(A)X = B$ for the unknown X, multiply the solution matrix X with the coefficient matrix (A) to check how closely the constant matrix B is reproduced. Similarly, in curve fitting, the computed function may be plotted along with the given data to determine if, indeed, a satisfactory fit has been achieved; see Figure 2.13. In root solving, the computed roots x are substituted into the given equation $f(x) = 0$ to ensure that the equation is satisfied. The physical nature of the problem is also used, wherever possible, to choose between multiple solutions and to determine if the numerical results show the expended trends.

2.3.4 Numerical Stability

Another important consideration, related to the errors and accuracy of a numerical solution, is that of *numerical stability*. It is of particular concern in the numerical

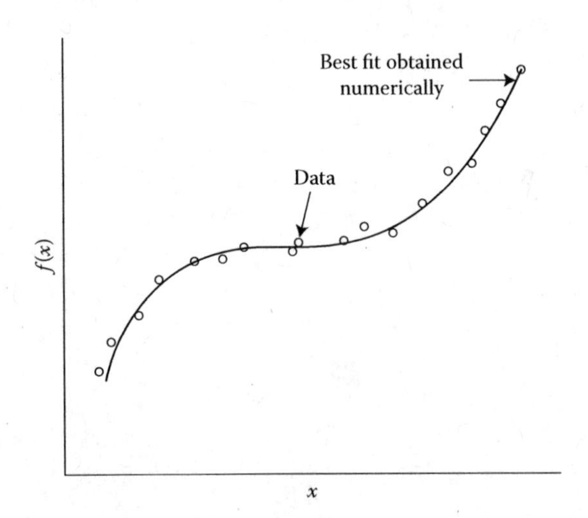

FIGURE 2.13 Comparison between the numerical results obtained for a best fit and the given data, for a check on the accuracy of the curve fit.

solution of ODEs and PDEs. Instability in a numerical scheme can lead to an unbounded growth of numerical errors that arise in the computation and thus can completely disrupt the numerical solution. If the scheme is stable, the errors are bounded and, although they accumulate as computation progresses, they do not grow to an unacceptably large level.

Let us consider, as an example, the simple ODE $dy/dx = f(x, y) = -cy$, where x is the independent variable, y the dependent variable, and c a positive constant. The analytical solution to this equation is $y/y_0 = e^{-cx}$, where $y = y_0$ at $x = 0$. This equation may be solved by any one of the several methods discussed later in this book. One of the simplest methods is Euler's method, which advances the solution from x_i to x_{i+1}, where $x_{i+1} = x_i + \Delta x$, by the recursion formula

$$y_{i+1} = y_i + \Delta x f(x_i, y_i) \tag{2.9}$$

Here, the subscript refers to the number of the computational step, starting with $i = 0$ at $x = 0$. Thus, $x_i = i\Delta x$, where Δx is the step size. This recursion formula is obtained by simply using the basic definition of a derivative to write $dy/dx = (y_{i+1} - y_i)/\Delta x = f(x_i, y_i)$. With $f(x, y) = -cy$,

$$y_{i+1} = y_i + \Delta x(-cy_i) = (1 - c\,\Delta x)y_i \tag{2.10}$$

The analytical solution decays exponentially with x, as sketched in Figure 2.14. However, the numerical solution will decay with x only if $c\,\Delta x < 1$. If the step size Δx is chosen large enough to make $c\,\Delta x > 1$, the solution becomes oscillatory. Thus, the difference between the numerical and analytical results increases as Δx increases. However, the oscillations obtained for $c\,\Delta x > 1$ decay with increasing x, provided $|1 - c\,\Delta x| < 1$. But if Δx is increased still further so that $|1 - c\,\Delta x| > 1$, the numerical solution grows with increasing x and ultimately becomes very large, as x is increased to large values; see Figure 2.14. The computer will then indicate that the solution becomes unbounded. Thus, an increasing solution is obtained instead of the decaying one given by analysis. This problem is an example of numerical instability, which must be avoided to obtain a physically realistic solution.

In this case, the scheme is conditionally stable, since if $|1 - c\,\Delta x| < 1$, an unbounded growth of the solution does not arise. Also, for the simple equation considered, a repeated application of Equation 2.10 gives the numerical solution as

$$y_i = y_0\,(1 - c\,\Delta x)^i \tag{2.11}$$

Thus, the errors in the solution accumulate as x increases, and for $|1 - c\,\Delta x| > 1$, they become unbounded at large x. Such a situation arises for some of the numerical methods used for the solution of differential equations. If the method is conditionally stable, the step size must be kept small enough so that instability does not arise. A good check for instability is to solve the problem for two values of the step size that are close to each other. If the results obtained differ tremendously, numerical instability may be present. Frequently, numerical instability can be avoided by reducing the step size. If the scheme continues to be unstable even with small step sizes, it is best to find some other method. For further details on numerical instability, advanced books such as those by Ferziger (1998) and Jaluria and Torrance (2003) may be consulted.

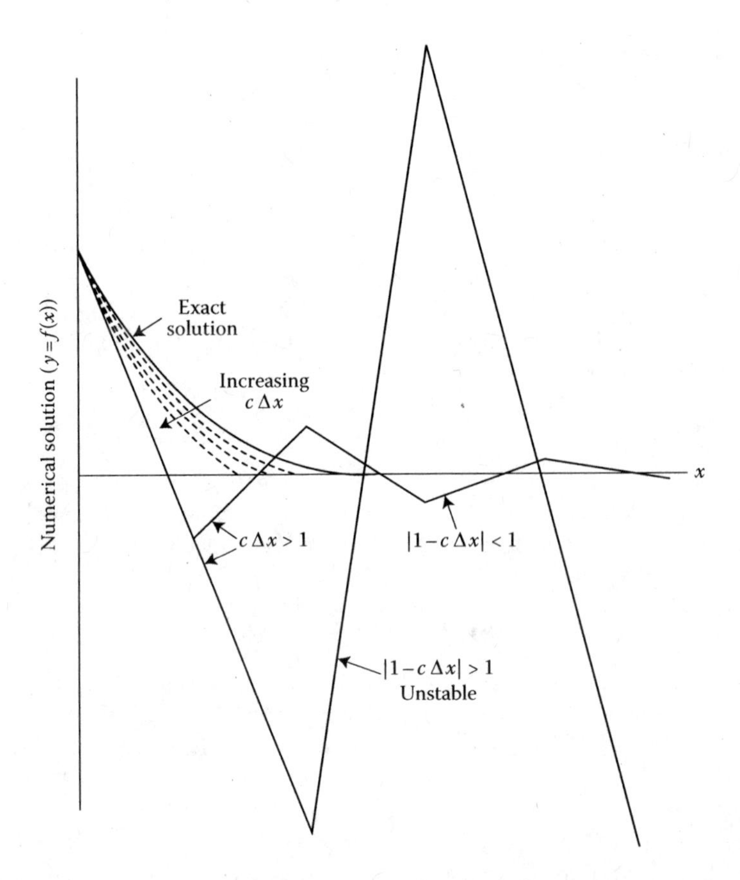

FIGURE 2.14 Increase in the numerical error and the onset of numerical instability as the step size Δx is increased in the solution of the differential equation $dy/dx = -cy$ by Euler's method.

2.4 ITERATIVE CONVERGENCE

Iteration is a numerical technique that is very commonly employed in the solution of a wide variety of problems. An approximation to the solution is assumed, and the approximation is gradually improved by iteration until the approximation to the solution does not vary significantly from one iteration to the next. The numerical method is then said to have *iteratively converged* to the desired numerical solution. However, convergence is not always obtained, and the conditions under which the scheme converges should be determined, whenever possible, before it is used in the solution of a given problem. Some of the important considerations related to iteration and convergence are outlined in this section.

The solution of nonlinear algebraic equations is usually based on systematic iteration methods, since except for a few special cases, such as quadratic equations, the solution cannot be obtained directly by algebra. For example, in the transcendental equation $\tan x = 2/x$, or in the polynomial equation $x^4 - 11x^3 + 41x^2 - 61x + 30 = 0$, the roots, which are the values of x that satisfy the equations, can be obtained by employing iteration. Similarly, systems of nonlinear equations are also generally solved by iterative methods. Even large systems of linear equations are often solved more effectively and more accurately by iteration than by direct algebraic methods. Such large systems frequently arise in the solution of ODEs and PDEs. There are several other circumstances where iterative procedures are employed to obtain the solution.

2.4.1 Conditions for Convergence

A very important consideration in the choice of an iterative method for a given problem is whether it would converge. As expected, convergence depends on the chosen, or guessed, initial approximation to the solution. A more rapid convergence usually results for an approximation that is closer to the actual solution than for one that is farther away. However, in many cases, the scheme diverges if the difference between the initial approximation and the actual solution is large. It is generally difficult to determine the region of convergence over which an arbitrary initial approximation would lead to convergence. Thus, the physical background of the problem and any available information on the solution must be used to approximate the solution as closely as possible. Still, several runs, with different starting approximations, may be needed before convergence is obtained. In some cases, the limiting values of the solution are known. Then numerical schemes that gradually reduce the region in which the solution lies and, thus, always converge may be developed.

In general, the conditions under which an iterative method converges must be determined. For many schemes, these conditions are known. For example, consider the following system of linear equations for the unknowns x, y, and z:

$$\begin{aligned} 2x + 4y + 8z &= 30 \\ 5x + y - 2z &= 4 \\ x + 5y - 3z &= 10 \end{aligned} \tag{2.12}$$

These equations can be solved by obtaining x in terms of y and z from the first equation, that is, $x = (30 - 4y - 8z)/2$, and similarly y in terms of x and z from the second, and z in terms of x and y from the last equation. We then assume starting values for x, y, and z, and solve for these variables iteratively in succession using the three equations for x, y, and z till the values do not change significantly from one iteration to the next. It is seen that, if this procedure is followed, the iterative process does not converge. However, if we solve for the variable with the largest coefficient in each equation, that is, for z in the first equation, x in the second, and y in the third and then carry out the iteration, it converges.

The condition for convergence in a system of linear equations, such as the one given by Equation 2.12, is expressed as

$$|a_{ii}| > \sum_{j=1, j \neq i}^{n} |a_{ij}| \tag{2.13}$$

where a_{ii} is the coefficient of the variable being solved for in a given equation and a_{ij} are the coefficients of the other variables. This condition requires each equation to have a dominant coefficient, which is greater in magnitude than the sum of the magnitudes of the other coefficients in the equation. Although convergence often occurs for weaker dominance than that given by Equation 2.13, this equation gives the condition under which convergence will occur. Further details are given in Chapter 6.

Similarly, the roots of a nonlinear equation $f(x) = 0$ may often be determined by rewriting the equation as $x = g(x)$ and using iteration, starting with an initial guess for x. This method, known as the *successive substitution* method, is convergent only if $|g'(\alpha)| < 1$, where $x = \alpha$ is the desired root, and the difference between the starting approximation and α is not too large. Again, it is difficult to quantify how close to the root the approximation must be for convergence to result. However, the condition $|g'(\alpha)| < 1$ may be used in formulating the function $g(x)$ before iteration is applied. Further details are given in Chapter 5.

2.4.2 Rate of Convergence

It is also important to determine the rate of convergence, if the scheme is confirmed to be convergent. If α is the desired solution and x_i is the ith approximation to the solution, the magnitude of the error after the ith iteration is $|(x_i - \alpha)|$. Similarly, the error after the $(i + 1)$th iteration is $|x_{i+1} - \alpha|$. Then the relation between these two errors indicates how rapidly the scheme is converging. First, for the scheme to be convergent,

$$|x_{i+1} - \alpha| < |x_i - \alpha| \quad \text{as } i \to \infty \tag{2.14}$$

Also, we may write the relationship between the errors as

$$|x_{i+1} - \alpha| \propto |x_i - \alpha|^n \tag{2.15}$$

where n is an exponent that depends on the numerical scheme. If $n = 1$, the scheme is said to have a *first-order convergence*, indicating that the error at a given iteration is proportional to that at the previous one. If $n = 2$, the scheme is said to have a *second-order*, or *quadratic*, *convergence*. Since the error is presumably small as i becomes large, this implies the squaring of a small quantity, resulting in a rapid reduction in error. This, in turn, results in a much more rapid convergence than that for a first-order convergence scheme. A still higher-order convergence will result in an even faster convergence.

2.4.3 Termination of Iteration

The next question is when and how an iterative process should be terminated. If x_i is the approximation to the solution after the ith iteration and x_{i+1} after the $(i + 1)$th iteration, a commonly employed criterion for deciding that convergence has been achieved and that the iteration should thus be terminated is

$$|x_{i+1} - x_i| \leq \varepsilon \tag{2.16}$$

where ε is a small quantity, known as the *convergence parameter* in the given convergence criterion. Unless the solution, or the approximation x_i, is zero, ε must be small compared to the solution. Thus, the relative convergence criterion given by

$$\left| \frac{x_{i+1} - x_i}{x_i} \right| \leq \varepsilon \tag{2.17}$$

is also very often employed. If x_i is expected to be close to zero, the absolute convergence criterion, given by Equation 2.16, is more appropriate, with $\varepsilon \ll 1.0$. Thus, ε is an arbitrarily chosen numerical parameter brought in to ascertain that the iteration has converged. However, if ε is too small, the computing time will be excessive; if ε is too large, the results may be in significant error. Also, it is necessary to ensure that the numerical results are essentially independent of the chosen value of ε. These considerations are discussed in greater detail in Section 2.5.

For an example on the use of such a convergence criterion, consider Example 2.1. We are interested in the sum S of the series. However, in a numerical scheme, we can sum only a finite number of terms. Then the error involved in neglecting the nth term, as compared to the sum S of the terms of the series up to this term, may be employed as the convergence criterion. Thus, if SN is the nth term, we have

$$\frac{SN}{S} \leq \varepsilon \tag{2.18}$$

as the condition for convergence. Similar considerations would apply for other iterative schemes. Unless the solution or its approximation could possibly be zero, the relative convergence condition is generally preferred, in comparison with the absolute condition, since the solution is generally not known, making it difficult to choose the value of ε in Equation 2.16. For the relative convergence condition, Equation 2.17, ε may be chosen to be around 10^{-4}, as the starting value, in order to obtain a reasonably small variation from one iteration to the next, in the approximation to the solution.

2.5 NUMERICAL PARAMETERS

The preceding sections have demonstrated that one must often introduce several arbitrarily chosen parameters into the numerical scheme in order to solve the problem. Among the most important of these chosen numerical variables are the step, or grid, size Δx, the convergence parameter ε, and the initial approximation to the solution. It is obvious that since such variables, or parameters, are chosen arbitrarily, it must be ensured that the numerical results obtained from the scheme are essentially independent of the chosen values.

2.5.1 Step Size

The effect of the step size Δx on the numerical solution has been considered earlier (see Figure 2.12). As Δx is reduced, starting with relatively large values, the TE is also reduced. The round-off error generally does not become significant unless very small Δx, which involves a very large amount of computation, is employed. Thus, TEs dominate over much of the commonly used range of Δx and, with decreasing Δx, the numerical results tend to approach essentially constant values. When this occurs, the effect of the step, or grid, size on the solution is negligible. Then the value of Δx may be chosen as the upper limit of the Δx range in which this effect is small; see Figure 2.15. The largest value of Δx for which the solution is essentially independent of Δx is chosen so that both the computational effort and the round-off error are minimized. Of course, at very small Δx, the round-off error becomes significant and may substantially affect the solution, as shown in Figure 2.15. Analytically, we allow Δx, or dx, to approach zero in order to determine, for instance, a derivative or an integral. However, numerically, this is not possible because of unacceptably high CPU times and large round-off errors.

2.5.2 Convergence Criterion

The convergence parameter ε must be similarly treated. A relatively large value of ε is initially employed so that a rapid convergence is achieved. Then ε is gradually reduced until the numerical results remain essentially unchanged if ε is reduced further. Since the computations involved increase with reducing ε, a continued reduction in ε will ultimately result in substantial round-off error. Thus, as before, the largest value of ε at which the dependence of the numerical solution on ε first disappears is chosen; see Figure 2.16. Also, the convergence criterion may be applied to different variables being computed in the solution to confirm that convergence has indeed occurred.

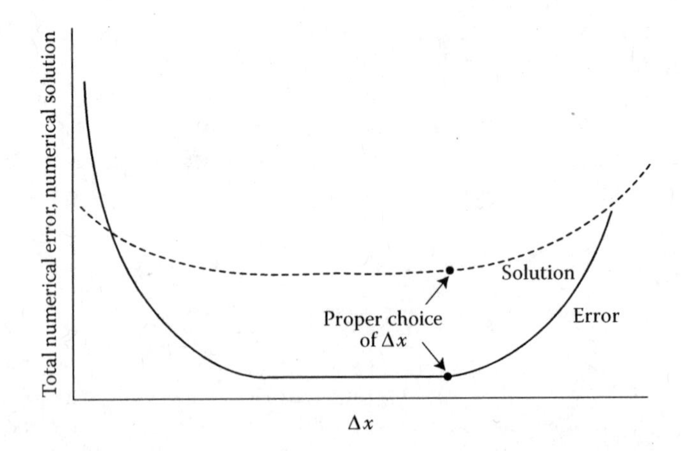

FIGURE 2.15 Sketch of the variation of the total numerical error and the solution with the step, or grid, size Δx. Also, indicated is the appropriate value of Δx that may be chosen for the computations.

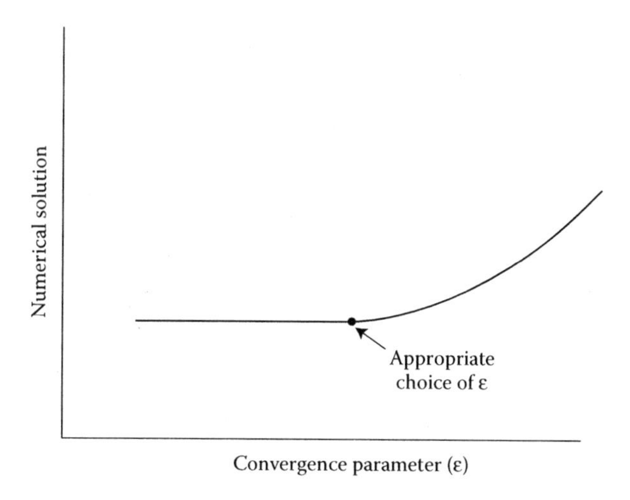

FIGURE 2.16 Sketch of the typical effect of a variation in the convergence parameter ε on the numerical solution.

2.5.3 Other Arbitrarily Chosen Variables

An initial approximation to the solution is needed in order to start an iteration scheme. Since convergence often depends strongly on the starting values, physical considerations and any available previous results on similar problems must be employed to choose the initial approximation. In root solving, for instance, the range of values in which the root lies is often known from the physical nature of the problem and may be used to obtain the first guess. Similarly, analytical or numerical results for similar problems are frequently used to obtain the starting values in iterative schemes for solving differential equations. However, it is important to ensure that the results are not significantly affected by the chosen initial guess. Thus, the initial approximation must be varied until the converged numerical solution is essentially independent of the starting values.

Example 2.3

In a chemical process, the concentration C in kg/m^3 of a given species decays with time t, in seconds, as follows:

$$C = 22.5 + 62.3\ \exp(-0.01t) \tag{2.19}$$

Thus, the concentration approaches a steady-state value of 22.5 kg/m^3 as time increases, that is, as $t \to \infty$. If the time t is increased with step size Δt, starting with $t = 0$, determine the dependence of the number of steps, the time t_{ss} required to attain steady state, and the concentration at steady state on the convergence parameter ε employed to indicate steady-state conditions.

SOLUTION

The initial concentration, at $t = 0$, is $22.5 + 62.3 = 84.8$ kg/m^3. As time $t \to \infty$, $C \to 22.5$ kg/m^3. However, we wish to terminate the computation as soon as C is close to the steady-state value of 22.5 kg/m^3, within a chosen convergence criterion. If such a criterion is not used, the computation will proceed until C is 22.5 kg/m^3, within the round-off error of the computer, and this would generally involve a considerable wastage of computer time. Thus, we may use a condition of the form

$$|C - 22.5| \leq \varepsilon \tag{2.20}$$

where ε is the convergence parameter, in order to decide that the steady-state value has been attained and that the computation may be terminated.

The given problem is employed to demonstrate the necessity of using a convergence criterion and the effect of ε on the results. The concentration C is computed at increasing time t, starting with $t = 0$, until Equation 2.20 is satisfied. The step size Δt determines only the values of t at which C is computed, and thus the time t_{ss} at which the computation is terminated is obtained within an accuracy of Δt. Since the exact, analytical expression for C is given, no TEs are involved, and round-off error arises only for each individual computation. There is no accumulation of error. Thus, the chosen value of Δt has a small effect on the solution and we may focus on the effect of ε.

A simple calculation may be carried out to increase t from 0, in steps of Δt, until Equation 2.20 is satisfied. The convergence criterion ε is varied from a high value of 100, at which convergence occurs at the very first step, to very low values, on the order of 10^{-9}. The value of Δt is chosen as 100 s. Thus, t_{ss} would be obtained to an accuracy of 100 s. A smaller value of Δt, $\Delta t = 10$ s, was also considered, and the effect of this change in Δt on the results at small values of ε was quite small. At steady state, as determined by Equation 2.20 being satisfied, the number of steps n, time t_{ss}, and concentration C_{ss} are obtained. Here, t_{ss} is related to n simply by $t_{ss} = n\,\Delta t$.

Figure 2.17 shows the dependence of the number of steps n and of the steady-state concentration C_{ss} on ε. The computational effort, as indicated by n, increases sharply as ε is reduced to very small values, whereas the solution is hardly affected as ε is reduced below about 10^{-2}. This figure indicates the importance of choosing the proper value of ε. A large value of ε results in considerable error, and a very small value leads to a very large, unnecessary computational effort. Here, a value of 10^{-2} may be chosen for ε. Such problems, in which the steady-state condition is to be determined, are frequently encountered in engineering problems. Although the first estimate of ε may be based on expected results or on previous experience with similar problems, ε must be varied to ensure that an appropriate value is chosen.

2.6 SUMMARY

This chapter discusses some of the important and fundamental considerations that form the basis for an efficient and accurate numerical scheme. The computational procedure is discussed in some detail, outlining method selection, programming language and computer system considerations, and program development. Besides indicating a systematic approach to the computational solution of a given problem, this

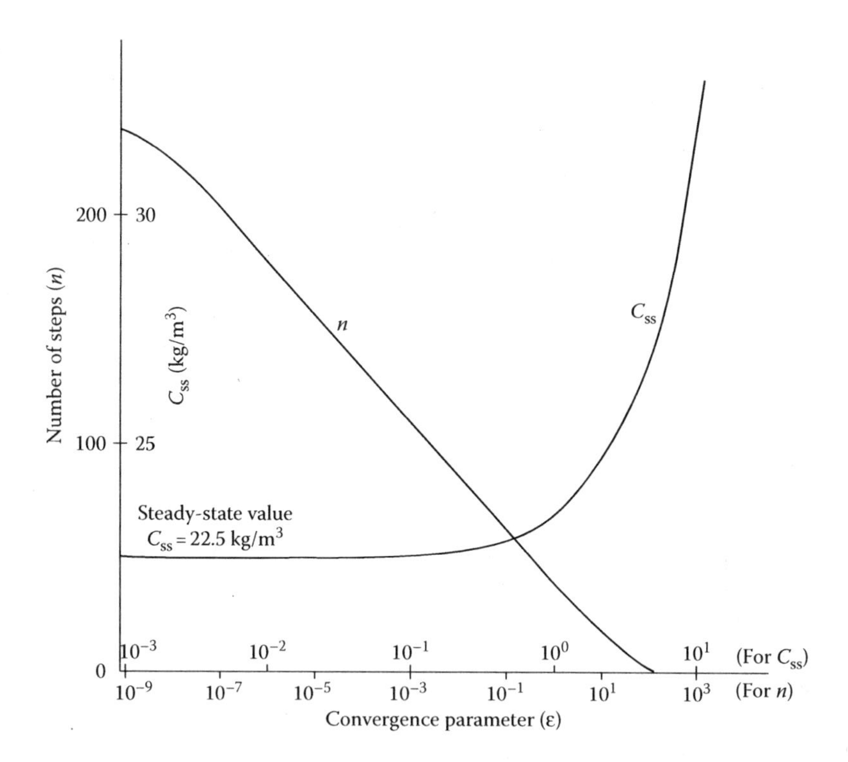

FIGURE 2.17 Dependence of the number of steps to convergence n and of the steady-state solution C_{ss} on the convergence parameter ε, for the problem considered in Example 2.3.

discussion also presents some of the recent trends in the area of numerical methods for engineering applications. Although fairly straightforward for most experienced users of the computer, this discussion nevertheless focuses on several relevant aspects that need to be considered before proceeding with the development of the computer code.

Numerical errors and accuracy are of crucial importance in any computational result. The nature and characteristics of errors that arise, particularly truncation and round-off errors, are discussed, along with the methods for evaluating and improving the accuracy of the numerical solution. Numerical instability is also considered. The convergence of iterative methods, which are frequently used for various types of problems of engineering interest, is discussed in terms of a few examples. The importance of a criterion for deciding if convergence has occurred and of determining the conditions under which the scheme is convergent are outlined. Finally, numerical variables and parameters, which are often introduced into the numerical method in order to obtain the solution, are considered. Since such parameters are chosen arbitrarily, it is important to ensure that the numerical results are not significantly affected by a variation in the chosen values. Methods to do so and the anticipated trends are outlined. The various considerations discussed in this chapter will arise in the following chapters, and the importance of these aspects will become quite apparent as we proceed with different types of problems and solution methods.

PROBLEMS

2.1. Calculate the number of arithmetic operations involved in solving the two simultaneous linear algebraic equations, $a_1x + b_1y = c_1$, and $a_2x + b_2y = c_2$, where x and y are the unknowns and a_1, a_2, b_1, b_2, c_1, and c_2 are given constants. Try different algebraic approaches to the solution, using elimination and substitution. Does your answer depend on the procedure adopted?

2.2. Write an algorithm to calculate the real or complex roots of the quadratic equation $ax^2 + bx + c = 0$, where a, b, and c are given constants. Use this algorithm with a calculator or computer to find the roots for (a) $a = 1, b = -3, c = 2$; (b) $a = 1, b = -5, c = 6$; and (c) $a = 2, b = 1, c = -1$.

2.3. Employing the binomial series generated by $1/(1 + x)$, where $|x| < 1$, compute the sum of the series, using a finite number of terms with a convergence criterion ε, as done in Example 2.1. Write an algorithm for the purpose and, using a calculator or a computer, study the effect of varying the convergence criterion on the numerical results.

2.4. Write an algorithm to determine the maximum of the function $f(x) = 12 + 18x - 3x^2$ in the range $0 \leq x \leq 4$. Starting with the lower limit on x, advance x with a step size $\Delta x = 0.1$ until the maximum is determined. Use a calculator or a computer and employ analytical expressions for the derivatives.

2.5. Repeat the preceding problem for determining the minimum of the function $f(x) = 7 - 12x^2 + 2x^3$ in the range $1 \leq x \leq 6$.

2.6. In Example 2.2, if the requirement is that the FW of the monthly deposits of $1000 must attain $200,000 at an interest rate of 7.5%, compute the number of months needed to achieve this FW. Also calculate the PW of the total money deposited. The given program may be suitably modified to solve this problem.

2.7. Employing a calculator or the computer program of Example 2.2, calculate the time needed for the repayment of the loan of $50,000 if the monthly payment is $1500 and the interest rate is 12%. Repeat the calculation for a monthly payment of $2000. In both cases, calculate the last payment if the loan is to be paid off exactly.

2.8. Write an algorithm to study round-off errors by adding 1/3 300 times and 1/6 600 times. Using a calculator or a computer, vary the number of decimal places retained in the calculations from 1 to 8, by appropriate programming statements. Compute the round-off error and show its dependence on the number of decimal places retained, in tabular or graphical form.

2.9. The second-order derivative may be written in finite-difference form as $\partial^2 f/\partial x^2 = [f(x + \Delta x) - 2f(x) + f(x - \Delta x)]/(\Delta x)^2$, where Δx is the step or grid size. Compute the resulting error if the round-off errors involved in the evaluation of the function at the three values of x are equal. Repeat this calculation if the round-off errors are equal in magnitude but alternating in sign from one grid point to the next. Comment on the significance of your results.

2.10. Employing Equation 2.9 for the numerical solution of the differential equation $dy/dx = -2y$, study the effect of varying Δx on the solution, including instability at large Δx. Confirm the trends shown in Figure 2.14.

2.11. Consider the functions $f(x) = 2 + 3/x$ and $g(x) = 5.2 + 2.4/x^2$. Both of these approach constant values as $x \to \infty$. Employing a convergence criterion, as illustrated in Example 2.3, determine the effect of the convergence parameter ε on the value of x, x_{ss}, at which the solution has essentially attained these constant values. Does the step size Δx have any significant effect on the results?

2.12. Determine the effect of varying Δx on the computed result for the second derivative, as given in Problem 2.9, for the function $f(x) = 5 + 10x - 4x^2 + 6x^3$. The second derivative is to be determined at $x = 1$. Using a computer or a calculator, calculate the second derivative at $x = 1$ with $\Delta x = 0.5$, 0.1, 0.05, and 0.01. Compare the results obtained with the exact value of 28.

2.13. In Example 2.3, employ a relative convergence criterion, as given by Equation 2.17, and choose the most appropriate value by varying ε and computing its effect on the numerical results.

2.14. The numerical integral I of a function $f(x)$ may be obtained by the simple expression $\int_a^b f(x)\mathrm{d}x \cong \sum_{i=0}^{n-1} f(x_i)\Delta x = I$, which involves summing the function values at various x values, $x_i = a + i\,\Delta x$, for $i = 0, 1, 2, \ldots$, so that $n\,\Delta x = b - a$, $x_0 = a$, and $x_{n-1} = b - \Delta x$. Using this formula, which is known as the *rectangular rule*, compute the integral $\int_0^2 x^2\,\mathrm{d}x$ for $\Delta x = 2$, 1, 0.5, 0.1, 0.05, and 0.01. Compare the results obtained with the exact value of 8/3. Plot the numerical error versus the step size Δx. What value of Δx will you choose for such computations, on the basis of the results obtained?

2.15. For the problem given in Example 2.2, with a loan of \$50,000 at 12% interest, consider reducing the monthly payments; that is, instead of \$1000, the payment is, say, \$950. Compute the time needed for repaying the loan if the monthly payment is \$950. Then recompute with a monthly payment of \$900, and so on. Is a limiting value, beyond which the monthly payment cannot be decreased further for repaying the loan, indicated from your results? If so, why does such a limitation arise?

2.16. The mass transfer rate $\dot{m}$, in kg/s, at the surface in a chemical reactor at a particular time is given by the series

$$\dot{m} = 556.3 \sum_{n=1,3,5,\ldots}^{\infty} \exp(-0.04n^2)$$

where n is an odd number. Using a suitable convergence criterion, determine the number of terms needed for the numerical evaluation of $\dot{m}$ and the resulting value of the mass transfer rate.

3 A Review of MATLAB® Programming

3.1 INTRODUCTION

In Chapter 2, we considered the main elements of a computer program, including the algorithm, programming language, and code development. Commercially available software, as well as computer programs that are available in the public domain, were discussed. It was mentioned that, in recent years, MATLAB® has become the most frequently used software for solving mathematical equations that arise in scientific and engineering problems. It provides a convenient and user-friendly environment to enter input data and obtain results in desired graphical, tabulated, or digital form. Fortran, which was probably the most common programming language used for engineering applications in the past and which continues to be important even today, *C++*, and other high-level programming languages are also frequently used for a variety of engineering systems. In many cases, programs in these languages and computed results are coupled with MATLAB programs for employing many attractive features, such as graphics and optimization, available in MATLAB.

In this chapter, the basic characteristics of the MATLAB environment for the numerical solution of mathematical problems are briefly outlined in order to discuss the development of an appropriate code as well as to present a few readily available commands to solve commonly encountered problems in engineering. Only the main features of the MATLAB environment are presented here for providing an appropriate, basic, background for presentations in the following chapters. Further details can be obtained from the references given at the end of the book, such as Rectenwald (2000), Chapra and Canale (2002), Matthews and Fink (2004), Palm (2005), Chapra (2005), Gilat (2008), Littlefield and Hanselman (2005), and Moore (2006).

3.2 MATLAB® ENVIRONMENT

3.2.1 Basic Commands

MATLAB provides a software environment in which a wide variety of mathematical operations can be carried out very easily. Though similar to the *C* programming language in some respects, it has its own style and format based on a large number of available commands, functions, and algebra built into the environment. Besides the usual mathematical functions such as sin(x), cos(x), tan(x), exp(x), log(x), abs(x), and so on, to represent sine, cosine, tangent, exponential, natural logarithm, and absolute value, respectively, of a variable x, many other specialized functions are available. Here, x is in radians for the trigonometric functions. Similarly, *a*sin, *a*cos, and *a*tan are the inverse of these functions and yield the angle in radians. MATLAB also has functions like *pi* to represent π, as in the circumference $2\pi r$ of a circle of radius r, sqrt(x) for square root of x, *eps* to denote spacing of floating point numbers and thus a small quantity, *inf* to denote infinity, *NaN* to denote *not a number* (such as 0/0), and *real* and *imag* to denote real and imaginary parts of a complex number. Help is easily available for various functions and commands. For instance, *help*('exp') or *help* exp will provide additional information on this function. The command *who* yields all the variables in the given session and *whos* gives the detailed information on the variables. All the definitions and variables remain in the current session till the command *clear* is typed.

Mathematical operations are denoted by +, –, *, /, and ^, where these represent addition, subtraction, multiplication, division, and raised to a power. Parentheses can be used to separate various operations for clarity and correctness. Then the operations within the parentheses are carried out first. The raised to a power operation is performed next, followed by multiplication and division, and then addition and subtraction, moving from left to right for the latter two sets of operations. The result of a calculation is printed immediately as the answer, denoted by *ans*, unless a semicolon is placed after the equation. Similarly, the value of a variable being defined is printed unless a semicolon is placed at the end of the expression.

Examples of such calculations given at the command (>>) level, are

```
>> 4 + 3/2 + 1 − 3 + 4^2 − 5
ans =
 14.5000
```

whereas

```
>> 4 + 3/2 + 1 − 3 + 4^2 − 5;
```

suppresses the printing. Similarly,

```
a = 3;
b = 4*8;
c = sqrt(55);
```

can be used to define variables a, b, and c, without printing. If the semicolon is not used, the result is printed, as, for example,

```
>> a = sqrt  (−25)
prints
a =
  0 + 5.000i
```

where i (or j) is used to indicate the imaginary part of the expression. Also,

```
>> a = sqrt  (2 + 3i)
prints
a =
  1.6741 + 0.8960i
```

This also indicates that complex algebra is built into the MATLAB environment and can be used easily to perform mathematical operations. Thus, *imag*, *real*, *conj*, *angle*, and *abs* can be used to obtain the imaginary part, real part, conjugate, angle in polar representation (in radians), and magnitude of a complex number. Thus,

$$\text{abs}(A) = \sqrt{(\text{real}(A))^2 + (\text{imag}(A))^2} \tag{3.1}$$

$$\text{angle}(A) = \tan^{-1}\left(\frac{\text{imag}(A)}{\text{real}(A)}\right) \tag{3.2}$$

Also,

$$\exp(\text{i}\theta) = \cos(\theta) + \text{i}\sin(\theta) \tag{3.3}$$

Addition, subtraction, multiplication, and division can be performed as done with real numbers, with MATLAB following the mathematical rules of complex algebra.

The semicolon ends the statement, so that the next definition can be given without going to the next line, for example,

```
a=2; b=3; c=4; d=exp(1.5);
```

Variables are case sensitive and must begin with a letter. Therefore, definitions of variables and mathematical operations can be carried out easily in the MATLAB environment.

3.2.2 Matrices

One of the major strengths of MATLAB lies in the definition, use, and algebra of matrices. Several functions, programs, and operations are built into the environment, so that many routine matrix calculations can be carried out easily and concisely. For instance, matrices with all the elements as one can easily be defined by

```
>> ones (2, 4)
```

which yields a 2×4 matrix with all the elements as 1. Similarly, *ones* (4, 2) yields a 4×2 matrix with all the elements as 1. We could also define a matrix A as A = *ones* (3, 3), A = *ones* (4, 4), A = *ones* (3, 4), and so on, with A being printed if a semicolon is not placed at the end of the equation. Similarly, *zeros* (2, 4) and *zeros* (4, 2) yield 2×4 and 4×2 matrices, respectively, with all the elements as 0. An identity matrix, with zeros every where except at the diagonal where the elements are 1, is obtained by the command *eye* (n), which gives an $n \times n$ identity matrix. In order to enter different values for the elements in a matrix, the rows are separated by a semicolon or by going to the next line by carriage return. The elements are separated by space and brackets denote a matrix. For instance, consider the following:

```
>> a=[1 2 3]; b=[2 4 9]; c=[3; 4; 7]; d=[1 2 3; 2 4 6; 3 5 8];
```

These statements yield row vectors a and b, with 1, 2, 3 and 2, 4, 9 as the element values, respectively, and c as a column vector with 3, 4, and 7 as the element values. The last statement gives d as a 3×3 matrix. Thus,

$$a = \begin{pmatrix} 1 & 2 & 3 \end{pmatrix} \quad b = \begin{pmatrix} 2 & 4 & 9 \end{pmatrix} \quad c = \begin{pmatrix} 3 \\ 4 \\ 7 \end{pmatrix} \quad d = \begin{pmatrix} 1 & 2 & 3 \\ 2 & 4 & 6 \\ 3 & 5 & 8 \end{pmatrix}$$

The transpose of a matrix is obtained by using an apostrophe, as a', b', c', and d'. Then, a' is just the column vector of the elements in matrix a and c' is the row vector consisting of elements in matrix c. Similarly, d' transposes the rows and columns in matrix d. Another apostrophe as $(d')'$ will yield the original matrix d.

A few other useful commands are mentioned here. Random numbers between 0 and 1.0 are generated by the command *rand* (n), which gives an $n \times n$ matrix of random numbers. The diagonal elements are given by *diag* (A), where A is a given square matrix. The command *diag* (c), where c is a vector, puts the elements on the diagonal with the other elements being zero. The number of elements in a vector is given by *length* (b), where matrix b is a vector. For a square matrix, it gives the number of rows or columns and for other matrices, $m \times n$, it gives the larger of the two parameters m and n. The command *size* (B) gives the number of rows m and number of columns n of an $m \times n$ matrix B.

Any desired element of a matrix A can be obtained by the statement A (m, n), where m is the row and n is the column of the element. For a row or column vector B, the command becomes B (k), where the kth element is desired. Similarly, the given element can be assigned a value, as, for instance,

```
>> A(2, 3) = 9; B(4) = 7;
```

A row of a given matrix A can be obtained by using the colon notation as A (2, :) and a column by A (:, 3), yielding the second row and the third column, respectively. A (1:2, 2:3) yields the elements in the first and second rows and the second and third columns. Similarly, A(1:5, 2) and A(1, 1:4) yield the specified elements in the second column and those in the first row, respectively. All these expressions can also be used for assigning values to the elements. Elements can be deleted by expressions like B (3) = [ ] and B (1:3) = [ ], where B is a vector. The third element is deleted in the first case and the first to the third one in the second case.

3.2.3 Arrays and Vectorization

A row vector of linearly spaced elements is given by the command *linspace*, whereas a logarithmic distribution to base 10 is given by *logspace*. Thus,

```
>> linspace (xmin, xmax, n);
```

generates n evenly spaced points between $x$min and $x$max and including both these boundary points. Thus, the region is divided into (n – 1) subdivisions. For example,

```
>> linspace (0, 4, 5)
```

gives 0, 1, 2, 3, and 4. If 5 is replaced by 10, we get 0, 0.4444, 0.8889, . . ., 3.5556, and 4.0000. If the number of points is not given, the default value is 100. Similarly, *logspace* (0, 2, 5) yields 1.0000, 3.1623, 10.0000, 31.6228, and 100.0000, since logarithm to base 10 is 0.0 for 1.0 and 2.0 for 100.0. A vector of evenly spaced points can also be generated by

```
>> x=0:5;
```

which yields six points as $x = 0, 1, 2, 3, 4, 5$, giving a default spacing of 1.0 between the points. The starting value is 0.0 and the ending value is 5.0, if the spacing yields it as appropriate point. Thus, if 5.5 were employed, instead of 5, in the above command, the six points will remain the same. However, the spacing can be changed by specifying the value as

```
>> x=0:0.5:5
```

which now yields 11 points separated by 0.5. If the spacing is given as 0.6, instead, 9 points are generated at a spacing of 0.6 with 4.8 as the last point. Different ways of distributing points over a given region are valuable in plotting the computed results obtained.

3.2.4 Matrix Algebra

Once the matrices are defined, MATLAB can be used effectively for various mathematical operations, such as addition, subtraction, and multiplication of matrices. For instance, if U and V are two matrices, we can use the commands

```
>> C=U+ V
>> D=U- V
>> E=U* V
```

for addition, subtraction, and multiplication of these two matrices, using the basic matrix algebra covered in mathematics courses. Thus, for addition and subtraction, the two matrices must have the same number of rows m and columns n, so that each element of V is added to or subtracted from the corresponding element in U. This implies that $C_{i,j} = U_{i,j} + V_{i,j}$ and $D_{i,j} = U_{i,j} - V_{i,j}$, where the subscripts i and j indicate the row and the column of an element. However, for multiplication, the number of rows in V must be equal to the number of columns in U, that is, if U is an $m \times n$ matrix, V must be an $n \times p$ matrix. For example,

```
>> U=[2 3 4; 1 1 2; 3 5 7];
>> V=[1 2; 4 3; 6 8];
>> E=U* V
```

yields $E = [38\ 45;\ 17\ 21;\ 65\ 77]$, with $2 \times 1 + 3 \times 4 + 4 \times 6 = 38$, $2 \times 2 + 3 \times 3 + 4 \times 8 = 45$, and so on, on the basis of the fundamental rules of matrix algebra. Clearly, $V*U$ is not defined since the rows in U are not equal to the columns in V. Also, even if square matrices are involved so that $U*V$ and $V*U$ are allowed, the two are generally not equal.

As seen in the preceding, arrays can be generated easily in MATLAB by simple commands such as

```
>> x=0: pi/4: pi;
```

which gives x as 0, 0.7854, 1.5708, 2.3562, and 3.1416, or 0, $\pi/4$, $\pi/2$, $3\pi/4$, and π. This distribution of points can be used to generate other arrays, or vectors, such as

```
>> y=cos(x);
```

which gives y as 1, 0.7071, 0, –0.7071 and –1. Thus, x represents an array, which can be used as a variable to generate arrays of functions like cosine, sine, exponential, logarithm, and so on. This process of generating and using arrays, or vectors, is known as *vectorization*. It provides a major advantage of MATLAB over many other languages and software.

Arrays can also be multiplied and divided by using the operator with a period preceding the operation as .* and ./, respectively, with no gap between the period and the operator. Thus, if x and y are two arrays given as

```
>> x=[1 2 3]; y=[4 5 6];
```

then, element-by-element multiplication or division can be achieved by

```
>> z1=x.*y; z2=x./y;
```

Here, $z1$ is obtained as [4 10 18] and $z2$ as [0.25 0.4 0.5], indicating element-by-element multiplication and division, respectively.

We can also raise all the elements to a power by the command

```
>> z3=x.^3; z4=y.^2;
```

which gives $z3$ = [1 8 27] and $z4$ = [16 25 36], indicating that each element has been raised to the given power. The same operators can be used for matrices, as long as the two matrices have the same number of rows m and columns n. Thus, A .* B and A ./ B give element-by-element multiplication and division for two $m \times n$ matrices A and B. We can also use A.^ 2 to square each element in A or A .^ (1/2) to take a square root of each element. Clearly, the placing of a dot before the operator changes the result and the operation does not follow matrix algebra, but carries out an element-by-element algebra. For instance, if x is an array, 1 ./ x and x .^ 2 can be used to obtain reciprocal and square of all the elements. This again provides a very effective tool to carry out a series of operations concisely and efficiently.

As an example of the use of arrays, consider the following simple program in MATLAB:

```
x=linspace (0, 0.9, 10);
y=1./(1-x);
z=1+x+x .^ 2+x .^ 3+x .^ 4;
error=100*(y-z) ./ y;
plot (x, error)
```

Thus, 10 evenly distributed points are obtained for x between 0 and 0.9, the corresponding values of the function $1/(1-x)$ are generated and the first five terms of the Taylor series for the function are added. The TE, in percentage, if only these terms are retained in the series, is determined and the results are plotted, as shown in Figure 3.1, indicating increasing error with x. Similarly, additional terms could be retained in the series and the effect on the error determined. Also, other functions of x such as sqrt(x), sin(x), and exp(x) can be used, instead of 1./(1 – x), to plot and study their variation with x. This simple example shows how arrays of the different quantities can be used effectively to solve problems.

3.2.5 Polynomials

MATLAB has an extensive library to formulate, evaluate, and perform mathematical operations on polynomials. An n-degree polynomial is defined as

$$P_n(x) = C_1x^n + C_2x^{n-1} + C_3x^{n-2} + \cdots + C_nx + C_{n+1} \tag{3.4}$$

Then the polynomial is given as $[C_1\ C_2\ C_3 \ldots C_n\ C_{n+1}]$, with the coefficients placed in descending order of the power of x. Thus, a third-order polynomial, $x^3 - 2x + 12$, is written as [1 0 –2 12]. The value of the polynomial at a given value of x can be obtained by the command *polyval*. For example,

```
>> c= [1 0 -2 12];
```

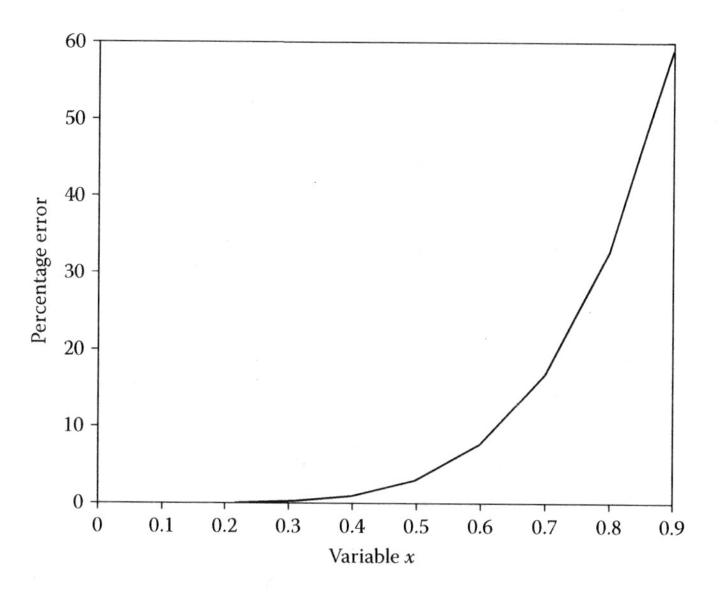

FIGURE 3.1 The truncation error (TE), in percentage, if only five terms in the Taylor series for the function $1/(1-x)$ are retained to approximate its value, as a function of x.

defines the given polynomial and the following command:

```
>> polyval (c, 1.5)
ans =
12.3750
```

yields the value of the polynomial at $x = 1.5$. Similarly, *conv* gives the product of two polynomials, *deconv* the division, *residue* the ratio, *poly* the polynomial from the roots, *polyfit* the least-squares fit to given data, and *roots* the n roots of an n-degree polynomial. These are discussed in greater detail in the following.

3.2.6 Root Solving

The n roots of a polynomial equation $P_n(x) = 0$ can be obtained easily by using the roots command. Consider a root-solving problem for temperature T obtained from the energy balance at a surface as

$$0.6 \times 5.67 \times 10^{-8} \times \left[(850)^4 - T^4 \right] = 40 \times (T - 350) \tag{3.5}$$

Although computer programs may be written in Fortran, C++, or other programming languages to solve this root-solving problem, the MATLAB environment provides a particularly simple solution scheme on the basis of the internal logic of the software. The polynomial p is given in terms of the coefficients a, b, c, d, and e, in descending powers of T, as

```
a=0.6*5.67*10^-8;
b=0;
c=0;
d=40.0;
e=-40.0*350.0-0.6*5.67*(10^-8)*(850^4);
p=[a b c d e];
```

Then the roots are simply obtained by using the command

$$r = \text{roots}(p) \tag{3.6}$$

This yields four roots since a fourth-order polynomial is being considered. It turns out, when the above scheme is used, that one negative and two complex roots are obtained in addition to one real root at 645.92, which lies in the appropriate physical range and is the correct solution.

Also, if the roots r of a polynomial are known, the polynomial may be formed by the command *poly* (r). For instance, the roots of the polynomial equation $x^3 - 6x^2 + 11x - 6 = 0$ can be obtained by the commands

```
>> c=[1 -6 11 -6];
>> r=roots (c)
```

which gives r as [3; 2; 1].

Then the command

```
>> p=poly (r)
```

yields the polynomial from the roots just obtained. It must be remembered that the coefficients are arranged in descending powers of the independent variable.

3.2.7 Linear Algebraic Equations

A system of linear equations $(a)\ (x) = (b)$ can be solved very easily in the MATLAB environment since it is particularly well suited to matrix algebra, as discussed earlier. Available commands and built-in functions may be used in a MATLAB environment to obtain the solution. For instance, let us enter the matrices (a) and (b) in MATLAB as

```
a= [2 1 0 6; 5 2 0 0; 0 7 2 2; 0 0 8 9];
b= [64; 37; 66; 104];
```

Then the solution (x) is obtained simply by using $(x) = (a)^{-1}\ (b)$ as

```
x=inv(a)*b
```

or as

```
x=a\b
```

The second approach uses the internal logic of the backslash, \, operator in MATLAB to indicate the left division of a into b. As discussed in Chapter 6, this operator uses a direct approach to the solution of the system of linear equation and requires fewer arithmetic operations compared to the preceding method based on matrix inversion, resulting in smaller computational time and smaller round-off error.

The solution can also be obtained by the decomposition of the matrix a into upper (u) and lower (l) triangular matrices, as discussed in detail in Chapter 6, by the commands

```
[l,u,p] =lu(a);
y=l\(p*b);
x=u\y
```

where p is the permutation matrix which stores the information on row exchanges during the computation process. When any of these approaches is used, the solution vector for the given problem is obtained as [5; 6; 4; 8].

3.2.8 Curve Fitting

A computer program may be developed to solve the system of linear algebraic equations generated by curve-fitting techniques, as discussed later in this book, using programming languages such as Fortran 90 and C++. However, MATLAB is particularly well suited for such problems since the command *polyfit* yields the best fit to a chosen order of the polynomial for curve fitting. For instance, the following commands may be used, with % representing the comment statement.

```
%Input Data
>> t=[0.2 0.4 0.6 0.8 1.0 1.2 1.8 2.2];
>> y=[4.29 3.78 3.54 3.59 4.0 4.8 10.11 16.53];
% Cutve Fit
>> y1=polyfit(t,y,1)
>> y2=polyfit(t,y,2)
>> y3=polyfit(t,y,3)
```

Then *y*1 yields the best fit with a first-order polynomial, that is, linear fit, *y*2 yields a best fit with a second-order polynomial, and *y*3 with a third-order polynomial. The results for the three cases are obtained as

```
5.9683 0.2135
5.5069 -7.3133 5.7585
1.0058 1.9845 -3.9929 4.9998
```

since only two coefficients are needed for a line, three for a parabola, or second-order polynomial, and four for a cubic, or third-order polynomial. The linear best fit is of particular interest in curve fitting with nonpolynomial forms, such as exponential, logarithm, and power-law variations. For instance, if the chosen exponential function for curve fitting is $y = A\ \exp(-ax)$, a logarithm is taken to yield $\log(y) = \log(A) - ax$. Then a new variable Y is defined as $Y = \log(y)$ and a linear fit is obtained from the given data for Y versus x. From this curve fit, the values of A and a can be determined. The use of the *polyfit* function is considered in greater detail in Chapter 7.

3.2.9 Flow Control

Many commands are used to control the flow of the program. These include *if...else...end*, *for...end*, and *while...end* commands. Relational expressions such as <, <=, >, >=, ==, and ~= refer to, respectively, less than, less than or equal to, greater than, greater than or equal to, equal to, and not equal to. Similarly, logical expressions such as &, |, and – refer to, respectively, and, or, and not.

If an expression is *false*, a zero number is assigned and if it is *true* a nonzero number is assigned. For instance, if we define *a* and *b* as

```
>> a=2; b=4;
```

Then the statement

```
>> c=a<b
```

is true and the result is given as

```
c=1
```

On the other hand, the statement

```
>> b_is_smaller = b < a
```

is false and yields

```
b_is_smaller = 0
```

Similarly, a command like

```
>> both_true = a_is_smaller & b_is_smaller
```

indicates that both statements are not true and yields zero. Also, the relational statement pertaining to *equal to* is written as

```
a == 3
```

Since this statement is false, the result is given as zero. Other such statements can be written to check if certain relationships are true or false and can thus be used for flow control.

The *if* ... *else* ... *end* commands are used as

```
if x >= y
c = 7
else
c = 6
end
```

If x and y are given as 3 and 5, respectively, the above command will yield c as 6, since x is not greater than or equal to y. Additional conditional statements can be introduced by using *elseif*. For example,

```
if x >= y
c = 7
elseif x == 2
c = 6
else
c = y/x
end
```

Since x is given as 3, the first two conditions are not satisfied and c is given as 5/3, or 1.6667. The commands after *else* or *elseif* can be a print statement, discussed later, a mathematical operation, a plotting command, or some other statement. For instance, it could be

```
else
fprintf ('Warning: either x and y are both negative or x < y\n')
fprintf ('x = %f  y = %f \n',x,y)
end
```

Here, \n moves the cursor to the next line after printing the results, %f is simply a default format for floating point data, and the apostrophe prints the string of words given. Print statements are discussed later.

The `for ... end` command gives loops for carrying out a series of repeated operations, such as

```
sum x=0;
for k=1:n
sum x=sum x+k;
end
```

Here, k takes the values of 1, 2, 3, ..., n. Then the result obtained is the sum of n consecutive numbers, that is, $n(n + 1)/2$. Similarly, *for... end* can be used to generate other loops in which a series of mathematical operations are to be performed. The *while ... end* command is similar in that the given series of mathematical operations are performed while a given statement, such as $x > 0.01$, is true. The series of operations and commands are carried out as long as the conditional statement is satisfied. Since the conditional statement may not be satisfied due to an error in the program or due to the operations being carried out, the calculations may go on for ever without stopping. The command *break* is often used to avoid such infinite loops. It takes the control out of the loop to the line just after the loop. If some undesirable result is obtained, such as an extremely high value of a variable beyond the overflow limit specified in the given software, or if allotted time expires, the loop is broken and the command shifts the operation to the end of the loop where it may be asked to print or stop.

3.3 ORDINARY DIFFERENTIAL EQUATIONS

MATLAB can be used to solve ODEs quite easily by employing standard commands available in the software. For example, consider the motion of a stone, which is thrown vertically at velocity V from the ground at $x = 0$ and at time $t = 0$ and which is governed by the differential equation

$$\frac{d^2x}{dt^2} = -g - 0.1\left(\frac{dx}{dt}\right)^2 \tag{3.7}$$

where g is the magnitude of gravitational acceleration, given as 9.8 m/s^2, and the velocity is dx/dt, denoted by V. We can solve this equation, if the initial conditions on x and V are given, to obtain displacement x and velocity V as functions of time t. As mentioned above, x is given as 0 at $t = 0$. Let us assume that the initial velocity V is given as 25 m/s.

Thus, the second-order differential equation in terms of the displacement x is given in Equation 3.7, with the initial conditions

$$t = 0: x = 0 \quad \text{and} \quad \frac{dx}{dt} = 25 \tag{3.8}$$

The corresponding differential equation in terms of the velocity V, where $V = dx/dt$, is obtained from Equation 3.7 as

$$\frac{dV}{dt} = -g - 0.1V^2 \tag{3.9}$$

with the initial condition

$$t = 0 : V = 25 \tag{3.10}$$

Both these cases are *initial-value* problems since all the necessary conditions are given at the initial time, $t = 0$. MATLAB can be used very easily for these problems by using *ode*23, *ode*45, and other built-in functions for the solution of ODEs. Both *ode*23 and *ode*45 are based on *Runge–Kutta* methods, which are discussed in Chapter 9, and use adaptive step sizes. Two solutions are obtained at each step, allowing the algorithm to monitor the accuracy and adjust the step size according to a given or default tolerance. The first method, *ode*23, uses second- and third-order Runge–Kutta formulas and the second one, *ode*45, uses fourth- and fifth-order formulas. Details on these methods are given later in this book. Only the appropriate commands are given here.

Considering first the equation for the velocity, the following MATLAB statements

```
dvdt=inline('(-9.8-.1*v.^2)', 't', 'v');
v0=25;
[t,v]=ode45(dvdt,1.4,v0)
```

yield the solution in terms of V. The first command defines the first-order differential equation (Equation 3.9), the second the initial condition on V, and the third allows time and velocity to be obtained till time $t = 1.4$. These can then be plotted, using MATLAB plotting routines, as shown in Figure 3.2. The velocity decreases from 25 m/s to zero with time. After the velocity becomes zero, the drag reverses direction and the differential equation changes, so the solution is valid only till $V = 0$. The final time may be varied according to the needs of the problem.

Similarly, the equation for x may be solved. However, this is a second-order equation, which is first reduced to two first-order equations as

$$\frac{dx}{dt} = V \tag{3.11}$$

$$\frac{dV}{dt} = -g - 0.1\,V^2 \tag{3.12}$$

First, the right-hand side of these two equations is defined as

```
function dydt=rhs(t,y)
dydt= [y(2);-9.8-0.1*y(2)^2];
```

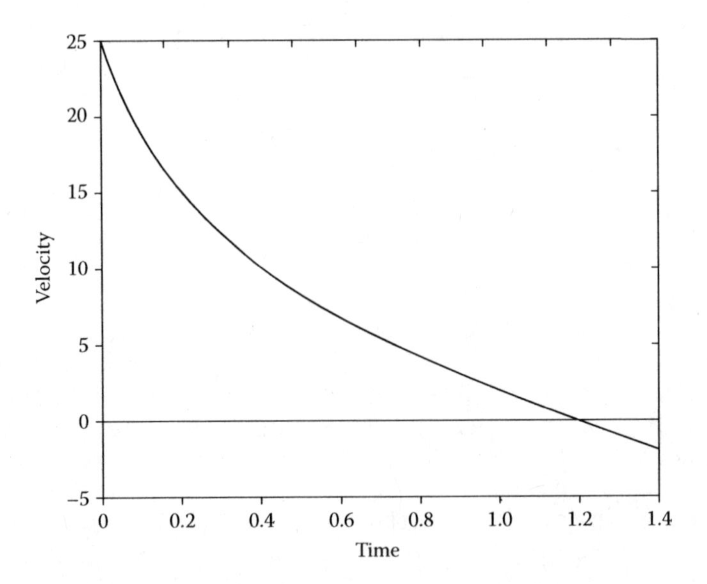

FIGURE 3.2 Calculated variation of velocity with time from the solution of Equation 3.9.

Thus, y is a taken as a vector with distance and velocity as the two components. Then the MATLAB commands are given as

```
y0 = [0;25];
[t,v] =ode45('rhs',1.4,y0)
```

Again, the initial conditions are given by the first line and the solution by the second. The results are obtained in terms of distance and velocity, which may be plotted, as shown in Figure 3.3. Here, the calculated distance x and the velocity V are plotted against time. Clearly, the results in terms of the velocity V are the same by the two approaches. Thus, MATLAB may be used effectively for solving such initial-value problems, considering single equations as well as multiple and higher-order equations. Other built-in functions for solving ODEs include those based on implicit and multistep methods, such as *ode*113 and *ode*15s. Further details on such functions and their usage can be obtained by using the *help* command in MATLAB. Several common MATLAB commands are given in Appendix A. Additional programs in MATLAB for such mathematical problems are given in Appendix B and discussed later in this book.

3.4 INPUT/OUTPUT

A fairly large variety of input and output commands are available in MATLAB to facilitate interaction with the computer. A common command for entering a given quantity or parameter is

```
>> x=input ('Enter the value of x, x=');
```

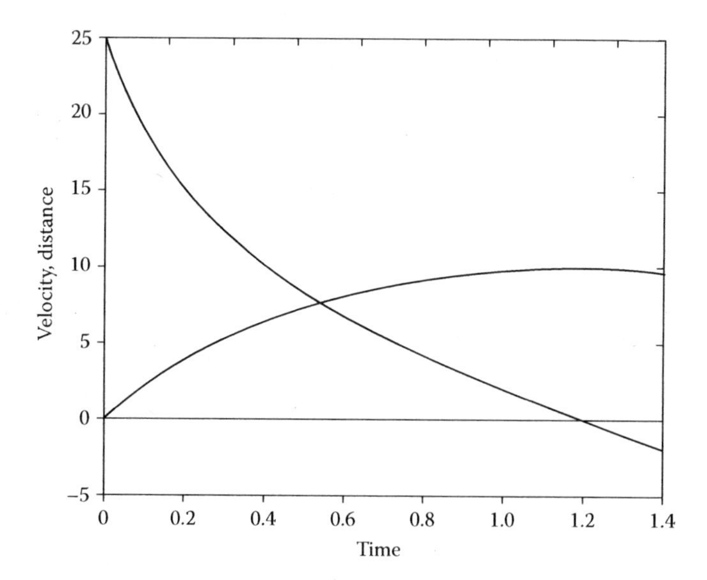

FIGURE 3.3 Variation of velocity and distance traveled as obtained from a solution of Equation 3.7.

This command allows the user to enter the value of the variable x, without printing it out. Just the command $x =$ *input*('*x*') can also be used instead. A string input, that is, a string of letters and numbers can also be given instead of a numeric value. For instance, the command

```
>> yourName = input ('Enter your name: ', 's');
```

prompts the user to enter his/her name. If the semicolon is not used, the computer prints out the name entered, say, John.

```
yourName =
= John
```

The output can be obtained in many different forms, from printing in different styles or formats to plotting. The command *disp* is good for simple tasks that have few requirements. The commands *fprintf* and *sprintf* have more control and greater options, including printing to a file. Also, *sprintf* is the same as *fprintf* except that it returns the data in a MATLAB string variable rather than writing it to a file. Thus, commands such as the following may be used.

```
>> disp ('My home is in New Jersey')
>> disp (['Your name is: ', yourName])
```

The first one just prints the given string *My home is in New Jersey* and the second one prints *Your name is: yourName*, where *yourName* is the name entered earlier. Strings are often used, along with numbers, to present the results in an easy-to-understand style. The command *num2str*, which allows a variation from a number to a string, is used as

```
>> outstring = ['x = ',num2str(x)];
>> disp (outstring)
```

which prints

```
x = 3
```

These two commands can also be combined to give the same result as

```
>> disp(['x = ',num2str(x)])
```

Similarly,

```
>> disp(['sin(x) = ',num2str(sin(pi/3))])
```

yields

```
sin(x) = 0.86603
```

Vectorization can also be used with *disp* as

```
>> x = 0 : pi/5 : 2*pi;
>> y = sin (x);
>> disp ([x 'y'])
```

which yields the x and the corresponding sin (x) results. The apostrophe is used in the *disp* command to obtain these two variables in the column format, as

```
 0         0
0.6283  0.5878
1.2566  0.9511
1.8850  0.9511
2.5133  0.5878
3.1416  0.0000
3.7699 -0.5878
4.3982 -0.9511
5.0265 -0.9511
5.6549 -0.5878
6.2832 -0.0000
```

There are several *format* statements that can be used to obtain the outputs in desired form. These include *short, long, short e, long e, and bank*, which give, respectively, a short (typically four decimal places), long (typically 15 decimal places), exponential short and long representations, and bank-style representation with two decimal places. Thus, if $x = 42.546314$, then it is given as 42.5463, 42.546314000000002, 4.2546e + 01, 4.254631400000000e + 01, and 42.55, respectively, in these formats.

Also, various formats such as %s, %d, %f, %e, and %g are used in statements for printing and refer to output as string, integer, floating-point, exponential notation, and the most compact format (%f or %e), using the default number of digits and decimals. More control is obtained by adding the field width, or total number of digits, and decimal places in terms of a floating-point number as %8.3f, %12.5e, %10g, and %7d. For example,

```
>> fprintf ('x=%f \n',x)
>> fprintf ('x=%8.3f \n',x)
>> fprintf ('x=%8.2e \n',x)
```

yield, respectively,

```
x=42.546314
x=42.546
x=4.25e+01
```

Similarly, other format statements can be used to obtain the output in desired form with chosen accuracy and strings.

As an example, let us consider the various commands considered above and see how the results are printed. Consider the commands

```
a=input('Enter the value of a, a=');
x=a^(1/3);
disp(['x=',num2str(x)])
disp(['The variable x =',num2str(x)])
format short
disp(x)
format long
disp(x)
fprintf('The variable x is %7.3f\n',x)
fprintf('The variable x is %7.5f\n',x)
fprintf('The variable x is %7.3e\n',x)
fprintf('The variable x is %7.5e\n',x)
fprintf('The variable x is %.3g\n',x)
fprintf('The variable x is %.5g\n',x)
```

Thus, a value of x is entered, it is raised to power 1/3 and the results are presented in different formats. For example, if x is entered as 0.4, the results are

```
Enter the value of a, a=0.4
x=0.73681
The variable x=0.73681
0.7368
0.736806299728077
The variable x is 0.737
The variable x is 0.73681
The variable x is 7.368e-01
The variable x is 7.36806e-01
The variable x is 0.737
The variable x is 0.73681
```

This indicates the use of various formats to obtain the results in the desired form.

3.5 SCRIPT *m*-FILES

These are analogous to computer programs in Fortran, *C* and other languages and involve a sequence of interactive statements stored in a file, so that the program can be employed to obtain results for different inputs and conditions. A plain text file can be generated, using the text editor available in MATLAB or any other text editor, and saved as a file with an extension of *m*, such as, *test.m*. Let us consider the following script *m*-file, saved as *test.m*,

```
% Summation of the Taylor Series for 1/(1-x)
%
for i=1:5
x=input('Enter the value of x, x =');
anal=1/(1-x);
sum=1+x+x^2+x^3+x^4;
diff=anal - sum;
error=diff *100/anal
end
```

Then, the following command

```
>> test
```

will run the program, which asks for the input value of x and then prints the percentage error if only the first five terms of the Taylor series are retained. The loop runs this sequence five times, so that results for five x values can be obtained. It can be shown that the error increases as x is increased. Similarly, other such programs are written and discussed in this book for the solution of various problems of engineering and scientific interest. Let us consider a simple script *m*-file to work with matrices, as discussed earlier.

Example 3.1

Write a script-*m* file to do the following:

1. Using the *rand* command, generate a 3×2 matrix A consisting of random numbers between 25 and 5.
2. Print the matrix generated, without printing "A =" or "ans =." But give a heading "MATRIX A."
3. Determine the smallest element in the matrix and its location.
4. Using the *sprintf* command, print the value of the smallest element and its location in terms of row and column numbers.
5. Obtain a new matrix *B* which flips the matrix, that is, the third row becomes the first row, the second remains the second, and the first row becomes the third row.
6. Append a third column to matrix *B*, with the elements as 4, 7, and 12.
7. Print a heading "MATRIX B" and print matrix *B*.

SOLUTION

This example demonstrates the use of various commands described earlier with respect to matrices and also the output of the results from a simple MATLAB program, as given in Figure 3.4.

The results obtained, when this program (Figure 3.4) is executed are given in Figure 3.5.

Therefore, this example illustrates the development and use of script *m*-files for solving mathematical and engineering problems. Once the program is developed, it can easily be used for solving other similar problems or modified according to

```
%       Example 3.1 Solution
%
format bank
%
%       Generate matrix
%
a=20*rand(3,2)+5;
%
%       Print Matrix A
%
disp(sprintf('MATRIX A'))
disp(a)
%
%       Determine smallest element and its location
%
amin=min(min(a));
[i, j]=find(a==amin);
%
%       Print results
%
disp(sprintf('The smallest element is %.5g.',amin))
disp(sprintf('The smallest element is at row %.0g ...
  and column %.0g.',i,j))
%
%       Obtain matrix B
%
b=flipud(a);
b(1,3)=4;
b(2,3)=7;
b(3,3)=12;
%
%       Print Matrix B
%
disp(sprintf('MATRIX B'))
disp(b)
```

FIGURE 3.4 MATLAB script *m*-file for the problem given in Example 3.1.

```
MATRIX A
      21.29       23.27
      23.12       17.65
       7.54        6.95
The smallest element is 6.9508.
The smallest element is at row 3 and column 2.
MATRIX B
       7.54        6.95        4.00
      23.12       17.65        7.00
      21.29       23.27       12.00
```

FIGURE 3.5 Results from the MATLAB program given in Figure 3.4.

changes in the requirements of the problem. The following example illustrates the use of MATLAB programming for a problem considered earlier and solved by using Fortran.

Example 3.2

For the problem considered in Example 2.1, write a MATLAB computer program, or script *m*-file, to obtain the sum of the series in the binomial expansion of $1/(1-x)$, the analytical result, the error and the number of terms needed to make the ratio of the *n*th term to the series sum up to this term less than 10^{-6} for x varying from 0.1 to 0.9.

SOLUTION

Using the algorithm presented in Example 2.1, a MATLAB computer program may easily be developed. A simple loop may be used to vary x from 0.1 to 0.9 in steps of 0.1. The corresponding MATLAB program is given in Figure 3.6, employing a simple *for... end* loop to go up to 1000 terms, if needed. As soon as the convergence check is satisfied, the loop breaks and the results are obtained in terms of the sum, error and number of terms. These are printed after all the x values have been considered.

Figure 3.7 shows the results obtained from the program. It is seen that the results agree with the earlier ones from Example 2.1. However, the error depends on the number of significant digits retained by the computer system and the software and, though small in both cases, is not expected to be equal. Here, the error varies from 0.01×10^{-3} for $x = 0.1$ to 0.8335×10^{-3} for $x = 0.9$ and is thus less than that obtained earlier in Example 2.1.

3.6 FUNCTION *m*-FILES

These are subroutines, or subprograms, that define functions, which can be used in a similar way to the MATLAB functions described earlier. Input/output parameters are given and all variables local to the function are defined. Then the function can be called to execute a specific mathematical operation or sets of operations. The basic format is *function [output parameter list] = functionName (input parameter list).*

```
% SERIES SUMMATION
%
% Enter Input Quantities
%
x=0.1;
for i=1:9
n=1;
sn=x^(n-1);
s=sn;
for k=1:1000
n=n+1;
%
% Sum the Series
%
sn=x^(n-1);
s=s+sn;
%
% Convergence Check
%
if sn/s<=10^(-6),break,end
end
      y(i)=x;
m(i)=n;
%
% Compute Analytical Value of the Sum and the Error
%
sum(i)=s;
sx(i)=1/(1-x);
er(i)=((sx(i)-sum(i))/sx(i))*100.0;
      x=x+0.1;
end
%
% Print Results
%
%
   disp(sprintf('The values of x are:'))
   disp(y)
   disp(sprintf('The number of terms needed are:'))
   disp(m)
   disp(sprintf('The sum of the series is:'))
   disp(sum)
   disp(sprintf('The error in percent is:'))
   disp(er)
```

FIGURE 3.6 MATLAB script *m*-file for the problem given in Example 3.2.

```
The values of x are
  0.1000  0.2000  0.3000  0.4000  0.5000  0.6000  0.7000  0.8000  0.9000

The number of terms needed are:
  7       10      13      16      20      27      37      56      111

The sum of the series is:
  1.1111  1.2500  1.4286  1.6667  2.0000  2.5000  3.3333  5.0000  9.9999

The error in percent is:
  1.0e-03 *
  0.0100  0.0102  0.0159  0.0429  0.0954  0.1023  0.1856  0.3741  0.8335
```

FIGURE 3.7 Numerical results obtained from the MATLAB program given in Figure 3.6.

The file is saved as *functionName.m*. Simple examples of function definition are

```
function [s, p] = addmult (x, y)
% Compute sum and product of two matrices
%
s = x + y;
p = x*y;
end
```

The file is stored as *addmult.m* and, when used, it yields the sum *s* and product *p* of the two variables *x* and *y*. Similarly, *fn1.m* represents the function file

```
function z = fn1 (x, y)
z = 0.5*y + x
end
```

It defines the function *fn1*(x, y), which can then be used in the computations similar to built-in functions like sin(x), exp(x), and log(x). Similarly, *absolute.m* represents the function

```
function y = absolute (x)
if x < 0
y = -x;
else
y = x;
end
```

which yields the absolute value of a given variable x.

Function *m*-files are very important in the development of MATLAB programs, since functions that are frequently needed are defined by these files. Functions can also be defined inline, as seen earlier for defining the ODE or as the following three functions:

```
f1 = inline('2.*(150.*x./(1 + exp(x)))', 'x');
f2 = inline('(2/(pi^0.5))*exp(-x.*x)', 'x');
f3 = inline('exp(x) + x^2', 'x');
```

The second entry within parentheses gives the independent variable and the first one the function definition. In the preceding cases, the second entry is not needed since it is implied by the function definition. However, if the independent variable were, say, y or z, then it would be needed. For instance, if 'y' is entered instead of 'x' in the third function definition, we would get f3(y) = exp(x) + x^2.

Once the function is defined using the *inline* command, given above, or as a function file, such as *f1.m*, the function can be evaluated at a given value of the independent variable, such as x, where x may be a scalar quantity or an array. However, if x is a vector, that is, an array, the function must be defined so that the mathematical operations such as multiplication and division can be performed on an array by using .* and ./, as discussed earlier. The command used for the evaluation of the function is *feval*, written as

```
feval(f,x) or feval('f',x)
```

where the first version is used for a function defined inline and the second for a function defined by a function file. Then, the value of the function f at given scalar x or the values of f corresponding to the components of an array x are given.

Function files and inline function definitions are extensively used in MATLAB programming, as seen later in various script files developed for different problems.

3.7 PLOTTING

One of the major strengths in MATLAB is the ease and variety with which outputs can be obtained in graphical form. A simple sequence of commands, such as

```
>> x=linspace (0, 2*pi);
>> y1=sin (x);
>> y2=cos (x);
>> y3=y1 .* y2;
>> plot (x, y1, '*', x, y2, '+', x, y3, '-')
```

uses vectorization to obtain the discrete values of sin (x), cos (x) and sin (x) cos (x), and plots these on a graph, with *, +, and – as the symbols to characterize the three variations. Legends and axes can also be defined, such as

```
legend ('sin(x)', 'cos(x)', 'sin(x) cos(x)')
axis ([0 2*pi -1.5 1.5])
```

which must be part of the script file or in one line with the plot command.

Different symbols, line formats, and colors are available. For instance, b, g, r, and k stand for blue, green, red, and black. Similarly, -, -., --, and : give solid, dash-dot, dashed, and dotted lines. Also, o, x, +, *, s, d, and h yield circle, cross, plus, asterisk, square, diamond, and hexagon markers. The command *hold on* allows a plot to be held so that other graphs can be added to the same figure. The *hold off* releases the hold on the figure.

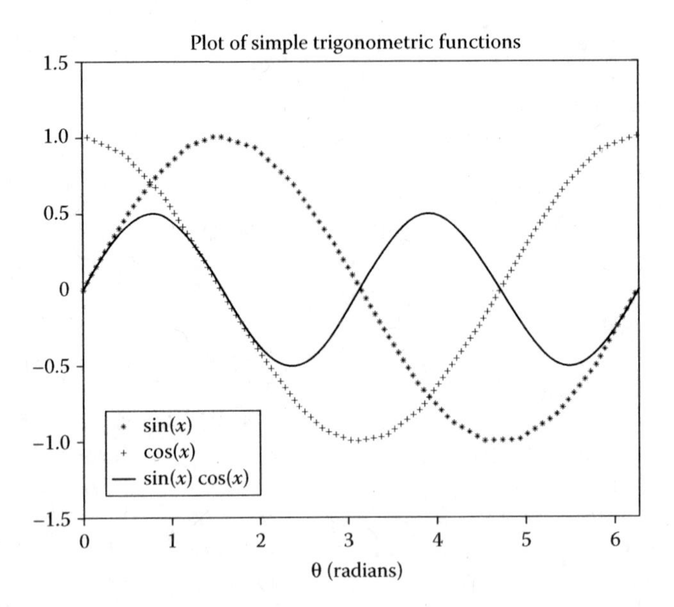

FIGURE 3.8 Plot of trigonometric functions sin(x), cos(x), and sin(x) cos(x) versus x, using MATLAB plotting commands.

Open figure windows can be closed with the command *close all*, so that other graphs can be displayed. The labels on the figure can similarly be specified, for instance,

```
xlabel ('\theta (radians)', 'Fontname', 'Times', ...
'Fontsize', 14)
```

This gives θ(radians) with the given font characteristics. For the y-label, "sin(*theta*)" may be used. The title of the figure can also be given easily, as, for instance,

```
Title ('Plot of simple trigonometric functions', ...
'FontName', 'Helvetica', 'FontSize', 16)
```

The results are shown in Figure 3.8. Similarly, various other possibilities are available to obtain the graphs in the desired form, including contour plots.

3.8 SUMMARY

This chapter presents a brief review of MATLAB programming, focusing on common commands and various features used in developing a MATLAB program in terms of a script *m*-file or a function *m*-file. The main strengths of using the MATLAB environment are discussed, particularly with respect to the generation, use, and manipulation of matrices. Simple commands that can be used for root solving of polynomials, curve fitting, solution of linear algebraic equations, ODEs, and so on, are outlined. Input and output commands, including entering variables and functions, as well as plotting, are briefly presented. A few examples are given in order to present the use of the MATLAB environment, and commands to solve mathematical problems. This brief review will serve to orient the students to the use of MATLAB for solving problems in science and engineering. However, for additional details and for more complex problems and situations, the various references mentioned here may be consulted. Also, it must be mentioned that, even though this book mainly considers the basic MATLAB computing environment generally available on common computer systems, a wide variety of *Toolboxes* are available and are used for specific and complex problems. These toolboxes include those for statistical analysis, economic analysis, finite-element methods, optimization, symbolic methods, solution of PDEs, visualization, data processing, and so on.

PROBLEMS

3.1. Become familiar with the MATLAB environment. Use commands to input data, employ mathematical operations, save variables, and vary format of outputs. Write a simple script *m*-file. Save and execute this file.

3.2. Try different mathematical operations in MATLAB, with and without semicolon. Print *eps*, *pi*, sin(x), and other such built-in functions. Also, calculate the inverse of sine, cosine, and tangent for common angles like 30°, 45°, 60°, and 90°. Remember that the angle x is in radians for such trigonometric functions.

3.3. Define different matrices by defining row and column elements in one command. Then obtain these matrices by defining the rows separately first and then using these rows to obtain the matrices.

3.4. Generate a 4×4 matrix of random numbers ranging from 0 to 8. Similarly, generate a 3×5 matrix of random numbers ranging from 2 to 12.

3.5. The following data are given for the independent variable x and dependent variable y:

```
x: 0 1 3 5 7 9 11
y: 2 5 14 47 128 281 530
```

Using the *polyfit* command, obtain the best fit to these data by increasing orders of the polynomial, from 1 to 5. From these results, is it possible to determine what order of polynomial is best suited to these data?

3.6. A polynomial equation is given as $x^3 - 7x^2 + 14x - 8 = 0$. Define this polynomial in the MATLAB environment and, using the *roots* function, obtain the three roots of the equation.

3.7. Using the approach discussed in the text, solve Equations 3.7 and 3.9 with the coefficient 0.1 replaced by 0.2 and the initial velocity given as 45 m/s.

3.8. Plot common functions exp(x), log(x), and x^n, for n = 1, 2, and 3, versus x, showing the appropriate title, legends, and labels.

3.9. Define the function $f(x) = x^3 + 2x^2 - 4x + 5$ by using the *inline* command as well as a function file. Then evaluate the function $f(x)$ at x values of 0, 1, 2, 3, . . ., 10. Finally, plot $f(x)$ versus x, with appropriate labels and ranges of the axes.

3.10. The charge q at a given time t in a capacitor in an electrical circuit is given by the expression:

```
q=E C+(Q-E C) exp(-t/RC)
```

where E is the source voltage, C the capacitance, Q the initial charge, and R the resistance. Two circuits are considered with $C = 1$, $Q = 10$, $R = 15$, and $E = 20$ in one case and $E = 7.5$ in the second case, all in SI units.

Do the following:

a. Write a script-*m* file to calculate the charge at $t = 0$, $t = 200$ (which is large time) and at any arbitrary time t.
b. Define the different variables for the two circuits.
c. Calculate charge at $t = 0$ and print it as *q_initial*.
d. Calculate charge at $t = 200$ and print it as *q_steady_state*.
e. Include an input statement to enter t.
f. Calculate the charge at time t and print it as *q_time_t*.
g. Run the program to ensure that all desired outputs are obtained.
h. Get the results for two times, $t = 5$ and $t = 10$ seconds.

3.11. Write a MATLAB script-*m* file to do the following:

a. Generate a 3×5 matrix a of random numbers between 30 and −5.
b. Determine the largest and smallest elements and their locations.
c. Print the information in (b) using the *sprintf* command.

3.12. Solve the following system of equations using the matrix algebra in MATLAB. Use both the invert matrix and backslash commands. Write a script-*m* file for this purpose.

$$x_1 + x_2 - x_3 = 0$$
$$x_2 - x_4 - x_5 = 0$$
$$x_1 + x_5 - x_6 = 0$$
$$2x_2 + 4x_3 + 6x_4 = 10$$
$$-6x_4 + 3x_5 + 5x_6 = -8$$
$$8x_1 - 2x_2 - 3x_5 = 0$$

where x_1, x_2, x_3, x_4, x_5, and x_6 are unknowns to be calculated.

3.13. For the function $f(v)$ given below,

$$f(v) = \frac{0.25v^2}{450 + [\log(v)]^{2.5}} - 0.023v - 9.8$$

write a script-*m* file to

a. Generate an array of velocities from 50 to 250 m/s.
b. Generate the corresponding array of $f(v)$ values.
c. Plot $f(v)$ versus v, using your choice of color and line style.
d. Label your axes and give a heading to the figure.
e. Run your program to ensure that the results are satisfactory.

3.14. Using *if*, *while*, and *for* loops, calculate the sum of n natural numbers, odd numbers (starting with 1), and even numbers (starting with 2). Choose and vary n. Validate your results.

4 Taylor Series and Numerical Differentiation

4.1 INTRODUCTION

In problems of engineering interest, the numerical solution is generally based on discrete values of a given function and its derivatives at a finite number of points in the computational domain. The need to discretize a function arises since a digital computer can usually carry out only the standard arithmetic operations, employing a finite number of discrete values. Also, in many cases, interest lies in estimating the derivatives from discrete numerical or experimental values of the function, given at specified data points. The derivatives are then computed at these data points or at a number of intermediate locations, employing only arithmetic operations. Similarly, the numerical integration of a function may be carried out, using the discrete values of the function. Of course, as mentioned earlier, symbolic algebra may also be used in a few limited cases to differentiate or integrate continuous functions, employing software such as *Maple* or *Mathematica*. This chapter discusses the basic concepts involved in discretization as well as in the computation of the derivatives of a given function from given discrete values.

Numerical differentiation refers to the computational procedure for evaluating the derivatives of a function, which is given as an analytical expression or in terms of discrete values at a finite number of points in the computational region. There are many diverse areas in engineering where numerical differentiation is needed. For example, in the dynamics of particles and systems, the time derivative of the displacement gives the velocity, and the second derivative gives the acceleration, which on multiplication with the mass of the body yields the force. In many engineering systems, such as robotics, the motion of the components is quite complex, and numerical differentiation is needed to determine the forces, velocities, and trajectories of the elements. Similarly, the heat transfer rate and the shear force at a surface due to fluid flow over the surface are obtained from the spatial derivatives of the temperature and the velocity, respectively. The distributions of temperature and velocity are often too complicated to permit use of the standard analytical methods for differentiation. The numerical values of the derivatives are also needed, for example, in optimization to obtain the best solution under given constraints, in economics to obtain the effect of a change in the interest rate or inflation on the financial dealings of a company, in electromagnetics to determine the wavelength at which maximum energy transfer arises, and in many other problems of practical interest.

Frequently, in engineering problems, one must solve an ODE or a PDE to obtain an unknown variable. Again, a numerical solution involves a finite number of locations or points where the value of the variable is computed. An important class of numerical methods for the solution of differential equations is based on replacing the derivatives by their discretized forms, known as *finite difference approximations*, and then solving the resulting algebraic equations. The numerical analysis that forms the basis of discretization of derivatives is often called *finite difference calculus*. Differential equations arise in many engineering areas, such as dynamics and vibrations, heat transfer and fluid flow, electronic circuitry, structures, mass transfer, and neutron diffusion in nuclear reactors. The numerical methods for the solution of ODEs and PDEs are discussed in Chapters 9 and 10, respectively.

In this chapter, we shall obtain finite difference formulations, which allow the computation of derivatives from discrete values of the function given at a finite number of points as well as the representation of derivatives in terms of discrete values. An important consideration in finite difference calculus is the error that arises due to the use of an approximation instead of the exact mathematical expression. Some discussion on the errors associated with discretization was included in Chapter 2. These errors are considered in greater detail here. The Taylor series forms the basis of many numerical techniques and also is used for estimating the errors involved. The general form of the series is presented, and the error resulting from the truncation of the series after a finite number of terms is determined. There are several approaches that may be adopted for deriving the finite difference approximation of the derivatives of a function. These approaches include the direct method, based on the definition of the derivative, the Taylor-series approach, and the use of a polynomial representation of the function. These three approaches are discussed, with particular emphasis on the derivation based on the Taylor series since it also yields quantitative information on the error. Finally, the corresponding approximations for partial derivatives are outlined.

4.2 TAYLOR SERIES

4.2.1 Basic Features

Let us consider a function $f(x)$ whose value at a given point $x = x_i$ is denoted by $f(x_i)$. The Taylor series is an infinite power series that expresses the value of the function in a region sufficiently close to $x = x_i$ as follows:

$$f(x) = f(x_i) + (x - x_i)f'(x_i) + \frac{(x - x_i)^2}{2!} f''(x_i) + \frac{(x - x_i)^3}{3!} f'''(x_i) + \cdots \quad (4.1)$$

or

$$f(x_i + \Delta x) = f(x_i) + \Delta x\, f'(x_i) + \frac{(\Delta x)^2}{2!} f''(x_i) + \frac{(\Delta x)^3}{3!} f'''(x_i) + \cdots \quad (4.2)$$

where Δx is the magnitude of a finite increment in the independent variable x, from the given value $x = x_i$, and the primes denote differentiation with respect to x. The

Taylor-series expansion has been taken about $x = x_i$ and, thus, all the derivatives are evaluated at $x = x_i$. Similarly, we may write

$$f(x_i - \Delta x) = f(x_i) - \Delta x\, f'(x_i) + \frac{(\Delta x)^2}{2!} f''(x_i) - \frac{(\Delta x)^3}{3!} f'''(x_i) + \cdots \tag{4.3}$$

It is assumed that all the derivatives of the function $f(x)$, at $x = x_i$, exist and are finite. Also, Δx must be sufficiently small so that the series is convergent. Such a power series has a radius of convergence, given in terms of the increment Δx, within which the series is convergent (Keisler, 1986; Larson and Edwards, 2009). Generally, the radius of convergence is finite, and if Δx is taken as larger than this value, the series is no longer convergent and the region is not sufficiently close to $x = x_i$. However, in finite difference computations, we do have the freedom to choose the value of Δx and thus control the convergence of the series and also the accuracy of the solution, as discussed below.

If an infinite number of terms is taken in the series given by Equations 4.2 and 4.3, the exact value of $f(x_i + \Delta x)$, or $f(x_i - \Delta x)$, may be computed, provided the series is convergent. However, it is obviously not possible to compute an infinite number of terms. The practical approach to such a computation is to retain only a few terms in the series for approximating the function and to estimate the error resulting from neglecting the remaining terms. If only the first term in the series of Equation 4.2 is retained, then $f(x_i + \Delta x) \simeq f(x_i)$, and the function $f(x)$, is taken as a constant. The retention of the first two terms gives

$$f(x_i + \Delta x) \cong f(x_i) + \Delta x f'(x_i) \tag{4.4}$$

Thus, a linear approximation of the function is employed over the region from x_i to $(x_i + \Delta x)$, and the slope is taken as constant. Similarly, if the first three terms in the series are retained,

$$f(x_i + \Delta x) \cong f(x_i) + \Delta x f'(x_i) + \frac{(\Delta x)^2}{2!} f''(x_i) \tag{4.5}$$

This expression allows a variation in the slope over the region and is, therefore, a more accurate approximation for $f(x_i + \Delta x)$ than that given by Equation 4.4.

Figure 4.1 shows the three circumstances of retaining one, two, or three terms in Equation 4.2 graphically. Equation 4.4 becomes exact only if $f(x)$ is a linear function of x. Similarly, Equation 4.5 is exact for a parabolic, or second-order, function. Therefore, for an arbitrary function $f(x)$, the accuracy of the representation by the Taylor series improves as additional terms are retained. Although an nth-order series expansion is exact for an nth-order polynomial, an infinite number of terms is, in general, needed for other differentiable and continuous functions.

4.2.2 Finite Difference Calculus

In finite difference calculus, the function $f(x_i + \Delta x)$ is generally written as $f(x_{i+1})$, indicating the value of the function at a neighboring point $x = x_{i+1}$, which is at an

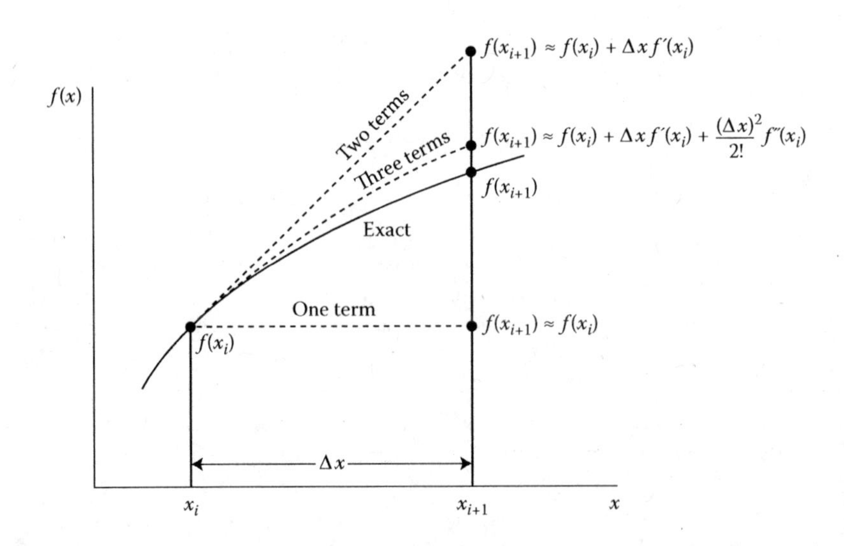

FIGURE 4.1 Approximation of a function $f(x)$ by a Taylor-series expansion, retaining one, two, or three terms in the series.

incremental distance Δx from the point $x = x_i$, about which the Taylor-series expansion has been taken. Then the series in Equations 4.2 and 4.3 may be written as

$$f(x_{i\pm1}) = f(x_i) \pm \Delta x\, f'(x_i) + \frac{(\Delta x)^2}{2!} f''(x_i) \pm \frac{(\Delta x)^3}{3!} f'''(x_i) + \cdots \tag{4.6}$$

This infinite series must be truncated after a finite number of terms in order to be useful in digital computation. If the series is truncated after the $(n + 1)$th term, that is, after the term containing the nth power of Δx, the neglected terms may be taken into account by means of a remainder term R_n, given for $f(x_{i+1})$ by

$$R_n = \frac{d^{n+1} f}{dx^{n+1}}(\xi) \frac{(\Delta x)^{n+1}}{(n+1)!}, \quad \text{where } x_i < \xi < x_{i+1} \tag{4.7}$$

The derivative in this expression is evaluated at a point $x = \xi$, which lies within the interval from x_i to x_{i+1}, and $(n + 1)!$ represents the factorial of $(n + 1)$. This remainder term is often known as the Lagrange form and its derivation is given in most textbooks on calculus; see, for instance, the books by Amazigo and Ruhenfeld (1980) and Larson and Edwards (2009).

The expression for the remainder given by Equation 4.7 can be employed for estimating the error, known as *truncation error* and briefly considered in Chapter 2, that results from a truncation of the series. Thus, the error when the series is truncated after the term containing $(\Delta x)^n$ is less than $\left|d^{n+1} f / dx^{n+1}\right|_{max} (\Delta x)^{n+1}/(n+1)!$, where the maximum magnitude of the derivative in the interval $x_i < x < x_i + 1$, is denoted by the subscript "max." The value of the $(n + 1)$th derivative of the given function, in the entire interval, is generally not known, since this would require an analytical expression for $f(x)$, which is assumed to be unknown. If $f(x)$ is known in the interval, the Taylor-series expansion is not needed for evaluating the function at $x = x_{i+1}$. Therefore, one cannot use the remainder term to determine the error exactly. However, the term does indicate the dependence of the truncation error on Δx, and we do have control over the value of Δx.

The remainder R_n, and thus the error, is usually written as

$$R_n = O\left[(\Delta x)^{n+1}\right] \tag{4.8}$$

where, as discussed in Chapter 2, this expression implies that the truncation error is on the order of $(\Delta x)^{n+1}$. Since the quantities that multiply $(\Delta x)^{n+1}$ in Equation 4.7 are constants over the given interval, the expression $O[(\Delta x)^{n+1}]$ may be taken to indicate that the error is proportional to the step size Δx raised to the $(n + 1)$th power. Then Equation 4.1 may be written, with the corresponding truncation error, as

$$f(x_{i+1}) = f(x_i) + \Delta x\, f'(x_i) + \frac{(\Delta x)^2}{2!} f''(x_i) + O\left[(\Delta x)^3\right] \tag{4.9a}$$

or

$$f(x_{i+1}) = f(x_i) + \Delta x\, f'(x_i) + \frac{(\Delta x)^2}{2!} f''(x_i) + \frac{(\Delta x)^3}{3!} f'''(x_i) + O\left[(\Delta x)^4\right] \tag{4.9b}$$

For an arbitrary function $f(x)$, Equation 4.9b yields a more accurate value of $f(x_{i+1})$, as seen in Figure 4.1. Thus, within the radius of convergence of the series, the error term due to truncation after n terms is related to that due to truncation after $(n + 1)$ terms as follows:

$$O[(\Delta x)^n] > O[(\Delta x)^{n+1}] \tag{4.10}$$

implying that the error is larger in the former case. We will assume this relationship to be valid, as long as the series is convergent.

The representation of the truncation error as $O[(\Delta x)^n]$ also indicates the behavior of the error as Δx is reduced. Thus, if Δx is halved, the error becomes $1/2^n$ of the previous error. The order of the Taylor-series approximation is given by the value of n. A higher value of n implies the retention of a larger number of terms in the series and, thus, a smaller truncation error. Usually, Δx is taken as sufficiently small, so that only the first few terms in the series are required to obtain a fairly accurate estimate of $f(x_{i+1})$. The characteristics of the Taylor series and of the error resulting from truncation are illustrated in the following example.

Example 4.1

a. Derive the Taylor-series expansions for e^x and $\log(1 - x)$, about $x = 0$. Employing the first six terms in the series, determine the values of these functions at $x = 0.1$, 0.2, 0.3, 0.4, and 0.5. Add the terms successively, indicating the effect of the number of retained terms on the accuracy of the numerical results.

b. The relationship between the pressure p and temperature T of a given fluid is

$$\log p = 19.2 - \frac{5301.4}{T} \tag{4.11}$$

where log represents the natural logarithm, p is in kilopascals, abbreviated as kPa (l kPa = 10^3 N/m^2), and T is in kelvins. Using the Taylor-series expansion for p, compute the pressure at $T = 351, 352, 355, 360$, and 370 K, given the value at 350 K, from Equation 4.11, and using only five terms in the expansion.

SOLUTION

4.1(a) Here, the functions $f(x) = e^x$ and $\log(1 - x)$ are to be expanded in Taylor series about $x = 0$, at which location $e^x = 1$ and $\log(1 - x) = 0$. The Taylor series about zero is also often referred to as the *Maclaurin* series. In order to obtain the series, as given by Equation 4.1, we need to evaluate the derivatives at $x = 0$. Thus, for $f(x) = e^x$,

$$\frac{d(e^x)}{dx} = \frac{d^2(e^x)}{dx^2} = \frac{d^3(e^x)}{dx^3} = \frac{d^4(e^x)}{dx^4} = \cdots = e^x$$

At $x = 0$, all these derivatives are 1.0. Therefore, the required series for e^x, about $x = 0$, is obtained from Equation 4.1 as

$$e^x = 1 + x + \frac{x^2}{2!} + \frac{x^3}{3!} + \frac{x^4}{4!} + \cdots + \frac{x^n}{n!} + \cdots \tag{4.12}$$

The second function, $\log(1 - x)$, yields the following derivatives:

$$\frac{d[\log(1-x)]}{dx} = -\frac{1}{1-x}$$

$$\frac{d^2[\log(1-x)]}{dx^2} = -\frac{1}{(1-x)^2}$$

$$\frac{d^3[\log(1-x)]}{dx^3} = -\frac{2}{(1-x)^3}$$

$$\frac{d^4[\log(1-x)]}{dx^4} = -\frac{6}{(1-x)^3}$$

$$\vdots$$

At $x = 0$, the derivatives of the function $f(x) = \log(1 - x)$ are

$$f'(0) = -1, \quad f''(0) = -1, \quad f'''(0) = -2, \quad f''''(0) = -6, \ldots$$

Therefore, the Taylor-series expansion for $\log(1 - x)$ about $x = 0$ is obtained from Equation 4.1 by setting $x_i = 0$ as

$$\log(1 - x) = 0 + x \cdot (-1) + \frac{x^2}{2!} \cdot (-1) + \frac{x^3}{3!} \cdot (-2) + \frac{x^4}{4!} \cdot (-6) + \frac{x^5}{5!} \cdot (-24) + \cdots$$

$$= -\left[x + \frac{x^2}{2} + \frac{x^3}{3} + \frac{x^4}{4} + \frac{x^5}{5} + \cdots + \frac{x^n}{n} + \cdots\right] \quad (4.13)$$

A calculator or a computer may be used to sum a finite number of terms in the two series given by Equations 4.12 and 4.13. A computer program may also be easily written for this purpose. Starting with the first term, additional terms can be included successively and the resulting sum determined. This process is then continued up to the sixth term, employing the various values of x given in the problem. For instance, a typical program in MATLAB® for e^x could include the following commands;

```
x=0.1;
s=0;
b=1;
for n=1:6
sn=(x^(n-1))/b;
s=s+sn;
b=b*n;
n,s
end
f=exp(x)
```

Appropriate input/output formats can be also included such as

```
>>fprintf('N=%1d: S=%.6f\n',n,s)
>>fprintf('EXACT VALUE= %.6f\n',f)
```

to obtain the results in a desired form. Similarly, a MATLAB program may be written for the function $\log(1 - x)$.

The numerical results obtained from such a computer program are shown in Figure 4.2. The exact values of the functions at the various x values considered are also computed and printed, along with the numerical results for comparison. Note that, as expected, the accuracy of the numerical evaluation of the functions from their respective Taylor-series expansions improves as the number of terms considered increases. Six terms are found to be quite adequate at smaller x values, although more terms should be employed for x equal to or larger than 0.5 for better accuracy. The convergence is thus slower at larger x, as expected and as shown in most calculus textbooks. Also, the series for e^x converges at all x, whereas that for $\log(l - x)$ converges only if $|x| < 1$. Therefore, at larger x, within the convergent space for the given function, additional terms should be included until the sum remains essentially unchanged with a further addition of terms.

(a) $f(x) = e^x$	(b) $f(x) = \log(1 - x)$
X = .1	X = .1
N = 1 : S = 1	N = 1 : S = −.1
N = 2 : S = 1.1	N = 2 : S = −.105
N = 3 : S = 1.105	N = 3 : S = −.105333
N = 4 : S = 1.105167	N = 4 : S = −.105358
N = 5 : S = 1.105171	N = 5 : S = −.105360
N = 6 : S = 1.105171	N = 6 : S = −.105361
EXACT VALUE = 1.105171	EXACT VALUE = −.105361
X = .2	X = .2
N = 1 : S = 1	N = 1 : S = −.2
N = 2 : S = 1.2	N = 2 : S = −.22
N = 3 : S = 1.22	N = 3 : S = −.222667
N = 4 : S = 1.221333	N = 4 : S = −.223067
N = 5 : S = 1.2214	N = 5 : S = −.223131
N = 6 : S = 1.221403	N = 6 : S = −.223141
EXACT VALUE = 1.221403	EXACT VALUE = −.223144
X = .3	X = .3
N = 1 : S = 1	N = 1 : S = −.3
N = 2 : S = 1.3	N = 2 : S = −.345
N = 3 : S = 1.345	N = 3 : S = −.354
N = 4 : S = 1.3495	N = 4 : S = −.356025
N = 5 : S = 1.349837	N = 5 : S = −.356511
N = 6 : S = 1.349858	N = 6 : S = −.356632
EXACT VALUE = 1.349859	EXACT VALUE = −.356675
X = .4	X = .4
N = 1 : S = 1	N = 1 : S = −.4
N = 2 : S = 1.4	N = 2 : S = −.48
N = 3 : S = 1.48	N = 3 : S = −.501333
N = 4 : S = 1.490667	N = 4 : S = −.507733
N = 5 : S = 1.491733	N = 5 : S = −.509781
N = 6 : S = 1.491819	N = 6 : S = −.510464
EXACT VALUE = 1.491825	EXACT VALUE = −.510826
X = .5	X = .5
N = 1 : S = 1	N = 1 : S = −.5
N = 2 : S = 1.5	N = 2 : S = −.625
N = 3 : S = 1.625	N = 3 : S = −.666667
N = 4 : S = 1.645833	N = 4 : S = −.682292
N = 5 : S = 1.648438	N = 5 : S = −.688542
N = 6 : S = 1.648698	N = 6 : S = −.691146
EXACT VALUE = 1.648721	EXACT VALUE = −.693147

FIGURE 4.2 Numerical results on the summation of the Taylor series expansions for e^x and $\log(1 - x)$ at various values of x, as given in Example 4.1a, along with the exact values of these functions.

The given relation between p and T, Equation 4.11, may be written as follows:

$$p = \exp\left[19.2 - \frac{5301.4}{T}\right] = \exp(19.2)\exp\left(-\frac{5301.4}{T}\right) \tag{4.14}$$

Therefore, a Taylor-series expansion for $\exp(-5301.4/T)$ is needed for computing the pressure p at temperatures close to $T = 350$ K, about which the expansion is to be carried out. We may write Equation 4.14 as

$$p = Af(T), \quad \text{where } f(T) = \exp\left(\frac{B}{T}\right) = \exp\left(-\frac{5301.4}{T}\right) \quad \text{and} \quad A = \exp(19.2) \tag{4.15}$$

The Taylor-series expansion for p is then given by

$$p = p_{350} + (T - 350)Af'(350) + \frac{(T-350)^2}{2!}Af''(350)$$
$$+ \frac{(T-350)^3}{3!}Af'''(350) + \frac{(T-350)^4}{4!}Af''''(350) + \cdots \tag{4.16}$$

where p_{350} refers to the pressure at 350 K, as calculated from Equation 4.14, and the quantity within the parentheses indicates the temperature T at which the evaluation is made. Now, f', f'', f''', and so on, are obtained by differentiation as

$$f'(T) = -\frac{B}{T^2}e^{B/T}$$

$$f''(\mathrm{T}) = \left(\frac{B^2}{T^4} + \frac{2B}{T^3}\right)e^{B/T}$$

$$f'''(\mathrm{T}) = \left(-\frac{B^3}{T^6} - \frac{6B^2}{T^5} - \frac{6B}{T^4}\right)e^{B/T}$$

$$\vdots$$

If T is replaced by 350 K in these expressions and substituted in Equation 4.16, the series for p is obtained.

The value of the pressure p_{350} at $T = 350$ is calculated from Equation 4.14 as 57.5781 kPa. Using this value, we calculate the pressures at $T = 351$, 352, 355, 360, and 370 K from the Taylor-series given by Equation 4.16. Again, a computer program similar to the ones for the preceding problem may be developed to sum the series. The numerical results obtained are shown in Figure 4.3. The sum of the series, considering one, two, and up to five terms, is given, along with the corresponding exact value from Equation 4.14. As expected, the accuracy of the numerical value for the pressure improves as the number of terms employed is increased. Five terms are found to be quite satisfactory, particularly for small temperature differences $\Delta T = T - 350$. However, as $(T - 350)$ increases

TEMPERATURE = 351
N = 1: P = 57.5781331 N = 2: P = 60.0699267
N = 3: P = 60.1167256 N = 4: P = 60.1172157
N = 5: P = 60.1172187
THE EXACT VALUE OF THE PRESSURE = 60.1172137

TEMPERATURE = 352
N = 1: P = 57.5781331 N = 2: P = 62.5617203
N = 3: P = 62.748916 N = 4: P = 62.7528363
N = 5: P = 62.7528845
THE EXACT VALUE OF THE PRESSURE = 62.7528848

TEMPERATURE = 355
N = 1: P = 57.5781331 N = 2: P = 70.0371011
N = 3: P = 71.2070744 N = 4: P = 71.2683794
N = 3: P = 71.2702111
THE EXACT VALUE OF THE PRESSURE = 71.2702421

TEMPERATURE = 360
N = 1: P = 57.5781331 N = 2: P = 82.496069
N = 3: P = 87.1759625 N = 4: P = 87.6660022
N = 5: P = 87.6961086
THE EXACT VALUE OF THE PRESSURE = 87.6971056

TEMPERATURE = 370
N = 1.: P = 57.5781331 N = 2: P= 107.414005
N = 3: P = 126.133579 N = 4 : P= 130.053896
N = 5: P = 130.535599
THE EXACT VALUE OF THE PRESSURE = 130.567704

FIGURE 4.3 Numerical values of the pressure p obtained from a summation of the Taylor series expansion for the function $p(T)$, as given in Example 4.1b, for various values of the temperature T.

to larger values, additional terms will be needed for accurate results. Obviously, the numerical sum of the series is in considerable error if only one or two terms are retained, particularly at large ΔT.

The simple problems given in Example 4.1 illustrate practical applications of the Taylor-series expansions and also bring out the important aspects that one needs to bear in mind in summing the series. It is important to ensure that the numerical results converge to an essentially constant value as the number of terms is increased. If the results diverge as additional terms are included, the series is not convergent, and an alternative approach to obtain the desired numerical results must be employed. In some engineering problems, the function $f(x)$ may be too complicated to be evaluated easily in the vicinity of the x value at which it is known. Then Taylor series may be used, as outlined above. The series is also sometimes employed for providing starting values in the solution of differential equations. However, the major interest in the

Taylor series is because of the approximation to the derivatives obtained by the use of the series, as discussed in Section 4.4.

4.3 DIRECT APPROXIMATION OF DERIVATIVES

In many diverse engineering problems, the accurate determination of derivatives from the measured or calculated values of the function $f(x)$ at a finite number of discrete points is needed. Consider, for example, an engineer, on a test track, involved in the measurement of the location of a moving body, such as a car, as a function of time. The velocity and acceleration of the object are given by the computed values of the first and second derivatives of the displacement. Similarly, a chemical or civil engineer may measure the concentration of a pollutant in a water body as a function of location and time and then use this information to obtain the rate of spread of chemical pollution. Heat and mass transfer processes are also concerned with the rates of transport, and measurements of temperature and concentration are often employed for developing models for predicting transport rates in several practical circumstances. Thus, the approximation of derivatives is important in many practical problems and also in the solution of differential equations by the finite difference approach, as presented in Chapters 9 and 10.

A simple approach to the derivation of the finite difference approximation to the derivatives of a function $f(x)$ is based on the replacement of infinitesimal differences by finite differences in the mathematical definition of differentiation. Finite differences are considered in the variation of the independent variable x, and the values of the function $f(x)$ at discrete points are employed in deriving the approximation. Consider the variation of $f(x)$ with x, as sketched in Figure 4.4. Three discrete points, denoted by subscripts $i - 1$, i, and $i + 1$, are shown along the x-axis. We may approximate the derivatives of $f(x)$, with respect to x, in terms of the corresponding discrete values.

The first derivative df/dx at $x = x_i$ can be approximated by $\Delta f/\Delta x$, where Δ denotes finite differences. Three approximations for $(df/dx)_i$ can be written by considering the differences between the values at the three discrete locations or nodes. These approximations are as follows:

$$\left(\frac{df}{dx}\right)_i \cong \frac{f_{i+1} - f_i}{\Delta x} \tag{4.17}$$

$$\left(\frac{df}{dx}\right)_i \cong \frac{f_i - f_{i-1}}{\Delta x} \tag{4.18}$$

$$\left(\frac{df}{dx}\right)_i \cong \frac{f_{i+1} - f_{i-1}}{2\Delta x} \tag{4.19}$$

where the subscripts denote the nodal location, in x, where the quantity is evaluated. The first approximation is known as the *two-point forward difference approxima-*

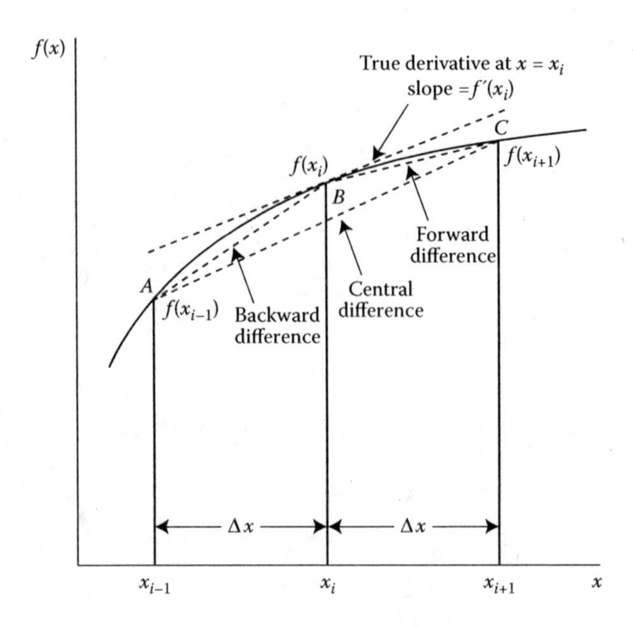

FIGURE 4.4 Graphical representation of the forward, backward, and central finite difference approximations of the first derivative of a function $f(x)$.

tion for $(df/dx)_i$, since only two nodes are involved and the value of the function in the forward, or the increasing x, direction is employed. Similarly, the second approximation, Equation 4.18, is known as the *two-point backward difference*, and the third approximation, Equation 4.19, as the *three-point central difference approximation.*

These approximations employ the slopes of the chords to the right of, to the left of, and centered on the node point at $x = x_i$, as shown in Figure 4.4, to approximate the gradient of the function at $x = x_i$. Since these are only approximations to the derivative $(df/dx)_i$, an approximate equality sign (≅) is used. The central difference may be interpreted in either of the following two ways:

$$\frac{f_{i+1} - f_{i-1}}{2\Delta x} = \frac{1}{2}\left[\frac{f_{i+1} - f_i}{\Delta x} + \frac{f_i - f_{i-1}}{\Delta x}\right] \quad (4.20a)$$

or

$$\frac{f_{i+1} - f_{i-1}}{2\Delta x} = \frac{1}{\Delta x}\left[\frac{1}{2}\left(f_{i+1} + f_i\right) - \frac{1}{2}\left(f_i + f_{i-1}\right)\right] \quad (4.20b)$$

The first equation represents an average of the two one-sided differences, and the second the difference based on the average values of the function at $(i + 1/2)$ and $(i - 1/2)$. Because the central difference averages out the variations on either side of the node $x = x_i$, it is expected to be a more accurate representation of the derivative. This is shown to be true on the basis of the Taylor-series approach presented in Section 4.4.

Similarly, the finite difference approximation for the second derivative d^2f/dx^2 at $x = x_i$ may be derived. Thus,

$$\frac{d^2 f}{dx^2} = \frac{d}{dx}\left(\frac{df}{dx}\right) \cong \frac{\Delta}{\Delta x}\left(\frac{\Delta f}{\Delta x}\right) \cong \frac{1}{\Delta x}\left[\frac{f_{i+1} - f_i}{\Delta x} - \frac{f_i - f_{i-1}}{\Delta x}\right]$$

or

$$\frac{d^2 f}{dx^2} \cong \frac{f_{i+1} - 2f_i + f_{i-1}}{\Delta x^2} \tag{4.21}$$

Here, the difference in $\Delta f/\Delta x$ is approximated by the difference between the slopes of the two chords, on either side of the node at $x = x_i$. In fact, $(f_{i+1} - f_i)/\Delta x$ represents the central difference approximation of the derivative at $x = x_i + (\Delta x/2)$, or $i + 1/2$, since it uses the values on either side of this location, with discrete differences of $\Delta x/2$ in x. Similarly $(f_i - f_{i-1})/\Delta x$ represents the central difference approximation of the derivative at $x = x_i - (\Delta x/2)$, or $i - 1/2$. Therefore,

$$\frac{d^2 f}{dx^2} \cong \frac{\Delta}{\Delta x}\left(\frac{\Delta f}{\Delta x}\right) \cong \frac{1}{\Delta x}\left[\left(\frac{\Delta f}{\Delta x}\right)_{i+1/2} - \left(\frac{\Delta f}{\Delta x}\right)_{i-1/2}\right]$$
$$\frac{d^2 f}{dx^2} \cong \frac{1}{\Delta x}\left[\frac{f_{i+1} - f_i}{\Delta x} - \frac{f_i - f_{i-1}}{\Delta x}\right] = \frac{f_{i+1} - 2f_i + f_{i-1}}{(\Delta x)^2} \tag{4.22}$$

The preceding finite difference approximation of the second derivative is known as the *three-point central second difference approximation.* Other approximations for the second derivative may also be obtained by employing other finite difference representations in the above derivation. However, the central second difference is the most frequently employed approximation. Similarly, finite difference approximations for higher-order derivatives may be derived. Again, several representations are usually possible, with central differences being more accurate than one-sided differences, if the same nodal points are used in the two cases.

The direct approximation of the derivatives thus allows one to derive the required finite difference representations. The approach is based on the mathematical interpretation of differentiation, and, therefore, it provides a physical background for the formulation of finite differences. However, it does not give any information on the accuracy of a particular representation. For an estimation of the error involved, we must employ the Taylor-series approach as described in Section 4.4.

4.4 TAYLOR-SERIES APPROACH AND ACCURACY

The Taylor-series expansions about a given nodal point $x = x_i$ may be employed to derive the finite difference approximations of the derivatives of a function $f(x)$. Since the error resulting from the truncation of the series after a finite number of terms can be estimated from the remainder term, given by Equation 4.7, the errors associated with the various finite difference approximations of the derivatives, obtained by the Taylor-series approach, may also be estimated. Using this approach, one can derive one-sided, forward and backward, and central difference approximations.

4.4.1 Finite Difference Approximation of the First Derivative

Consider the variation of the function $f(x)$ with x, as shown in Figure 4.4. If the function $f(x)$ is sufficiently smooth, it can be expanded in a Taylor series in the neighborhood of $x = x_i$. Assuming that the points x_{i-1} and x_{i+1} lie within the region of convergence of the series, the function $f(x)$ at these points is given by Equation 4.6 as

$$f_{i+1} = f_i + \Delta x\, f_i' + \frac{(\Delta x)^2}{2!} f_i'' + \frac{(\Delta x)^3}{3!} f_i''' + \cdots \tag{4.23}$$

$$f_{i-1} = f_i - \Delta x\, f_i' + \frac{(\Delta x)^2}{2!} f_i'' - \frac{(\Delta x)^3}{3!} f_i''' + \cdots \tag{4.24}$$

where the subscripts again denote the nodal locations, in x, where the function is evaluated, and the primes denote differentiation with respect to x.

If Equation 4.23 is solved for the first derivative f_i', we obtain

$$f_i' = \frac{f_{i+1} - f_i}{\Delta x} - \frac{\Delta x}{2} f_i'' - \frac{(\Delta x)^2}{6} f_i''' + \cdots$$

When the infinite series is replaced by the remainder term, this equation becomes

$$\begin{aligned} f_i' &= \frac{f_{i+1} - f_i}{\Delta x} - \frac{\Delta x}{2} f''(\xi), \quad \text{where } x_i < \xi < x_{i+1} \\ &= \frac{f_{i+1} - f_i}{\Delta x} + O(\Delta x) \end{aligned} \tag{4.25}$$

This equation gives the forward difference approximation of the first derivative, given by Equation 4.17, along with the truncation error in the approximation.

Similarly, the two-point backward difference for the first derivative may be obtained by solving Equation 4.18 for f_i' as follows:

$$f_i' = \frac{f_i - f_{i-1}}{\Delta x} + \frac{\Delta x}{2} f_i'' - \frac{(\Delta x)^2}{6} f_i''' + \cdots$$

which gives

$$f_i' = \frac{f_i - f_{i-1}}{\Delta x} + \frac{\Delta x}{2} f''(\xi), \quad \text{where } x_{i-1} < \xi < x_i$$
$$= \frac{f_i - f_{i-1}}{\Delta x} + O(\Delta x) \tag{4.26}$$

The truncation error is of the same order as that in the forward difference approximation. Since the error terms are included in Equations 4.25 and 4.26, the approximate equality signs of Equations 4.17 and 4.18 are not needed here.

A more accurate finite difference approximation of the first derivative is obtained by subtracting Equation 4.24 from Equation 4.23, to yield

$$f_i' = \frac{f_{i+1} - f_{i-1}}{2\Delta x} - \frac{(\Delta x)^2}{6} f'''(\xi), \quad \text{where } x_{i-1} < \xi < x_{i+1}$$
$$= \frac{f_{i+1} - f_{i-1}}{2\Delta x} + O\left[(\Delta x)^2\right] \tag{4.27}$$

The result is the three-point central difference approximation of the first derivative, as given earlier in Equation 4.19. The truncation error is on the order of $(\Delta x)^2$, and, therefore, this representation is more accurate than the forward and backward differences. Graphically, this expression approximates the derivative of the function $f(x)$ at x_i as the slope of the line AC in Figure 4.4. The forward and backward differences approximate the derivative by the slopes of the chords BC and AB, respectively. Also note from the preceding expressions that if the step size Δx is halved, the truncation error is also approximately halved for the forward and backward differences, whereas the error becomes one-fourth for the central difference.

4.4.2 Second Derivative

A finite difference approximation for the second derivative f_i'' may be derived by adding Equations 4.23 and 4.24, yielding

$$f_i'' = \frac{f_{i+1} - 2f_i + f_{i-1}}{(\Delta x)^2} - \left[\frac{1}{12}(\Delta x)^2 f_i'''' + \cdots\right]$$

Therefore,

$$f_i'' = \frac{f_{i+1} - 2f_i + f_{i-1}}{(\Delta x)^2} - \frac{(\Delta x)^2}{12} f''''(\xi), \quad \text{where } x_{i-1} < \xi < x_{i+1}$$
$$= \frac{f_{i+1} - 2f_i + f_{i-1}}{(\Delta x)^2} + \left[O(\Delta x)^2\right] \tag{4.28}$$

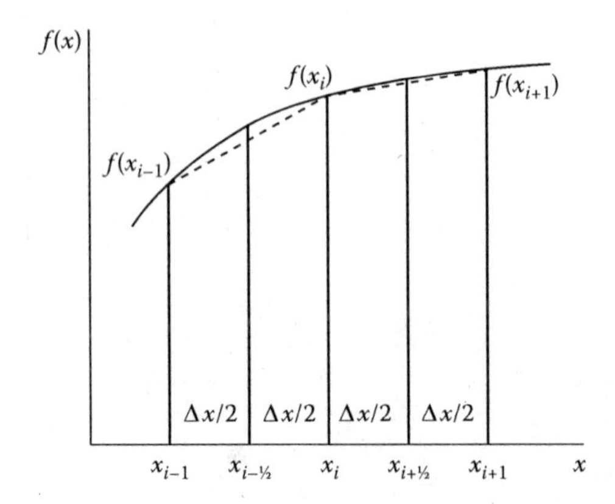

FIGURE 4.5 Graphical depiction of the finite difference approximation of the second derivative.

The result is the second central difference, which was derived in the preceding section by the direct approximation approach. The truncation error is $O[(\Delta x)^2]$, and, therefore, this finite difference approximation is of second order. Graphically, this expression approximates the second derivative by dividing the difference in the slopes of the chords that approximate the first derivatives at $x_{i+1/2}$ and $x_{i-1/2}$ by Δx; see Figure 4.5. The slopes of these chords are approximated in the central difference formulation as follows:

$$f'\left(x_i + \frac{\Delta x}{2}\right) = f'_{i+1/2} = \frac{f_{i+1} - f_i}{\Delta x} + O\left[(\Delta x)^2\right] \tag{4.29}$$

and

$$f'\left(x_i - \frac{\Delta x}{2}\right) = f'_{i-1/2} = \frac{f_i - f_{i-1}}{\Delta x} + O\left[(\Delta x)^2\right] \tag{4.30}$$

Thus,

$$\begin{aligned} f_i'' &= \frac{f'_{i+1/2} - f'_{i-1/2}}{\Delta x} + O\left[(\Delta x)^2\right] \\ &= \frac{f_{i+1} - 2f_i + f_{i-1}}{(\Delta x)^2} + O\left[(\Delta x)^2\right] \end{aligned} \tag{4.31}$$

Similarly, one-sided forward or one-sided backward differences may be derived for f_i'' by employing points on only one side of $x = x_i$, rather than on both sides, as done for the central difference. Let us consider, for example, the three points at x_i, x_{i+1}, and x_{i+2}, as shown in Figure 4.6. The Taylor-series expansion for $f(x_{i+1})$ is given by Equation 4.23, and the expansion for $f(x_{i+2})$ is

$$f_{i+2} = f_i + (2\Delta x) f_i' + \frac{(2\Delta x)^2}{2!} f_i'' + \frac{(2\Delta x)^3}{3!} f_i''' + \frac{(2\Delta x)^4}{4!} f_i'''' + \cdots \tag{4.32}$$

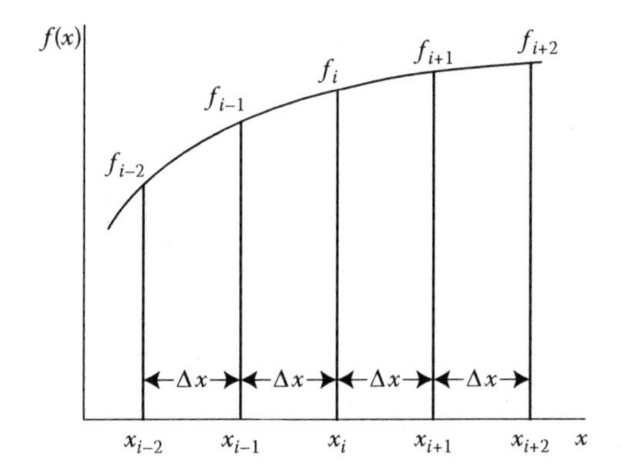

FIGURE 4.6 Distribution of the nodal points for deriving forward, backward, and central difference approximations for higher-order derivatives and also for higher-accuracy formulas.

Now, the first derivative f_i' may be eliminated from Equations 4.23 and 4.32 to yield an expression for f_i''. Thus, multiplying Equation 4.23 by 2 and subtracting the resulting equation from Equation 4.32 gives

$$
\begin{aligned}
f_i'' &= \frac{f_{i+2} - 2f_{i+1} + f_i}{(\Delta x)^2} - (\Delta x\, f_i''' + \cdots) \\
f_i'' &= \frac{f_{i+2} - 2f_{i+1} + f_i}{(\Delta x)^2} + O(\Delta x)
\end{aligned}
\tag{4.33}
$$

This result is the forward difference approximation of the second derivative. This approximation is accurate to within an error of order Δx.

Similarly, the backward difference approximation may be obtained by employing the Taylor-series expansions for $f(x_{i-1})$ and $f(x_{i-2})$ as follows:

$$
f_i'' = \frac{f_i - 2f_{i-1} + f_{i-2}}{(\Delta x)^2} + O(\Delta x) \tag{4.34}
$$

Note again that the forward and backward difference approximations are less accurate than the central difference approximation if the same number of nodal points is used in all three cases. Higher-order approximations may be derived by employing additional points, as shown later. Even though the one-sided differences are less accurate than the central difference, they are often employed for approximating the derivatives, particularly near the boundaries of a computational domain since nodal points may be available on only one side of the boundary.

4.4.3 Higher-Order Derivatives

The finite difference approximations of higher-order derivatives may be derived by the use of Taylor-series expansions, as outlined earlier for the first and second derivatives.

However, the derivation becomes more involved as one proceeds to successively higher derivatives since an increasingly larger number of simultaneous equations must be solved. The larger number of equations is obtained by employing expansions at a larger number of nodal points. Thus, the formulas for the third and fourth derivatives may be obtained by employing the expansions for f_{i+1} and f_{i-1}, given by Equations 4.23 and 4.24, along with those for f_{i+2} and f_{i-2}. The expansion for f_{i+2} is given by Equation 4.32, and that for f_{i-2} is

$$f_{i-2} = f_i - (2\Delta x) f_i' + \frac{(2\Delta x)^2}{2!} f_i'' - \frac{(2\Delta x)^3}{3!} f_i''' + \frac{(2\Delta x)^4}{4!} f_i'''' + \cdots \quad (4.35)$$

Subtracting Equation 4.35 from Equation 4.32, we obtain

$$f_{i+2} - f_{i-2} = 4\Delta x\, f_i' + \frac{8(\Delta x)^3}{3!} f_i''' + O\left[(\Delta x)^5\right]$$

The substitution of the finite difference expression for f_i' from Equation 4.27 gives

$$f_{i+2} - f_{i-2} = 4\Delta x \left\{ \frac{f_{i+1} - f_{i-1}}{2\Delta x} - \frac{(\Delta x)^2}{6} f_i''' + O\left[(\Delta x)^4\right] \right\} + \frac{8(\Delta x)^3}{3} f_i''' + O\left[(\Delta x)^5\right]$$

This yields

$$f_i''' = \frac{f_{i+2} - 2f_{i+1} + 2f_{i-1} - f_{i-2}}{2(\Delta x)^3} + O\left[(\Delta x)^2\right] \quad (4.36)$$

Similarly, the finite difference approximation for the fourth derivative f_i'''' may be derived by adding Equations 4.32 and 4.35 and then substituting the approximation for f_i''. The resulting approximation is

$$\begin{aligned} f_i'''' &= \frac{f_{i+2} - 4f_{i+1} + 6f_i - 4f_{i-1} + f_{i-2}}{(\Delta x)^4} - \frac{(\Delta x)^2}{6} \frac{d^6 f}{dx^6}(\xi), \quad \text{where } x_{i-2} < \xi < x_{i+2} \\ &= \frac{f_{i+2} - 4f_{i+1} + 6f_i - 4f_{i-1} + f_{i-2}}{(\Delta x)^4} + O\left[(\Delta x)^2\right] \end{aligned} \quad (4.37)$$

Thus, the five grid points shown in Figure 4.6 are involved in the finite difference expression for the fourth derivative f_i'''', with a truncation error of order $(\Delta x)^2$. The corresponding expression for f_i''' also involves these points, except for f_i which drops out in the derivation. By employing a still larger number of points, one may derive expressions for the fifth and sixth derivatives to the same accuracy.

However, these derivations are quite involved because of the large number of equations to be solved. Another method, which is based on difference and derivative operators, may often be employed more easily for the derivation of higher-order derivatives. This approach is discussed by Salvadori and Baron (1961) and Hornbeck (1975).

Equations 4.36 and 4.37 give the central difference approximations of the third and fourth derivatives of $f(x)$, respectively. Similarly, one-sided forward or one-sided backward differences may be derived. As mentioned earlier, one-sided differences are of interest in only a few cases, such as near the boundaries of the computational region. The central differences are much more important and are employed for the approximation of the derivatives in a wide variety of engineering problems. Several of the commonly used finite difference formulations are given in Figures 4.7 through 4.9, including higher-accuracy formulas discussed in Section 4.4.4.

4.4.4 Higher-Accuracy Approximations

The Taylor-series approach for the derivation of finite difference formulas may be employed for obtaining approximations of higher accuracy, that is, smaller truncation

Forward Difference Approximations of $O(\Delta x)$

$$f_i' = \frac{f_{i+1} - f_i}{\Delta x}$$

$$f_i'' = \frac{f_{i+2} - 2f_{i+1} + f_i}{(\Delta x)^2}$$

$$f_i''' = \frac{f_{i+3} - 3f_{i+2} + 3f_{i+1} - f_i}{(\Delta x)^3}$$

$$f_i'''' = \frac{f_{i+4} - 4f_{i+3} + 6f_{i+2} - 4f_{i+1} + f_i}{(\Delta x)^4}$$

Forward Difference Approximations of $O[(\Delta x)^2]$

$$f_i' = \frac{-f_{i+2} + 4f_{i+1} - 3f_i}{2\Delta x}$$

$$f_i'' = \frac{-f_{i+3} + 4f_{i+2} - 5f_{i+1} + 2f_i}{(\Delta x)^2}$$

$$f_i''' = \frac{-3f_{i+4} + 14f_{i+3} - 24f_{i+2} + 18f_{i+1} - 5f_i}{2(\Delta x)^3}$$

$$f_i'''' = \frac{-2f_{i+5} + 11f_{i+4} - 24f_{i+3} + 26f_{i+2} - 14f_{i+1} + 3f_i}{(\Delta x)^4}$$

FIGURE 4.7 Forward finite difference formulas, along with the truncation errors.

Backward Difference Approximations of $O(\Delta x)$

$$f_i' = \frac{f_i - f_{i-1}}{\Delta x}$$

$$f_i'' = \frac{f_i - 2f_{i-1} + f_{i-2}}{(\Delta x)^2}$$

$$f_i''' = \frac{f_i - 3f_{i-1} + 3f_{i-2} - f_{i-3}}{(\Delta x)^3}$$

$$f_i'''' = \frac{f_i - 4f_{i-1} + 6f_{i-2} - 4f_{i-3} + f_{i-4}}{(\Delta x)^4}$$

Backward Difference Approximations of $O[(\Delta x)^2]$

$$f_i' = \frac{3f_i - 4f_{i-1} + f_{i-2}}{2\Delta x}$$

$$f_i'' = \frac{2f_i - 5f_{i-1} + 4f_{i-2} - f_{i-3}}{(\Delta x)^2}$$

$$f_i''' = \frac{5f_i - 18f_{i-1} + 24f_{i-2} - 14f_{i-3} + 3f_{i-4}}{2(\Delta x)^3}$$

$$f_i'''' = \frac{3f_i - 14f_{i-1} + 26f_{i-2} - 24f_{i-3} + 11f_{i-4} - 2f_{i-5}}{(\Delta x)^4}$$

FIGURE 4.8 Backward finite difference formulas, along with the truncation errors.

error than given thus far. All the finite difference representations derived earlier had a truncation error of order Δx or $(\Delta x)^2$. Although an accuracy of $O[(\Delta x)^2]$ is adequate for most problems of practical interest, since we can choose smaller Δx to improve the accuracy, higher-accuracy formulas are often employed if a given circumstance demands very accurate numerical results. Such a requirement arises, for instance, in the determination of the displacement and velocity of a projectile or of a robotic arm.

Higher-accuracy formulas can be developed by including additional terms in the Taylor-series expansions. However, a larger number of grid points will be required to generate the additional equations needed for eliminating the higher-order derivatives that arise due to the retention of additional terms. Consider, for example, the forward difference expression for the first derivative f_i'. As obtained earlier,

$$f_i' = \frac{f_{i+1} - f_i}{\Delta x} - \frac{\Delta x}{2} f_i'' - \frac{(\Delta x)^2}{6} f_i''' + \cdots$$

If instead of truncating the series after the first term, as done earlier, we retain the term of order Δx and substitute the forward finite difference expression for f_i'', we

Central Difference Approximations of $O[(\Delta x)^2]$

$$f_i' = \frac{f_{i+1} - f_{i-1}}{2\Delta x}$$

$$f_i'' = \frac{f_{i+1} - 2f_i + f_{i-1}}{(\Delta x)^2}$$

$$f_i''' = \frac{f_{i+2} - 2f_{i+1} + 2f_{i-1} - f_{i-2}}{2(\Delta x)^3}$$

$$f_i'''' = \frac{f_{i+2} - 4f_{i+1} + 6f_i - 4f_{i-1} + f_{i-2}}{(\Delta x)^4}$$

Central Difference Approximations of $O[(\Delta x)^4]$

$$f_i' = \frac{-f_{i+2} + 8f_{i+1} - 8f_{i-1} + f_{i-2}}{12\Delta x}$$

$$f_i'' = \frac{-f_{i+2} + 16f_{i+1} - 30f_i + 16f_{i-1} - f_{i-2}}{12(\Delta x)^2}$$

$$f_i''' = \frac{-f_{i+3} + 8f_{i+2} - 13f_{i+1} + 13f_{i-1} - 8f_{i-2} + f_{i-3}}{8(\Delta x)^3}$$

$$f_i'''' = \frac{-f_{i+3} + 12f_{i+2} - 39f_{i+1} + 56f_i - 39f_{i-1} + 12f_{i-2} - f_{i-3}}{6(\Delta x)^4}$$

FIGURE 4.9 Central difference approximations, with the associated truncation errors.

will obtain a higher-accuracy forward difference expression for f_i'. Thus, from Equation 4.33,

$$\begin{aligned} f_i' &= \frac{f_{i+1} - f_i}{\Delta x} - \frac{\Delta x}{2}\left[\frac{f_{i+2} - 2f_{i+1} + f_i}{(\Delta x)^2} - \Delta x\, f_i''' + \cdots\right] - \frac{(\Delta x)^2}{6} f_i''' + \cdots \\ &= \frac{-f_{i+2} + 4f_{i+1} - 3f_i}{2\Delta x} + \frac{(\Delta x)^2}{3} f_i''' + \cdots \\ f_i' &= \frac{-f_{i+2} + 4f_{i+1} - 3f_i}{2\Delta x} + O\left[(\Delta x)^2\right] \end{aligned} \tag{4.38}$$

Similarly, a backward difference expression of $O[(\Delta x)^2]$ may be obtained by retaining an additional term in the backward difference expression for f_i' and substituting the backward difference formula of $O(\Delta x)$ for f_i''.

Formulations of still higher accuracy can be obtained by retaining additional terms in the series. As seen in Equation 4.38, the value of the function at an additional grid point, x_{i+2}, is brought in to obtain the higher accuracy. Similarly, finite difference expressions for f_i' with truncation errors of order $(\Delta x)^3$ and $(\Delta x)^4$ are obtained as follows:

$$f_i' = \frac{1}{6\Delta x}(-2f_{i-3} + 9f_{i-2} - 18f_{i-1} + 11f_i) + \frac{(\Delta x)^3}{4} f_i''''(\xi) \tag{4.39a}$$

$$f_i' = \frac{1}{12\Delta x}(3f_{i-4} - 16f_{i-3} + 36f_{i-2} - 48f_{i-1} + 25f_i) + \frac{(\Delta x)^4}{5} f_i'''''(\xi) \quad (4.39b)$$

$$f_i' = \frac{1}{12\Delta x}(f_{i-2} - 8f_{i-1} + 8f_{i+1} - f_{i+2}) + \frac{(\Delta x)^4}{30} f_i'''''(\xi) \quad (4.39c)$$

where ξ is within the range of the appropriate expansion. The first two equations are third- and fourth-order correct backward differences. The third equation is a fourth-order accurate, five-point central difference approximation for the first derivative at $x = x_i$. With these five points, a higher-order approximation for the second derivative is

$$f_i'' = \frac{1}{12(\Delta x)^2}\left[-f_{i-2} + 16f_{i-1} - 30f_i + 16f_{i+1} - f_{i+2}\right] + \frac{(\Delta x)^4}{90}\frac{d^6 f}{dx^6}(\xi), \quad \text{where } x_{i-2} < \xi < x_{i+2} \quad (4.40)$$

It is evident that finite difference approximations of desired accuracy may be derived by the use of Taylor-series expansions. As shown later in Chapters 9 and 10, most finite difference solutions of ODEs and PDEs are based on expressions of accuracy $O[(\Delta x)^2]$. However, finite difference representations of higher accuracy are also employed, depending on the special needs of a given problem.

The accuracy of the numerical results may be improved either by employing a higher-accuracy formula or by reducing the grid spacing Δx. As discussed in Chapter 2, both of these approaches are employed in practice. Grid refinement, or reducing Δx, is generally carried out until the numerical results are essentially unaffected by a further reduction. At this stage, the numerical results are as accurate as can be obtained with the chosen finite difference expression. A continued reduction in grid spacing will lead to increasing round-off error and, thus, less accurate results. Then the accuracy of the results can be increased by using a higher-accuracy formulation.

An interesting point that may be observed from all the finite difference expressions given here is that the sum of all the coefficients, which multiply the function values in the numerator, is always zero. This result arises because the derivatives must become zero if $f(x)$ is a constant. Also, if Δx approaches zero, the numerator must also approach zero so that the limiting result obtained as $\Delta x \to 0$ yields a finite value for an arbitrary continuous function $f(x)$.

Example 4.2

An engineer involved in the design of automobiles uses an experimental system for studying the motion of a wide variety of vehicular devices in a full-scale laboratory environment. One particular test involves an accurate measurement of the displacement x of the vehicle as a function of time t. This information is then used to determine velocity V, acceleration A, and rate of change of acceleration F as functions of time. In a given experiment, the displacement x was measured over a time range of 0–10 s, at steps of 0.1 s. Some of the results obtained are given as follows:

t(s)	0.0	0.1	0.2	0.3	0.4	0.5
x(m)	0.0	0.8733	1.8224	2.8611	4.0032	5.2625
t(s)		0.6	0.8	1.0	1.2	
x(m)		6.6528	9.8816	13.80	18.5184	

From these data, compute V, A, and F at $t=0$ s, employing forward differences, and at $t=0.3$, employing central differences, with a step size Δt of 0.1 s. Repeat these calculations for $t=0$ s and $t=0.6$ s, with a step size Δt of 0.2 s.

SOLUTION

The velocity V, the acceleration A, and the rate of change of acceleration F are given in terms of the displacement x and time t by

$$V = \frac{dx}{dt}, \quad A = \frac{d^2x}{dt^2}, \quad F = \frac{d^3x}{dt^3} \tag{4.41}$$

Since measurements are available only for $t \geq 0$, the values at $t=0$ s can be computed only by forward differences. At $t=0.3$ s, central differences can be employed with a step size of 0.1 s, and at $t=0.6$ s, central differences with a step size of 0.2 s can be employed, according to the data given.

Various orders of approximation may be considered. The formulas needed for forward differences of $O(\Delta t)$ and $O[(\Delta t)^2]$ are given in Figure 4.7. In addition, formulas of $O[(\Delta t)^2]$ and $O[(\Delta t)^4]$ may be obtained, where t is the independent variable, instead of x, in Figures 4.7 through 4.9. For the first derivative, Equations 4.39a and 4.39b give the formulas for backward differences. Similarly, for forward differences, the first derivative may be approximated, for a function $f(t)$, by

$$f_i' = \frac{1}{6\Delta t}(2f_{i+3} - 9f_{i+2} + 18f_{i+1} - 11f_i) + O\left[(\Delta t)^3\right] \tag{4.42a}$$

$$f_i' = \frac{1}{12\Delta t}(-3f_{i+4} + 16f_{i+3} - 36f_{i+2} + 48f_{i+1} - 25f_i) + O\left[(\Delta t)^4\right] \tag{4.42b}$$

These two formulas are employed, in addition to those given in Figure 4.7, in order to demonstrate the effect of higher-order forward difference approximations on the numerical results for the velocity V. It is seen from Equation 4.42b that five points, including the one at which the derivative is sought, are needed in the forward direction to obtain an accuracy of $O[(\Delta t)^4]$. For computing A and F by forward differences, only formulas of $O(\Delta t)$ and $O[(\Delta t)^2]$ are used. The central differencing formulas of $O[(\Delta t)^2]$ and $O[(\Delta t)^4]$ are given in Figure 4.9. These may be employed for the computation of the first, second, and third derivatives needed in the present case.

A calculator may be used to carry out these calculations or a computer program may be written in MATLAB or Fortran for solving this problem. For example, the time

t at which the derivatives are to be computed can be entered in terms of the integer variable I, where $I = 1$ at $t = 0$ s. Then, I is taken as 4 at $t = 0.3$ s, with $\Delta t = 0.1$ s, and at $t = 0.6$ s, with $\Delta t = 0.2$ s. The step size Δt also can be entered interactively, as are the data values needed for the computations. For forward differences, the values of x at five points, $I, I+1, \ldots, I+4$, are to be entered. Similarly, for central differences, the values of x at six points, $I-3, I-2, \ldots, I, \ldots, I+3$, are needed. The program would first employ forward differencing to compute the derivatives and then central differencing.

Typical numerical results obtained from such a computer program are shown in Figures 4.10 and 4.11. Here, V, A, and F represent the velocity, acceleration, and rate of change of acceleration. The numbers after these variables indicate the order of the approximation and the numbers within the parentheses the index I that labels the time at which the quantity is computed. Also, T represents the time t and DT the time step Δt. First, consider forward difference results, shown in Figure 4.10, note that the velocity V converges to 8.4 m/s as the order of the approximation is increased. A considerable error is observed for the first-order approximation, particularly for the larger Δt (0.2 s), as expected. However, the third-order approximation is adequate for this problem, since essentially no change is observed by going to the fourth-order approximation. The acceleration A is given as 6.2 m/s^2 by the second-order approximation at both the mesh sizes considered. Again, the first-order approximation is in considerable error, particularly at $\Delta t = 0{,}2$ s. The computed value of F is found to be 13.8 m/s^3 at $\Delta t = 0.2$ s, although a variation is observed from the first-order to the second-order approximation at $\Delta t = 0.1$ s. In this problem, F is a constant at 13.8 m/s^3, as illustrated by the remaining results, discussed below. Thus, a higher-order approximation will not improve the accuracy

```
FORWARD DIFFERENCES
INPUT DATA:
ENTER THE VALUES OF I, T AND DT
0      0.0      0.1
ENTER THE MEASURED VALUES OF X(I) TO X(I + 4)
0.0         0.8733         1.8224         2.8611         4.0032
CALCULATED RESULTS:
TIME = 0.0000        TIME STEP = 0.1000
Vl = 8.7330  V2 = 8.3540  V3 = 8.4000  V4 = 8.4000
Al = 7.5800  A2 = 6.2000  Fl = 13.7998  F2 = 13.7997

INPUT DATA:
ENTER THE VALUES OF I, T AND DT
0      0.0      0.2
ENTER THE MEASURED VALUES OF X(I) TO X(I + 4)
0.0         1.8224         4.0032         6.6528         9.8816
CALCULATED RESULTS:
TIME = 0.0000        TIME STEP = 0.2000
Vl = 9.1120  V2 = 8.2160  V3 = 8.4000  V4 = 8.4000
Al = 8.9600  A2 = 6.2000  Fl = 13.8001  F2 = 13.8001
```

FIGURE 4.10 Numerical results obtained by employing forward differences for the problem given in Example 4.2.

CENTRAL DIFFERENCES
INPUT DATA:
ENTER THE VALUES OF I, T AND DT
4 0.3 0.1
ENTER THE VALUES OF X(I – 3) TO X(I + 3)
0.0 0.8733 1.8224 2.8611 4.0032 5.2625 6.6528
CALCULATED RESULTS:
TIME = 0.3000 TIME STEP = 0.1000
V2 = 10.9040 A2 = 10.3399 F2 = 13.8004
V4 = 10.8810 A4 = 10.3399 F4 = 13.8004

INPUT DATA:
ENTER THE VALUES OF I, T AND DT
4 0.6 0.2
ENTER THE VALUES OF X(I – 3) TO X(I + 3)
0.0 1.8224 4.0032 6.6528 9.8816 13.8 18.5184
CALCULATED RESULTS:
TIME = 0.6000 TIME STEP = 0.2000
V2 = 14 6960 A2 = 14.4800 F2 = 13.8000
V4 = 14.6040 A4 = 14.4799 F4 = 13.7999

FIGURE 4.11 Numerical results obtained by employing central differences for the problem given in Example 4.2.

if the function being considered is a polynomial of lower order. Of course, for an arbitrary function, accuracy is generally improved by employing a higher-order approximation.

The results from central differencing, shown in Figure 4.11, indicate only small changes from the second-order to the fourth-order approximations. Thus, the second-order formulas are adequate for this problem, as is often the case in most engineering problems. At $t = 0.3$ s, V, A, and F are obtained as 10.881 m/s, 10.34 m/s^2, and 13.8 m/s^3, respectively. Similarly, at $t = 0.6$ s, V, A, and F are obtained as 14.604 m/s, 14.48 m/s^2, and 13.8 m/s^3, respectively. Again, the second-order approximations are found to be adequate.

Example 4.2 has illustrated the use of numerical differentiation in a practical circumstance. The displacement x can generally be measured very accurately as a function of time t, and finite difference formulas can then be employed to yield velocity, acceleration, and so on. Forward and backward differences are generally used only at the start and the termination of the measurements, central differences being appropriate for other times. Although higher-order approximations may be used, second-order formulas often yield satisfactory accuracy in most problems of engineering interest.

4.5 POLYNOMIAL REPRESENTATION

Another frequently employed approach for the derivation of the finite difference approximations to the derivatives of a given function $f(x)$ is based on a polynomial fit to the values at the given grid points. Depending on the order of the derivative whose approximation has to be obtained and the desired accuracy, the order of the polynomial may be chosen. For an nth-order polynomial, $(n + 1)$ grid points are needed to evaluate all the coefficients that appear in the polynomial. Curve fitting is discussed in detail in Chapter 7, and only a few simple aspects are presented here in order to obtain the finite difference approximations.

By way of illustration, let us consider fitting a second-order polynomial to the three grid points shown in Figure 4.12. The function $f(x)$ is taken as

$$f(x) = A_0 + A_1 x + A_2 x^2 \tag{4.43}$$

Fitting this parabola to the three points yields

$$f_i = A_0 + A_1 x_i + A_2 x_i^2 \tag{4.44a}$$

$$f_{i+1} = A_0 + A_1(x_i + \Delta x) + A_2(x_i + \Delta x)^2 \tag{4.44b}$$

$$f_{i+2} = A_0 + A_1(x_i + 2\Delta x) + A_2(x_i + 2\Delta x)^2 \tag{4.44c}$$

From these equations, the coefficients A_0, A_1, and A_2 may be determined. The first and second derivatives of the function are given by

$$f_i' = A_1 + 2A_2 x_i \tag{4.45}$$

$$f_i'' = 2A_2 \tag{4.46}$$

Since the finite difference expression should depend only on the relative positions of the grid points, that is, it should be independent of the absolute location of the points, any arbitrary value of x_i may be taken. The algebra is simplified if x_i is taken

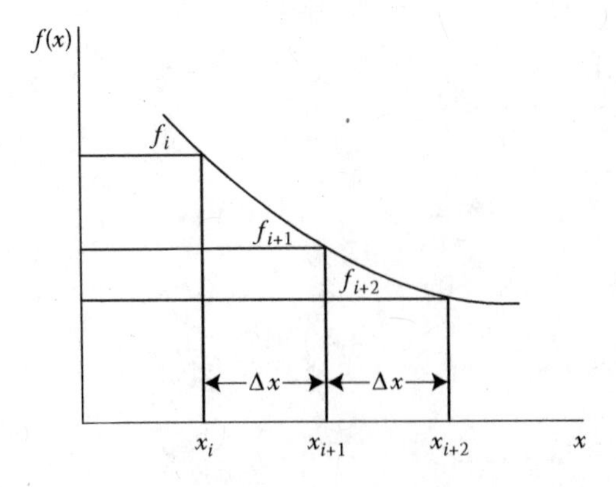

FIGURE 4.12 Uniform one-sided distribution of grid points used for illustrating the derivation of the finite difference approximations of the first and second derivatives by means of a second-order polynomial representation of the function $f(x)$.

as 0 so that $x_{i+1} = \Delta x$ and $x_{i+2} = 2\Delta x$. However, the resulting expressions for f_i' and f_i'' are the same whatever the value of x_i. These expressions are obtained from Equations 4.44 through 4.46 as follows:

$$f_i' = \frac{-f_{i+2} + 4f_{i+1} - 3f_i}{2\Delta x} \tag{4.47}$$

and

$$f_i'' = \frac{f_{i+2} - 2f_{i+1} + f_i}{(\Delta x)^2} \tag{4.48}$$

The approximations are identical to those derived earlier from the Taylor series; see Equations 4.33 and 4.38. The first expression was shown to have a truncation error of $O[(\Delta x)^2]$ and the second one of $O(\Delta x)$. Both the equations give forward difference approximations because of the chosen grid points, which are on one side of $x = x_i$ in the direction of increasing x. The error term is not explicitly given by this approach. For an accurate evaluation of the error, one must resort to the Taylor-series approach.

The polynomial representation is particularly useful in the derivation of finite difference expressions for grid points that are located at nonuniform distances from each other. For instance, consider the distribution shown in Figure 4.13. Taking $x_i = 0$, $x_{i+1} = \Delta x$, and $x_{i+2} = 3\ \Delta x$, the parabola of Equation 4.43 gives

$$f_i = A_0 \tag{4.49a}$$

$$f_{i+1} = A_0 + A_1\Delta x + A_2(\Delta x)^2 \tag{4.49b}$$

$$f_{i+2} = A_0 + A_1(3\Delta x) + A_2(3\Delta x)^2 \tag{4.49c}$$

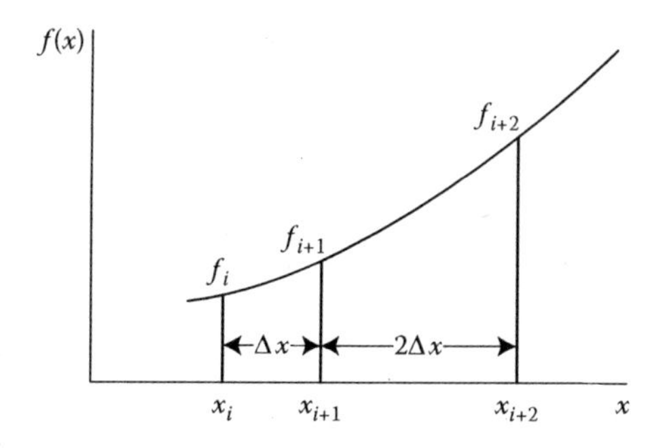

FIGURE 4.13 Nonuniform distribution of grid points employed for illustrating the use of a polynomial representation of $f(x)$ to derive the finite difference approximations of the derivatives.

with

$$f'(0) = A_1 \quad \text{and} \quad f''(0) = 2A_2 \tag{4.50}$$

From these equations, we obtain the finite difference expressions for f_i' and f_i'' as follows:

$$f_i' = \frac{-f_{i+2} + 9f_{i+1} - 8f_i}{6\Delta x} \tag{4.51}$$

and

$$f_i'' = \frac{f_{i+2} - 3f_{i+1} + 2f_i}{3(\Delta x)^2} \tag{4.52}$$

Similarly, expressions for other arbitrary distributions of grid points may be derived. This approach is frequently employed for determining the derivatives from experimental data. Examples of such data are those pertaining to the variation of material properties and of physical quantities such as pressure and density with an independent variable, such as temperature. Such data are generally available at non-uniformly distributed values of the independent variable, and the polynomial approach provides a simple method for computing the derivatives.

4.6 PARTIAL DERIVATIVES

In the preceding sections, we have considered the numerical differentiation of an arbitrary function $f(x)$ that depends on a single independent variable x. The finite difference representations of the ordinary derivatives of the function were derived by considering the variation with x. However, in engineering problems, we frequently encounter circumstances where the dependent variable is a function of two or more independent variables. In such cases, partial derivatives arise, and the finite difference approximations of these derivatives are of interest. Since a partial derivative is defined in terms of the variation of the function with a given independent variable while the others are held constant, the finite difference approximations are analogous to those for the ordinary derivatives.

Consider, for instance, a function $f(x, y)$. Then the partial first derivatives of the function are $(\partial f/\partial x)$ and $(\partial f/\partial y)$, where y is kept constant in the first case and x in the second. The variables that are held constant for a particular differentiation are sometimes indicated by means of subscripts as, for instance, $(\partial f/\partial x)_y$ and $(\partial f/\partial x)_x$. However, it is understood that for the partial differentiation $\partial f/\partial x$, only the variation of $f(x, y)$ with x is under consideration, y being kept unchanged. Since two independent variables, x and y, are involved, a location in the computational domain is represented by two subscripts, instead of only one needed for ordinary derivatives. Thus, the value of the function $f(x, y)$ at a grid point represented by indices (i, j) may be denoted as $f_{i,j}$, where $x = i\Delta x$ and $y = j\Delta y$, as shown in Figure 4.14. Such a grid is employed in the solution of PDEs by finite difference methods, as discussed in Chapter 10.

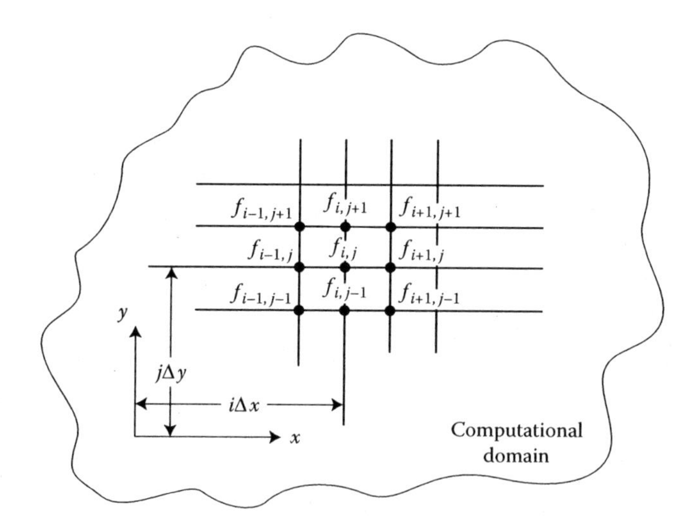

FIGURE 4.14 A two-dimensional grid indicating the finite number of locations at which the function $f(x, y)$ is evaluated in the computational domain.

Considering the variation of the function $f(x, y)$ with x alone, we may write the finite difference approximations of the first and second derivatives, in a manner analogous to that outlined in Section 4.3:

$$\left(\frac{\partial f}{\partial x}\right)_{i,j} \cong \frac{f_{i+1,j} - f_{i,j}}{\Delta x} \quad \text{Forward difference} \tag{4.53}$$

$$\left(\frac{\partial f}{\partial x}\right)_{i,j} \cong \frac{f_{i,j} - f_{i-1,j}}{\Delta x} \quad \text{Backward difference} \tag{4.54}$$

$$\left(\frac{\partial f}{\partial x}\right)_{i,j} \cong \frac{f_{i+1,j} - f_{i-1,j}}{2\Delta x} \quad \text{Central difference} \tag{4.55}$$

$$\left(\frac{\partial^2 f}{\partial x^2}\right)_{i,j} \cong \frac{f_{i+1,j} - 2f_{i,j} + f_{i-1,j}}{(\Delta x)^2} \quad \text{Second central difference} \tag{4.56}$$

Therefore, the subscript j is not varied in these expressions. Its presence indicates that the function also depends on another independent variable.

Similarly, the partial derivatives with respect to y may be obtained. Thus, the central differences yield

$$\left(\frac{\partial f}{\partial y}\right)_{i,j} \cong \frac{f_{i,j+1} - f_{i,j-1}}{2\Delta y} \tag{4.57}$$

$$\left(\frac{\partial^2 f}{\partial y^2}\right)_{i,j} \cong \frac{f_{i,j+1} - 2f_{i,j} + f_{i,j-1}}{(\Delta y)^2} \tag{4.58}$$

In this case, the subscript i is not varied. Thus, all the expressions derived in the preceding sections may easily be extended to partial derivatives.

The derivation of the finite difference approximations for partial derivatives may, again, be based on direct approximation, the Taylor series, or polynomial representation. The Taylor-series expansions about the point (i,j) may be written as

$$f_{i\pm1,j} = f_{i,j} \pm \Delta x\left(\frac{\partial f}{\partial x}\right)_{i,j} + \frac{(\Delta x)^2}{2!}\left(\frac{\partial^2 f}{\partial x^2}\right)_{i,j} \pm \frac{(\Delta x)^3}{3!}\left(\frac{\partial^2 f}{\partial x^3}\right)_{i,j} + \cdots \tag{4.59}$$

$$f_{i,j\pm1} = f_{i,j} \pm \Delta y\left(\frac{\partial f}{\partial y}\right)_{i,j} + \frac{(\Delta y)^2}{2!}\left(\frac{\partial^2 f}{\partial y^2}\right)_{i,j} \pm \frac{(\Delta y)^3}{3!}\left(\frac{\partial^2 f}{\partial y^3}\right)_{i,j} + \cdots \tag{4.60}$$

The remainder terms may also be obtained for truncation after a finite number of terms, as outlined earlier for ordinary derivatives. The remainder term for truncation after the term containing the mth power of Δx, that is, after $(m + 1)$ terms in Equation 4.59 is

$$R_{m,x} = (-1)^{m+1}\frac{(\Delta x)^{m+1}}{(m+1)!}\left(\frac{\partial^{m+1} f}{\partial x^{m+1}}\right)_{\xi,j}, \quad \text{where } i < \xi < i+1 \tag{4.61}$$

Similarly, the remainder term $R_{m,y}$ is obtained as $O[(\Delta y)^{m+1}]$ for truncation after $(m + 1)$ terms in Equation 4.60. The total remainder term is the sum of $R_{m,x}$ and $R_{m,y}$ as discussed in Chapter 10.

Thus, the finite difference approximations may be derived from the Taylor-series expansions, as given earlier. Sometimes cross derivatives such as $\partial^2 f/\partial x \partial y$ have to be evaluated. The corresponding finite difference representations may be derived by applying the approximation twice for the two differentiations with different independent variables. A two-variable Taylor-series expansion for $f(x, y)$ may also be employed for the purpose, as outlined by Jaluria and Torrance (2003).

Partial derivatives are of interest in many important engineering applications, such as those that involve fluid flow, heat and mass transfer, thermodynamics, chemical reactions, structural vibrations, and electrical fields. Obviously, these topics encompass a wide range of engineering problems, extending from aerospace and environmental problems to nuclear and chemical reactors and power plants. Partial differential equations, which are discussed in detail in Chapter 10, govern such physical phenomena. The finite difference approximations of the partial derivatives are then employed for developing the numerical procedure for solving these problems. An example on the evaluation of partial derivatives is given in the following.

Example 4.3

Planck's law for blackbody radiation is given as

$$E_{b\lambda}(\lambda, T) = \frac{C_1}{\lambda^5[e^{C_2/\lambda T} - 1]} \tag{4.62}$$

where $C_1 = 3.7413 \times 10^8$ W μm^4/m^2, $C_2 = 1.4388 \times 10^4$ μm K, λ is wavelength in μm, and T is temperature in K. $E_{b\lambda}$ is a function of λ and T, and is known as the *monochromatic emissive power* of a blackbody (see Figure 1.3). Thus, $E_{b\lambda}$ is a function of λ and T. Partial derivatives of $E_{b\lambda}$ are of interest in areas such as physics and heat transfer. Numerically determine $\partial E_{b\lambda}/\partial\lambda$ and $\partial E_{b\lambda}/\partial T$ at $\lambda = 4$ μm and $T = 1600$ K. Repeat the calculation for $\lambda = 2$ μm and $T = 1000$ K. Use different values of $\Delta\lambda$ and ΔT to ensure accuracy of your results, employing the second-order formula for the derivative. Compare your results with analytical ones.

SOLUTION

Energy transfer by radiation is of importance in several areas of engineering and physical sciences. Planck's law is of considerable value in the calculations for energy transfer since it gives the characteristics and magnitude of energy lost through thermal radiation by an idealized surface, termed *blackbody*, as functions of temperature T and wavelength λ. Our interest here lies in numerically evaluating the rate of change of the emissive power $E_{b\lambda}$ with these two independent variables. The results are to be obtained at the two sets of values given for λ and T. Also, the step sizes $\Delta\lambda$ and ΔT are to be varied so that we may study their effect on the results and choose the most appropriate values, as outlined earlier in Section 2.5.

The problem is fairly straightforward, and a calculator may be used for the calculations or a simple computer program may be written for obtaining the required numerical results. In such a program, the function $E_{b\lambda}(\lambda,T)$ needs to be defined and the constants C_1 and C_2 specified. The input values for λ and T, at which the gradients $\partial E_{b\lambda}/\partial\lambda$ and $\partial E_{b\lambda}/\partial T$ are to be determined, can be entered interactively. The starting values of the step sizes $\Delta\lambda$ and ΔT can be taken as 1 μm and 500 K. These can then be successively halved in the program, and the corresponding derivatives computed by the following central differencing formulas for a function $f(x,y)$:

$$\frac{\partial f}{\partial x} = \frac{f(x + \Delta x, y) - f(x + \Delta x, y)}{2\Delta x} \tag{4.63a}$$

$$\frac{\partial f}{\partial y} = \frac{f(x, y + \Delta y) - f(x, y + \Delta y)}{2\Delta y} \tag{4.63b}$$

Therefore, an appropriate computer program may be written as :

```
ebl=inline('3.7413*10^8/((x^5)*(exp(1.4388*10^4/(x*t))...
     -1))'); x=4;t=1600;
fprintf('WAVELENGTH= %.2f TEMPERATURE= %.2f\n',x,t)
fprintf('DX DE/DX DT DE/DT\n')
for i=0:12
dx=1/(2^i);dt=500/(2^i);
dxebl=(ebl(t,x+dx)-ebl(t,x-dx))/(2*dx);
dtebl=(ebl(t+dt,x)-ebl(t-dt,x))/(2*dt);
fprintf('%.5f %.2f %.3f %.2f\n',dx,dxebl,dt,dtebl)
end
```

where the given function is defined as ebl (t,x), the two independent variables, temperature t and wavelength x, being assigned alphabetically. The derivatives with respect to x and t are represented by d*xebl* and d*tebl*, respectively.

The numerical results obtained are presented in Figure 4.15. The wavelength step size $\Delta\lambda$ is varied from 1.0 to 0.00024 μm, and the temperature step size ΔT from 500 to 0.122 K. The step sizes are halved in each successive computation. Here, X and T refer to λ and T, respectively. DX refers to $\Delta\lambda$ and DT to ΔT, while DE/DX and DE/DT refer to the derivatives $\partial E_{b\lambda}/\partial\lambda$ and $\partial E_{b\lambda}/\partial T$, respectively. Note that $\partial E_{b\lambda}/\partial T$ approaches a constant value, as ΔT is reduced, much more rapidly than $\partial E_{b\lambda}/\partial\lambda$, with reduction in $\Delta\lambda$. This indicates that the results are more sensitive to variations in λ, as is also evident from the λ^5 dependence in the denominator of the function $E_{b\lambda}$; see Equation 4.62. From the numerical results presented in Figure 4.15, the computed values of both $\partial E_{b\lambda}/\partial T$ and $\partial E_{b\lambda}/\partial\lambda$ vary monotonically from the starting value of $\Delta\lambda$ until they reache constant values at small $\Delta\lambda$. For still smaller $\Delta\lambda$, the values may oscillate due to the appearance of significant round-off error. Thus, $\partial E_{b\lambda}/\partial T$ is evaluated as 67.77 W/m^2 μm K at $\lambda = 4$ μm and $T = 1600$ K and as 63.27W/m^2 μm K at $\lambda = 2$ μm and $T = 1000$ K. Similarly, $\partial E_{b\lambda}/\partial\lambda$ is evaluated as –26,813.97 W/m^2 μm^2 at $\lambda = 4$ μm and $T = 1600$ K and as 9664.07 W/m^2 μm^2 at $\lambda = 2$ μm and $T = 1000$ K.

The corresponding analytical values may also be determined by using mathematics to differentiate $E_{b\lambda}$ successively with respect to the two independent variables λ and T. The given sets of input values may then be substituted into the mathematical expressions obtained. The analytical results thus obtained are as follows:

	$\partial E_{b\lambda}/\partial\lambda$	$\partial E_{b\lambda}/\partial T$
$\lambda = 4$, $T = 1600$	–26,813.970	67.765
$\lambda = 2$, $T = 1000$	9664.069	63.268

Therefore, the numerical results obtained are essentially identical to the analytical results at the chosen values of the step sizes. Obviously, at large values of the step sizes, the numerical results are in considerable error; see Figure 4.15. An appropriate reduction in step sizes is, therefore, needed until the change in the numerical results is small.

WAVELENGTH = 4.00	TEMPERATURE = 1600.00		
DX	DE/DX	DT	DE/DT
1.00000	–28565.47	500.000	65.96
0.50000	–27287.70	250.000	67.32
0.25000	–26934.25	125.000	67.65
0.12500	–26844.15	62.500	67.74
0.06250	–26821.52	31.250	67.76
0.03125	–26815.86	15.625	67.76
0.01562	–26814.44	7.812	67.77
0.00781	–26814.09	3.906	67.77
0.00391	–26814.00	1.953	67.77
0.00195	–26813.98	0.977	67.77
0.00098	–26813.97	0.488	67.77
0.00049	–26813.97	0.244	67.77
0.00024	–26813.97	0.122	67.77

WAVELENGTH = 2.00	TEMPERATURE = 1000.00		
DX	DE/DX	DT	DE/DT
1.00000	6308.24	500.000	97.40
0.50000	8804.91	250.000	72.68
0.25000	9456.31	125.000	65.68
0.12500	9612.72	62.500	63.87
0 06250	9651.27	31.250	63.42
0.03125	9660.87	15.625	63.31
0.01562	9663.27	7.812	63.28
0 00781	9663.87	3.906	63.27
0.00391	9664.02	1.953	63.27
0.00195	9664.06	0.977	63.27
0.00098	9664.07	0.488	63.27
0.00049	9664.07	0.244	63.27
0.00024	9664.07	0.122	63.27

FIGURE 4.15 Numerical results obtained for Example 4.3, indicating the dependence of the computed derivatives on the step sizes $\Delta\lambda$ and ΔT.

4.7 SUMMARY

In this chapter, we have considered the basic concepts underlying numerical differentiation and finite difference calculus. Three different approaches, namely, direct approximation, Taylor series, and polynomial representation, are presented for the derivation of the finite difference approximations to the various derivatives of an arbitrary function $f(x)$. The truncation error resulting from the retention of a finite number of terms in the Taylor series is considered in detail and is related to the accuracy of the various finite difference approximations. The general procedures for deriving finite difference expressions of higher accuracy and those for higher-order derivatives are outlined. The Taylor-series approach is the preferred one since it also yields the error, which the other two methods, although relatively simpler to employ, do not. In addition, the Taylor-series expansions may be successively applied to improve the accuracy of the finite difference approximation, if a higher level of accuracy is desired in a given application. The polynomial representation approach is particularly useful if a nonuniform distribution of grid points is employed. Finally, the chapter discusses partial derivatives and shows how the finite difference approximations may easily be obtained from those for ordinary derivatives.

PROBLEMS

4.1. Derive the Taylor-series expansions for $\sin x$, $(1-x)^{-2}$, and e^{-x}, about $x = 0$. Are there any constraints on $|x|$ for the series to be convergent? Why does the series for $\log(1-x)$, derived in Example 4.1, converge only if $|x| < 1$?

4.2. Using the Taylor series for e^x and $\sin x$, obtain the series for $e^x \sin x$, about $x = 0$. Compute the value of $e^{0.2} \sin 0.2$ by a summation of the series, retaining terms so that the first neglected term is $O[(\Delta x)^4]$. Compare your result with the true value of the function $e^x \sin x$ at $x = 0.2$.

4.3. Show that the Taylor-series expansion for x^5 about $x = 0$ is x^5 itself.

4.4. Calculate $e^{0.3}$ by employing the Taylor series for e^x. How many terms are needed if the error from the true value of the quantity $e^{0.3}$ is to be less than 0.01%?

4.5. The pressure p of a gas is given by the expression $\log p = 21.6–2420/T$, where T is the temperature in kelvins. Using the exact value of p from this expression, at $T = 400$ K, and the Taylor-series expansion for $p(T)$, compute the pressures at 410, 420, and 450 K. Compare these values with the exact ones obtained from the given expression. Refer to Example 4.1b.

4.6. Compute the value of $e^{\sin(x)}$ at $x = 0.25$, employing the corresponding Taylor-series expansion, and compare the result with the exact value.

4.7. Consider the function $f(x) = 2x^{1/2} + 3x$. The derivatives of the function at $x = 0$ are all infinite, and therefore the Taylor-series expansion about $x = 0$ cannot be obtained. Instead, obtain the series about $x = 0.1$ and also about $x = 0.2$. Compare the two and comment on the difference.

4.8. Compute the first and second derivatives of $\sin x$ and e^x at $x = 0$, employing forward and central differencing formulas of $O[(\Delta x)^2]$. Consider three values, 0.2, 0.1, and 0.01, of the step size Δx. Compare the numerical results obtained with the exact, mathematical values of the derivatives. Discuss the effect of Δx on the accuracy of the numerical results.

4.9. Calculate the numerical value of $d[(\sin x)^{e^x}]/dx$ at $x = 1$, using central difference approximations of $O[(\Delta x)^2]$. Start with $\Delta x = 0.2$ and reduce it until the numerical result remains essentially unaffected by a further reduction in Δx.

4.10. Consider the expressions for computing the PW and FW of a series of equal monthly payments, given by Equations 2.4 and 2.5, respectively. It is important for the economic planning of an engineering system to determine the effect of a change in the interest rate x on PW and FW. Compute the rate of change of these quantities with x, at $x = 10\%$, for $R = \$1000$ and $m = 240$ months. Also, refer to Example 2.2 for details on this problem. Employ an appropriate value of Δx, as discussed in Section 2.5.

4.11. In Problem 4.10, compute d(FW)/dm and d(PW)/dm, where m is the number of months, at $x = 10\%$, $m = 240$, and $R = \$1000$.

4.12. The measured temperature distribution in a solar energy heating system may be represented by the equation

$$T(x) = 15.5 + 68.2\left[1 - \frac{\exp(-x/2.7)}{(1 + x^2)}\right]$$

where x is the distance away from the surface being heated and T is the temperature. Compute the temperature gradient dT/dx and the second derivative d^2T/dx^2 at $x = 0$. The heat transfer rate is proportional to the temperature gradient at $x = 0$. The second derivative is related to energy lost or gained by radiative transport. Employ forward differences to $O[(\Delta x)^2]$ and reduce Δx to obtain numerical results that are largely independent of the value of Δx chosen.

4.13. For the problem considered in Example 4.2, if, in addition to the data given, the displacements at $t = 0.7$ s and 0.9 s are given as 8.1879 m and 11.7477 m, respectively, compute the velocity and acceleration at $t = 0.5$ s, using forward, backward, and central differences of $O[(\Delta x)^2]$. Employ a step size Δt of 0.1 s.

4.14. In a periodic mass transfer process in a chemical plant, the concentration C of the moisture is obtained from analysis as

$$C = 9.5\left[\exp(-0.75x)\right]\cos(t - 0.75x - 2)$$

where x is distance in meters, t is the time in seconds, and C is in kg/m^3. The mass transfer rate is proportional to the gradient $\partial C/\partial x$ at $x = 0$. The second derivative is related to the rate of moisture addition per unit volume. Compute both $\partial C/\partial x$ and $\partial^2 C/\partial x^2$ at $t = 1$ s and $x = 0$, using forward differences. Start with $\Delta x = 0.1$ m and reduce it to 0.01 m to see whether there is any significant effect on the numerical results.

4.15. In fluid mechanics, the stream function ψ, which is related to the flow rate, is frequently employed for analysis. The distribution of ψ for a particular problem is obtained as

$$\psi = 1.1\left[\exp\left(-y^{1.3}\right)\right]\cos\left(\frac{y^2}{3\pi}\right) + 0.9\left[\exp\left(-y^{0.7}\right)\right]\sin\left(\frac{y^2}{3\pi}\right)$$

The velocity V is given by $d\psi/dy$, and the shear force generated by the flow is proportional to $d^2\psi/dy^2$. Compute both these derivatives at $y = 0.5$, with $\Delta y = 0.2$, 0.1, and 0.05. Employ second-order central differences.

4.16. It is known from analysis that the distribution of $E_{b\lambda}$, given in Example 4.3, has a maximum at $\lambda T = 2897.6$ µm K. Confirm this by using numerical differentiation to obtain the first and second derivatives, with respect to λ, at $T = 1000$ K and $\lambda T = 2897.6$ µm K.

Remember that, at a maximum, the first derivative is zero and the second derivative is negative.

4.17. The hot-wire anemometer is an instrument used for measuring velocities or temperatures. If, during its calibration, the output signal E is measured as 0, 1.7, 3.3, and 5.6 V at velocities V of 0, 1, 1.5, and 2 m/s, obtain the gradient dE/dV at $V = 0$ m/s, using the polynomial representation of the function $E(V)$.

4.18. Obtain the finite difference approximation for df/dx, where $f(x)$ is a given function of x, using four uniformly spaced grid points and the polynomial representation of $f(x)$.

5 Roots of Equations

5.1 INTRODUCTION

In a wide variety of engineering problems, there is a task of determining the values of the variable x that would satisfy a given algebraic equation, such as $x^3 - 4x^2 + 5x = 2$, or $x \tan x = 1$. Depending on the problem, defined by the equation and the range of x under consideration, these values of x, which are termed as *roots* of the equation, may be real or complex and may be finite or infinite in number. Root solving is needed, for instance, in determining the terminal velocity of a falling body, the concentration of a chemical species at a surface subjected to mass transfer, the time needed to repay a loan at a given interest rate and monthly payment, and the natural frequencies of vibration of a beam.

The algebraic equation to be solved is represented by the general form

$$f(x) = 0 \tag{5.1}$$

where the function $f(x)$ may designate a polynomial or a transcendental expression, such as $x \tan x - 1$, from the equation given above. The problem of finding the roots of Equation 5.1, therefore, involves obtaining the values of x at which the function $f(x)$ is zero. Consequently, the roots are also often referred to as *zeros* of the function. Although interest usually lies in determining the real roots of equations with real coefficients, we do encounter problems, such as periodic processes, in which complex roots are of interest or in which the equation has complex coefficients. Therefore, the discussion in this chapter is initially directed at obtaining the real roots of equations with real coefficients, the other circumstances being considered later in Section 5.5.1.

There are several methods available for finding the roots of algebraic equations. Some of these are applicable only to polynomial equations, which are obtained when $f(x)$ represents a polynomial to yield an equation of the form

$$f(x) = x^n + a_1 x^{n-1} + a_2 x^{n-2} + \cdots + a_{n-1} x + a_n = 0 \tag{5.2}$$

where n is the degree of the polynomial equation and $a_1, a_2, \ldots, a_n$ are real coefficients. This equation has n roots, which may be real or complex. Some of the real roots may be equal. Also, the complex roots occur in conjugate pairs, that is, a complex root $a + ib$, where a and b are real and $i = \sqrt{-1}$, will occur in conjunction with another root $a - ib$. For a linear equation, $n = 1$, the root may be found directly as $x = -a_1/a_0$ from the equation $a_1 + a_0 x = 0$. For a quadratic equation, $n = 2$, the roots may again be determined by using the following well-known expression for the two roots, α_1 and α_2:

$$\alpha_1, \alpha_2 = \frac{-b \pm \sqrt{b^2 - 4ac}}{2a} \tag{5.3}$$

where $f(x) = ax^2 + bx + c = 0$ is the quadratic equation. Depending on whether the determinant $(b^2 - 4ac)$ is positive, zero, or negative, the roots are, respectively, real and distinct, equal, or complex. In a few limited cases, such formulas are available for higher-order equations too. Although these formulas provide a direct analytical method for finding the exact solution, they are usually very restrictive in their applicability and are also often quite complicated. Therefore, it is generally easier or necessary to use indirect, or *iterative*, methods to find the roots of a nonlinear equation, for which $n \neq 1$, numerically. Transcendental equations, which involve trigonometric and other special functions such as exponentials and logarithms, also arise in engineering problems. These equations are also generally nonlinear, and the number of roots is often unknown. Several of the methods considered in this chapter are applicable to both polynomial and transcendental equations.

In many cases of practical interest, the approximate variation of the function $f(x)$ with x, the nature of the roots, and the interval over which these are to be determined are known. For instance, if the surface temperature of a pond, resulting from the various heat transfer processes operating at the surface, is to be determined, the energy balance equation must be solved to yield a single, real, positive root in a given range of temperature. Similarly, if the terminal velocity of a particle moving under the action of various forces or the lowest natural frequency of vibration of a dynamic system were to be determined, one would search for real, positive roots of the corresponding equations over specified ranges.

Real, negative roots may also be obtained, for instance, when considering temperature and concentration differences, account balances, weight changes, forces, or velocities that may be positive or negative depending on the direction of motion, and price changes. If no prior information is available on the function and on the roots, a rough plot of the variation of $f(x)$ with x may be obtained numerically, for real roots, to determine the behavior of the function and the approximate location of the roots. This information may then be used in the choice of the method and the interval over which the roots are sought. In some engineering problems, such as those concerned with the stability of systems and with periodic processes, complex roots are of interest. Some of the techniques discussed here may also be employed for finding complex roots. Again, a prior knowledge of the approximate value and nature of the roots would be useful. Therefore, the basic nature of the problem, which gives rise to the equation whose roots are to be determined, is often important in the solution of the equation.

This chapter discusses several numerical methods for finding the real roots of polynomial and transcendental equations. These include the search method, the *bisection* method, the *regula falsi* method (also known as the method of *linear interpolation*), the *secant* method, *Newton's* method, *Newton's second-order* method, and the method of *successive substitution*. Some of these are also applicable to complex roots, and the corresponding procedure is outlined. There are several other available methods, such as *Muller's* method, *Brent's* method, *Graeffe's* root-squaring method, and *Bairstow's* iterative factorization of polynomials. These are also briefly discussed. Since, in many engineering applications, information on the basic background of the problem may be used effectively in the choice of the method and of the interval of interest, several examples of engineering problems are taken in order to illustrate the importance of the characteristics of the problem in the solution of the equations. Different types of equations are considered in order to present the various methods considered here.

5.2 SEARCH METHOD FOR REAL ROOTS

The search method is a very simple method, which is based on the change in the sign of the function $f(x)$ as x is incremented, starting with an initial value x_0, to determine the zero crossings of the function, that is, the locations where the plot of the function $f(x)$ crosses the x-axis, as shown in Figure 5.1. An increment Δx is chosen, and x is successively increased by this value. If the function $f(x)$ changes sign between two successive values x_i and x_{i+1}, then $[f(x_i) \cdot f(x_{i+1})] < 0$, and the presence of a real root in the interval between these values of x is indicated, as shown in Figure 5.1. This process is repeated by starting with $x = x_i$ and taking a smaller increment to narrow the interval containing the root. Therefore, one can make the interval in which the root lies as small as desired by successively reducing the increment Δx to search for the zero crossing of the function. Generally, one starts with a large step size Δx and successively reduces it to a small fraction, say, one-tenth of the previous value, for locating the root more accurately in the reduced interval obtained from a sign change of the function. If the initial increment is chosen as Δx and the subsequent reduced increments are $\Delta x/n$, $\Delta x/n^2$, $\Delta x/n^3$, and so on, where n is a constant, it is evident that the root may be obtained to the desired accuracy in only a few incremental searches.

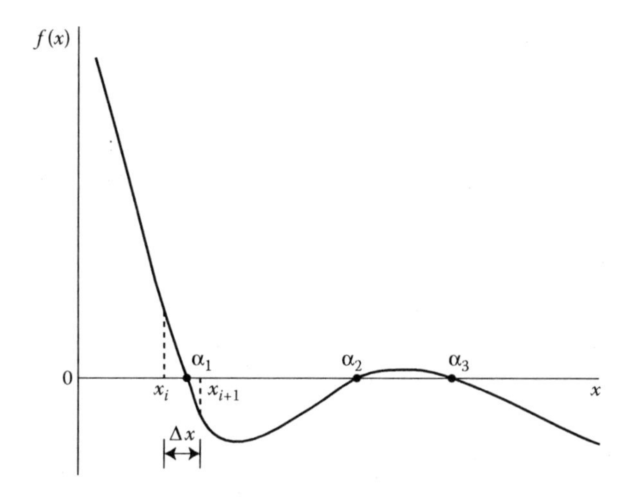

FIGURE 5.1 Search method for the real roots of the equation $f(x) = 0$.

When one root, $x = \alpha_1$, has been determined and if other roots are sought, the incremental search proceeds to larger values of x, taking $x > \alpha_1$; until another sign change in $f(x)$ occurs and the preceding process is repeated. Generally, one takes the starting value x_0 as the smallest value in the range of interest and proceeds to larger x until all the real roots in the given range are found. The method can be used to find positive or negative real roots of a given polynomial or transcendental equation, if the function $f(x)$ crosses the x-axis. It will fail to find a zero resulting from the function being tangent to the x-axis, since no sign change occurs in this case.

This method, which is sometimes known as the *incremental search method*, is particularly suitable for obtaining the various intervals in which the real roots of the given equation $f(x) = 0$ are located. Once an interval containing a root has been found, various other, more efficient methods, discussed later, may be employed for a faster convergence to the root. Consequently, the incremental search method often precedes other methods and is thus employed in conjunction with them. Frequently, the starting point, $x = x_0$, and the incremental step size Δx are chosen and $f(x)$ is determined for successively incremented values of x over the entire range of interest, thus yielding the various intervals in which real roots are located. Such a search for roots over the entire range is often known as *exhaustive search*. This procedure also allows one to obtain a rough plot of $f(x)$ versus x and thus determine the approximate behavior of the function. This information is useful in dealing with problem spots such as equal, or multiple, roots, obtained if $f(x)$ does not cross the x-axis but is tangent to it, and roots that are very close to each other in value. If no prior information is available on the nature of the roots and the behavior of the function, a small value of the increment Δx may be taken at the onset to ensure that no roots are missed in the search.

The incremental search method is frequently employed with an interactive computer program so that one might choose the increment and the starting point as the interval containing the root is successively reduced, thus coupling one's previous experience and knowledge with the program. The function $f(x)$ may also be plotted, using available software for graphics, for a visual study of the behavior of the function and the approximate location of the roots. This is particularly convenient in MATLAB® due to the availability of plotting software. Search methods, such as the one outlined here, are frequently employed in optimization of systems, where one seeks to maximize or minimize a given function. The incremental search method can easily be extended to find the maxima or minima of a function by seeking the zeros of the derivative of the function instead of those of the function $f(x)$.

As outlined here, the search method provides an iterative procedure for determining the roots of the given algebraic equation $f(x) = 0$. Iteration is terminated when the root has been determined to the desired accuracy level. A commonly employed criterion for convergence is specified as

$$\left| \frac{x^{(n)} - x^{(n-1)}}{x^{(n)}} \right| \leq \varepsilon \tag{5.4a}$$

where $x^{(n)}$ and $x^{(n-1)}$ represent the approximations to the root, that is, the values of x at which the function changes sign, for the nth and the $(n-1)$th iterations, respectively, and ε is a specified small quantity, often taken in the range 10^{-5}–10^{-3}. The preceding convergence criterion is thus based on the magnitude of the relative change in the approximation to the root from one iteration to the next. Another criterion, based on the increment $\Delta x^{(n)}$ for the nth iteration, is also frequently used. This criterion may be written as follows:

$$\left| \frac{\Delta x^{(n)}}{x^{(n)}} \right| \le \varepsilon \tag{5.4b}$$

The above conditions for convergence, therefore, imply that the fractional error in the computed root is less than or equal to ε. The first criterion, Equation 5.4a, does not explicitly indicate the accuracy of the root, and, therefore, the second one, which does give the accuracy, is preferable in most cases. The preceding criteria are also often replaced by

$$| x^{(n)} - x^{(n-1)} | \le \varepsilon \tag{5.5a}$$

or

$$| \Delta x^{(n)} | \le \varepsilon \tag{5.5b}$$

where the actual values of the quantities are considered, instead of the relative magnitudes. This form is particularly useful if the expected value of the root is equal to or close to zero, since, in this case, the forms in Equation 5.4 cannot be used because of the denominator becoming zero. The accuracy of the calculated root may be improved by taking it as the average of the two final x values, x_i and x_{i+1}, between which a sign change occurs, that is, $\alpha = (x_i + x_{i+1})/2$. The following example illustrates the use of the search method in finding the real roots of an algebraic equation.

Example 5.1

The surface of a furnace wall is exposed to radiative, convective, and conductive heat transfer, as shown in Figure 5.2. Under steady-state conditions, the surface temperature T is obtained from an energy balance, which yields the equation

$$\varepsilon\sigma(T_h^4 - T^4) = h(T - T_a) + \frac{k}{d}(T - T_2) \tag{5.6}$$

where the three terms, from the left, represent heat transfer by radiation, convection, and conduction, respectively. Here, T_h is the temperature of the hot environment radiating to the surface, T_a is the air temperature, and T_2 is the temperature at the outer surface of the wall. Also, h is known as the *convective heat transfer coefficient*, k is the thermal conductivity of the wall material, d is the wall thickness, ε is the emissivity of the surface, and σ is the Stefan–Boltzmann constant, given as 5.67×10^{-8} W/m² K⁴. Find the wall temperature by the search method,

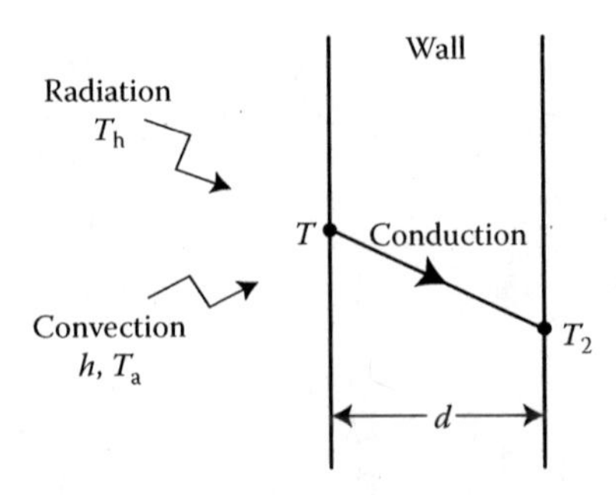

FIGURE 5.2 Heat transfer at the surface of the furnace wall considered in Example 5.1.

if T_h =1000 K, T_a = 500 K, T_2 = 300 K, h = 50 W/m² K, k = 25 W/m K, ε = 0.8, and d = 0.15 m.

SOLUTION

The given problem reduces to finding the roots of the equation

$$f(x) = 0.8 \times 5.67 \times 10^{-8}[(1000)^4 - x^4] - 50(x - 500) - \frac{25}{0.15}(x - 300) = 0 \tag{5.7}$$

where x is the unknown temperature. Since this is a fourth-order polynomial, there are four roots that will satisfy the equation. However, a consideration of the physical problem, described here, indicates that a unique, positive value of the temperature must be obtained in the range 300–1000 K, these being the two extreme temperatures in the problem. Therefore, we expect only one real and positive root to lie in this range. The others will be physically unacceptable, being negative, complex, or beyond the indicated range.

The general behavior of the function $f(x)$ can be first studied by obtaining a rough plot of $f(x)$ versus x. This plot may be obtained simply by incrementing x, starting with the lower limit of x = 300 K, and computing $f(x)$ until the upper limit of x = 1000 K is reached. In MATLAB, the following commands will yield the desired graph, with a line at $f(x)$ = 0 to indicate the sign change in $f(x)$.

```
x=linspace(300,1000,20);
f1=inline(' 0.*x ');
f=inline(' (0.8*5.67*10^(-8))*(10^12- ...
   x.^4)-50.*(x-500)-(25/0.15).*(x-300)');
plot(x,f(x),'k-',x,f1(x),'k--')
xlabel('x (K)','fontsize',14);ylabel('f(x)','fontsize',14)
```

Figure 5.3 shows this plot, and a single, real, positive root is observed to lie between 500 and 550 K.

We may now proceed with the incremental search method, starting with x = 300 K or with a value close to 500 K, and narrow in on the root. A maximum value of x = 1000 K may also be specified to avoid going beyond the range. In the present case, the function is well behaved and we do not expect to encounter

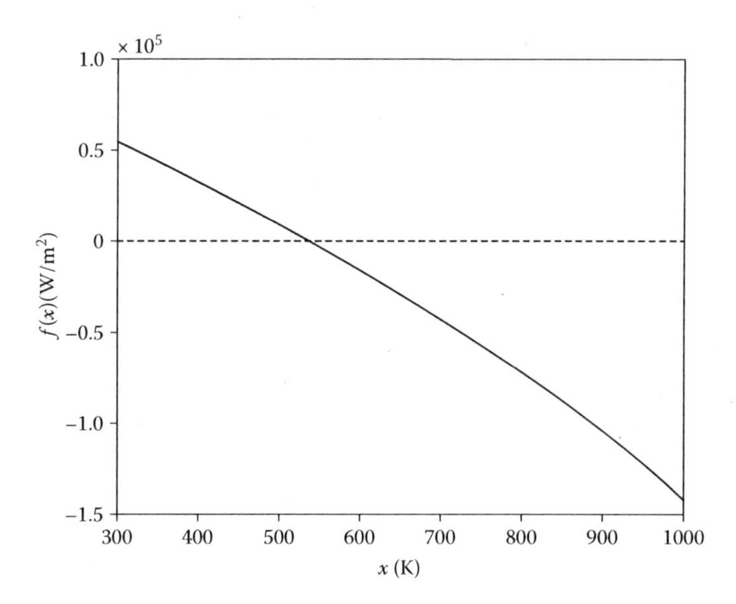

FIGURE 5.3 A rough plot of $f(x)$ versus x to determine the approximate value of the root in Example 5.1.

any problems. A computer program in MATLAB is given in Appendix B.1. The corresponding program in Fortran is given in Appendix C.1 to show a comparison between the two. The logic is quite similar, but the commands to implement the algorithm are different, with MATLAB being more convenient to use. However, for more involved problems, Fortran, C++, or other high-level languages may be more efficient, as discussed earlier in Chapter 2. The program increments x, with a chosen increment of 50 K and an initial value of 300 K, until the function $f(x)$ changes sign. The increment is reduced to one-tenth of the earlier value, and the initial value of x is now taken as the beginning of the step in which the sign change occurs. This process is repeated until the unknown temperature x is obtained to the desired accuracy level, given by the convergence parameter *eps*.

The numerical results from the MATLAB program at various values of the convergence parameter *eps*, which is applied to Δx, as given by Equation 5.5b, are shown in Figure 5.4. These results demonstrate that, as expected, the number of iterations increases as *eps*, printed as EPS here, is decreased. Also, the value of the function $f(x)$ at the estimated root decreases toward zero. Here, F1 and F2 represent $f(x)$ at the two ends of the subinterval. Note that a change only in the third decimal place occurs when EPS is reduced from 0.01 to 0.001, indicating that the first value will give adequate accuracy. The desired accuracy will generally be governed by the engineering application being considered. However, it is important to vary the convergence criterion EPS to ensure that the numerical results are not significantly affected by the value chosen, as discussed in Section 2.5. The convergence criterion may also be applied to the function $f(x)$, which represents the net energy gain at the surface. For further details on the physical problem considered here, books on heat transfer or on college physics, such as Young et al. (2000), may be consulted.

```
EPS = 10.00000
X = 550.00000        F1 = 9191.6667       F2 = –2957.3902
X = 540.00000        F1 = 727.2266        F2 = –496.9862
TEMPERATURE = 535.50000                   F(X) = 727.2266

EPS = 1.00000
X = 550.00000        F1 = 9191.6667       F2 = –2957.3902
X = 540.00000        F1 = 727.2266        F2 = –496.9862
X = 538.00000        F1 = 115.6117        F2 = –6.8290
TEMPERATURE = 537.55000                   F(X) = 115.6117

EPS = 0.10000
X = 550.00000        F1 = 9191.6667       F2 = –2957.3902
X = 540.00000        F1 = 727.2266        F2 = –496.9862
X = 538.00000        F1 = 115.6117        F2 = –6.8290
X = 538.00000        F1 = 5.4168          F2 = –6.8290
TEMPERATURE = 537.95500                   F(X) = 5.4168

EPS = 0.01000
X = 550.00000        F1 = 9191.6667       F2 = –2957.3902
X = 540.00000        F1 = 727.2266        F2 = –496.9862
X = 538.00000        F1 = 115.6117        F2 = –6.8290
X = 538.00000        F1 = 5.4168          F2 = –6.8290
X = 537.97500        F1 = 0.5186          F2 = –0.7060
TEMPERATURE = 537.97050                   F(X) = 0.5186

EPS = 0.00100
X = 550.00000        F1 = 9191.6667       F2 = –2957.3902
X = 540.00000        F1 = 727.2266        F2 = –496.9862
X = 538.00000        F1 = 115.6117        F2 = –6.8290
X = 538.00000        F1 = 5.4168          F2 = –6.8290
X = 537.97500        F1 = 0.5186          F2 = –0.7060
X = 537.97250        F1 = 0.0287          F2 = –0.0937
TEMPERATURE = 537.97205                   F(X) = 0.0287

EPS = 0.00010
X = 550.00000        F1 = 9191.6667       F2 = –2957.3902
X = 540.00000        F1 = 727.2266        F2 = –496.9862
X = 538.00000        F1 = 115.6117        F2 = –6.8290
X = 538.00000        F1 = 5.4168          F2 = –6.8290
X = 537.97500        F1 = 0.5186          F2 = –0.7060
X = 537.97250        F1 = 0.0287          F2 = –0.0937
X = 537.97215        F1 = 0.0042          F2 = –0.0080
TEMPERATURE = 537.97210                   F(X) = 0.0042
```

FIGURE 5.4 Numerical results obtained from the MATLAB program for the search method in Example 5.1.

As presented earlier in Chapter 3, the given problem can also be solved very easily by using the software available in MATLAB for finding the roots of polynomial equations. The given polynomial p is defined by specifying its coefficients and the *roots*(p) command is used to obtain all the roots of the equation. The coefficients are given in descending powers of the independent variable x. Since this is a fourth-order polynomial, four roots are obtained. For example, the following simple program can be used to specify the polynomial from Equation 5.7 in descending powers of x and obtain the roots.

```
format short e
a=0.8*5.67*10^(-8);
b=0;c=0;
d=50+25/0.15;
e=-0.8*5.67*10^(-8)*1000^(4)-50*500-25*300/0.15;
p=[a b c d e];
r=roots(p);
disp(r)
```

This program yields the results

```
-1.8390e+03
6.5052e+02+1.5030e+03i
6.5052e+02-1.5030e+03i
5.3797e+02
```

It is seen that only one root, 537.97, is positive and lies within the acceptable range of temperature. Two are complex and one is negative, making them unacceptable. Also, the root obtained is close to that obtained earlier by the search method.

Similarly, the *fzero* command, available in MATLAB, can be used to obtain the location, in x, where the graph of $f(x)$ versus x crosses the x-axis. A search is carried out for the *zero* of the function close to a specified location or over a given range. The function may be defined by an inline statement as

```
f=inline('0.8*5.67*10^(-8)*(1000^4-x^4)-50* ...
   (x-500)-(25/0.15)*(x-300)');
```

or a function file *f.m* may be created to define the function as

```
function z=f(x)
z=0.8*5.67*10^(-8)*(1000^4-x^4)-50* ...
   (x-500)-(25/0.15)*(x-300);
```

where the 3 periods indicate continuation of the command.
For the former case, the *fzero* command may be given, with a given x, as

```
>> root=fzero(f,300)
```

or, with the given range specified, as

```
>> root = fzero(f,300,1000)
```

yielding the resulting root as

```
root =
    537.9721
```

Similarly, for the second case, employing the *f.m* function file, the command is given as

```
>> fzero('f',300)
```

or

```
>> root = fzero('f',300,1000)
```

This gives the root as

```
root =
    537.9721
```

Thus, the root is close to that obtained earlier and may easily be determined by using the *fzero* command if the $f(x)$ versus x graph crosses the x-axis, yielding a real root.

5.3 BISECTION METHOD

The *bisection*, or *half-interval*, method may be used for a rapid convergence to the root once the interval containing a real root has been determined by the incremental search method or by plotting $f(x)$ versus x. Consider the function $f(x)$, shown in Figure 5.5, which is known to have only one real root in the interval $x_1 \leq x \leq x_2$. The interval is bisected, and the function is computed at the midpoint x_0, where

$$x_0 = \frac{x_1 + x_2}{2} \tag{5.8}$$

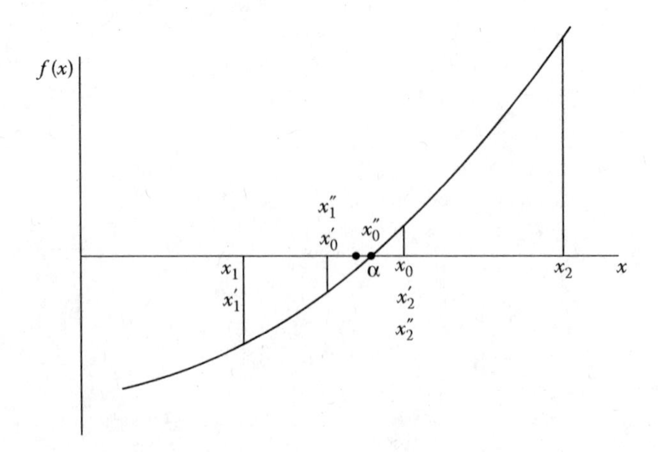

FIGURE 5.5 Sketch illustrating the computational procedure for the bisection method.

Now the product $f(x_0) \cdot f(x_1)$ is calculated. If $f(x_0) \cdot f(x_1) < 0$, then the root lies in the interval $x_1 < x < x_0$, since the function has changed sign in this half-interval. If the product is positive, then the root must be in the other half-interval $x_0 < x < x_2$. The interval containing the root is, therefore, reduced by half, and the preceding procedure is next applied to the reduced interval. The process is repeated until the location of the root is obtained to the desired accuracy. Since the interval containing the root is halved in each bisection, the original interval is reduced by a factor of 2^n after n bisections. The computed root is taken at the midpoint of the interval obtained after n bisections. Then, the maximum error in the calculated root equals half the size of this interval. Therefore, the error ε in the root is given by $I_0/2^{n+1}$, where I_0 is the starting interval containing the root. The number of bisections needed to reduce the maximum error to ε is obtained by taking the logarithm of the equation $\varepsilon = I_0/2^{n+1}$ as follows:

$$n = \frac{\log(I_0/\varepsilon)}{\log 2} - 1 \tag{5.9}$$

where log represents the natural logorithm. Therefore, if we wish to reduce the error to less than 0.1% of the original interval I_0, that is, $\varepsilon/I_0 = 0.001$, the above equation yields $n = 8.97$, implying that only nine bisections are needed to achieve this level of accuracy.

After each interval-halving operation, the new interval containing the root is determined, and the designation of the endpoints of the reduced interval is changed, with x_0 replacing the appropriate endpoint x_1 or x_2. Figure 5.5 shows a few steps in the computational procedure, denoting the successive values of x_1, x_2, and x_0 by means of primes. If, at any stage in the computation, the function $f(x_0)$ is found to be zero or close to it, within a specified error, the root of the equation is taken as the corresponding value of x_0, and further computation is stopped. Otherwise, the computation may be carried out for a specified number of bisections n or until the root is obtained to the desired accuracy, the location of the root being taken as the midpoint of the successively reducing interval. Frequently, the computation is terminated when the change in the root from one bisection to the next is less than a specified error tolerance ε. In most practical problems, the value of ε may be obtained from a consideration of the accuracy needed in the determination of the physical quantity represented by the root. For example, if the pressure of a given volume of gas is to be obtained by solving its equation of state, ε may be chosen as a small fraction of the acceptable error in the pressure.

The bisection method will always yield a root of the given equation $f(x) = 0$ if $f(x)$ changes sign in the interval. However, if the interval contains more than one real root, one must determine the subintervals containing the roots before proceeding to bisection. An odd number of roots in the interval $x_1 < x < x_2$ will give $f(x_1) \cdot f(x_2) < 0$. If there are no roots in the interval or if there are an even number of roots, then $f(x_1) \cdot f(x_2) > 0$. Several bisections may be needed to obtain the subintervals in which $f(x)$ changes sign, and the method may converge to the same root more than once. In such cases, the search method, exhaustive or incremental, may be employed effectively to determine the approximate location of the roots and the subintervals where bisection may then be used. Similarly, bisection will not locate multiple roots that arise due to the plot of the function $f(x)$ being tangent to the x-axis, without crossing it, since a sign change does not occur at the root. Again, the approximate behavior of the function may be determined graphically or from search methods. If such multiple roots arise, other methods, discussed later in this chapter, will be needed. If the multiple root arises at a location where the function $f(x)$ crosses the axis, bisection can be used to find the root. The numerical process will always converge if the interval containing a zero crossing of the function $f(x)$ is known. The convergence is faster than that obtained by the search method.

A MATLAB program is given in Appendix B.2 in order to illustrate the algorithm discussed here. A similar program in Fortran is given in Appendix C.2 for finding the root of a given equation $\log_{10}(x) + x^2 - 6 = 0$ to show the similarities and differences between the two. In both cases, the function $f(x)$ in the equation $f(x) = 0$, whose roots are to be determined, can be employed to solve a given problem. If the MATLAB program shown is to be used for the problem of Example 5.1, the appropriate function $f(x)$ has to be defined. It can be defined within the program as an inline statement, as done earlier in Appendix B.1, or a function file can be saved as *f.m* in the same subdirectory as the main program. For the problem in Example 5.1, the function file was given earlier and is rewritten as

```
function z=f(x)
z=0.8*5.67*10^(-8)*(1000^4-x^4)-50* ...
   (x-500) - (25/0.15)*(x-300);
```

Then, if the given program is executed, the program asks for the end points of the interval and yields the approximations to the root as

```
Enter lowest value of interval, a=300
Enter highest value of interval, b=1000
Iteration converged
 650.0000
 475.0000
 562.5000
 518.7500
 540.6250
 529.6875
 535.1562
 537.8906
 539.2578
 538.5742
 538.2324
 538.0615
 537.9761
 537.9333
 537.9547
 537.9654
 537.9707
 537.9734
 537.9721
```

Therefore, the root obtained as the same as that calculated earlier by the search method. The convergence criterion is based on the absolute value of the function $f(x)$ at the approximation to the root, which is given by Equation 5.8, becoming less than the convergence parameter, which is taken as 0.02. Again, this parameter may be varied to ensure that the results are independent of the value chosen. The convergence, though generally faster than the search method, is still seen to be quite slow and several methods with faster convergence are discussed in Sections 5.4 and 5.5.

5.4 REGULA FALSI AND SECANT METHODS

5.4.1 Regula Falsi Method

The *regula falsi*, or *false-position*, method is similar to the bisection method in that it will always yield a real root in the interval in which the function $f(x)$ changes sign. However, the convergence to the root is generally more rapid. Let us again consider an interval $x_1 < x < x_2$ found graphically or from the search method to contain a real root of the equation $f(x) = 0$. Therefore, $f(x_1)$ and $f(x_2)$ are opposite in sign. A chord is drawn joining the two endpoints $[x_1, f(x_1)]$ and $[x_2, f(x_2)]$. The intersection of the chord, which represents a linear approximation to the function $f(x)$, with the x-axis, $x = x_3$, is taken as the first estimation of the root of the given equation. From Figure 5.6, x_3 may be obtained by using the geometrical relationship between the two triangles formed as follows:

$$x_3 = \frac{x_1 f(x_2) - x_2 f(x_1)}{f(x_2) - f(x_1)} \tag{5.10}$$

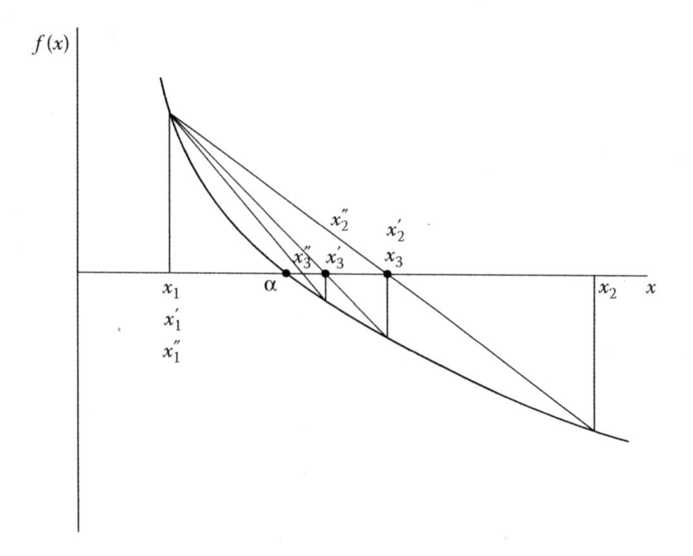

FIGURE 5.6 The regula falsi method for root solving.

If $f(x_3)$ is zero or close to zero, within a specified convergence criterion, the process is terminated, and the root is located at $x = x_3$. If $f(x_3)$ has the same sign as $f(x_1)$, then $f(x_3) \cdot f(x_1) > 0$, and the root lies between x_2 and x_3. Then x_2 remains unchanged and x_3 becomes the new value of x_1, thus giving the reduced interval containing the root as $x_3 < x < x_2$. Similarly, if $f(x_3) \cdot f(x_1) < 0$, then the root is located in the interval $x_1 < x < x_3$. In this case, x_1 remains unaltered, and x_3 becomes the new value of x_2. Figure 5.6 shows a few steps in the computational process, denoting new values by primes.

The new values of x_1 and x_2 are employed in Equation 5.10 to yield an improved approximation to the root. The process is continued, successively reducing the interval containing the root until $|f(x_3)|$ or the change in the root, which is approximated by x_3, from one computational step to the next is less than a specified small quantity ε. The value of the error tolerance ε is chosen on the basis of the accuracy desired in the evaluation of the root. Since, at each step, a subinterval containing the root is considered, the method will always converge if a sign change in $f(x)$ occurs in the initial interval. Multiple roots due to the plot of $f(x)$ being tangent to the x-axis cannot be located by this method, which requires a sign change in $f(x)$. The rate of convergence to the root depends on the nature of the function $f(x)$ and the initial interval. Although convergence is often faster than that obtained by the bisection method, examples can be found where such is not the case. The procedure outlined above is also sometimes known as the *linear interpolation method.*

5.4.2 Secant Method

The preceding methods always considered the subinterval that enclosed the root and will, therefore, always yield the solution if a sign change in $f(x)$ occurs in the interval. The secant method is similar to the regula falsi, or false-position, method, but it does not always consider subintervals containing the root. This method is, therefore, not guaranteed to converge, unlike the enclosure methods discussed so far. However, when it does converge, it does so more rapidly than the previous methods. Instead of using the two values of x that bound a subinterval containing a real root, the method uses the two most recent values of x in the iterative procedure. Therefore, it employs both interpolation and extrapolation to approximate the root by the intersection of the line joining the two points $[x_1, f(x_1)]$ and $[x_2, f(x_2)]$ with the x-axis. Figure 5.7 shows a few iterative steps for the secant method, using primes to denote the new values. As in the regula falsi method, a chord is drawn between the two endpoints of the initial interval, and the intersection with the x-axis is obtained from Equation 5.10. However, in the next step, the interval containing the root is not considered and the two most recent values of x, x_2 and x_3, are taken. The former thus becomes x_1 and the latter x_2. These are again substituted in Equation 5.10 to yield the new intersection point, which is the next approximation to the root. Therefore, the general expression for iteration by the secant method may be written from Equation 5.10 as

$$x_{i+1} = \frac{x_{i-1} f(x_i) - x_i f(x_{i-1})}{f(x_i) - f(x_{i-1})} \tag{5.11}$$

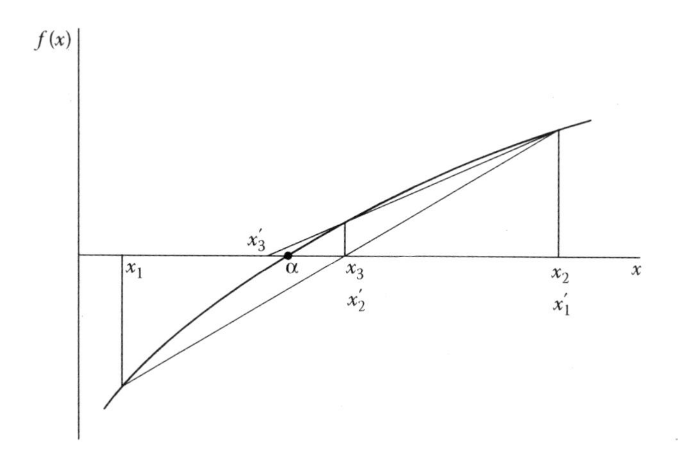

FIGURE 5.7 The secant method for finding the real roots of an algebraic equation.

where the subscript represents the order of the iteration, starting with $x_{i-1} = x_1$ and $x_i = x_2$ as the first two approximations to the root.

The above process is continued until $|f(x_i)| \leq \varepsilon$, where ε is the specified convergence parameter. Since the method employs linear extrapolation for subintervals that do not contain the root, the process may not converge. Convergence depends on the nature of the function and on the limits of the initial interval. If the initial guesses x_1 and x_2 are chosen sufficiently close to the root, $x = \alpha$, convergence of the iterative process to the root can be expected. The method may diverge if the initial guesses are not well chosen. Therefore, this method is suitable if the interval containing the root is known to a fairly good approximation from a rough plot of the function or from the search method. Both the regula falsi and the secant methods may be employed for the real roots of polynomial as well as transcendental equations. The following examples illustrate the use of these methods for finding the roots of algebraic equations.

Example 5.2

A flat plate falling freely in air is subjected to a downward gravitational force and an upward frictional drag due to air. This drag force D_f is given by the expression

$$D_f = \frac{0.2275V^2}{465.9 + (\log V)^{2.58}} - 0.017V \quad \text{for } V \geq 1\,\text{m/s} \tag{5.12}$$

where V is the vertical velocity of the plate and $\log V$ is the natural logarithm of V. A terminal velocity is attained when this drag force equals the gravitational force. The net force F acting on the plate is given by

$$F = D_f - mg \tag{5.13}$$

where m is the mass of the plate and g the magnitude of the gravitational acceleration. Find the terminal velocity by the regula falsi method if $m = 1$ kg and

$g = 9.8$ m/s^2. Vary the convergence parameter ε, applied to F, from 1.0 to 10^{-4}, and study the effect on the numerical value obtained for the terminal velocity.

SOLUTION

The terminal velocity is given by the root of the equation

$$F(V) = \frac{0.2275V^2}{465.9 + (\log V)^{2.58}} - 0.017V - 9.8 = 0 \qquad (5.14)$$

Since the expression for the drag force D_f is valid for $V \geq 1$ m/s, we may take the value of 1 m/s as the lower limit, x_1, of the range in which the root is located. The upper limit, x_2, may be taken as a large value, say, 500 m/s. If the root is not found within this range, a still larger value of x_2 may be employed. A rough plot of $F(V)$ versus V may also be obtained to guide the choice of the initial range of values. The root is expected to be real, positive, and unique.

A computer program may easily be written for solving this problem by the regula falsi method. The algorithm is very similar to that for the bisection method, discussed in the preceding section. The approximation to the root x_3 is given by Equation 5.10, instead of Equation 5.8 which was used for the bisection method. As before, depending on the sign of $f(x_1) \cdot f(x_3)$, the new values of x_1 and x_2 are chosen. If $f(x_1) \cdot f(x_3) < 0$, then the root lies between x_1 and x_3. Thus, x_3 becomes the new x_2, and the procedure is repeated. Similarly, if $f(x_1) \cdot f(x_3) > 0$, x_3 becomes the new x_1. Iteration is terminated when $|f(x_3)| \leq$ EPS, where EPS is the chosen convergence parameter. The numerical results obtained for various values of EPS are shown in Figure 5.8. A velocity V of 173.0431 m/s is obtained for EPS = 10^{-4}, and a change of less than 0.003% is observed when EPS is varied from 10^{-3} to this value. The net force F on the plate is essentially zero, within the chosen convergence criterion. Depending on the desired accuracy of the terminal velocity, the corresponding value of EPS may be chosen.

Example 5.3

Solve the problem of Example 5.2 by the secant method, and compare the results with those obtained by the regula falsi, or false-position, method.

SOLUTION

In this case, the two most recent values of the unknown x are employed, with interpolation and extrapolation, to find the root. Equation 5.11 is used instead of

```
EPS = 1.00000     TERMINAL VELOCITY = 167.1907     FUN(X) = -0.7216
EPS = 0.10000     TERMINAL VELOCITY = 172.2868     FUN(X) = -0.0948
EPS = 0.01000     TERMINAL VELOCITY = 172.9755     FUN(X) = -0.0086
EPS = 0.00100     TERMINAL VELOCITY = 173.0377     FUN(X) = -0.0008
EPS = 0.00010     TERMINAL VELOCITY = 173.0431     FUN(X) = -0.0001
```

FIGURE 5.8 Computed results for various values of the convergence parameter EPS, using the regula falsi method for Example 5.2.

Equation 5.10 to find the next approximation. Therefore, x_1 and x_2 are the first two approximations to the root, followed by x_2 and x_3 as the approximations in the next step. A MATLAB program for the secant method is given in Appendix B.3. The program is given as a function *m*-file, where the file, saved as *secant.m*, is given as *function [p1,err,k] = secant(f,p0,p1,delta,max1)*. Thus, the function *f*(*x*) has to be stored separately as a function file *f.m* and specified as a string, such as 'f,' when calling the secant function *m*-file. Also, p0 and p1 are the starting values for the unknown *x*, *delta* is the convergence parameter and *max1* is the given maximum number of iterations before the execution is stopped if convergence is not achieved. The results from this program will yield the resulting approximation to the roots after each iteration and the final result if convergence is obtained. If convergence is not achieved, it will indicate that the maximum number of iterations has been reached. The print out commands may be suitably modified to obtain desired format for the results. Similarly, different starting values p0 and p1, as well as different convergence parameters may be used. The corresponding program in Fortran is given in Appendix C.3 for comparison.

The initial values of x_1 and x_2 may be taken as 150 m/s and 200 m/s, respectively. The function *F*(*V*) is well behaved, with no discontinuities or sharp changes, and convergence is obtained even with a much larger initial range. For example, with x_1 taken as 1 m/s and x_2 as 500 m/s, convergence is again achieved and a smaller computer (CPU) time than that needed for the regula falsi method is required. However, a narrower initial interval for the unknown root would generally be needed for the secant method, as compared to the regula falsi method, in order to obtain convergence. The secant method is not guaranteed to converge since the interval being considered at any given stage does not necessarily contain the root.

The numerical results for the terminal velocity in the given problem at various values of the convergence criterion EPS are shown in Figure 5.9. It is interesting to note that the value obtained, for $x_1 = 150$ m/s and $x_2 = 200$ m/s, does not change when EPS is varied from 10^{-2} to 10^{-4}, for the four significant decimal places printed. The value itself is within 0.001% of that obtained in Example 5.2, and a relatively large value of EPS, 0.01 from the results shown, is found to be satisfactory. Therefore, if the appropriate range or interval containing the root is known, a rapid convergence to the root may be obtained by the secant method. The results

```
INITIAL X1 = 150.00              INITIAL X2 = 200.00
EPS = 1.00000     TERMINAL VELOCITY = 171.1760     FUN(X) = –0.2331
EPS = 0.10000     TERMINAL VELOCITY = 172.9019     FUN(X) = –0.0178
EPS = 0.01000     TERMINAL VELOCITY = 173.0445     FUN(X) = 0.0001
EPS = 0.00100     TERMINAL VELOCITY = 173.0445     FUN(X) = 0.0001
EPS = 0.00010     TERMINAL VELOCITY = 173.0445     FUN(X) = 0.0001

INITIAL X1 = 2.00                INITIAL X2 = 500.00
EPS = 1.00000     TERMINAL VELOCITY = 170.7368     FUN(X) = –0.2876
EPS = 0.10000     TERMINAL VELOCITY = 173.1468     FUN(X) = 0.0129
EPS = 0.01000     TERMINAL VELOCITY = 173.0430     FUN(X) = –0.0001
EPS = 0.00100     TERMINAL VELOCITY = 173.0430     FUN(X) = –0.0001
EPS = 0.00010     TERMINAL VELOCITY = 173.0430     FUN(X) = –0.0001
```

FIGURE 5.9 Computed results from the solution of the problem in Example 5.2 by the secant method.

for the starting values of x_1 and x_2 taken as 2 and 500 m/s, respectively, are also shown in Figure 5.9. The results are very slightly different from those obtained when these are taken as 150 and 200 m/s, respectively, and are also very close to those obtained in Example 5.2.

5.5 NEWTON–RAPHSON METHOD AND MODIFIED NEWTON'S METHOD

The *Newton–Raphson* method, or simply *Newton's* method, for finding the roots of polynomial and transcendental equations is very widely used because of its versatility and generally rapid convergence. The method employs an initial approximation to the root of a given equation and iteratively improves the root until convergence to the desired accuracy is achieved. However, as in the secant method, convergence is not assured. The method is not based on the plot of the function $f(x)$ crossing the x-axis, and therefore it may be used for complex and multiple roots as well. The *modified Newton's method* is an extension of the conventional method and has certain important advantages, as discussed later in Section 5.5.2.

5.5.1 Newton–Raphson Method

If $x = x_1$ is the first approximation to the root of the equation $f(x) = 0$, the function $f(x)$ may be expanded in a Taylor series about x_1, for x close to x_1, as given in Chapter 4 to yield

$$f(x) = f(x_1) + (x - x_1)f'(x_1) + \frac{(x - x_1)^2}{2!} f''(x_1) + \cdots \tag{5.15}$$

where the primes denote the order of the differentiation with respect to x. In order to determine the root, $x = \alpha$, we set $f(x)$ equal to zero and then solve the resulting equation for the root. However, this gives a polynomial equation of order infinity. If only the first two terms are retained, the next approximation to the root, x_2, may be obtained. Therefore, setting $f(x) = 0$, we get

$$0 = f(x_1) + (x_2 - x_1)f'(x_1)$$

or

$$x_2 = x_1 - \frac{f(x_1)}{f'(x_1)}$$

where x_2 represents an improved estimate of the root. In the next iteration, x_1 is replaced by this new approximation x_2, and a further improved approximation to the root x_3 is obtained. The general expression for iteration by Newton's method is, therefore, written as follows:

$$x_{i+1} = x_i - \frac{f(x_i)}{f'(x_i)} \tag{5.16}$$

where x_i and x_{i+1} are the values obtained after the ith and the $(i + 1)$th iterations, respectively.

This iterative process is continued until the approximation to the root from one iteration to the next changes by less than a specified small quantity ε. As discussed for the previous methods, the convergence criterion ε is often chosen on the basis of the nature of the physical problem under consideration and may be applied to the approximate root or to the function $f(x)$, as $|f(x_i)| \leq \varepsilon$. Note that Newton's method is similar to the secant method, discussed in Section 5.4.2. In the secant method, the slope $f'(x_i)$ is approximated by $[f(x_i) - f(x_{i-1})]/(x_i - x_{i-1})$. If this approximation is substituted in Equation 5.16, the formula for iteration by the secant method, Equation 5.11, is obtained. Consequently, the convergence characteristics of the secant method and the Newton–Raphson method are quite similar.

The derivative $f'(x_i)$ of the function is needed for using this method. In many cases, particularly for polynomial equations, the derivative may be obtained easily. However, there are problems, such as those involving transcendental functions, in which the differentiation of the function $f(x)$ may be quite complicated. The derivative may then be computed numerically by a finite difference approximation, as outlined in Chapter 4. The iterative procedure converges very rapidly, as shown in terms of a few iterative steps in Figure 5.10. It is seen graphically that the intersection of the tangent, to the curve of $f(x)$ versus x at a given approximation, with the x-axis, where $f(x) = 0$, gives the next approximation. From the figure, the slope $f'(x_i)$ is given by $f(x_i)/(x_i - x_{i+1})$, which yields the iterative formula for Newton's method.

However, the method may not converge if the initial guess is too far from the root and also if the derivative is close to zero or varies substantially near the root. A few cases in which the iteration does not converge are shown in Figure 5.11. The computer program should include this possibility so that, if the method diverges or if the root is not obtained in a specified number of iterations, a new initial guess is chosen and the numerical procedure for finding the root is carried out again. Any information on the value and characteristics of the root, from the physical background of the

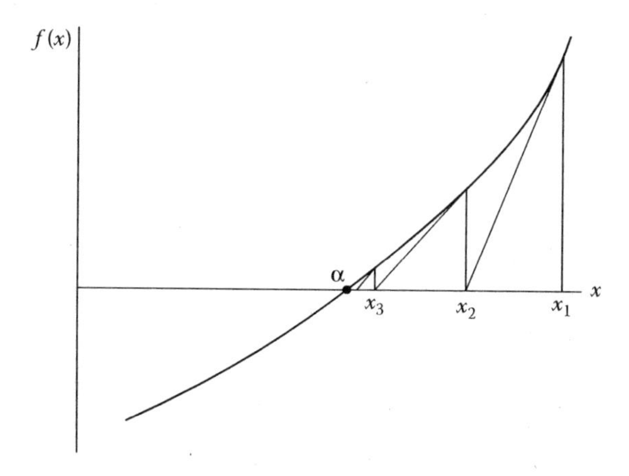

FIGURE 5.10 A graphical representation of the Newton–Raphson iterative procedure for solving an algebraic equation.

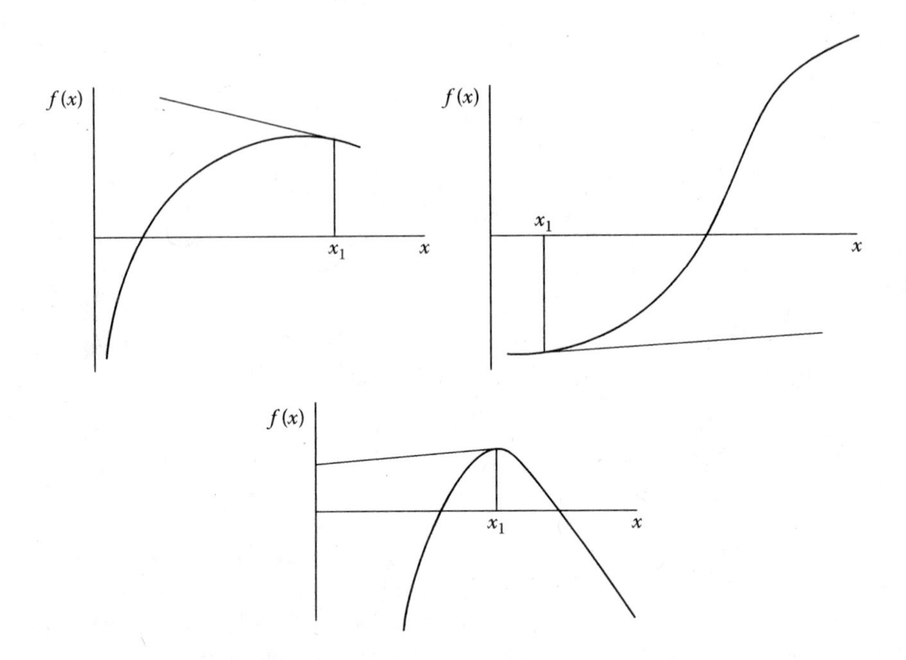

FIGURE 5.11 A few cases in which the Newton–Raphson method diverges.

problem, from the variation of the function $f(x)$ with x, or from the search method, will be useful in the choice of the initial approximation, so that a rapid convergence to the root may be obtained.

Newton's method may also be used for determining the complex roots of the equation $f(z) = 0$, where the complex variable $z = x + iy$, i being $\sqrt{-1}$ and x and y real quantities. The function $f(z)$ may be written as $f(z) = u(x, y) + iv(x, y)$, where u and v are the real and imaginary parts of the function. Then the iterative process for Newton's method is given by the complex expression

$$z_{i+1} = z_i - \frac{f(z_i)}{f'(z_i)} \tag{5.17}$$

where z_i represents the approximation to the root after the ith iteration. Therefore, if complex algebra is available on the computer, complex roots may be determined by the procedure outlined above for real roots. If complex variables cannot be used, the following procedure may be used. The real and imaginary parts on the two sides of the above equation may be equated to obtain the following (Carnahan et al., 1969):

$$x_{i+1} = x_i + \left(\frac{vu_y - uu_x}{u_x^2 + u_y^2} \right)_{x_i, y_i} \tag{5.18}$$

$$y_{i+1} = y_i + \left(\frac{-vu_x - uu_y}{u_x^2 + u_y^2} \right)_{x_i, y_i} \tag{5.19}$$

where u_x and u_y denote the partial derivatives $\partial u/\partial x$ and $\partial u/\partial y$, respectively. The subscripts x_i, y_i outside the parentheses indicate that the values are to be determined at $x = x_i$ and $y = y_i$. For obtaining the above equations, the Cauchy–Riemann equations, $u_x = v_y$ and $u_y = -v_x$, have been employed. Therefore, this method may be used for finding the zeros of complex functions whose real and imaginary parts can be separated easily. In most computer systems, complex algebra is available, and Equation 5.17 may be employed directly to determine the complex roots, as outlined in Example 5.5. In MATLAB, if the initial guess is given as a complex quantity, complex algebra is automatically employed to yield the complex roots. Of course, if the *roots*(*p*) command is used for a polynomial equation, complex roots, if any, are given by the results. A similar treatment is applicable if the coefficients of the equation are not real but complex.

As mentioned earlier, complex functions arise in several engineering problems, such as those concerned with vibrations, stability of systems, electrical circuits with alternating current sources, periodic processes, wave phenomena, and flow fields that may be represented by a complex potential. For further details on complex algebra, see any textbook on calculus, such as Thomas and Finney (1999). It may be mentioned that though z is used to denote an independent complex variable here for clarity, the independent variable x will, in general, be used in the following to denote a real or complex variable.

5.5.2 Modified Newton's Method

The Newton–Raphson method can also be used for multiple roots corresponding to points where the function $f(x)$ becomes tangent to the x-axis. However, since $f'(x)$ also goes to zero, as $f(x)$ approaches zero at the root, the convergence is slow, and computational difficulties may arise. For such cases and for achieving a faster rate of convergence, the above procedure may be modified to obtain Newton's second-order method, which employs the second derivative $f''(x)$ of the function in the computation of the root. If three terms are retained from the Taylor series given in Equation 5.15, instead of the two used for the Newton–Raphson method, we obtain

$$0 = f(x_1) + (x_2 - x_1)\left[f'(x_1) + \frac{(x_2 - x_1) f''(x_1)}{2} \right]$$

To avoid solving this quadratic equation for x_2, one may substitute the approximation for $(x_2 - x_1)$ from the Newton–Raphson method in the brackets above. The resulting linear equation may be solved to obtain the next approximation to the root. This method is often known as the *modified Newton's method*. Therefore,

$$0 = f(x_1) + (x_2 - x_1)\left[f'(x_1) - \frac{f(x_1)}{f'(x_1)} \cdot \frac{f''(x_1)}{2} \right]$$

or

$$x_2 = x_1 - \frac{f(x_1)}{f'(x_1) - (f(x_1) f''(x_1) / 2 f'(x_1))} \tag{5.20}$$

This equation gives the general expression for iteration by this method as

$$x_{i+1} = x_i - \frac{f(x_i)}{f'(x_1) - \dfrac{f(x_i)f''(x_i)}{2f'(x_i)}} \tag{5.21}$$

Newton's second-order method, therefore, requires the value of the second derivative of the function. Similarly, higher-order modifications of the conventional Newton's method may be derived for better convergence characteristics. However, the applicability of the method is limited by the computational difficulty in obtaining the derivatives. If they are obtained easily from the given function $f(x)$, the method is advantageous to use. However, if the derivatives are not easy to obtain, one may need to compute the finite difference approximations of the derivatives, leading to a considerable increase in the computational effort. In most problems of engineering interest, the Newton–Raphson method is employed, instead of its higher-order modifications, because of the programming and computational simplicity of the method.

5.5.3 Convergence

As mentioned earlier, the Newton–Raphson method may not converge. But, if it does converge, it does so very rapidly. It can be shown that for nonzero $f'(\alpha)$, where α is a real root, convergence is guaranteed if the starting value x_1 is close enough to α (Carnahan et al., 1969). Also, once the approximation x_i to the root is close to the exact value α, the error after the next iteration, $x_{i+1} - \alpha$, can be shown to be proportional to the square of the error in the present step, $x_i - \alpha$. The relationship between the two is obtained as follows (Carnahan et al., 1969; Atkinson, 1989)

$$x_{i+1} - \alpha = (x_i - \alpha)^2 \frac{f''(\alpha)}{2f'(\alpha)} \tag{5.22}$$

where in the limit $i \to \infty$, $x_i \to \alpha$. Therefore, the error, which is assumed to be small near the root α, reduces very rapidly with the number of iterations. The resulting convergence is termed *quadratic*, or *second order*, and is more rapid than that for the other methods discussed earlier in this chapter. Most of these methods have a linear convergence, that is, the error after a given iteration is proportional to that obtained after the preceding iteration. Thus, they have *first-order* convergence. The order of convergence for the secant method can similarly be shown to be 1.62, implying that its convergence is faster than methods like search and bisection, but not as fast as Newton's method.

Because of its high rate of convergence, applicability to a variety of equations, and simplicity in programming, the Newton–Raphson method is used extensively in engineering applications. It is also used as a correction scheme in the solution of ODEs, for satisfying the boundary conditions, and in the iterative solution of a system of nonlinear equations. These applications are considered in Chapters 6 and 9. The modified Newton's method is generally employed if the derivative $f'(\alpha)$

goes to zero or becomes very small in the vicinity of the root. The following examples illustrate the use of these methods.

Example 5.4

The water mass flow rate *w*, in kg/s, in a heating equipment that transfers energy from condensing steam to water, is to be obtained from energy balance considerations. If 250 kW of thermal energy are to be exchanged between the two fluids, the equation for the conservation of energy is given as

$$250 = 294w\left[1 - \exp\left(\frac{-1000}{21(5 + 20w)}\right)\right]$$

Find the root of this equation by the Newton–Raphson method. The flow rate is known to be less than 5 kg/s.

SOLUTION

It is evident from the above outline of the physical problem under consideration that the root to be obtained is real and positive, being in the range 0 to 5 kg/s. The given equation may be written as

$$f(x) = 294x\left[1 - \exp\left(\frac{-1000}{21(5 + 20x)}\right)\right] - 250 = 0 \tag{5.23}$$

where x is the unknown flow rate in kg/s. To apply the Newton–Raphson method, we need a starting guess for the unknown and the value of the derivative df/dx at each approximation to the root. Although the derivative may be obtained analytically in this case, there are several problems where the differentiation may be quite involved. In such cases, numerical differentiation may be employed, using finite-difference approximations of Chapter 4. Therefore, numerical differentiation is used here. The function $f(x)$ is determined numerically at two values of x, which are close to each other and are represented by x and x_N, with $x_N > x$. Then the derivative of the function at x is approximated by

$$\frac{df}{dx} \cong \frac{f(x_N) - f(x)}{x_N - x} \tag{5.24}$$

Once the derivative $f'(x)$ has been evaluated, we use Equation 5.16 to determine the next approximation to the root.

Appendix B.4 gives a MATLAB script *m*-file for the Newton–Raphson method. The difference between x_N and x is taken arbitrarily as 0.001, considering the expected value of the root. A smaller value may be chosen for greater accuracy of the derivative. The initial guess for x is chosen as 0.1 kg/s. Various values of the convergence criterion EPS, as applied to the function $f(x)$, are considered, and the results for EPS = 10^{-3} are shown in Figure 5.12. A rapid convergence to the root, which is obtained as 0.9987 kg/s, is observed. Convergence was found to occur if the starting value is taken in the range 0 to around 2.5, but divergence

Enter the convergence parameter, eps = 0.001
EPS = 0.0010
Enter the initial guess, x(1) = 0.1
X = 0.1000 FUNCTION F(X) = −220.632656
X = 0.8529 FUNCTION F(X) = −28.191628
X = 0.9916 FUNCTION F(X) = −1.309440
X = 0.9987 FUNCTION F(X) = −0.002785
Iterations Converged
FLOW RATE X = 0.9987 FUNCTION F(X) = 0.000001

FIGURE 5.12 Numerical results by the Newton–Raphson method at EPS = 10^{-3} for the problem in Example 5.4.

of the scheme occurred at higher values. Therefore, a statement for terminating the computation is included in the program if the approximation to the root becomes very large, this being specified by its numerical value becoming greater than 1/EPS. If the iterations diverge, a new starting value is chosen and the computational scheme repeated. It was found that a smaller value of EPS gave essentially the same flow rate. A Fortran program is given in Appendix C.4 for comparison and it is seen that the logic is very similar in the two cases and may be used for other languages, such as C++, as well.

If the derivative can easily be obtained analytically, the scheme may be modified to evaluate the derivative directly, instead of using the numerical differentiation procedure given here. For instance, if the equation to be solved is $\exp(x) - x^2 = 0$, the derivative $f'(x) = \exp(x) - 2x$ and thus the finite-difference approximation is not needed and the calculated value of $f'(x)$ for each iteration may be obtained from the preceding expression.

Example 5.5

Several fluid flow circumstances of interest in mechanical and civil engineering problems can be represented in terms of a complex variable x, known as the *complex potential.* The complex potential x for a flow is governed by the polynomial equation

$$f(x) = x^4 - 4x^3 + 7x^2 - 6x + 2 = 0 \tag{5.25}$$

Using the Newton–Raphson method, find the complex roots of this equation. The zeros of the polynomial represent certain locations of symmetry in the flow.

SOLUTION

The given polynomial equation is of fourth order and thus has four roots, which may be real or complex. The complex roots arise in conjugate pairs. A rough plot of the function $f(x)$, shown in Figure 5.13, indicates the possibility of a multiple real root around $x = 1$, as confirmed later in Example 5.6. Therefore, a conjugate pair of complex roots is sought in the present problem. The mathematics for complex variables available on the computer is employed, with the function $f(x)$, the

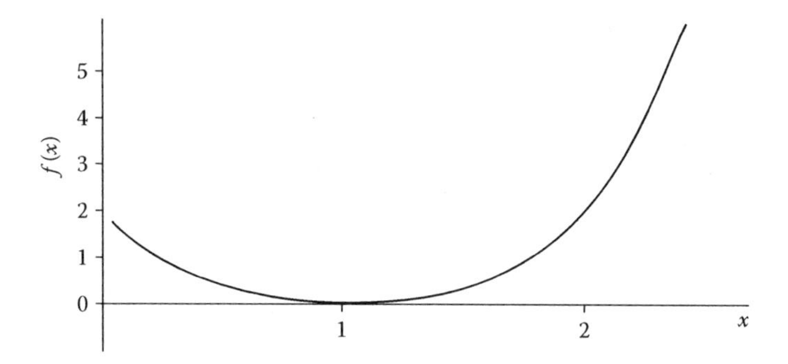

FIGURE 5.13 A plot of the function $f(x)$ versus x for the problem in Example 5.5.

unknown x, the derivative, and the increment Δx, in x, for the next iteration being defined as complex. The derivative is obtained simply as

$$\frac{df}{dx} = 4x^3 - 12x^2 + 14x - 6 \tag{5.26}$$

The convergence criterion is applied to the magnitude of the function, given by $\sqrt{u^2 + v^2}$, where u and v are the real and imaginary parts of the function $f(x)$. The real part represents the velocity potential, and the imaginary part the stream function which is related to the flow rate. The convergence criterion used ensures that, at convergence, the magnitudes of both of these are less than a given small quantity *eps*.

The MATLAB computer program given in Appendix B.4 may be used, with the definition of the appropriate function $f(x)$ and the calculation of the derivative $f'(x)$ from Equation 5.26, for finding the complex roots of the given polynomial equation, Equation 5.25. If the starting value of x is taken as a complex number, such as $1.5 + 2.0i$, where $i = \sqrt{-1}$, the complex algebra in MATLAB is automatically employed. Equation 5.17 is then used to determine the next approximation to the root and the process is repeated until convergence is achieved. The results for the convergence parameter eps, which is printed as EPS here, of 10^{-3} are shown in Figure 5.14, indicating a complex root at $(1 + i)$.

Therefore, the conjugate root is $(1 - i)$. It must be noted that, depending on the initial guess, or starting value for the iteration, the scheme may diverge or converge to another root. In this case, it could converge to the multiple real roots at $x = 1.0$, even if the initial guess is given as a complex number, or to the conjugate root $(1 - i)$.

Again, the convergence to the root is quite rapid, although each iteration involves a larger computational effort as compared to that for real variables. The roots may also be found analytically in this case. The values are found to be identical to those obtained numerically. Similarly, the roots of the polynomial equation may simply be found by the following commands in MATLAB:

```
>>p = [1 -4 7 -6 2];
>>disp(roots(p);
```

```
EPS = 0.0010
X   = 1.5000 + 2.0000i
X   = 1.3546 + 1.5644i
X   = 1.2366 + 1.2623i
X   = 1.1334 + 1.0732i
X   = 1.0427 + 0.9904i
X   = 0.9986 + 0.9952i
X   = 1.0000 + 1.0001i
THE SOLUTION IS X = 1.0000 + 1.0001i
```

FIGURE 5.14 The numerical results obtained at EPS = 10^{-3} for finding the complex roots of the polynomial equation of Example 5.5 by the Newton–Raphson method.

The results are printed as

```
1.0000 + 1.0000i
1.0000 - 1.0000i
1.0000 + 0.0000i
1.0000 - 0.0000i
```

indicating a multiple root at x = 1 and complex roots 1 + i and 1 – i.

Example 5.6

Find the real roots of the polynomial equation given in Equation 5.25 by the modified Newton's method, and compare the convergence to the root with that obtained by the Newton–Raphson method.

SOLUTION

As seen from the rough plot of the given function, shown in Figure 5.13, a multiple root is expected in the neighborhood of $x = 1$. The second derivative of the function is obtained as

$$\frac{d^2 f}{dx^2} = 12x^2 - 24x + 14 \tag{5.27}$$

An initial guess, say at $x = 0.1$, is taken. Then, the function and the first and second derivatives are computed at this value of x. Equation 5.21 is then employed to obtain the next approximation to the root. The convergence parameter *eps* is applied to the function $f(x)$ and is taken as 10^{-5}, although only a small difference in the results was observed when *eps* was varied between 10^{-5} and about 10^{-3}. The computer program given in Appendix B.4 for the Newton–Raphson method may be easily modified for applying the modified Newton's method. The numerical results are shown in Figure 5.15. As mentioned earlier, the analytical solution yields a multiple root at $x = 1$, and the computed value is obtained as 0.998419.

```
MODIFIED NEWTON'S METHOD
EPS  = 0.000010
X    = 0.100000                  FUNCTION= 1.466100
X    = 0.606555                  FUNCTION= 0.176761
X    = 0.863423                  FUNCTION= 0.019002
X    = 0.954440                  FUNCTION= 0.002081
X    = 0.984829                  FUNCTION= 0.000231
X    = 0.994977                  FUNCTION= 0.000026
X    = 0.996419                  FUNCTION= 0.000003
THE SOLUTION IS X = 0.998419     FUNCTION= 0.000003

NEWTON-RAPHSON METHOD

EPS  = 0.000010
X    = 0.100000                  FUNCTION= 1.466100
X    = 0.410876                  FUNCTION= 0.467519
X    = 0.645094                  FUNCTION= 0.141824
X    = 0.604693                  FUNCTION= 0.039600
X    = 0.898886                  FUNCTION= 0.010329
X    = 0.948940                  FUNCTION= 0.002614
X    = 0.974405                  FUNCTION= 0.000656
X    = 0.987205                  FUNCTION= 0.000165
X    = 0.993650                  FUNCTION= 0.000040
X    = 0.996804                  FUNCTION= 0.000010
X    = 0.998446                  FUNCTION= 0.000003
THE SOLUTION IS X = 0.998446     FUNCTION= 0.000003
```

FIGURE 5.15 The numerical results for Example 5.6 from the modified Newton's method and the Newton–Raphson method.

Because the plot of the function is tangent to the x-axis at this value of x, both $f(x)$ and $f'(x)$ go to zero at the root, resulting in a slower convergence as compared to the case where the function crosses the x-axis.

The numerical results obtained by the application of the Newton–Raphson method to this problem are also shown in Figure 5.14. The values of the root obtained in the two cases are quite close, but the convergence for the Newton–Raphson method is much slower. The retention of the additional term, involving a nonzero $f''(x)$, in the modified Newton's method accelerates the convergence. Therefore, for multiple roots, arising due to the plot of the function $f(x)$ being tangential to the x-axis, the modified Newton's method gives a faster convergence to the root.

5.6 SUCCESSIVE SUBSTITUTION METHOD

The *successive substitution*, or *fixed point*, method is an important, but simple, approach to determine the roots of an equation. Any type of equation, such as polynomial or transcendental, can be considered and the algorithm is quite straightforward. However, the scheme may not converge and various strategies are often employed to obtain convergence. In this method, the equation $f(x) = 0$ is rewritten to obtain an equation for the independent variable x as

$$x = g(x) \tag{5.28}$$

so that $\alpha = g(\alpha)$ if $f(\alpha) = 0$, where α is a root of the original equation. Thus, if x_1 is an initial approximation to a root, the successive approximations to the root may be obtained from the recursion relation

$$x_{i+1} = g(x_i) \tag{5.29}$$

Therefore, a successive substitution of the approximation x_i to a root into the function $g(x)$ yields a sequence of iterations that may converge to the root. The equation $x = g(x)$ can be obtained from the original equation $f(x) = 0$ in an unlimited number of ways. In many cases, the equation may contain a linear expression in x. Consider, for instance, $f(x) = x^4 - 6.5x^3 + 7x^2 - 11.5x + 3 = 0$. Then the equation for the successive substitution method may be obtained simply by isolating the linear expression to give $x = g(x) = (x^4 - 6.5x^3 + 7x^2 + 3)/11.5$. Similarly, we may rewrite the equation as $x = g(x) = (6.5x^3 - 7x^2 + 11.5x - 3)^{1/4}$. Convergence is often sensitive to the choice of $g(x)$ and may not occur for a chosen form of the function. In the preceding equation, for instance, the first formulation is appropriate if the root is less than 1.0, and the second formulation is better if the root is larger than 1.0, as seen in Example 5.7.

In order to modify the convergence characteristics of the method, we can also employ the following recursion equation:

$$x_{i+1} = (1 - \beta)x_i + \beta g(x_i) \tag{5.30}$$

where β is a constant and may be chosen to improve convergence. This equation is obtained from the consideration that if $\alpha = g(\alpha)$, then α also satisfies the equation $\alpha = (1 - \beta)\alpha + \beta g(\alpha)$. The choices for $g(x)$ and β are dependent on the behavior of the function $f(x)$. Because of the arbitrariness in the choice of $g(x)$ and the usual strong dependence of convergence on the function $g(x)$, the method is not used as frequently for root solving as other methods like Newton–Raphson. However, in many practical circumstances, the method is employed for the solution of simultaneous, nonlinear algebraic equations governing the performance of engineering systems. The successive substitution method then provides a relatively simple computational technique for obtaining the values of the physical variables that satisfy the given system of equations, as discussed later in Chapter 6.

It can be shown that the successive substitution method will converge if for $|x - \alpha| < |x_1 - \alpha|$, the function of $g(x)$ possesses a derivative $g'(x)$ such that $|g'(x)| < 1$. Here, x_1 is the initial approximation to the root. This condition implies that the magnitude of the derivative is less than 1.0 in the computational region. When x_i is close to the root α, the next approximation x_{i+1} can be shown to be given by the approximate relation

$$x_{i+1} - \alpha \cong g'(\alpha)(x_i - \alpha) \tag{5.31}$$

Therefore, if $|g'(\alpha)| < 1$, the method converges to the root in a region near the root. The derivative $g'(\alpha)$ is often termed the *asymptotic convergence factor*. The Newton–Raphson method, discussed in Section 5.5.1, may also be considered in terms of the successive substitution method to obtain the convergence characteristics, as outlined by Carnahan et al. (1969).

The major problem with the successive substitution method is the frequent divergence of the iteration for a given choice of the function $g(\alpha)$. The condition for convergence given by Equation 5.31, $|g'(x)| < 1$, can sometimes be employed in the formulation of the function $g(x)$, as indicated for the polynomial equation considered earlier. Frequently, convergence occurs over a very narrow range of the starting value, and one may need to try several values before the iteration converges. The method is very easy to program, and, as seen from Equation 5.31, a linear convergence is obtained when x_i is close to the root. The following example illustrates the use of the successive substitution method for finding the roots of an algebraic equation and also demonstrates the convergence characteristics of the method.

Example 5.7

The gas flow rate R, in m^3/s, through a duct in a chemical reactor due to a fan is given in terms of the pressure P, in N/m^2, by the equation

$$R = 15 - 75 \times 10^{-6} \times P^2 \tag{5.32}$$

where

$$P = 80 + 10.5R^{5/3} \tag{5.33}$$

Employing the successive substitution method, find the gas flow rate at which the system operates.

SOLUTION

The problem involves finding the roots of the equation

$$R = 15 - 75 \times 10^{-6} \times (80 + 10.5R^{5/3})^2 \tag{5.34}$$

Both the pressure P and the flow rate R are real and positive quantities. It is also obvious from Equation 5.32 that R must be less than 15 m^3/s, since P is zero if $R = 15$ m^3/s and imaginary if R is larger than this value. Thus, R lies between 0 and 15 m^3/s, the two extreme values being excluded, since nonzero values of R and P are expected.

Equation 5.34 is already in the form of Equation 5.28, and the successive substitution method may be applied to this equation. However, it is found that the method does not converge, mainly because of the large exponent of R on the right-hand side which makes even a small error in the numerical solution grow from one iteration to the next. In fact, $|g'(\alpha)|$, defined in Equation 5.31, is found to be larger than 1.0 for R larger than about 3.5. Consequently, the problem may be reformulated in terms of a smaller exponent of R as

$$R = \left(\frac{P - 80}{10.5}\right)^{3/5} \quad \text{where } P = \left(\frac{15 - R}{75 \times 10^{-6}}\right)^{1/2} \tag{5.35}$$

This gives the equation for R as

$$R = \left[\frac{\left((15 - R)/75 \times 10^{-6}\right)^{1/2} - 80}{10.5}\right]^{3/5} \tag{5.36}$$

The successive substitution method is now applied to this formulation of the problem. Since $0 < R < 15$, the minimum and maximum values of the unknown may be suitably specified. A simple MATLAB program may be written for this problem as given in Appendix B.5. Here, z represents the function $g(x)$ and the absolute value of $(z - x)$ is used with a specified convergence parameter *conv* to check for convergence. A fixed number of iterations are specified. A condition for divergence may also be used for termination of the iterations if the scheme does not converge. Figure 5.16 shows the numerical results obtained from such a program for two starting values, 0.5 and 1.0, of the flow rate, denoted by X, for convergence parameter *conv*, denoted by CONV, of 10^{-3} and 10^{-4}.

The flow rate is computed as 6.732 m^3/s, this value being only slightly changed by a variation in the convergence parameter, CONV. A larger number of iterations are needed at the smaller value of CONV, as expected, and the starting value has a negligible effect on the converged solution. As shown by this example, convergence may often be achieved in successive substitution by rewriting the algebraic equation in a different way, if the method diverges when applied to the given equation. It can be verified that $|g'(\alpha)|$ is indeed less than 1.0 near the root for the formulation given in Equation 5.36.

5.7 OTHER METHODS

So far, we have discussed many important methods for root solving and have considered their applicability, limitations, convenience, and convergence characteristics. There are several other methods that are available for finding the roots of certain types of equations and that are sometimes preferred due to superior convergence, ease in programming or wider range of applicability. Some of these methods are based on the techniques and algorithms for the methods discussed earlier and try to improve the earlier methods. Others employ different approaches to root solving. Some of these methods are presented in this section.

```
X = 0.50    CONV = 0.0010                 X = 1.00    CONV = 0.0010
X = 0.5000    Z = 8.3339                  X = 1.0000    Z = 8.2271
X = 8.3339    Z = 6.1733                  X = 8.2271    Z = 6.2136
X = 6.1733    Z = 6.9075                  X = 6.2136    Z = 6.8951
X = 6.9075    Z = 6.6752                  X = 6.8951    Z = 6.6792
X = 6.6752    Z = 6.7504                  X = 6.6792    Z = 6.7491
X = 6.7504    Z = 6.7262                  X = 6.7491    Z = 6.7266
X = 6.7262    Z = 6.7340                  X = 6.7266    Z = 6.7338
X = 6.7340    Z = 6.7315                  X = 6.7338    Z = 6.7315
X = 6.7315    Z = 6.7323                  X = 6.7315    Z = 6.7323
THE REQUIRED ROOT IS X = 6.7315           THE REQUIRED ROOT IS X = 6.7315
X = 0.50    CONV = 0.0001                 X = 1.00    CONV = 0.0001
X = 0.5000    Z = 8.3339                  X = 1.0000    Z = 8.2271
X = 8.3339    Z = 6.1733                  X = 8.2271    Z = 6.2136
X = 6.1733    Z = 6.9075                  X = 6.2136    Z = 6.8951
X = 6.9075    Z = 6.6752                  X = 6.8951    Z = 6.6792
X = 6.6752    Z = 6.7504                  X = 6.6792    Z = 6.7491
X = 6.7504    Z = 6.7262                  X = 6.7491    Z = 6.7266
X = 6.7262    Z = 6.7340                  X = 6.7266    Z = 6.7338
X = 6.7340    Z = 6.7315                  X = 6.7338    Z = 6.7315
X = 6.7315    Z = 6.7323                  X = 6.7315    Z = 6.7323
X = 6.7323    Z = 6.7320                  X = 6.7323    Z = 6.7320
X = 6.7320    Z = 6.7321                  X = 6.7320    Z = 6.7321
THE REQUIRED ROOT IS X = 6.7320           THE REQUIRED ROOT IS X = 6.7320
```

FIGURE 5.16 Computed results for the problem in Example 5.7 by the successive substitution method, for two values of the convergence parameter and two initial estimates of the unknown root.

5.7.1 Müller's Method

This method is based on the secant method, which employs the intersection of a line through two points on the graph of $f(x)$ with the x-axis to approximate the root for the next iteration. *Müller's method* uses three points, instead of two, obtains the parabola through these three points, and takes the intersection with the x-axis as the next approximation, as shown in Figure 5.17. If the three initial function values are f_0, f_1, and f_2 corresponding to the x values of x_0, x_1, and x_2, the parabola going through these points is determined and its intersection with the x-axis is obtained by solving the quadratic equation $ax^2 + bx + c = 0$. Using the alternative form of the solution, we have

$$z = \frac{-2c}{b \pm \sqrt{b^2 + 4ac}} \tag{5.37}$$

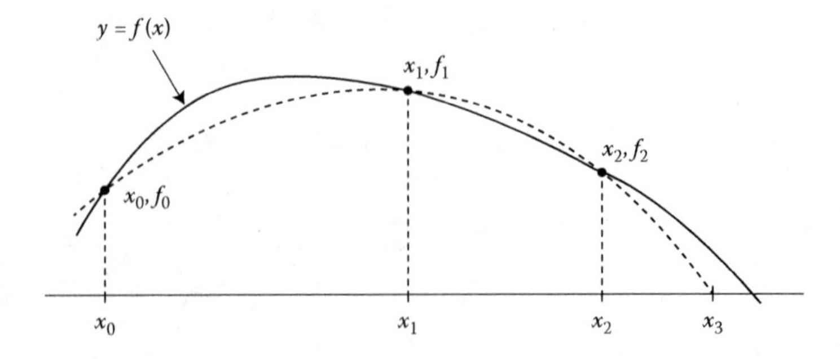

FIGURE 5.17 A sketch showing the starting approximations x_0, x_1, and x_2 to the root and the application of Müller's method.

where

$$
\begin{aligned}
a &= \frac{(f_0 - f_2)(x_1 - x_2) - (f_1 - f_2)(x_0 - x_2)}{(x_1 - x_2)(x_0 - x_2)^2 - (x_0 - x_2)(x_1 - x_2)^2} \\
b &= \frac{(f_1 - f_2)(x_0 - x_2)^2 - (f_0 - f_2)(x_1 - x_2)^2}{(x_1 - x_2)(x_0 - x_2)^2 - (x_0 - x_2)(x_1 - x_2)^2} \\
c &= f_2
\end{aligned}
\tag{5.38}
$$

The derivation is simplified by assuming $x = x_2$ to be the best approximation to the root and changing the independent variable to $x - x_2$. Then, the next approximation to the root x_3 is obtained as

$$x_3 = x_2 + z \tag{5.39}$$

To ensure the stability of the method, the root with the smallest absolute value, that is, the one closest to x_2 is chosen. Therefore, for $b > 0$, the positive sign is used, otherwise the negative sign is taken. The iterative process is continued till an appropriate convergence criterion applied to the root or the function is satisfied. Note that a particular approximation during the iteration can be complex, even if the previous values were all real. This is in contrast with other root-finding algorithms like the secant or Newton's method, whose iterates will remain real if one starts with real numbers. Having complex iterates can be an advantage if one is looking for complex roots or a disadvantage if it is known that all roots are real.

The order of convergence of Müller's method can be shown to be approximately 1.84. This can be compared with 1.62 for the secant method and 2 for Newton's method. So, the secant method makes less progress per iteration than Müller's method and Newton's method makes more progress. However, three starting approximations to the root are needed to initiate the iterative scheme. After each iteration, the latest three approximations may be employed to generate the next value. Thus, the algorithm is quite similar to that for the secant method and the programs given earlier may be modified to obtain one for Müller's method.

5.7.2 Iterative Factorization of Polynomials

Analytically, the roots of a polynomial equation can often be obtained by factorization and equating each factor to zero. A similar approach may be employed for root solving by numerical methods. Several methods are based on the iterative factorization of the given polynomial and can be used to obtain factors of arbitrary degree. Generally, linear or quadratic factors are determined so that the roots may be obtained directly from these factors. Let us first consider *Bairstow's* method, which iteratively determines quadratic factors of the form $x^2 + bx + c$, and so the roots are given by Equation 5.3 as

$$\alpha_1, \alpha_2 = \frac{-b \pm \sqrt{b^2 - 4c}}{2}$$

The polynomial given in Equation 5.2 can be written as

$$f(x) = (x^2 + bx + c)(d_0 x^{n-2} + d_1 x^{n-3} + d_2 x^{n-4} + \cdots + d_{n-3} x + d_{n-2}) + \text{remainder} \tag{5.40}$$

where the d's are functions of b and c and are obtained from a comparison with the original polynomial of Equation 5.2 as

$$\begin{aligned} d_0 &= 1 \\ d_1 &= a_1 - b \\ d_2 &= a_2 - d_1 b - c \\ d_3 &= a_3 - d_2 b - d_1 c \\ &\vdots \\ d_i &= a_i - d_{i-1} b - d_{i-2} c \end{aligned} \tag{5.41}$$

and the remainder is $(x + b)d_{n-1} + d_n$.

To extract the quadratic factor from the polynomial, we must reduce the remainder to zero, within a specified error tolerance. We do so by iteratively reducing d_{n-1} and d_n to zero. Since both of these are functions of b and c, we may use Taylor's expansion for a function of two variables. If only the linear terms are retained, we obtain

$$\begin{aligned} d_n(b + \Delta b, c + \Delta c) &\cong d_n(b, c) + \frac{\partial d_n}{\partial b}\Delta b + \frac{\partial d_n}{\partial c}\Delta c = 0 \\ d_{n-1}(b + \Delta b, c + \Delta c) &\cong d_{n-1}(b, c) + \frac{\partial d_{n-1}}{\partial b}\Delta b + \frac{\partial d_{n-1}}{\partial c}\Delta c = 0 \end{aligned} \tag{5.42}$$

where Δb and Δc are increments in b and c. We set the equations equal to zero in order to obtain the next approximation to b and c so that d_n and d_{n-1} become zero.

The set of equations for the d's may be differentiated to obtain a similar sequence of equations for their partial derivatives. The corresponding expressions are as follows:

$$
\begin{aligned}
&\frac{\partial d_1}{\partial b} = -1 = e_0 && \frac{\partial d_1}{\partial c} = 0 \\
&\frac{\partial d_2}{\partial b} = b - d_1 = e_1 && \frac{\partial d_2}{\partial c} = -1 = e_0 \\
&\frac{\partial d_3}{\partial b} = -d_2 - e_1 b - c e_0 = e_2 && \frac{\partial d_3}{\partial c} = b - d_1 = e_1 \\
&\vdots && \vdots \\
&\frac{\partial d_{n-1}}{\partial b} = -d_{n-2} - e_{n-3} b - e_{n-4} c = e_{n-2} && \frac{\partial d_{n-1}}{\partial c} = e_{n-3} \\
&\frac{\partial d_n}{\partial b} = -d_{n-1} - e_{n-2} b - e_{n-3} c = e_{n-1} && \frac{\partial d_n}{\partial c} = e_{n-2}
\end{aligned}
\tag{5.43}
$$

which give

$$e_i = -d_i - e_{i-1} b - e_{i-2} c \quad \text{for } i = 2, 3, \ldots, (n-1)$$

Also,

$$e_0 = -1 \quad \text{and} \quad e_1 = b - d_1$$

Therefore, the partial derivatives for the remainder terms may be obtained. From Equations 5.42,

$$
\begin{aligned}
d_n &= -e_{n-1}\Delta b - e_{n-2}\Delta c \\
d_{n-1} &= -e_{n-2}\Delta b - e_{n-3}\Delta c
\end{aligned}
\tag{5.44}
$$

Solving these simultaneous linear equations, we find that Δb and Δc are given by

$$
\begin{aligned}
\Delta b &= \frac{d_{n-1} e_{n-2} - d_n e_{n-3}}{e_{n-1} e_{n-3} - (e_{n-2})^2} \\
\Delta c &= \frac{d_n e_{n-2} - d_{n-1} e_{n-1}}{e_{n-1} e_{n-3} - (e_{n-2})^2}
\end{aligned}
\tag{5.45}
$$

To apply Bairstow's method, initial guessed values of b and c are taken and the corresponding d's and e's are determined. The increments Δb and Δc are obtained for the next approximation of b and c. The recursion formula is

$$b_{i+1} = b_i + \Delta b \quad \text{and} \quad c_{i+1} = c_i + \Delta c \tag{5.46}$$

where i is the iteration number. The iterative process for the determination of b and c is continued until $|\Delta b|$ and $|\Delta c|$ are less than a specified convergence criterion. The quadratic factor thus obtained yields two roots of the equation. The reduced polynomial of degree $(n - 2)$ is next considered to obtain the remaining roots. The algorithm is shown in terms of a flow chart in Figure 5.18.

Bairstow's method can be used for finding real, equal, or complex roots of a polynomial. Although the analysis appears to be complicated, the method may be programmed for the computer without too much difficulty. However, convergence cannot be guaranteed for an arbitrary choice of initial values. If there is some prior information available on the roots or on the coefficients of a factor, the method may be used very effectively to improve the accuracy of the roots. Since divergence may occur with an arbitrary choice of the initial values of b and c, one may restrict the total number of iterations in the program and choose the starting values again if divergence occurs.

Several other methods have been developed based on the extraction of factors from polynomials. Synthetic division by a linear or quadratic factor allows one to obtain equations for the coefficients of the reduced polynomial and for the remainder, as discussed above. Bairstow's method uses the Newton–Raphson method for the solution of simultaneous nonlinear equations to iteratively reduce the remainder to zero (see Equation 5.42). If the successive substitution method for simultaneous nonlinear equations is employed, instead of the Newton–Raphson method, to reduce the remainder to zero, the procedure is known as *Lin's method*. This method provides a simpler, although less efficient, iterative procedure for obtaining the quadratic factors of a polynomial of degree greater than two. The extraction of linear factors from the polynomial may also be carried out by using synthetic division. However, quadratic factors are the most desirable ones since they allow the direct determination of real and complex roots. The methods based on the iterative factorization of polynomials also have the attractive feature of obtaining all the real, multiple, and complex roots. Therefore, despite their complexity, they are frequently used, particularly for problems of engineering interest in which the nature and magnitude of the roots are not known. The following example illustrates the use of Bairstow's method.

Example 5.8

Use Bairstow's method for the iterative factorization of polynomials to find a quadratic factor and the roots of the characteristic equation given as

$$\lambda^4 - 10\lambda^3 + 35\lambda^2 - 50\lambda + 24 = 0 \tag{5.47}$$

SOLUTION

The quadratic factor to be determined is taken as $x^2 + bx + c$, where b and c are to be obtained from Bairstow's method. Equations 5.41 and 5.43 give the coefficients of the remaining polynomial and the derivatives of these coefficients with respect to b and c. The iterative procedure given by Equations 5.45 and 5.46 is employed to converge to the desired values of the constants b and c, using a convergence

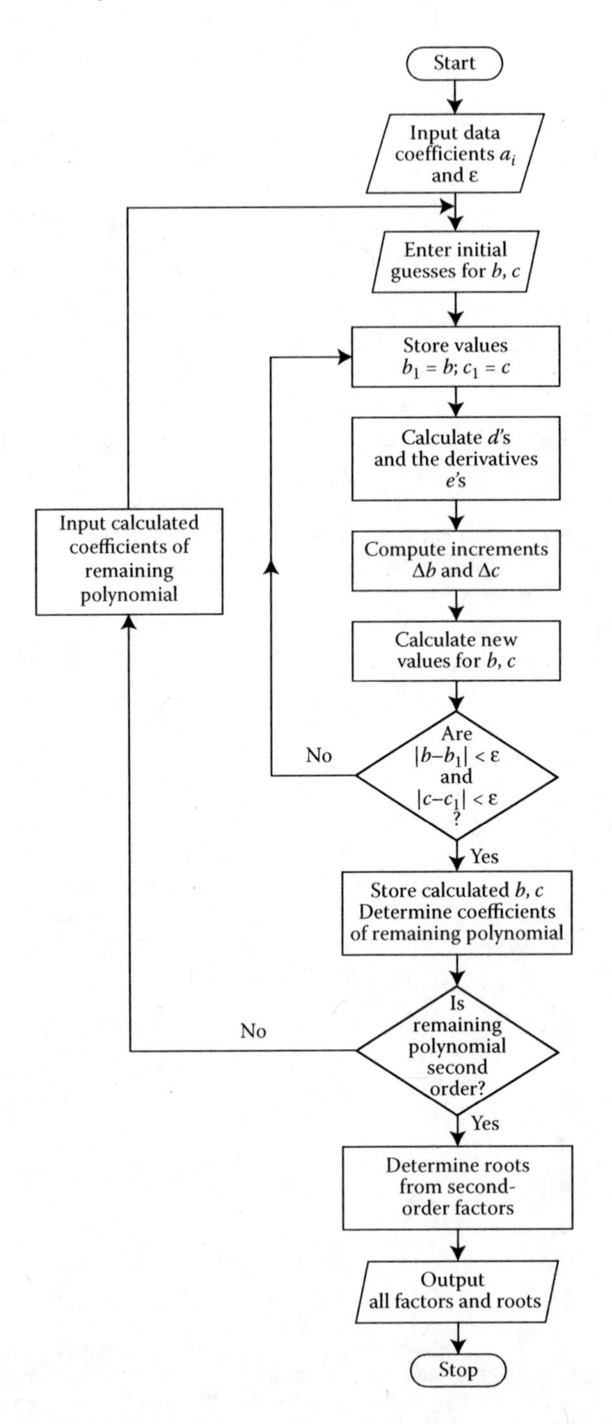

FIGURE 5.18 Flow chart for the solution of a polynomial equation by Bairstow's method.

```
THE INITIAL GUESS: B = -2.0000                    C = 1.0000
                   B = -2.776470                  C = 1.352941
                   B = -3.162417                  C = 1.922668
                   B = -2.825542                  C = 1.902719
                   B = -2.969001                  C = 1.973022
                   B = -2.999051                  C = 1.998892
                   B = -3.000000                  C = 1.999998
                   B = -3.000003                  C = 2.000003
THE QUADRATIC FACTOR:B = -3.000003                C = 2.000003
REMAINING POLYNOMIAL:D1 = -7.000000               D2 = 12.000000

THE INITIAL GUESS: B = 0.0000                     C = 5.0000
                   B = -1.120000                  C = 2.200001
                   B = -2.010739                  C = 1.453996
                   B = -2.633198                  C = 1.577660
                   B = -2.957543                  C = 1.885094
                   B = -3.005317                  C = 1.999321
                   B = -2.999916                  C = 1.999945
                   B = -2.999983                  C = 1.999984
THE QUADRATIC FACTOR:B = -2.999983                C = 1.999984
REMAINING POLYNOMIAL:D1 = -7.000084               D2 = 12.000400

THE INITIAL GUESS: B = 2.0000                     C = -2.0000
                   B = -1.967654                  C = -13.789750
                   B = -4.887068                  C = -14.755200
                   B = -4.943721                  C = -4.912420
                   B = -4.972162                  C = -0.009018
                   B = -4.986652                  C = 2.395084
                   B = -4.994331                  C = 3.505467
                   B = -4.998444                  C = 3.918154
                   B = -4.999864                  C = 3.996880
                   B = -4.999969                  C = 3.999967
                   B = -4.999969                  C = 3.999973
THE QUADRATIC FACTOR:B = -4.999969                C = 3.999973
REMAINING POLYNOMIAL:D1 = -5.000042               D2 = 6.000034
```

FIGURE 5.19 The numerical results for Example 5.8, employing three different starting values.

criterion of 10^{-3}. A computer program may easily be written for this problem, using the earlier program for Newton's method. The *d*'s and *e*'s are obtained and employed to determine the increments in b and c for the next iteration. The starting values are taken as $b = -2$ and $c = 1$. The iterative process converges to $b = -3$ and $c = +2$, which gives the quadratic factor as $x^2 - 3x + 2$. The remaining polynomial is $x^2 - 7x + 12$; see Figure 5.19.

Therefore, the roots of the given equation, Equation 5.47, may be obtained by solving the two quadratic equations

$$x^2 - 3x + 2 = 0 \quad \text{and} \quad x^2 - 7x + 12 = 0 \tag{5.48}$$

The four roots of the given equation are found to be 4, 3, 2, and 1. It is evident that the quadratic factor derived from the given polynomial equation will depend on the starting values of b and c. Six different quadratic factors are possible, since the first root may be combined with any one of the three remaining ones, the second root with two remaining ones, and the third with the fourth one, to yield a quadratic factor each. The convergence of the method is quite rapid, as shown in Figure 5.19. The convergence criterion must also be varied to ensure a negligible dependence of the results on the value chosen. Results are also shown for different sets of starting values and convergence to a different quadratic factor is observed in the last case.

5.7.3 Graeffe's Method

Graeffe's root-squaring method is suitable for polynomials and can be used to determine both real and complex roots, as well as multiple roots. It is based on obtaining a new polynomial, which is of the same degree as the original polynomial and whose roots are some large, even power of the roots of the original equation. The roots of the derived equation are first obtained, and these then yield the required roots of the given equation.

A given polynomial equation may be written as

$$f(x) = (x - \alpha_1)(x - \alpha_2)\cdots(x - \alpha_n) \tag{5.49}$$

where $\alpha_1, \alpha_2, \alpha_3, \ldots, \alpha_n$ are the roots. A new function $F(x)$ may be defined as

$$F(x) = (-1)^n f(x) f(-x) \tag{5.50}$$

which gives

$$F(x) = (x^2 - \alpha_1^2)(x^2 - \alpha_2^2)\cdots(x^2 - \alpha_n^2) \tag{5.51}$$

Therefore, $F(x)$ contains only even powers of x, and a function $f_2(x)$ may be defined as

$$f_2(x) = F(\sqrt{x}) = (x - \alpha_1^2)(x - \alpha_2^2)\cdots(x - \alpha_n^2) \tag{5.52}$$

Therefore, the roots of the derived equation $f_2(x) = 0$ are squares of the roots of the original equation. The process may be repeated to obtain a sequence of polynomials $f_4, f_6, f_8, \ldots$, so that a derived polynomial $f_m(x)$ is obtained, where

$$f_m(x) = (x - \alpha_1^m)(x - \alpha_2^m)\cdots(x - \alpha_n^m) = 0 \tag{5.53}$$

The roots of the above equation are a large, even power m of the roots of the original equation. If $|\alpha_1| > |\alpha_2| > \cdots |\alpha_n|$, then the ratios of the roots of the derived equation,

$|\alpha_2^m/\alpha_1^m|, |\alpha_3^m/\alpha_2^m|, \ldots, |\alpha_n^m/\alpha_{n-1}^m|$, may be made as small as desired by making m large. The derived polynomial $f_m(x)$ may be written as follows:

$$f_m(x) = x^n - (\alpha_1^m + \alpha_2^m + \cdots)x^{n-1} + (\alpha_1^m\alpha_2^m + \alpha_1^m\alpha_3^m + \alpha_2^m\alpha_3^m \cdots)x^{n-2} - (\alpha_1^m\alpha_2^m\alpha_3^m + \alpha_1^m\alpha_2^m\alpha_4^m + \cdots)x^{n-3} + \cdots + (-1)^n \alpha_1^m\alpha_2^m \ldots \alpha_n^m \quad (5.54a)$$

or

$$f_m(x) = x^n - A_1x^{n-1} + A_2x^{n-2} + \cdots + (-1)^n A_n \quad (5.54b)$$

Then the magnitude of the roots may be approximated by

$$\alpha_1^m \cong (A_1), \alpha_2^m \cong \frac{A_2}{A_1}, \ldots, \alpha_n^m \cong \frac{A_n}{A_{n-1}} \quad (5.55)$$

if only the leading, or dominant, terms within the parentheses in Equation 5.54a are retained. The values of the roots α_1, α_2, ..., α_n of the original equation may be obtained by taking the mth root of the above equations, that is, $\alpha_1 \cong \pm A_1^{1/m}, \alpha_2 \cong \pm(A_2 / A_1)^{1/m}$, and so on. The signs of the roots are not determined and must be obtained by substitution in the original equation or from any previous information on the roots, based on the physical nature of the problem.

If the original polynomial equation has real and equal roots, the regular relationship between the coefficients of successive polynomials, as mentioned above, is not obtained. Since the method does not determine the sign of the root, equal roots, in Graeffe's method, are those that have the same absolute value. If the roots α_i and α_{i+1} are taken as equal, it can be shown from the above analysis that the coefficient A_i of the polynomial $f_m(x)$ is essentially equal to half the square of the corresponding coefficient in the polynomial $f_{m-1}(x)$ for large m. The other coefficients are squares of the corresponding preceding values if the remaining roots are real and distinct. If three equal roots are present, say, α_i, α_{i+1}, and α_{i+2}, the coefficients A_i and A_{i+1} become one-third of the corresponding preceding values. The corresponding relationship between the roots and the coefficients of the polynomial may be obtained from Equation 5.54.

Graeffe's method may also be used for complex roots, which appear in conjugate pairs. The conjugate pair may be taken as $(u + iv)$ and $(u + iv)$, and the above analysis may be applied to such roots. It can be shown that, at large m, the real and distinct roots give rise to coefficients of the polynomial $f_m(x)$ that are essentially squares of the corresponding coefficients of $f_{m-1}(x)$. The presence of complex roots is indicated by a fluctuation in the sign of a coefficient, since a trigonometric function cos $m\theta$, where m is a constant and the complex roots are written as $Re^{i\theta}$ and $Re^{-i\theta}$, appears in the relationships. If the sign of the coefficient A_i fluctuates, the roots α_i and α_{i+1} are a conjugate pair of complex roots. The magnitude R of the roots is determined from the coefficients A_{i-1} and A_{i+1} as before. If more than one conjugate pair of complex roots is present, correspondingly more coefficients in the polynomial $f_m(x)$ fluctuate in sign.

Therefore, Graeffe's method provides a means of determining all the roots of a polynomial equation, whether they are real, equal, or complex. However, despite this attractive feature of the method, it has not become very popular mainly because of the need to make decisions that considerably complicate the programming. The round-off error introduced at any stage of the process accumulates in the computation and affects the accuracy of the roots obtained. Also, the coefficients frequently exceed the floating-point range of the computer, particularly if there are two roots which are close to each other and which, therefore, require a large value of m for the separation of the roots. However, this last problem may be avoided by the scaling of the polynomial, which involves dividing the roots by a scale factor. The roots of the modified equation yield the scaled roots, from which the desired roots are obtained.

Probably the best procedure for employing Graeffe's method is to work interactively with the computer. Such an interactive program would allow one to make decisions as the computation proceeds and make the necessary changes in the process. Although Graeffe's method is not widely used, it does have the attractive aspect of evaluating all the roots of a polynomial equation. The method is discussed here since it indicates a different approach, as compared to the methods outlined earlier in this chapter, to root solving and may form the basis for solving certain complicated equations of engineering interest that cannot be solved by other methods. For further details on the method and examples illustrating its use, see Jaluria (1996).

5.7.4 Additional Methods

There are obviously many more methods available in the literature for root solving. Many of these are based on considerations and techniques quite similar to those discussed here. Among those that may be mentioned are *Brent's* method, *Laguerre's* method, *Householder's* method, *Horner's* method, *Bernoulli's* method, and *Ward's* method.

Brent's method is a combination of the bisection method, the secant method and inverse quadratic interpolation. Before each iteration, the method decides which of these three is likely to perform the best, and proceeds by doing a step according to that method. This gives a robust and fast method, which, despite being complicated, enjoys considerable popularity. Laguerre's method uses second-order derivatives and complex algebra to obtain cubic convergence for simple roots whenever the initial guess x_0 is close enough to the root x. For a multiple root, the convergence is only linear. A major advantage of this method is that it is almost guaranteed to converge to a root of the polynomial. It may even converge to a complex root of the polynomial. Since failure to converge is extremely rare, this method is a good candidate for a general purpose polynomial root finding algorithm. The algorithm is fairly simple to use and the speed at which the method converges implies that only a few iterations are generally needed to get high accuracy. However, theoretical understanding of the algorithm is rather limited and this has made the method not as popular as one may expect.

Householder's methods lead to a class of methods used for functions of one real variable with continuous derivatives up to some order $m + 1$, where m will then be the order of the Householder's method, as well as the rate of convergence. The Householder's method of order 1 is just Newton's method and the method of order 2 yields another method, known as *Halley's* method, which has a cubic order of convergence but involves more operations per iteration. *Horner's* method can be used for finding the roots of a polynomial equation. It employs an iterative sequence of translations to place the root at the origin. The sum of these translations is the root of the original equation. Then, synthetic division is employed to reduce the equation by this root and search for the next root is carried out. The method becomes quite complicated if the degree of the polynomial equation is high or if complex roots arise. Similarly, Bernoulli's method is applicable for polynomial algebraic equations and obtains the real root with the largest absolute value.

An important class of methods has been developed using minimization principles. *Ward's* method uses these principles to find the roots of a complex polynomial equation $f(z) = 0$, where z is a complex variable and

$$f(z) = u(x, y) + \mathrm{i}v(x, y) \quad \text{and} \quad z = x + \mathrm{i}y \tag{5.56}$$

Ward's method seeks to minimize the function $p(x, y)$, where

$$p(x, y) = |u(x, y)| + |v(x, y)| \tag{5.57}$$

The method is iterative, and at each step the value of $p(x, y)$ is compared with the values at $(x + \Delta x, y)$, $(x - \Delta x, y)$, $(x, y + \Delta y)$, and $(x, y - \Delta y)$, where Δx and Δy are increments in x and y. If a smaller value of $P(x, y)$ is found at any of these four points, that point becomes the new (x, y) location. If the four points do not give a smaller value of p, Δx, and Δy may be reduced until they do. Otherwise, the minimum has been reached. Therefore, the method simply moves in the direction of decreasing $p(x, y)$. The root is obtained when u and v become zero, or smaller than a chosen convergence parameter. The search for a minimum value of p is, therefore, expected to lead to the root. However, convergence does not necessarily occur. Similarly, other functions, such as $(u^2 + v^2)$, may be minimized. Other minimization techniques are also available from mathematical procedures that have been developed for the optimization of systems. Search methods are frequently used in optimization, the above procedure being generally termed as *lattice search* (Stoecker, 1989; Jaluria, 2008). In engineering problems, one may encounter equations that are so complicated that the various methods discussed in this chapter may not be convenient to use. In such cases, one may resort to minimization methods to obtain the desired roots.

5.8 SUMMARY

In this chapter, several methods for finding the roots of a nonlinear algebraic equation, including polynomial and transcendental equations, have been presented. The discussion considers some of the most important and widely used methods, outlining their limitations and advantages. The selection of the method for solving a given problem depends on the nature of the equation and of the roots. The physical characteristics of the problem, if known, are useful in choosing the method and in determining the interval over which the roots are sought. The incremental search method may be used to yield the approximate nature of the function $f(x)$ and the approximate location of the roots. Once this information has been obtained, one may switch to a method, such as the Newton–Raphson method, that converges more rapidly. In most problems of practical interest, prior information on the nature and location of roots is available. This information should be built into the computer program to obtain rapid convergence to the roots and to reject unacceptable values. MATLAB is particularly convenient to use for polynomial equations since the available software can be directly used to obtain all the real and complex roots of the equations.

The convergence criterion, the convergence parameter ε and the initial guess must be varied to ensure that the results obtained are not significantly affected by the values chosen. Several of these considerations, related to the computational procedure, were also discussed earlier in Chapter 2 and may be employed in developing the computer program. Some of the methods presented here can also be extended to the solution of simultaneous equations, as discussed in the next chapter. Further details on the various methods considered in this chapter may be obtained from the discussions given by Traub (1964), Ostrowski (1966), Carnahan et al. (1969), Householder (1970), Brent (1973), and Atkinson (1989). Ralston and Rabinowitz (1978), Rice (1983), and Gerald and Wheatley (2003) may also be consulted for the mathematical background of some of the methods outlined here.

PROBLEMS

5.1. We wish to find the cube root of 17, that is, $17^{1/3}$, by root solving. Set up the equation to be solved, and outline a method to compute the desired value.

5.2. For the following equation, find the first two positive roots, which represent the lowest frequencies of natural vibration of a mechanical system:

$$\tan x = \tanh x$$

Use the search method to obtain an accuracy of order 10^{-3} on the roots.

5.3. The root of an algebraic equation is known to be between 0 and 800 m/s. This root is to be determined to an accuracy of ±0.1 m/s by the bisection method. Derive Equation 5.9 and use it to determine the number of bisections needed to achieve this accuracy.

5.4. The voltage x at a given junction in an electrical circuit is given by the first positive root of the equation

$$f(x) = \log_{10} x + x^2 - 6 = 0$$

Employ the bisection method, following the determination of the interval containing the root by the search method, to obtain the voltage. Also, use the *fzero* function in MATLAB to obtain the solution and compare the result with that obtained earlier.

5.5. The root of the following equation is to be determined.

$$x\left[1 - \exp\left(-\frac{10}{1+4x}\right)\right] - 1 = 0$$

Write a function-m file for the secant root solving method. Then, write a script m-file to use the secant function file to calculate the root. Using these script files, obtain the results for 5 values of the convergence parameter *delta* (*delta* = 1.0, 0.1, 0.01, 0.001, and 0.0001).

5.6. The equation that governs the frequency of vibration of a cantilever beam is of the form

$$\cos x \cosh x = -1$$

Use the search method to obtain the approximate locations of the first two positive roots of this equation. Then use the regula falsi method to converge to the roots.

5.7. If the derivative $f'(x_i)$ in Equation 5.16 is replaced by its backward finite-difference approximation, Equation 4.18, obtain the resulting recursive formula, and compare it with that for the secant method.

5.8. The roots of the equation

$$\tan x = \frac{B}{x}$$

where B is a constant, are needed in a series representation of the temperature field in a conduction heat transfer problem. Use the search method to obtain the approximate values of the first three positive roots of this equation, and then use the Newton–Raphson method to obtain these more accurately. Consider two values, 1.0 and 2.0, of B. Also use the successive substitution method, and compare the results with those obtained by Newton's method.

5.9. For the physical problem discussed in Example 5.1, take the values of T_h, h, and k as 1500 K, 10 W/m^2 K, and 50 W/mK. Using the Newton–Raphson method, find the resulting surface temperature. The temperature is to be determined to an accuracy of 0.01 K.

5.10. Explain what is meant by the statement that the Newton–Raphson method has second-order convergence. Obtain the general form of the corresponding convergence formulas for search, bisection, and successive substitution methods. Compare these with that for the

Newton–Raphson method, and discuss the resulting difference in convergent rate.

5.11. Use the Newton–Raphson method for root solving to find the nth root of a number N, that is, $x^n = N$, in order to derive the formula

$$x_{i+1} = \frac{x_i}{n}\left[n - 1 + \frac{N}{x_i^n}\right]$$

5.12. The temperature of an electrically heated wire is to be determined from its energy balance. If the energy input per unit surface area into the wire due to the electric current is 1000 W/m^2, the resulting equation is obtained as

$$1000 = 0.5 \times 5.67 \times 10^{-8} \times [T^4 - (300)^4] + 10 \times (T - 300)$$

Determine the temperature T of the wire by employing the search method. Since the equation is a fourth-order polynomial equation, there are four roots. How would you choose the correct solution? Also, use the *fzero* function in MATLAB to obtain the root and compare it with that obtained earlier.

5.13. In a manufacturing process, a spherical piece of metal is subjected to radiative and convective heat transfer, resulting in the energy balance equation

$$f(T) = 0.6 \times 5.67 \times 10^{-8} \times [(850)^4 - T^4] - 40 \times (T - 350) = 0$$

Obtain a rough plot of the function $f(T)$ versus T, and use the secant method to find the real root in the range $350 < T < 850$. Also, use the *roots* function in MATLAB to obtain all the roots and compare the results with the root obtained earlier. Comment on the choice of the correct root.

5.14. The Planck distribution for the emission of radiation from a blackbody is given by

$$E = \frac{c_1}{\lambda^5\left[\exp(c_2 / \lambda T) - 1\right]}$$

where E is termed the *monochromatic emissive power*, λ is the wavelength of radiation, T is the temperature, and c_1 and c_2 are constants. We wish to find the value of λ at which E is a maximum. Therefore, the first positive real root of the equation

$$\frac{dE}{d\lambda} = 0$$

is to be determined. Find this value and compare it with the result given in the literature as $\lambda_{max} = 2897.6/T$, where λ is in μm and T is in K. Take the constants c_1 and c_2 as 3.741×10^8 and 1.439×10^4 in the appropriate units.

5.15. In the turbulent flow of a fluid through a smooth pipe, the frictional force on the fluid is represented in terms of a friction factor f which is positive and less than 0.1. The equation for f is

$$\frac{1}{f^{1/2}} = 2\log_{10}(\mathrm{Re}\, f^{1/2}) - 0.8$$

where Re is a constant, termed the *Reynolds number*, which varies with the fluid properties, flow rate, and tube diameter. Obtain the approximate value of the friction factor by the search method, and then use the Newton–Raphson method to converge to the root for Re = 10^4 and 10^6.

5.16. If the fluid, in the physical circumstance of the above problem, flows through a rough pipe, whose roughness is given by a parameter ε/D, where ε is the physical size of the surface protrusions and D is the pipe diameter, the friction factor is given by

$$\frac{1}{f^{1/2}} = -2\log_{10}\left(\frac{\varepsilon/D}{3.7} + \frac{2.51}{\mathrm{Re}\, f^{1/2}}\right)$$

Obtain the friction factor f for $\varepsilon/D = 10^{-4}$ at Re = 10^6 and also for $\varepsilon/D = 4 \times 10^{-4}$ at Re = 10^7, using the search method. Also, use the *fzero* function in MATLAB to obtain the root and compare the result with that obtained earlier.

5.17. The equation of state for a gas is given by the *van der Waals equation*

$$\left(P + \frac{a}{v^2}\right)(v - b) = RT$$

where P is the pressure, v is the specific volume, T is the temperature, R is the gas constant, and a, b are constants that depend on the gas. For $P = 70$ atm, $T = 200$ K, $R = 0.08205$ liter atm/mole K, $a = 3.59$, and $b = 0.0427$, the specific volume is given in liters/mole. Find this value using the Newton–Raphson method, after obtaining the approximate value by the search method. Also, use the *roots* function in MATLAB to obtain the solution and compare the result with that obtained earlier.

5.18. Use the Successive Substitution method to determine the variable v from the equation

$$v = \left\{\left[\left(\frac{14 - v}{7 * 10^{-5}}\right)^{0.55} - 90\right] \Big/ 11\right\}^{0.63}$$

5.19. Use Newton's method or the Secant method to solve the equation

$$\exp(x) - x^2 = 0$$

5.20. Use Newton's method to find the real roots of the equation in Example 5.5, given as

$$x^4 - 4x^3 + 7x^2 - 6x + 2 = 0$$

Also, use the *roots* function in MATLAB to obtain the roots and compare these with those obtained earlier.

5.21. Solve the following nonlinear system by Newton's method

$$X^3 + 3Y^2 = 21$$

$$X^2 + 2Y + 2 = 0$$

Try to solve these equations by the successive substitution method as well.

5.22. (a) Using the Newton's method, solve the following equation for the value of x, which is known to be positive,

$$x^5 = [10\,(10 - x)^{0.5} - 8]^3$$

(b) Plot the appropriate function $f(x)$ versus x to get an approximate value of the root.

(c) Compare the solution and the convergence of the numerical scheme for starting guesses of 1 and 2.0 for the root.

(d) Can the bisection method be used for this problem? Explain your answer.

5.23. Use any suitable method to obtain all the roots of the following polynomial equations:

$$x^4 - 10x^3 + 35x^2 - 50x + 24 = 0$$

$$x^4 - 5x^3 + 5x^2 + 5x - 6 = 0$$

Also use the search method for the real roots. Compare the values obtained by the two methods. Then, employing the *roots* function in MATLAB, obtain the roots of these equations and compare the results with those obtained earlier.

5.24. Solve the problem discussed in Example 5.4 by the regula falsi and secant methods. Compare the results obtained and the iterations needed for convergence with those for the Newton–Raphson method.

5.25. Solve the problem considered in Example 5.7 by the Newton–Raphson method, and compare the value of the root and the convergence characteristics with those discussed for the successive substitution method. Comment on the observed differences.

5.26. A cylindrical probe of diameter D is placed in a stream of air, and the energy transfer from it is measured as 100 W. If the energy balance equation is obtained as

$$\left[\frac{60}{D} D^{0.466} + 50\right] \pi D = 100$$

Find the diameter D of the probe using the bisection method. Also write the equation as $f(D) = 0$, draw an approximate plot of $f(D)$ versus D, and discuss the behavior of the function as D increases from zero to 0.01 m. Also, use the *fzero* function in MATLAB to obtain the root and compare it with that obtained earlier.

5.27. A loan of \$5000 is taken from a bank that charges a nominal annual interest rate i, compounded monthly. A payment of \$200 is made each month, starting at the end of the first month, toward the loan. If it takes 36 months to pay off the loan, the rate of interest i may be determined from the following equation, which is obtained by summing the present worth of the monthly payments (see Example 2.2):

$$\frac{(1+i/12)^{36}-1}{\frac{i}{12}(1+i/12)^{36}}\times 200=5000$$

Find this interest rate, by any method of your choice. An accuracy of 10^{-3} on i is adequate.

5.28. A bond of \$1000 yields 8% interest annually and has 7 years to maturity. It is sold for \$500 due to the prevailing higher interest rate i. If the buyer achieves the current interest rate on his investment, the equation governing the transaction is obtained by equating the monetary value of the bond before and after the sale (Stoecker, 1989) as follows:

$$1000+1000\times 0.08\times\frac{(1+i)^{7}-1}{i}=500\times(1+i)^{7}$$

Find the prevailing interest rate i from this equation, using any suitable method.

5.29. A function $y(x)$ is given as

$$y(x)=\frac{\log x\cdot\sin(x^{2}/25)}{x}$$

where log represents the natural logarithm. Determine the maximum value of y for $x>1.5$, using the search method.

5.30. Use the Newton–Raphson and the second-order Newton's methods for finding the nonzero real roots of the equations

$$f(x)=e^{x}-5-x^{2}=0$$
$$f(x)=e^{-x}+1-x^{2}=0$$

Compare the results and the rate of convergence obtained by the two methods. Obtain the roots for two values of ε, 10^{-3} and 10^{-5}, where ε is the convergence criterion applied to the two functions represented by $f(x)$, and comment on the difference, if any.

5.31. Find all the roots of the polynomial obtained in Example 5.1 by Bairstow's method, and show that only one is acceptable because of the physical considerations of the problem.

5.32. Using any suitable method, obtain the four roots corresponding to the polynomial in Problem 5.12. Again, show that three of them are not acceptable.

5.33. For the following polynomial equations, use Bairstow's method to determine all the roots:

$$x^4 - 8x^3 + 22x^2 - 24x + 9 = 0$$

$$x^5 - 5x^4 - 16x + 80 = 0$$

Also, obtain the roots by employing the MATLAB function *roots* and compare the results with those obtained earlier.

5.34. Use the search method to determine the approximate location of the real roots and the behavior of the polynomial functions in the preceding problem. Once this information has been obtained, how would you choose the method for finding the real roots more accurately?

5.35. Use the Newton–Raphson method for finding the real roots, in the range $0 < x < 1.5$, of the following polynomial equation, which represents the variation of the force on a vertical structure with distance x:

$$f(x) = x^4 - 3x^3 + x^2 + 3x - 2 = 0$$

Also use the modified Newton's method for the problem and compare the convergence in the two cases. Obtain a rough plot of $f(x)$ versus x to guide the choice of the starting value.

5.36. Using Bairstow's method, find all the roots of the following polynomial equations:

$$x^7 - 3x^6 + 2x^5 - 32x^2 + 96x - 64 = 0$$

$$x^5 - 15x^4 + 85x^3 - 225x^2 + 274x - 120 = 0$$

$$x^6 - 21x^5 + 175x^4 - 735x^3 + 1624x^2 - 1764x + 720 = 0$$

Also use the Newton–Raphson method to find the real roots in the range $0 < x < 1.5$ in these three cases. Compare the convergence and the accuracy obtained by the two methods. The real roots give the frequencies of vibration of systems represented by these equations.

5.37. Use any two applicable methods for finding the first positive real root of the following transcendental equations:

$$\cosh x = 4x$$

$$\sin x = \cos^2 x$$

$$2e^x + x - 4 = 0$$

Compare the convergence of these methods to the root and the computational effort involved.

5.38. The calibration curve for a temperature-measuring device is given by

$$T = 15 + 3.5V + 0.6V^2 + 0.5V^3 + 0.1V^4$$

where T is the temperature in °C and V is the voltage signal. Determine the voltage output at $T = 30°C$ and 60°C. Use any suitable method.

5.39. Obtain a rough plot of the function $f(x)$ where

$$f(x) = e^{-x/3}(4 - x) - 2 - x$$

and determine the real roots of $f(x) = 0$ by the bisection method. How many bisections are needed to locate the roots with a convergence criterion of $\varepsilon = 10^{-5}$, where ε is the change in the root from one iteration to the next?

5.40. Determine the effect of a variation in the convergence criterion ε on the value of the root obtained in Problem 5.9. Take ε varying from 1 to 10^{-5}.

5.41. The critical load for the buckling of a vertical column is governed by the transcendental equation

$$\tan x = x$$

where $\sqrt{x}$ represents the critical load. Solve this equation by the modified Newton's method to obtain a real positive root, starting with an initial guess of $x = 4.0$. Also try to solve it by the successive substitution method. Discuss your results.

5.42. The real and complex roots of the following polynomial equation are related to the stability of a body subjected to a system of forces:

$$x^4 + 3x^3 + 6x^2 + 7x + 3 = 0$$

where x represents the complex amplification factor for the disturbance. Find these roots, using the Newton–Raphson method. Also use Newton's second-order method for the real root, which is a multiple root at $x = -1$. Compare the convergence by the two methods.

5.43. The decomposition of carbon dioxide into oxygen and carbon monoxide is governed by the equation

$$\left(\frac{P}{E^2} - 1\right)x^3 + 3x - 2 = 0$$

where P is the pressure in atmospheres, E is the temperature-dependent equilibrium constant, and x is the fractional decomposition of CO_2. Using any suitable method, find x for $P = 1$ atm and $E = 1.65$.

5.44. The vapor pressure P of a material is given in terms of the temperature T as

$$\log P = a + \frac{b}{T} + c \log T$$

where log is the natural logarithm and a, b, and c are constants that depend on the material. If their values are given as 17.5, -2.2×10^4, and -0.9, respectively, find the temperatures at pressures of 0.01 and

0.1 atmospheres, using the search method to determine the approximate value of the root, followed by the Newton–Raphson method to converge to the root.

5.45. When water vapor is heated to very high temperatures, it dissociates to give oxygen and hydrogen. Then the mole fraction, x, of water that dissociates is given by the equation

$$S = \frac{x}{1-x}\sqrt{\frac{2P}{2+x}}$$

where S is the equilibrium constant of the reaction and P is the pressure of the mixture. If the pressure P is given as 3.2 atmospheres and S as 0.055, compute the value of x that satisfies the above equation, using any suitable method.

5.46. The current I in an electrical circuit containing resistances R_1 and R_2, inductance L, and voltage source E is given by the equation

$$I = \frac{E}{R_1}\left[1 - \frac{R_2}{R_1 + R_2}e^{(-R_1/L)t}\right]$$

where t is the time. If $R_2 = 10$ ohms, $L = 10$ henries, and $E = 20$ V, find the resistance R_1 that gives a current of 1.4 amperes at $t = 0.5$ s.

6 Numerical Solution of Simultaneous Algebraic Equations

6.1 INTRODUCTION

Systems of simultaneous algebraic equations are frequently encountered in engineering applications such as those concerned with electrical networks, structural analysis, heat transfer, fluid flow, optimization, vibrations, chemical reactions, and data analysis. The numerical solution of an ODE or a PDE also often reduces to the solution of a set of algebraic equations, as discussed later in Chapters 9 and 10. A system of n simultaneous equations, with $x_1, x_2, \ldots, x_n$ as the n unknowns, may be written as

$$\begin{aligned} f_1(x_1, x_2, \ldots, x_n) &= 0 \\ f_2(x_1, x_2, \ldots, x_n) &= 0 \\ \vdots \qquad & \quad \vdots \\ f_n(x_1, x_2, \ldots, x_n) &= 0 \end{aligned} \tag{6.1}$$

where $f_1, f_2, \ldots, f_n$ denote n different functions of the n independent variables. Various methods have been developed to solve this system of equations to obtain the values of the variables $x_1, x_2, \ldots, x_n$. The choice of a particular method for a given problem generally depends on the nature of the equations and the number of unknowns n.

In many circumstances, the equations are linear in the unknown variables. Such a system of linear equations has the general form

$$\begin{aligned} a_{11}x_1 + a_{12}x_2 + \cdots + a_{1n}x_n &= b_1 \\ a_{21}x_1 + a_{22}x_2 + \cdots + a_{2n}x_n &= b_2 \\ \vdots \qquad\qquad & \quad \vdots \\ a_{n1}x_1 + a_{n2}x_2 + \cdots + a_{nn}x_n &= b_n \end{aligned} \tag{6.2}$$

where the a's represent n^2 coefficients and the b's similarly represent n constants. In matrix notation, this system may be written more concisely as

$$AX = B \tag{6.3}$$

where A is a square matrix of the coefficients, X is a column matrix, or vector, of the unknowns, and B is a column matrix, or vector, of the constants that appear on the right-hand side of the equations. From Equation 6.2, a_{ij} represents an element of the matrix A and b_i an element of the vector B. If the column matrix B is zero, that is, if the b's are all zero, the set of equations is said to be *homogeneous*, and nontrivial solutions can be obtained only if all the equations are not independent, as discussed later in this chapter. If the b's are not all zero, the set of equations is nonhomogeneous. In this case, all the equations must be independent in order to yield unique values of the unknowns.

A system of linear algebraic equations may be solved by employing *Cramer's rule* which gives the unknown x_i, as

$$x_i = \frac{\text{Det } A_i}{\text{Det } A} \tag{6.4}$$

where A_i is the matrix A with its ith column replaced by the column vector B and Det represents the determinant of the corresponding matrices. It can be shown that the number of basic arithmetic operations needed to solve for all the unknowns in a set of equations by Cramer's rule, employing expansion by minors for obtaining the determinants, is $(n + 1)!$ Therefore, this method is satisfactory only for a small number of equations, generally less than 5. For the large sets of equations generally encountered in engineering problems, the time required to solve the equations using Cramer's rule is very large and the method is quite impractical, as compared to other methods discussed in this chapter. However, Equation 6.4 indicates some important points regarding the solution of linear simultaneous algebraic equations.

If Det $A = 0$, the matrix A is termed *singular*, and no unique solution can be obtained if the numerator is nonzero. However, if the column matrix B is also zero, then Det $A_i = 0$, since one entire column of the matrix A_i is zero. In this case, nontrivial solutions can be obtained. This is the circumstance of homogeneous equations, which give rise to *eigenvalue* problems, as discussed later in this chapter. If Det $A \neq 0$, the equations are said to be *all independent*, and unique solutions may be obtained for nonhomogeneous equations. A brief outline of the matrix algebra needed for the following discussion is given here. For further details, textbooks on matrices such as Lancaster and Tismenetsky (1985) and Bronson and Costa (2008) may be consulted.

The solution to the set of linear equations given by Equation 6.3 may also be written as

$$X = A^{-1}B \tag{6.5}$$

where A^{-1} is the inverse of the matrix A, which must be *nonsingular* for the inverse to exist. Then $A^{-1} A = I$, where I is the *identity*, or unit, matrix, which is a square matrix consisting of zeros everywhere except at the diagonal, where all the elements are unity. Several methods for the solution of simultaneous linear equations are based on obtaining A^{-1} as an intermediate step. This is particularly advantageous if solutions are to be obtained for many systems of equations in which A is unchanged and only the column vector B varies. Also, many computer systems have available programs to invert a matrix. For example, matrix inversion is quite straightforward in MATLAB®. However, most methods, which are used for solving simultaneous algebraic equations, do not solve for A^{-1} as an intermediate step and solve only for X, unless there is a particular advantage in obtaining A^{-1}, such as those mentioned above, or unless the information is needed to study the nature of the equations.

There are two different types of methods, *direct* and *iterative*, that may be adopted for solving Equation 6.3 for the unknown X. Direct methods solve the equations exactly, except for the computational round-off error, in a finite number of operations. The methods based on finding the inverse matrix A^{-1} to obtain the solution from Equation 6.5 fall under direct methods, as do several other methods discussed in this chapter. These direct techniques are particularly useful when the number of equations to be solved is typically less than 20. However, a few special methods have been developed for particular types of equations. These methods may be used advantageously even for a much larger number of equations. Among these are the *Tridiagonal Matrix Algorithm (TDMA)*, *Fast Fourier Transform* method, and the *Cyclic Reduction* method, all of which are particularly suited for the large number of algebraic equations obtained from finite difference approximations of PDEs. The second class of methods is based on iteration. Iterative methods are appropriate for large systems of algebraic equations, typically of the order of 100 or more equations, in which the sparseness of the unknowns in the equations often makes iterative computation more efficient. Again, these methods are of particular interest in the finite difference and finite element solutions of PDEs.

In this chapter, both direct and iterative methods for the numerical solution of systems of linear algebraic equations are discussed. Most of the direct methods are based on matrix inversion or on elimination and reduction so that the given set of equations is obtained in a form that is amenable to a direct solution. Among those discussed here are the *Gaussian elimination*, *Gauss–Jordan elimination*, *L U decomposition*, *Crout's decomposition*, and *matrix inversion* methods. The iterative methods discussed here include the *Jacobi*, *Gauss–Seidel*, and *relaxation* methods. The solution of homogeneous linear equations, which often result in eigenvalue problems, is also discussed. The methods outlined for obtaining the eigenvalues and the corresponding eigenvectors include the *Gauss–Jordan* method, the *power* method, the *Jacobi* method, and *Householder's* method, used in conjunction with the *LR*, *QR*, and *QL* algorithms. Nonlinear algebraic equations are also of interest in many engineering applications, and the solution of these equations is outlined. In most cases, the equations are linearized in order to employ the methods applicable to linear equations. For small systems of nonlinear equations, methods based on those discussed in Chapter 5 for a single nonlinear algebraic equation, such as the Newton–Raphson and the successive substitution methods, may be employed.

MATLAB is particularly useful for the solution of a system of algebraic equations because of the advantage it has regarding the definition and manipulation of matrices. Matrix multiplication and inversion can be easily obtained by simple MATLAB commands. Similarly, the elements of a given matrix can be easily defined and augmented matrices can easily be formed from given matrices and vectors, making it easy to define the given system of equations and solve them to obtain the unknown variables. This advantage of MATLAB over other computational environments, particularly for direct methods, is clearly demonstrated in this chapter.

6.2 GAUSSIAN ELIMINATION

Gaussian elimination is a direct method for solving a system of linear algebraic equations and is frequently employed in a wide variety of engineering problems. By a process of elimination of the unknowns, the method reduces the given set of n equations to a triangular set, so that one of the equations has only one unknown. This unknown is determined and the remaining unknowns are obtained by the process of back-substitution. This method is of particular interest since several other direct methods are based on it.

6.2.1 Basic Approach

Let us consider a general system of three linear equations, given as

$$\begin{aligned} a_{11}x_1 + a_{12}x_2 + a_{13}x_3 &= b_1 \\ a_{21}x_1 + a_{22}x_2 + a_{23}x_3 &= b_2 \\ a_{31}x_1 + a_{32}x_2 + a_{33}x_3 &= b_3 \end{aligned} \tag{6.6}$$

As a first step, eliminate x_1 from the second equation by adding it to the equation obtained by multiplying the first equation by $-a_{21}/a_{11}$. Similarly, multiply the first equation by $-a_{31}/a_{11}$ and add the third equation to it to eliminate x_1 from the third equation as well. The resulting system of equations is

$$\begin{aligned} a_{11}x_1 + a_{12}x_2 + a_{13}x_3 &= b_1 \\ a_{22}^{(1)}x_2 + a_{23}^{(1)}x_3 &= b_2^{(1)} \\ a_{32}^{(1)}x_2 + a_{33}^{(1)}x_3 &= b_3^{(1)} \end{aligned} \tag{6.7}$$

where the superscripts indicate new values of the coefficients after the first step.

The first equation, which has been used to eliminate the unknown x_1 from the equations that follow, is known as the *pivot equation*, and the coefficient a_{11} of the eliminated unknown is the *pivot coefficient* or *pivot element*. In the next step, multiply the second equation by $-a_{32}^{(1)}/a_{22}^{(1)}$ and add it to the third equation, in order to eliminate x_2 from the latter. The result is an upper triangular set, given by

$$\begin{aligned} a_{11}x_1 + a_{12}x_2 + a_{13}x_3 &= b_1 \\ a_{22}^{(1)}x_2 + a_{23}^{(1)}x_3 &= b_2^{(1)} \\ a_{33}^{(2)}x_3 &= b_3^{(2)} \end{aligned} \tag{6.8}$$

where $a_{33}^{(2)}$ and $b_3^{(2)}$ arise from the second step in the elimination process. Now, x_3 is directly obtained as $b_3^{(2)}/a_{33}^{(2)}$ from the last equation. This value may be substituted in the second equation to obtain x_2, which, along with x_3, is substituted in the first equation to obtain x_1. This process, known as *back-substitution*, can be employed easily with a triangular set of equations to obtain the unknowns.

The preceding process can easily be extended to n equations, with n unknowns. Employing successive pivot equations, the elimination procedure, outlined here, is carried out until the original system of equations is reduced to an upper triangular set. Back-substitution then yields the unknowns. To illustrate the Gaussian elimination method, let us consider the following set of equations:

$$\begin{aligned} 3x+5y+z&=16 \\ x+4y+2z&=15 \\ 2x+2y+3z&=15 \end{aligned} \tag{6.9}$$

where x, y, and z are the unknowns. With the first equation as the pivot equation, x is eliminated from the other two equations to obtain

$$\begin{aligned} 3x+5y+z&=16 \\ \frac{7y}{3}+\frac{5z}{3}&=\frac{29}{3} \\ -\frac{4y}{3}+\frac{7z}{3}&=\frac{13}{3} \end{aligned} \tag{6.10}$$

Now, we use the second equation as the pivot equation to eliminate y from the third equation. This results in the following triangular set of equations:

$$\begin{aligned} 3y+5y+z&=16 \\ \frac{7y}{3}+\frac{5z}{3}&=\frac{29}{3} \\ \frac{69z}{21}&=\frac{207}{21} \end{aligned} \tag{6.11}$$

The fractional coefficients in the resulting equations may be avoided by multiplying on both side by the largest denominator. The value of z is obtained as 3 from the third equation. Back-substitution then yields the values of the remaining unknowns. The result is

$$x=1, \quad y=2, \quad z=3 \tag{6.12}$$

6.2.2 Computational Procedure

The Gaussian elimination method is very well suited for digital computation. The system of linear equations is written in matrix form as

$$AX=B \tag{6.3}$$

We then consider an augmented matrix C of this set, defined as follows:

$$C = \begin{bmatrix} a_{11} & a_{12} & \cdots & a_{1n} & a_{1,n+1} \\ a_{21} & a_{22} & \cdots & a_{2n} & a_{2,n+1} \\ \vdots & & & & \vdots \\ a_{n1} & a_{n2} & \cdots & a_{nn} & a_{n,n+1} \end{bmatrix} \tag{6.13}$$

where the $(n + 1)$th column consists of the b's, with $a_{1,n+1} = b_1$, $a_{1,n+2} = b_2$, and so on. The computational procedure is then concerned with reducing this matrix to the following upper triangular augmented matrix after $(n - 1)$ elimination steps:

$$C^{(n-1)} = \begin{bmatrix} a_{11} & a_{12} & a_{13} & \cdots & a_{1n} & a_{1,n+1} \\ 0 & a_{22}^{(1)} & a_{23}^{(1)} & \cdots & a_{2n}^{(1)} & a_{2,n+1}^{(1)} \\ \vdots & & & & & \vdots \\ 0 & 0 & 0 & \cdots & a_{nn}^{(n-1)} & a_{n,n+1}^{(n-1)} \end{bmatrix} \tag{6.14}$$

To help with the visualization of these matrices, Figure 6.1 shows qualitatively a few special types of matrices that are of particular interest in the numerical solution

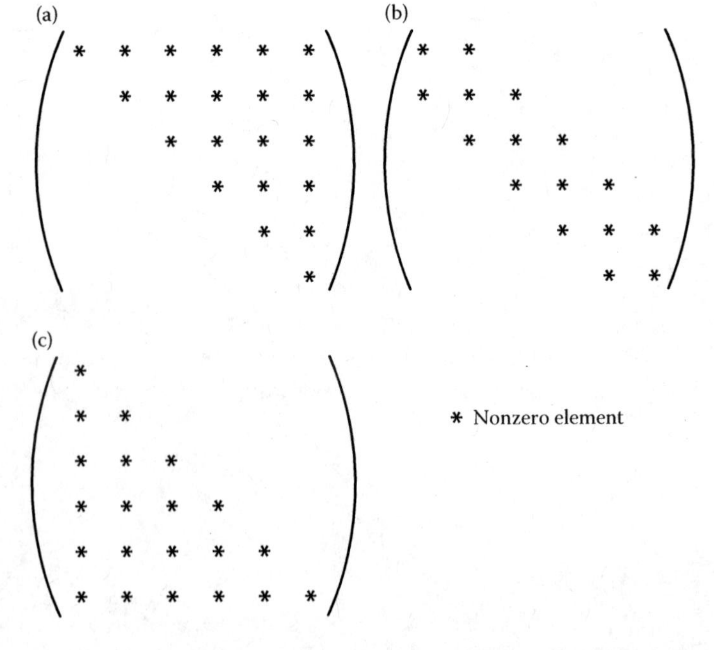

FIGURE 6.1 Sketch of a few special types of matrices that are of interest in the solution of simultaneous algebraic equations. (a) Upper triangular matrix, (b) tridiagonal matrix, and (c) lower triangular matrix.

of linear systems. This includes the upper and lower triangular matrices and the *tridiagonal* matrix, which is discussed later.

The first step in the reduction of the matrix C to the augmented triangular matrix $C^{(n-1)}$ is obtained from a generalization of the procedure outlined earlier. Therefore, the elements $a_{ij}^{(1)}$ of the matrix obtained after the first elimination step are given by

$$a_{ij}^{(1)} = a_{ij} - \frac{a_{i1}}{a_{11}}(a_{1j}), \quad \text{where } 2 \le i \le n \quad \text{and} \quad 1 \le j \le n+1 \tag{6.15}$$

Here, a_{ij} are the elements of the original matrix C. The first row is the pivot row and a_{11} is the pivot element. Note that the first element of each row, except the first one, becomes zero after this computational step.

Similarly, the complete elimination procedure may be generalized by recognizing that the pivot row varies from the first to the $(n-1)$th row as elimination proceeds and that the pivot element for each step is a_{rr}, where r denotes the number of the pivot row. Therefore, the general procedure is written as

$$a_{ij}^{(r)} = a_{ij}^{(r-1)} - \frac{a_{ir}^{(r-1)}}{a_{rr}^{(r-1)}}\left[a_{rj}^{(r-1)}\right], \quad \text{where } r+1 \le i \le n \quad \text{and} \quad r \le j \le n+1 \tag{6.16}$$

The superscripts within parentheses indicate the elimination step, with r varying from 1 to $(n-1)$. For $r = 1$, Equation 6.15 is obtained, with the superscript (0) simply denoting the coefficients of the original matrix. Again, note that the elements in the rth column of all the rows that follow the pivot row become zero. Therefore, an upper triangular matrix, augmented with the modified values of the b's, which were initially the given constants on the right-hand side of the given system of equations, is obtained in the form represented by the matrix $C^{(n-1)}$ after successively applying the preceding procedure to all the pivot rows, going from the first to the $(n-1)$th row. Once this reduced matrix has been obtained, as shown in Figure 6.1a, the unknown x_n is obtained from the elements of the matrix as

$$x_n = \frac{a_{n,n+1}}{a_{nn}} \tag{6.17}$$

The other unknowns are obtained from back-substitution as

$$x_i = \frac{a_{i,n+1} - \sum_{j=i+1}^{n} a_{ij}x_j}{a_{ii}}, \quad \text{where } i = n-1,\ n-2, \ldots, 2, 1 \tag{6.18}$$

Here the superscripts have been dropped for convenience. In a computer program, the old elements may be replaced by the new ones as elimination proceeds. No superscripts are needed if j is varied from $(r+1)$ to $(n+1)$ in Equation 6.16 and the elements in the rth column are simply replaced by zero. Back-substitution yields the unknowns in the reverse order, starting with x_n and ending with x_1, as given by Equation 6.18.

Let us apply the generalized procedure outlined above to the example given earlier, whose augmented matrix C is

$$C = \begin{bmatrix} 3 & 5 & 1 & 16 \\ 1 & 4 & 2 & 15 \\ 2 & 2 & 3 & 15 \end{bmatrix} \quad (6.19)$$

If Equation 6.16 is successively applied, with the first row as the pivot row in the first step and then the second row, the matrices obtained in the two steps are

$$\begin{bmatrix} 3 & 5 & 1 & 16 \\ 0 & \frac{7}{3} & \frac{5}{3} & \frac{29}{3} \\ 0 & \frac{-4}{3} & \frac{7}{3} & \frac{13}{3} \end{bmatrix} \text{ and } \begin{bmatrix} 3 & 5 & 1 & 16 \\ 0 & \frac{7}{3} & \frac{5}{3} & \frac{29}{3} \\ 0 & 0 & \frac{69}{21} & \frac{207}{21} \end{bmatrix}$$

Therefore, Equation 6.17 yields $x_3 = 3$, and back-substitution, given by Equation 6.18, gives $x_2 = 2$ and $x_1 = 1$. The generalized procedure can easily be applied on the computer, as discussed in Example 6.1.

It may be mentioned here that the solution of n linear equations by Gaussian elimination can be shown to require about $n^3/3$ arithmetic operations, as compared to n^3 for the multiplication of two $n \times n$ matrices (Atkinson, 1989). The method can also be used for matrix inversion and for the evaluation of a determinant, as outlined in Section 6.2.4. We now proceed to a consideration of the accuracy of the results obtained by Gaussian elimination.

6.2.3 Solution Accuracy

In the preceding example, the Gaussian elimination method yielded the exact solution of the given system of linear equations, because the number of equations was small and only whole numbers and exact fractions were involved. On the computer, however, fractions are replaced by decimals, retaining a limited number of significant places. As a consequence, a round-off error is introduced in dealing with quantities that have a larger number of significant decimal digits than those retained in the computation. The round-off error and its effect on accuracy and convergence were considered in detail in Chapter 2. Note from the earlier discussion that the round-off error may substantially affect the accuracy of the solution if a large number of equations are involved. Consequently, Gaussian elimination is generally used if the number of equations is typically less than 20 if most of the unknowns arise in each equation. Such a system gives rise to a *dense* coefficient matrix and, consequently, each element has to be considered at each elimination step. If, however, only a few unknowns are present in each equation, a *sparse* coefficient matrix is obtained. In certain special cases, several elimination steps are avoided and a larger number of equations may be solved by Gaussian elimination, while obtaining an acceptable accuracy level. An important example of such a sparse system is the *tridiagonal matrix*, which has nonzero elements only at the diagonal and on either side of the diagonal; see Figure 6.1b. This system arises in the numerical solution of ODEs and PDEs and is discussed in greater detail in Example 6.2.

6.2.3.1 Ill-Conditioned Set

The accuracy of the solution is also substantially influenced by the conditioning of the given system of linear equations. The system is said to be *ill-conditioned* if a relatively small change in one of the coefficients results in a relatively large change in the solution. Similarly, if there are elements in the inverse A^{-1} of the matrix that are several orders of magnitude larger than those in the original matrix A, then the matrix is probably ill-conditioned. The main problem with ill-conditioning is that the round-off error may cause slight changes in the coefficients which may, in turn, result in a large variation in the solution. To test whether the round-off error is significant for a given problem, one may use the solution vector computed to determine the constant vector B from the equation $AX = B$ and compare it with the original column matrix. Also, A may be inverted twice and compared with A, or the product $A^{-1}A$ may be compared with the identity matrix I to determine whether the round-off error is large for a given problem. An example of an ill-conditioned set of equations is

$$\begin{aligned} x - 1.9999y &= 0 \\ x - 1.9998y &= 1.0 \end{aligned} \tag{6.20}$$

for which the values of x and y are 19,999 and 10^4, respectively. A small change in the coefficients of y can result in a large effect on the solution. Since the round-off error can cause such a small change, the solution obtained may be quite inaccurate. To keep the error small, one may use double precision on the computer. Double precision reduces the speed of computation but is necessary for an ill-conditioned system.

6.2.3.2 Error Correction

We may also improve the accuracy of the solution obtained by applying an error correction. If X' is the solution vector obtained from numerical computation, the constant vector B' may be found by the substitution of X' into the given system of equations, as $B' = AX'$. Since the solution obtained is not exact due to the round-off error, B' will differ from the original constant matrix B. If X is the exact solution, the error vector E is defined as $X - X'$. Therefore, from the original system of equations, the following equations, known as *error equations*, are obtained:

$$A(X - X') = B - B' \quad \text{or} \quad AE = B - B' \tag{6.21}$$

Thus, the error vector may be computed from this set, which differs from the original system of equations only in the constant vector. For applying this error correction, Gaussian elimination is employed to solve the given system of equations in order to obtain X', from which B' is computed. The multipliers in the elimination process are $a_{ir}^{(r-1)}/a_{rr}^{(r-1)}$. We may retain these to solve other systems of equations with the same coefficient matrix A but a different constant vector B, by applying Equation 6.16 to the $(n + 1)$th column, which represents the new constant vector. Therefore, the error correction vector E may be obtained. Then a more accurate solution $\bar{X}$ to the set of equations is obtained from

$$\bar{X} = X' + E \tag{6.22}$$

The process may be repeated with the improved solution to obtain a still greater accuracy. However, the exact solution is not obtained at any iterative stage because of the presence of round-off errors.

6.2.3.3 Pivoting

In the solution of a system of linear equations by Gaussian elimination, we may encounter steps for which the pivot element is zero, or close to zero. In some cases, the pivot element may theoretically be zero but may acquire a small, non-zero value in the computational process due to the round-off error. The use of such a pivot element would lead to inaccurate results. In fact, the accuracy of the solution is considerably affected by the magnitude of the pivot element, which is employed in all the arithmetic operations for elimination in a given step. Greater accuracy is obtained if reduction is carried out with the row that contains the largest pivot element. A process, known as *partial pivoting*, in which the rows are interchanged at each step, to employ the row with the largest pivot element as the pivot row, is very commonly employed for more accurate results, particularly for large systems of equations. This procedure also avoids the problem with a zero pivot element during the elimination process. It is, therefore, necessary to include partial pivoting in the computational scheme. It can be easily incorporated in the computer program, as demonstrated in Example 6.1, given at the end of this section. In some cases, *complete pivoting*, with both rows and columns being interchanged to obtain the largest pivot element, is employed, as outlined later for homogeneous equations.

6.2.4 Matrix Inversion and Determinant Evaluation

Gaussian elimination may also be employed to obtain the inverse of a matrix. The inverse A^{-1} is generally not needed in the solution of a set of linear equations. However, as discussed later, the matrix may be needed for studying the nature of the equations and for solving several systems of equations that have the same coefficient matrix A but different constant vector B. Finding A^{-1} is equivalent to solving the equation $AX = I$, where X is now an $n \times n$ unknown matrix. An augmented matrix C is formed with the elements of A and I, placing the elements of I on the right-hand side of matrix A, as those of vector B were placed earlier. By applying Gaussian elimination to A, the solution matrix can be determined, which is the required inverse A^{-1}. The computation of the inverse by this method requires about $(4/3)n^3$ arithmetic operations, which is approximately four times the number of operations required for solving a set of linear equations (Atkinson, 1989). Other methods for calculating the inverse of a matrix are discussed in Section 6.5.

In some engineering problems, such as those that arise in vibrations and in stability analysis, it may be necessary to evaluate the determinant of a matrix. Although not needed for solving linear equations, since Cramer's rule is rarely applied, the value of the determinant may be required for studying the nature of the equations and for determining whether the inverse of a matrix exists. Gaussian elimination may again be used for the evaluation of a determinant. The value of the determinant of a matrix of an upper triangular form is simply the product of the diagonal elements. Therefore, Gaussian elimination may be used, as discussed earlier for a matrix, to obtain a given determinant in this form. This method is applicable since the value of a determinant is not altered if a constant multiple of the elements of a row or column are added to or subtracted from the elements of another row or column. However, if any two rows or columns are interchanged, the sign of the determinant is changed. Therefore, if partial pivoting is used, to achieve greater accuracy or to avoid a zero pivot element, this change in sign must be taken into account.

Let us consider the determinant of the coefficient matrix of the system of equations given by Equation 6.9. Then the determinant is

$$\begin{vmatrix} 3 & 5 & 1 \\ 1 & 4 & 2 \\ 2 & 2 & 3 \end{vmatrix}$$

We apply Gaussian elimination to this determinant by using Equation 6.16 with $r = 1$ and then with $r = 2$ to obtain the upper triangular form

$$\begin{vmatrix} 3 & 5 & 1 \\ 0 & \frac{7}{3} & \frac{5}{3} \\ 0 & 0 & \frac{69}{21} \end{vmatrix}$$

The value of this determinant is given by a product of the diagonal elements and is, therefore, obtained as

$$3 \times \frac{7}{3} \times \frac{69}{21} = 23$$

6.2.5 Tridiagonal Systems

The direct solution of systems of linear equations that have tridiagonal, or banded, coefficient matrices, as shown in Figure 6.1b, is important in many practical problems, particularly in the numerical solution of PDEs. The set of equations, in this case, may be written as

$$\begin{bmatrix} b_1 & c_1 & 0 & 0 & & \cdots & & & 0 \\ a_2 & b_2 & c_2 & 0 & & \cdots & & & 0 \\ 0 & a_3 & b_3 & c_3 & 0 & 0 & \cdots & & 0 \\ \vdots & & & & & & & & \vdots \\ 0 & 0 & \cdots & & 0 & 0 & a_{n-1} & b_{n-1} & c_{n-1} \\ 0 & 0 & & \cdots & & & 0 & a_n & b_n \end{bmatrix} \begin{bmatrix} x_1 \\ x_2 \\ \vdots \\ x_{n-1} \\ x_n \end{bmatrix} = \begin{bmatrix} d_1 \\ d_2 \\ \vdots \\ d_{n-1} \\ d_n \end{bmatrix} \tag{6.23}$$

Therefore, the matrix A is tridiagonal if only a_{ii}, $a_{i,i-1}$, and $a_{i,i+1}$ are nonzero. This implies that $a_{ij} = 0$ for $|i - j| > 1$. If Gaussian elimination is applied to this system, only one of the a's is eliminated from the column containing the pivot element in each step, since the remaining elements below the diagonal are zero. Therefore, only one elimination process is employed at each step. The original zero elements are kept unchanged, and the resulting system, after completion of the elimination procedure, which is often known as the *Thomas algorithm*, is of the form

$$\begin{bmatrix} b_1 & c_1 & 0 & 0 & \cdots & & & 0 \\ 0 & b_2' & c_2' & 0 & \cdots & & & 0 \\ 0 & 0 & b_3' & c_3' & \cdots & & & \vdots \\ \vdots & & & & & & & \\ 0 & & \cdots & & 0 & 0 & b_{n-1}' & c_{n-1}' \\ 0 & \cdots & & & 0 & 0 & 0 & b_n' \end{bmatrix} \begin{bmatrix} x_1 \\ x_2 \\ \vdots \\ x_{n-1} \\ x_n \end{bmatrix} = \begin{bmatrix} d_1 \\ d_2' \\ \vdots \\ d_{n-1}' \\ d_n' \end{bmatrix} \tag{6.24}$$

where the primes indicate new values. From this system of equations, the unknowns may easily be obtained by back-substitution, since the last equation has only one unknown and the others have two, including one which is obtained by solving the equation below a given equation.

The recursion formulas for the above system may be written in terms of the elements a_i, b_i, and c_i, where i denotes the row in the coefficient matrix. Therefore, the new elements are given by

$$\begin{aligned} a_i' &= 0, \quad b_i' = b_i - c_{i-1}\frac{a_i}{b_{i-1}'} \\ c_i' &= c_i, \quad d_i' = d_i - d_{i-1}'\frac{a_i}{b_{i-1}'} \end{aligned} \tag{6.25}$$

with the calculations being carried out for increasing values of i, from $i = 2$ to n. Once the reduced system, given by Equation 6.25, has been obtained, back-substitution may be applied as follows:

$$x_n = \frac{d'_n}{b'_n}, \quad x_i = \frac{d'_i - c'_i x_{i+1}}{b'_i} \tag{6.26}$$

On the computer, the new elements simply replace the old ones, and primes are not needed. The number of operations needed for solving a tridiagonal system is of order n, $O(n)$, as compared to $O(n^3/3)$ for Gaussian elimination applied to a system with a dense coefficient matrix. Therefore, much smaller computing times and, consequently, much smaller round-off errors arise in the solution of such systems. Thus, large tridiagonal systems are generally solved by this method. Example 6.2 illustrates the use of Gaussian elimination for solving systems of equations of the tridiagonal form.

In MATLAB, the solution of a system of linear equations can easily be achieved by defining the matrices a and b and applying the backslash operator, \, described in Chapter 3. For instance, considering the simple problem given earlier, a and b are defined as

```
a=[3 5 1; 1 4 2; 2 2 3];
b=[16; 15; 15];
```

Then, the desired solution x is given by

```
x=a\b
```

which uses the internal logic of the \ operator in MATLAB to obtain the left division of a into b. It uses the Gaussian elimination approach to achieve this, without finding the inverse of (A). It requires fewer arithmetic operations compared to methods based on matrix inversion and thus requires less CPU time and has smaller round-off error. The results are obtained as $x = [1.0; 2.0; 3.0]$, as before.

Example 6.1

The specific volume v of saturated steam in m³/kg is given at six dimensionless temperature T values of 1, 2, 3, 4, 5, and 6, where 1 represents 10°C in physical terms, as, respectively, 106.4, 57.79, 32.9, 19.52, 12.03, and 7.67 by Reynolds and Perkins (1977). Using the Gaussian elimination method, obtain a fifth-order polynomial that passes through these data points.

SOLUTION

The required polynomial is of the form

$$v = x_1 + x_2T + x_3T^2 + x_4T^3 + x_5T^4 + x_6T^5 \tag{6.27}$$

where T is the dimensionless temperature and $x_1, x_2, \ldots, x_6$ are the unknown coefficients to be determined. The following system of six equations is obtained if the given data are substituted in this equation:

$$x_1 + x_2 + x_3 + x_4 + x_5 + x_6 = 106.4$$

$$x_1 + 2x_2 + 2^2x_3 + 2^3x_4 + 2^4x_5 + 2^5x_6 = 57.79$$

$$x_1 + 3x_2 + 3^2x_3 + 3^3x_4 + 3^4x_5 + 3^5x_6 = 32.9$$

$$x_1 + 4x_2 + 4^2x_3 + 4^3x_4 + 4^4x_5 + 4^5x_6 = 19.52$$

$$x_1 + 5x_2 + 5^2x_3 + 5^3x_4 + 5^4x_5 + 5^5x_6 = 12.03$$

$$x_1 + 6x_2 + 6^2x_3 + 6^3x_4 + 6^4x_5 + 6^5x_6 = 7.67$$

This set of linear equations is to be solved to obtain the unknowns x_i, where $i = 1, 2, \ldots, 6$. The augmented matrix consists of six rows and seven columns, where the seventh column contains the constants on the right-hand side of the equations. It may be mentioned here that the dimensionless temperature T is employed simply for convenience. The actual temperatures may also be used if so desired.

A computer program in Fortran is shown in Appendix C.5 to present the logic for solving this problem by the Gaussian elimination method. The problem is written for a system of up to 10 equations. The number of equations and the augmented matrix are given as input data. At each elimination step, the row with the largest pivot element is found, considering the rows below and including the pivot row. If another row has a pivot element larger than that in the pivot row, it is interchanged with the pivot row, making it the pivot row for the next elimination step. This partial pivoting improves the accuracy of the solution and also avoids problems if a zero pivot element arises. Gaussian elimination is applied to reduce the given matrix to an upper triangular one. Then *X*(6), which represents the unknown x_6, is computed directly from Equation 6.17. The other unknowns are determined by back-substitution, using Equation 6.18.

Figure 6.2 shows the computed results. The program also computes the constants B_i, where $i = 1, 2, \ldots, 6$, using the equation $B = AX$, where X is the computed vector of the unknowns. These computed constants, denoted by *B*(*I*) in the program, may be compared with the constants in the given equations to determine the accuracy of the numerical results. It is seen from the results presented that the computed values of the constants are close to the given values. In fact, they are identical if we retain the same number of significant figures as those in the given data. Therefore, a high level of accuracy in the computed results is indicated.

This problem can be solved in MATLAB by using the backslash operator, as discussed earlier. The matrix *a* and the vector *b* is defined from the given equations as

```
>>a= [1.0 1.0 1.0 1.0 1.0 1.0;1.0 2.0 4.0 8.0 16.0 32.0;1.0 ...
     3.0 9.0 27.0 81.0 243.0;1.0 4.0 16.0 64.0 256.0 ...
     1024.0;1.0 5.0 25.0 125.0 625.0 3125.0;1.0 6.0 36.0 ...
     216.0 1296.0 7776.0];
>>b= [106.4;57.79;32.9;19.52;12.03;7.67];
```

THE SOLUTION TO THE EQUATIONS IS
$X(1) = 201.26010$
$X(2) = -128.82130$
$X(3) = 40.67448$
$X(4) = -7.42302$
$X(5) = 0.74085$
$X(8) = -0.03108$

THE CONSTANT VECTOR OF THE EQUATIONS IS
$B(1) = 106.39990$
$B(2) = 57.79021$
$B(3) = 32.90009$
$B(4) = 19.52042$
$B(5) = 12.03053$
$B(6) = 7.67023$

FIGURE 6.2 Numerical results obtained from the solution of the system of linear equations in Example 6.1 by the Gaussian elimination method.

Matrix a may also be defined by defining the six rows separately as, say, a_1, a_2, a_3, a_4, a_5, and a_6, and then specifying the matrix a as $a = [\, a_1; a_2; a_3; a_4; a_5; a_6]$. Then the command

```
>>x = a\b
```

Yields the results as

```
x =
201.2600
-128.8210
40.6742
-7.4229
0.7408
-0.0311
```

which are close to those obtained from the Fortran program earlier. But, clearly, this is a much simpler approach, which uses MATLAB advantageously by employing the built in backslash operator.

However, a MATLAB program may also be written using the computational scheme outlined here. Appendix B.6 gives the corresponding MATLAB program as a function *m*-file, *gauss.m*, which reduces the given matrix to an upper triangular matrix and uses another function *m*-file, *backsub.m*, for back substitution. The two matrices a and b are given and the function file is invoked as *gauss* (a, b). This yields the desired results as identical to those given above from the backslash operator. The polynomial obtained for the specific volume v, as given by Equation 6.27, may also be plotted, along with the given data, to evaluate the accuracy of the numerical results obtained. Figure 6.3 shows the resulting graph and the given data, indicating the high level of accuracy achieved in the computed results.

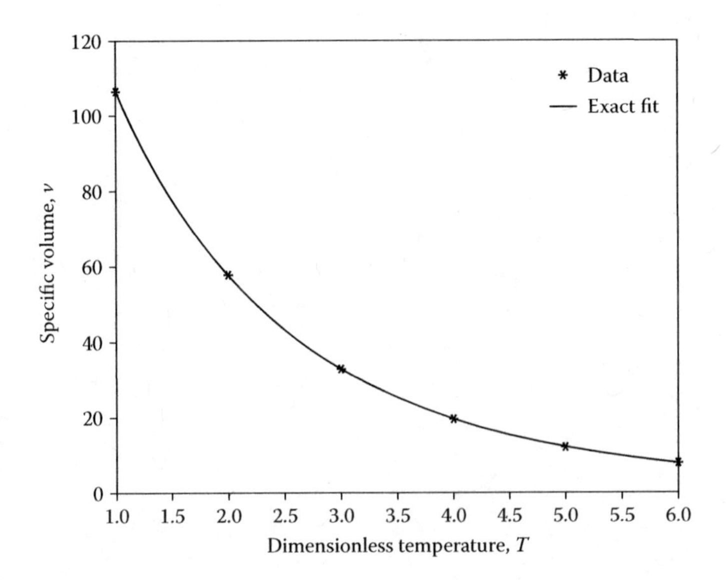

FIGURE 6.3 Graph of the polynomial given in Equation 6.27, as obtained from the computed results, along with the given data in Example 6.1.

Example 6.2

The temperature $T(x)$ that arises due to steady-state heat conduction in a bar 30 cm long is governed by the following ODE, if uniform temperature is assumed across any cross section:

$$\frac{d^2T}{dx^2} - GT = 0 \tag{6.28}$$

where T is the temperature difference from the ambient medium, which is at 20°C, x is the axial coordinate distance, and G is a constant that depends on the surface heat transfer rate. As discussed in greater detail in Chapter 9, this equation may be replaced by a finite difference approximation, using the second central difference given in Chapter 4, as

$$\frac{T_{i+1} - 2T_i + T_{i-1}}{(\Delta x)^2} - GT_i = 0 \tag{6.29}$$

where $x = i\,\Delta x$. Considering 30 subdivisions of the length of the rod, with $\Delta x = 1$ cm, as shown in Figure 6.4, find the temperature differences T_i, where $i = 1, 2, \ldots, 29$.

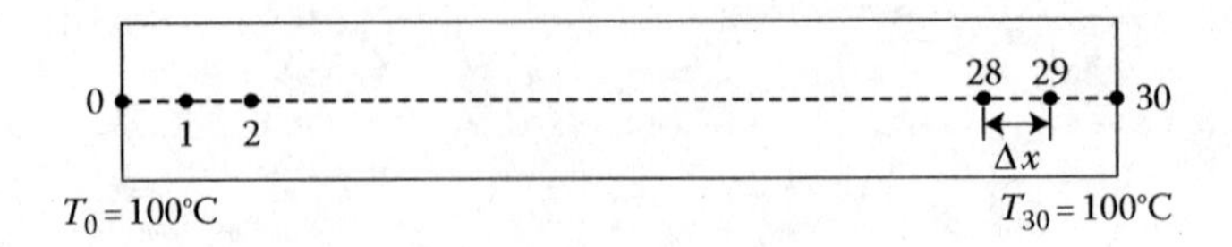

FIGURE 6.4 Physical problem considered in Example 6.2.

The temperatures differences T_0 and T_{30}, at $x = 0$ and $x = 30$ cm, respectively, are given as 100°C, and the constant G as $(0.071)^2$ cm^{-2}. Use Gaussian elimination.

SOLUTION

The system of equations to be solved by Gaussian elimination is

$$T_{i+1} - [2 + G(\Delta x)^2]T_i + T_{i-1} = 0 \tag{6.30}$$

or

$$-T_{i+1} + (2 + S)T_i - T_{i-1} = 0 \quad \text{for } i = 1, 2, \ldots, 29 \tag{6.31}$$

where

$$S = G(\Delta x)^2 = (0.071)^2(1.0)^2$$

The constants F_i on the right-hand side of Equation 6.31 are all zero except in the two equations corresponding to $i = 1$ and 29, where T_0 and T_{30} appear and yield $F_1 = 100$ and $F_{29} = 100$. A tridiagonal system is obtained, which may be written as follows:

$$\begin{bmatrix} 2+S & -1 & 0 & 0 & \cdots & & 0 \\ -1 & 2+S & -1 & 0 & \cdots & & 0 \\ 0 & -1 & 2+S & -1 & \cdots & & 0 \\ \vdots & & & & & & \vdots \\ 0 & 0 & \cdots & & 0 & -1 & 2+S \end{bmatrix} \begin{bmatrix} T_1 \\ T_2 \\ \vdots \\ T_{29} \end{bmatrix} = \begin{bmatrix} 100 \\ 0 \\ \vdots \\ 0 \\ 100 \end{bmatrix} \tag{6.32}$$

This tridiagonal system of equations can easily be solved by the Gaussian elimination method, as presented earlier. The three nonzero elements in each row are denoted by $A(I)$, $B(I)$, and $C(I)$, where $I = 1, 2, \ldots, 29$. The constants on the right-hand side are denoted by $F(I)$. The coefficients and constants are given as input data. Gaussian elimination is used to eliminate the left-most element in each row in one traverse from the top row to the bottom row. The temperature difference T_{29} is then computed as $F(29)/B(29)$, where both F and B are the new values after reduction. Back-substitution then yields the remaining temperature differences.

Appendix C.6 gives the Tridiagonal Matrix Algorithm (TDMA), also known as *Thomas algorithm*, as a subroutine in Fortran. Thus, the matrix coefficients and the constants F_i may be given, and the unknown temperature differences T_i computed by this subroutine. Finally, the ambient temperature of 20°C is added to T_i to yield the physical temperatures T_p, where $T_p = T_i + 20$. It is evident that the program is much simpler than the corresponding program for a system that is not tridiagonal. Since tridiagonal systems arise frequently in the numerical solution of differential equations, the above algorithm is of considerable importance. Because of the associated small number of arithmetic operations, the Gaussian elimination method of solving a tridiagonal system results in smaller computer time and smaller round-off error than most of the other methods discussed in this chapter.

THE REQUIRED TEMPERATURES ARE
TP(1) = 114.6583
TP(2) = 109.7937
TP(3) = 105.3816
TP(4) = 101.3998
TP(5) = 97.8283
TP(6) = 94.6491
TP(7) = 91.8481
TP(8) = 89.4051
TP(9) = 87.3139
TP(10) = 85.5620
TP(11) = 84.1405
TP(12) = 83.0422
TP(13) = 82.2616
TP(14) = 81.7948
TP(15) = 81.6395
TP(16) = 81.7948
TP(17) = 82.2616
TP(18) = 83.0421
TP(19) = 84.1404
TP(20) = 85.5620
TP(21) = 87.3140
TP(22) = 89.4052
TP(23) = 91.8462
TP(24) = 94.6493
TP(25) = 97.8286
TP(26) = 101.4001
TP(27) = 105.3819
TP(28) = 109.7939
TP(29) = 114.6584

FIGURE 6.5 Computed temperatures obtained by solving the tridiagonal system of equations in Example 6.2.

The computed temperature distribution is shown in Figure 6.5. The temperatures at the two ends of the rod, *TP*(0) and *TP*(30), are 120°C. The problem is symmetric about *TP*(15), and the computational procedure may be simplified by employing only 16 points and taking *TP*(16) = *TP*(14) from symmetry.

Similarly, the tridiagonal system may be solved using MATLAB. Again, the elements *a*, *b*, and *c* of the coefficient matrix, as well as the constant vector *f*, are given and the computational procedure outlined earlier for tridiagonal systems is applied to obtain the solution. Appendix B.7 gives this algorithm, for Example 6.2, in MATLAB as an *m*-file in (a) and as a function *m*-file in (b). The matrix and the constant vector are easily defined and element-by-element operations can be used advantageously to solve the problem. The results obtained are very close to those given in Figure 6.5. The resulting temperature distribution is also plotted using MATLAB plotting commands and is shown in Figure 6.6.

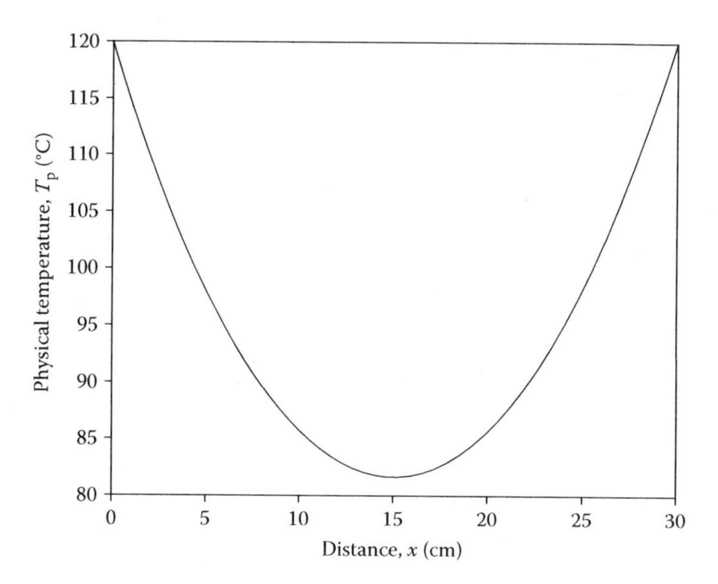

FIGURE 6.6 Temperature distribution from the calculated temperatures in Example 6.2.

6.3 GAUSS–JORDAN ELIMINATION

In the Gauss–Jordan elimination method, which is a variation of Gaussian elimination, the original matrix A of the coefficients is reduced to an identity matrix I so that the unknowns $x_1, x_2, \ldots, x_n$ are found directly, without back-substitution. At each step involved in the elimination of an unknown using a pivot equation, the unknown is eliminated from the equations above the pivot equation as well as from those below it. The pivot equation is normalized by dividing it throughout by the pivot element, so that the diagonal elements are finally obtained as unity, resulting in a reduced system of equations of the form $IX = B'$, where B' now gives the solution vector. Since the procedure is similar to Gaussian elimination, partial or complete pivoting may be employed to improve the accuracy of the solution and to avoid the use of a pivot element, that is, zero, or close to it.

6.3.1 Mathematical Procedure

Let us consider the following set of linear equations for solving by the Gauss–Jordan elimination method:

$$\begin{aligned} 2x + y + 3z &= 9 \\ 3x - 4y + 4z &= 7 \\ x + 4y - z &= 3 \end{aligned} \tag{6.33}$$

We normalize the first equation by dividing it by 2, which is the pivot element. Then we eliminate x from the other two equations by multiplying the normalized pivot equation by the coefficients of x in the other equations and subtracting the equations thus obtained from the original equations, to yield

$$\begin{aligned} x+\frac{1}{2}y+\frac{3}{2}z&=\frac{9}{2} \\ -\frac{11}{2}y-\frac{1}{2}z&=-\frac{13}{2} \\ \frac{7}{2}y-\frac{5}{2}z&=-\frac{3}{2} \end{aligned} \tag{6.34a}$$

We repeat the above step with the second equation as the pivot equation, which we first normalized by dividing it by the pivot element –11/2. Thus, y is eliminated from the first and third equations to give

$$\begin{aligned} x+\frac{16}{11}z&=\frac{43}{11} \\ y+\frac{1}{11}z&=\frac{13}{11} \\ -\frac{31}{11}z&=-\frac{62}{11} \end{aligned} \tag{6.34b}$$

In the final step, z is eliminated from the first two equations, to give

$$\begin{aligned} x&=1 \\ y&=1 \\ z&=2 \end{aligned} \tag{6.34c}$$

6.3.2 Computational Scheme

To implement this method on the computer, we follow the same procedure as that outlined earlier for Gaussian elimination. The augmented matrix C of a system of linear equations is given by Equation 6.13. The first row is the pivot equation for the first elimination step, and the elements of the matrix obtained after this step are

$$a_{1j}^{(1)}=\frac{a_{1j}}{a_{11}}, \quad \text{where } 1\le j\le n+1 \tag{6.35a}$$

$$a_{ij}^{(1)}=a_{ij}-a_{i1}a_{1j}^{(1)}, \quad \text{where } 1\le j\le n+1 \quad \text{and} \quad 2\le i\le n \tag{6.35b}$$

We can easily generalize this procedure by noting that Equation 6.35b is not applied to the pivot row, in a given step, and that the columns to the left of the one containing the pivot element are not affected. Therefore, if r denotes the pivot row and, hence, the elimination step, the Gauss–Jordan elimination method is given by the following recursion formulas:

$$a_{rj}^{(r)} = \frac{a_{rj}^{(r-1)}}{a_{rr}^{(r-1)}}, \quad \text{where } r \le j \le n+1 \tag{6.36a}$$

$$a_{ij}^{(r)} = a_{ij}^{(r-1)} - a_{ir}^{(r-1)} a_{rj}^{(r)}, \quad \text{where } r \le j \le n+1 \text{ and } 1 \le i \le n, \text{ except } i = r \tag{6.36b}$$

In most computational schemes, the new elements simply replace the old ones as they are computed, to avoid additional storage. However, the $j = r$ column must be stored separately at each step to provide the $a_{ir}^{(r-1)}$ needed in Equation 6.36b. This situation may be avoided by varying j from right to left or from $(r + 1)$ to $(n + 1)$. In the latter case, the $j = r$ column is not computed but may simply be inserted after the completion of the step, with 1 at the pivot element and zeros above and below it in the column. Then the storage required is the same as that for the original system. This approach is demonstrated in Example 6.3, using partial pivoting.

Let us now apply the above generalized procedure to the system of equations considered earlier, with the augmented matrix C given by

$$C = \begin{bmatrix} 2 & 1 & 3 & 9 \\ 3 & -4 & 4 & 7 \\ 1 & 4 & -1 & 3 \end{bmatrix} \tag{6.37}$$

Applying the recursion formulas of Equation 6.36, successively, for $r = 1$, 2, and 3, we find that the resulting matrices are, respectively,

$$\begin{bmatrix} 1 & \frac{1}{2} & \frac{3}{2} & \frac{9}{2} \\ 0 & -\frac{11}{2} & -\frac{1}{2} & -\frac{13}{2} \\ 0 & \frac{7}{2} & -\frac{5}{2} & -\frac{3}{2} \end{bmatrix} \begin{bmatrix} 1 & 0 & \frac{16}{11} & \frac{43}{11} \\ 0 & 1 & \frac{1}{11} & \frac{13}{11} \\ 0 & 0 & -\frac{31}{11} & -\frac{62}{11} \end{bmatrix} \begin{bmatrix} 1 & 0 & 0 & 1 \\ 0 & 1 & 0 & 1 \\ 0 & 0 & 1 & 2 \end{bmatrix}$$

Thus the required solution is $x = 1$, $y = 1$, and $z = 2$. As shown above, the columns to the left of the pivot element, at each step, are unaffected, and the elements above and below the pivot element, in the same column, become zero. The solution is given by the $(n + 1)$th column. No back-substitution is needed.

If several systems of equations with the same coefficient matrix A and different constant vector B are given, they can all be solved by application of the above procedure once. The various constant vectors are simply added as columns to the augmented matrix, and the reduction process is carried out. At the completion of the elimination, the solutions for the different systems are given by the corresponding columns in the reduced matrix. It can be shown that the solution of a system of n linear equations by the Gauss–Jordan elimination method requires about $n^3/2$ arithmetic operations. Therefore, this method takes a somewhat larger computing time and has a larger round-off error than Gaussian elimination and is not preferred for solving linear systems. However, it can be used to develop a method for matrix inversion employing minimum storage. For inverting a matrix A, the equation to be solved is $AX = I$, where X becomes A^{-1} when Gauss–Jordan elimination transforms A into I. This procedure is discussed in greater detail in Section 6.5. Also, if several systems of equations with the same coefficient matrix A and different constant vector B are to be solved, as is the case, for instance, in engineering problems where the boundary conditions are varied while the governing equations remain unchanged, Gauss–Jordan elimination is more advantageous to use than Gaussian elimination.

Example 6.3

Consider the electrical network shown in Figure 6.7a and compute the electric currents $I_1, I_2, \ldots, I_6$ through the six resistances. Also, solve the problem, employing three loop currents I_1, I_2, and I_3 in the three closed circuits shown in Figure 6.7b. Use Gauss–Jordan elimination for this problem.

SOLUTION

A system of six linear equations may be written for the six unknowns x_i, where x_i represents the current through a given resistance and $i = 1, 2, \ldots, 6$. By Kirchhoff's laws, the sum of the currents entering a node is equated to the sum of the currents leaving it. Thus,

$$x_1 + x_2 - x_3 = 0 \tag{6.38a}$$

$$x_2 - x_4 - x_5 = 0 \tag{6.38b}$$

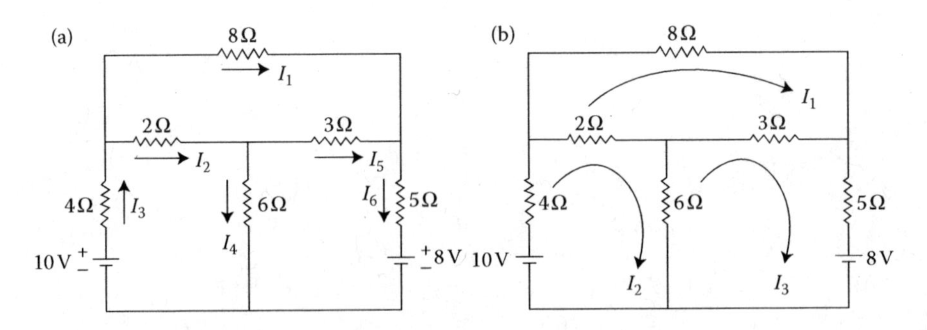

FIGURE 6.7 Electrical network considered in Example 6.3.

$$x_1 + x_5 - x_6 = 0 \tag{6.38c}$$

Also, the voltage change as one goes around each loop is equated to zero to yield

$$2x_2 + 4x_3 + 6x_4 - 10 = 0 \tag{6.38d}$$

$$-6x_4 + 3x_5 + 5x_6 + 8 = 0 \tag{6.38e}$$

$$8x_1 - 2x_2 - 3x_5 = 0 \tag{6.38f}$$

Therefore, six independent linear equations are obtained for determining the six currents x_i. If other nodes are considered, only linear combinations of the first three equations will be obtained; see, for instance, Young et al. (2000).

The Gauss–Jordan elimination method, with partial pivoting, is used for solving this system of equations. The computer program is quite similar to that for Gaussian elimination. However, in this case, the elements in the column containing the pivot element are reduced to zero in rows both above and below the pivot row. Also, at each step, the pivot row is taken as the row with the largest pivot element, considering the rows below the pivot row for the preceding step. The elements of this row are normalized by dividing throughout by the pivot element. Therefore, the reduced matrix is an identity matrix, instead of the upper triangular matrix obtained in Gaussian elimination. The augmented matrix, with the constant vector B taken as the last column, is supplied to the program as input data. This column vector becomes the solution after the Gauss–Jordan elimination process has been completed. Therefore, no back-substitution is necessary, and the last column yields the solution.

Appendix B.8 shows the MATLAB program for the Gauss–Jordan method as a function *m*-file, *jordan.m*. The given matrix *a* and constant vector *b* are given to form the augmented matrix *aug* = [*a b*]. The earlier function *m*-file, *gauss.m*, for the Gaussian elimination method is easily modified to obtain the function *m*-file for the Gauss–Jordan method. The back-substitution function *m*-file is not needed and the commands for generating an upper triangular matrix are modified to obtain an identity matrix instead. The last column then yields the solution. Similarly, Appendix C.7 shows the computer program in Fortran. Again, the program given earlier for Gaussian elimination is modified to include division of the pivot row by the pivot element to obtain 1.0 at the diagonal and elimination of elements both above and below the pivot element to reduce the coefficient matrix to an identity matrix.

Figure 6.8 shows the results obtained in terms of the six currents, which are the unknowns x_i, denoted here by $X(I)$, where $I = 1, 2, \ldots, 6$. Also shown is the reduced matrix achieved at the end of the calculations, indicating that, as expected, an identity matrix is obtained. As done, for Example 6.1, the constant vector B may also be computed from the obtained values of the currents and compared with the given values, in order to evaluate the accuracy of the solution. This method can also be applied to the problem given in Example 6.1 and the same computed results as before are obtained.

THE SOLUTION TO THE EQUATIONS IS

$X(1) = 0.05135$
$X(2) = 0.66351$
$X(3) = 0.71486$
$X(4) = 0.96892$
$X(5) = -0.30541$
$X(6) = -0.25405$

THE REDUCED MATRIX IS

1.000	0.000	0.000	0.000	0.000	0.000
0.000	1.000	0.000	0.000	0.000	0.000
0.000	0.000	1.000	0.000	0.000	0.000
0.000	0.000	0.000	1.000	0.000	0.000
0.000	0.000	0.000	0.000	1.000	0.000
0.000	0.000	0.000	0.000	0.000	1.000

FIGURE 6.8 Computed results from the solution of the linear system of equations in Example 6.3 by the Gauss-Jordan elimination method.

If the loop currents of Figure 6.7b are considered instead, the voltage change around each loop may be equated to zero:

$$-2I_1 + 12I_2 - 6I_3 - 10 = 0 \tag{6.39a}$$

$$-3I_1 - 6I_2 + 14I_3 + 8 = 0 \tag{6.39b}$$

$$13I_1 - 2I_2 - 3I_3 = 0 \tag{6.39c}$$

This system may also be solved, using the algorithm discussed above. The resulting values of I_1, I_2, and I_3 were obtained as 0.0514, 0.7149, and –0.2541 amperes, which are almost identical to the values of $X(1)$, $X(3)$, and $X(6)$, as expected from the nomenclature of Figure 6.7. Once these currents have been obtained, the other currents and desired voltages may be computed from Kirchhoff's laws.

6.4 COMPACT METHODS

6.4.1 Matrix Decomposition

There are several numerical methods for the solution of simultaneous linear equations that are based on the decomposition of the coefficient matrix A into an upper triangular matrix U and a lower triangular matrix L, as shown in Figure 6.1, such that

$$A = LU \tag{6.40}$$

In Gaussian elimination, we obtain an upper triangular matrix U of the form

$$U = \begin{bmatrix} a_{11} & a_{12} & \cdots & a_{1n} \\ 0 & a_{22}^{(1)} & \cdots & a_{2n}^{(1)} \\ \vdots & & & \vdots \\ 0 & \cdots & 0 & a_{nn}^{(n-1)} \end{bmatrix} \tag{6.41}$$

In order to obtain the above form, we use Equation 6.16, with the multipliers m_{ir} at each elimination step being given by

$$m_{ir} = \frac{a_{ir}^{(r-1)}}{a_{rr}^{(r-1)}} \quad \text{for } i = r+1, \ldots, n \tag{6.42}$$

These multipliers, if stored, can be used for solving different systems of equations, $AX = B$, which have the same coefficient matrix A but a different constant vector B. Usually, these multipliers m_{ij} are stored in place of the zero elements below the diagonal of the matrix U, that is, in the space originally employed for a_{ij} for $i > j$.

Let us consider the lower triangular matrix L, defined as follows:

$$L = \begin{bmatrix} 1 & 0 & 0 & \cdots & 0 \\ m_{21} & 1 & 0 & \cdots & 0 \\ \vdots & & & & \vdots \\ m_{n1} & m_{n2} & \cdots & \cdots & 1 \end{bmatrix} \tag{6.43}$$

We can show that $A = LU$, by using the preceding definitions of L and U and carrying out the matrix multiplication of Equation 6.40. Therefore, the coefficient matrix A may be decomposed into the two triangular matrices L and U. If the system of equations given by Equation 6.9 is considered, the multipliers m_{ir} may be retained to yield the matrices

$$L = \begin{bmatrix} 1 & 0 & 0 \\ \frac{1}{3} & 1 & 0 \\ \frac{2}{3} & -\frac{4}{7} & 1 \end{bmatrix} \quad \text{and} \quad U = \begin{bmatrix} 3 & 5 & 1 \\ 0 & \frac{7}{3} & \frac{5}{3} \\ 0 & 0 & \frac{69}{21} \end{bmatrix} \tag{6.44}$$

It can be easily verified that $A = LU$.

Recursive formulas for the elements of L and U may be obtained directly from Equation 6.40 and employed in the solution of a system of linear equations. If the above form for L, with all the diagonal elements equal to 1, is considered, the decomposition is the one obtained from Gaussian elimination. For this circumstance, *Doolittle's* method gives the corresponding explicit recursive formulas for the elements l_{ij} and u_{ij} and employs them in the solution of linear systems (Atkinson, 1989).

Another method, known as *Crout's method*, employs a U matrix whose diagonal elements are all equal to 1. This method is discussed in some detail here, since it is widely used in many engineering problems. It is also generally more efficient than the other elimination methods discussed earlier. Consequently, it requires less computer time and generates smaller round-off error.

6.4.2 Matrix Decomposition in MATLAB®

In MATLAB, as mentioned in Chapter 3, the $L\,U$ decomposition is easily obtained by the command

$$[l,u,p] = lu(a); \quad (6.45a)$$

Here, p is the permutation matrix which stores the information on row exchanges for partial pivoting. Then the solution of the system of equations $AX = B$ is obtained by the commands

$$y = l\backslash(p*b); \quad (6.45b)$$

$$x = u\backslash y \quad (6.45c)$$

These commands can be employed for the problem in Example 6.3 as

```
>>a= [1 1 -1 0 0 0; 0 1 0 -1 -1 0; 1 0 0 0 1 -1;...
     0 2 4 6 0 0;  0 0 0 -6 3 5;  8 -2 0 0 -3 0];
>>b= [0 0 0 10 -8 0]';
>>[l,u,p] = lu(a);
>>y=l\(p*b);
>>x=u\y
```

This yields the results

```
x=
  0.0514
  0.6635
  0.7149
  0.9689
 -0.3054
 -0.2541
```

which are identical to the results from the Gauss–Jordan method presented earlier.

Similarly, three linear equations given by

```
>>a= [3 5 1;1 4 2;2 2 3];
>>b= [16;15;15];
```

can be entered and the preceding $L\,U$ decomposition used to obtain the solution as

```
x=
  1.0000
  2.0000
  3.0000
```

6.4.3 Crout's Method

In *Crout's* method, Gaussian elimination is written in a more compact form, using the following decomposition. Then, the recursive relations for the matrix elements are obtained using matrix algebra.

$$A = LU \quad \text{where } L = \begin{bmatrix} l_{11} & 0 & \cdots & \cdots & 0 \\ l_{21} & l_{22} & 0 & \cdots & 0 \\ \vdots & \vdots & \vdots & & \vdots \\ l_{n1} & l_{n2} & \cdots & \cdots & l_{nn} \end{bmatrix}$$
$$\text{and} \quad U = \begin{bmatrix} 1 & u_{12} & u_{13} & \cdots & u_{1n} \\ 0 & 1 & u_{23} & \cdots & u_{2n} \\ \vdots & & & & \vdots \\ 0 & \cdots & \cdots & 0 & 1 \end{bmatrix} \tag{6.46}$$

From the above decomposition, the product rule of the determinants may be employed to obtain

$$\mathrm{Det}(A) = \mathrm{Det}(L)\,\mathrm{Det}(U) \tag{6.47}$$

For independent equations, the determinant of matrix A is nonzero, as discussed in Section 6.1. Therefore, the determinant of matrix L is also nonzero. Since L is a lower triangular matrix, its determinant is the product of its diagonal elements. This implies that if Det(L) is nonzero, all the diagonal elements l_{ii} of this matrix are nonzero. It is shown below that, if l_{ii} and a_{11} are nonzero, L and U matrices exist and their elements can be determined uniquely.

To solve a system of linear equations by Crout's method, we write A and U as augmented matrices of the form

$$\begin{bmatrix} l_{11} & 0 & \cdots & 0 \\ l_{21} & l_{22} & 0 & 0 \\ \vdots & & & \vdots \\ l_{n1} & l_{n2} & \cdots & l_{nn} \end{bmatrix} \begin{bmatrix} 1 & u_{12} & u_{13} & \cdots & u_{1n} & u_{1,n+1} \\ 0 & 1 & u_{23} & \cdots & u_{2n} & u_{2,n+1} \\ \vdots & & & & & \vdots \\ 0 & 0 & \cdots & \cdots & 1 & u_{n,n+1} \end{bmatrix}$$
$$= \begin{bmatrix} a_{11} & a_{12} & a_{13} & \cdots & a_{1n} & a_{1,n+1} \\ a_{21} & a_{22} & a_{23} & \cdots & a_{2n} & a_{2,n+1} \\ \vdots & & & & & \vdots \\ a_{n1} & a_{n2} & a_{n3} & \cdots & a_{nn} & a_{n,n+1} \end{bmatrix} \tag{6.48}$$

Recursive formulas for l_{ij} and u_{ij} may be developed by application of matrix multiplication to the above equation. Therefore,

$$l_{11} = a_{11}, l_{11}u_{12} = a_{12}, l_{11}u_{13} = a_{13}, \ldots, l_{11}u_{1,n+1} = a_{1,n+1}$$

which gives

$$l_{i1} = a_{i1} \quad \text{for } i = 1, 2, \ldots, n \tag{6.49a}$$

$$u_{1j} = \frac{a_{1j}}{l_{11}} \quad \text{for } j = 2, 3, \ldots, n+1 \tag{6.49b}$$

Similarly,

$$l_{21}u_{12} + l_{22} = a_{22}, \quad {}_{31}u_{12} + l_{32} = a_{32}$$

or

$$l_{22} = a_{22} - l_{21}u_{12}, \quad l_{32} = a_{32} - l_{31}u_{12} \tag{6.49c}$$

Also,

$$l_{21}u_{13} + l_{22}u_{23} = a_{23}, \quad l_{21}u_{14} + l_{22}u_{24} = a_{24}$$

or

$$u_{23} = \frac{a_{23} - l_{21}u_{13}}{l_{22}}, \quad u_{24} = \frac{a_{24} - l_{21}u_{14}}{l_{22}} \tag{6.49d}$$

Therefore, we may proceed as given above to obtain all the elements of the matrices L and U. General equations may also be developed for l_{ij} and u_{ij} as follows:

$$l_{i1} = a_{i1} \quad \text{for } i = 1, 2, \ldots, n \tag{6.50a}$$

$$u_{1j} = \frac{a_{1j}}{l_{11}} \quad \text{for } j = 2, 3, \ldots, n+1 \tag{6.50b}$$

$$l_{ij} = a_{ij} - \sum_{m=1}^{j-1} l_{im}u_{mj} \quad \text{for} \begin{bmatrix} j = 2, 3, \ldots, n \\ i = j, j+1, \ldots, n \end{bmatrix} \tag{6.50c}$$

$$u_{ij} = \frac{a_{ij} - \sum_{m=1}^{i-1} l_{im}u_{mj}}{l_{ii}} \quad \text{for} \begin{bmatrix} i = 2, 3, \ldots, n \\ j = i+1, i+2, \ldots, n+1 \end{bmatrix} \tag{6.50d}$$

We can easily verify that these general equations yield the elements of the two matrices, given by Equations 6.49, by employing the corresponding values of i and j. Also, note from the above relations that if a_{11} and l_{ii} are nonzero, the two matrices exist and can be uniquely determined. As shown earlier, the diagonal elements l_{ii} are all nonzero if matrix A is nonsingular.

It is evident from the recursive formulas given for l_{ij} and u_{ij}, in Equation 6.50, that the first column of L and the first row of U are determined first. Then the second column of L is determined from the third equation, and the second row of U from the fourth equation. We then proceed to the third column of L and the third row of U, and continue this process until both matrices are determined. The unknowns $x_1, x_2, \ldots, x_n$ are determined by back-substitution, as before, from

$$
\begin{aligned}
x_n &= u_{n,n+1} \\
x_i &= u_{i,n+1} - \sum_{j=i+1}^{n} u_{ij} x_j \quad \text{for } i = n-1, n-2, \ldots, 1
\end{aligned}
\tag{6.51}
$$

The main advantage of compact methods, such as Crout's method outlined above, is that a smaller number of arithmetic operations are needed, as compared to those for the Gaussian and the Gauss–Jordan elimination methods. Besides requiring less computing time, it also results in smaller round-off error. The algebra for determining l_{ij} and u_{ij} may be carried out in double precision for greater accuracy and then rounded off to single precision to reduce computer storage. This limited use of double precision is not possible in regular elimination methods, which would then require all operations and storage to be done in double precision. Since several elements in the L and U matrices are 1 or 0, considerable reduction in computer storage may be accomplished by storing both the matrices in the storage locations for the original augmented matrix. Then both l_{ij} and u_{ij} are termed a_{ij}, and the general equations may be suitably modified. The first equation, Equation 6.50a, is automatically satisfied. For the other elements, the old a_{ij} values are replaced by new ones, as computation proceeds. Figure 6.9 shows the algorithm for Crout's method, in terms of a flow chart.

Other methods based on matrix decomposition and factorization have been developed. For symmetric matrices, which often arise in many engineering problems such as those related to the analysis of structures, an important method for factorization is *Cholesky's* method. This method is based on finding a lower triangular matrix L such that

$$
A = LL^{\mathrm{T}} \tag{6.52}
$$

where L^{T} is the transpose of the matrix L. This factorization is possible for matrices that are symmetric and positive definite, a necessary and usually sufficient condition for which is that the eigenvalues of the matrix (Section 6.7) be positive. These properties of matrices are discussed in most books on matrices, such as Reiner (1971) and Bronson and Costa (2008). Once L has been determined, one obtains the solution x_i

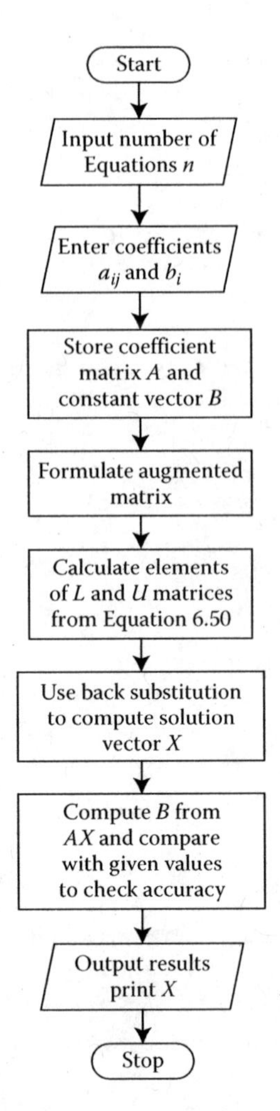

FIGURE 6.9 Flow chart for the use of Crout's method for solving a system of linear equations.

by computing the first unknown x_1 directly from the resulting linear equation and the remaining by substitution of the computed values of the preceding unknowns into the reduced equations, with increasing i. The Cholesky decomposition requires about $(1/6)n^3$ operations, instead of $(1/3)n^3$ needed for the Gaussian elimination.

Thus, matrix decomposition methods for solving systems of linear equations are very efficient. However, the computer programming is generally much more involved than the elimination methods, such as Gaussian or Gauss–Jordan elimination. Consequently, engineers frequently use available programs in engineering applications, rather than develop the necessary software, for methods such as those discussed here. However, following the approach discussed earlier for elimination methods, one may also write computer programs for these methods without too much difficulty. For further details on these methods, refer to Carnahan et al. (1969), James et al. (1985), and Atkinson (1989), listed among the references at the end of this book.

6.5 NUMERICAL SOLUTION OF LINEAR SYSTEMS BY MATRIX INVERSION

In the methods discussed so far, we solved the system of linear equations, given by $AX = B$, by applying various elimination procedures directly to the given system, without finding the inverse A^{-1} of the coefficient matrix. However, if $\text{Det}(A) \neq 0$, the inverse A^{-1} exists, and the solution of the set of linear equations may be obtained as

$$X = A^{-1}B \tag{6.5}$$

If a given set of equations is to be solved, it is generally advantageous to solve the system directly, without computing the inverse matrix A^{-1}. However, as mentioned earlier, the matrix itself may be needed in the problem in order for us to study the behavior of the mathematical or physical system. Also, if several sets of equations with the same coefficient matrix A but different constant vectors B are to be solved, it is often more efficient to compute A^{-1} and to employ it with the different constant vectors B to obtain the corresponding solutions from Equation 6.5. Another important consideration is that many computer systems have programs available for matrix inversion. These prepared programs may often be employed to solve systems of linear equations.

In MATLAB, the inverse of a given matrix a is obtained by the command

$$c = \text{inv}(a) \tag{6.53a}$$

Then the solution to the given system of linear equations, $ax = b$ is obtained by

$$x = c * b \tag{6.53b}$$

or, simply,

$$x = \text{inv}(a) * b \tag{6.53c}$$

Again, this approach may be applied to the examples presented earlier and results essentially identical to those presented earlier are obtained.

6.5.1 Computational Procedure

In view of the preceding discussion, it is evident that matrix inversion, as an intermediate step in solving linear systems, may be desirable in some cases. To invert a square matrix, such as the coefficient matrix A, we use the following definition of the inverse A^{-1}

$$AA^{-1} = I \tag{6.54}$$

where I is the identity or unit matrix. Therefore, the inverse may be obtained by solving the equation

$$AX = I \tag{6.55}$$

where X assumes the role of the column vector of unknowns and I assumes that of the constant vector employed earlier. This equation may be solved by applying methods such as Gaussian elimination and Gauss–Jordan elimination. Then the matrix of the unknowns yields the inverse A^{-1}. Matrix inversion requires about $(4/3)n^3$ arithmetic operations, while only $(1/3)n^3$ are needed for directly solving a set of linear equations by Gaussian elimination. Gauss–Jordan elimination is particularly suitable for matrix inversion, since it transforms the matrix A into the identity matrix I, which also constitutes the right-hand side of Equation 6.54, and the inverse A^{-1} is obtained directly. This method is outlined below.

The augmented matrix C for Equation 6.55 is obtained as follows:

$$C = \begin{bmatrix} a_{11} & a_{12} & \cdots & a_{1n} & 1 & 0 & \cdots & \cdots & 0 \\ a_{21} & a_{22} & \cdots & a_{2n} & 0 & 1 & 0 & \cdots & 0 \\ \vdots & & & & & & & & \vdots \\ a_{n1} & a_{n2} & \cdots & a_{nn} & 0 & 0 & \cdots & 0 & 1 \end{bmatrix} \tag{6.56}$$

Now, if Gauss–Jordan elimination is applied to this matrix, using Equation 6.36, until the a's are replaced by the elements of an identity matrix, the identity matrix in the augmented matrix above is transformed into the inverse A^{-1}. For illustration, let us consider the system of equations given by Equation 6.33. The augmented matrix C for matrix inversion is

$$C = \begin{bmatrix} 2 & 1 & 3 & 1 & 0 & 0 \\ 3 & -4 & 4 & 0 & 1 & 0 \\ 1 & 4 & -1 & 0 & 0 & 1 \end{bmatrix}$$

As before, the first row is divided by the pivot element, which is 2. Then it is multiplied by 3 and subtracted from the second row to yield the second row of the reduced matrix. The new third row is similarly obtained by subtraction of the normalized first row from the third row. This process is continued, using Equation 6.36, to yield the following matrices during reduction:

$$\begin{bmatrix} 1 & \frac{1}{2} & \frac{3}{2} & \frac{1}{2} & 0 & 0 \\ 0 & -\frac{11}{2} & -\frac{1}{2} & -\frac{3}{2} & 1 & 0 \\ 0 & \frac{7}{2} & -\frac{5}{2} & -\frac{1}{2} & 0 & 1 \end{bmatrix}, \begin{bmatrix} 1 & 0 & \frac{16}{11} & \frac{4}{11} & \frac{1}{11} & 0 \\ 0 & 1 & \frac{1}{11} & \frac{3}{11} & -\frac{2}{11} & 0 \\ 0 & 0 & -\frac{31}{11} & -\frac{16}{11} & \frac{7}{11} & 1 \end{bmatrix},$$

$$\text{and} \begin{bmatrix} 1 & 0 & 0 & -\frac{12}{31} & \frac{13}{31} & \frac{16}{31} \\ 0 & 1 & 0 & \frac{7}{31} & -\frac{5}{31} & \frac{1}{31} \\ 0 & 0 & 1 & \frac{16}{31} & -\frac{7}{31} & -\frac{11}{31} \end{bmatrix}$$

As discussed before, we must employ partial pivoting, or row interchange, to avoid a zero pivot element or to increase the accuracy, by considering the pivot row and all the rows below it at each step and exchanging the rows to employ one with the largest pivot element as the pivot row for the elimination process.

Once the coefficient matrix has been transformed into I, the original identity matrix should become A^{-1}. Therefore,

$$A^{-1} = \begin{bmatrix} -\frac{12}{31} & \frac{13}{31} & \frac{16}{31} \\ \frac{7}{31} & \frac{-5}{31} & \frac{1}{31} \\ \frac{16}{31} & -\frac{7}{31} & -\frac{11}{31} \end{bmatrix}$$

We can easily verify that the above is true by multiplying the original matrix A by this matrix to obtain I.

$$\begin{bmatrix} 2 & 1 & 3 \\ 3 & -4 & 4 \\ 1 & 4 & -1 \end{bmatrix} \begin{bmatrix} -\frac{12}{31} & \frac{13}{31} & \frac{16}{31} \\ \frac{7}{31} & -\frac{5}{31} & \frac{1}{31} \\ \frac{16}{31} & -\frac{7}{31} & -\frac{11}{31} \end{bmatrix} = \begin{bmatrix} 1 & 0 & 0 \\ 0 & 1 & 0 \\ 0 & 0 & 1 \end{bmatrix} = I$$

The solution vector X may now be obtained by applying Equation 6.5. Therefore,

$$X = A^{-1}B = [1,\ 1,\ 2]$$

Similarly, the solution of sets of equations with the same A but different B may be obtained easily once A^{-1} has been determined. For instance, if B is given as [11, 10, 4], X is computed as

$$X = A^{-1}B = [2,\ 1,\ 2]$$

6.5.2 Additional Considerations

Partial pivoting is generally incorporated in the program for matrix inversion. Besides avoiding problems with a zero or relatively small pivot element, it improves the accuracy of the computational results obtained. Complete pivoting, with both row and column interchanges, may also be employed to obtain the largest pivot element at each step and thus increase the accuracy. Another improvement in matrix inversion by Gauss–Jordan elimination is obtained by storing the inverted matrix in the same location as the original matrix. The identity matrix is seldom stored, although its transformed columns, which finally give the inverse, are stored in place of the columns in the coefficient matrix, that have been reduced to the diagonal form by Gauss–Jordan elimination. Most commercially available computer programs for matrix inversion incorporate these features for accuracy and reduction in storage.

If the storage-saving feature, outlined above, is employed, the general equations for matrix inversion may be obtained from Equation 6.36. Since the transformed elements of the identity matrix are stored in place of the diagonalized columns, as the inversion proceeds, we obtain

$$
\begin{aligned}
a'_{mj} &= \frac{a_{mj}}{a_{mm}} && \text{for } j = 1,2,\ldots,n \text{ and } j \neq m \\
a'_{ij} &= a_{ij} - a'_{mj}a_{im} && \text{for } \begin{bmatrix} i = 1,2,\ldots,n \text{ and } i \neq m \\ j = 1,2,\ldots,n \text{ for each } i \text{ and } j \neq m \end{bmatrix} \\
a'_{mm} &= \frac{1}{a_{mm}} \\
a'_{im} &= -a_{im}a'_{mm} && \text{for } i = 1,2,\ldots,n \text{ and } i \neq m
\end{aligned}
\tag{6.57}
$$

where the prime denotes the new elements which replace the old ones after each cycle. The first two equations are the same as those given earlier in Equation 6.36. The last two are obtained from the transformation of an appended identity matrix, whose elements are zero everywhere except at the diagonal, where they are unity. If partial pivoting is used, without storing the appended matrix, the matrix obtained after the reduction process must be reordered, in the same sequence as the row interchanges, in order to obtain the inverse of the original matrix. However, each row interchange during the inversion corresponds to a column interchange in the identity matrix. Therefore, in the reordering of the final matrix, column interchanges are performed corresponding to each row interchange in the computation process and in the reverse sequence, starting with the last row interchange; see James et al. (1985).

Example 6.4

Solve the equations obtained in Example 6.3 for the six currents in the electrical network of Figure 6.7a by matrix inversion.

SOLUTION

The Gauss–Jordan elimination method may be used for inverting the coefficient matrix A in the system of equations $AX = B$. Then the unknown X is obtained from the inverse of the matrix A^{-1} as $X = A^{-1}B$. The augmented matrix consists of the coefficient matrix A with an identity matrix appended to it. Gauss–Jordan elimination is applied to the coefficient matrix so that it is reduced to an identity matrix. When this is accomplished, the original identity matrix is transformed into the inverse of the matrix A^{-1}, since this amounts to solving the equation $AY = I$, where Y is the unknown matrix inverse.

The MATLAB function *m*-file given in Appendix B.8 for solving a system of equations by the Gauss–Jordan method may easily be modified to solve the given system by matrix inversion, as shown in Appendix B.9. The augmented matrix is formed from the given coefficient matrix a and the identity matrix I, and Gauss–Jordan elimination is performed on the augmented matrix to reduce the $n \times n$ coefficient matrix to an identity matrix, where n is the number of equations. The inverse of the matrix may be printed and the unknowns computed from the equation $x = a^{-1}b$. In the function *m*-file, the matrix multiplication command in MATLAB is used, though commands may be written to achieve this as well without involving the multiplication software. The computed values, shown in Figure 6.10 are identical to those obtained in Example 6.3. The matrix a may be multiplied with its inverse to check whether the identity matrix is obtained. From the results obtained, it is found that $A\,A^{-1}$ was very close to the identity matrix I. It can also be confirmed that the use of Equation 6.53c, in a MATLAB environment, yields the same results.

```
THE INVERSE OF THE MATRIX
 0.1243  -0.0081   0.1622  0.0311   0.0324   0.0892
 0.3432   0.3689   0.1216  0.0858   0.0243  -0.0581
-0.5324   0.3608   0.2838  0.1169   0.0568   0.0311
 0.2405  -0.3635  -0.2297  0.0601  -0.0459  -0.0014
 0.1027  -0.2676   0.3514  0.0257   0.0703  -0.0568
 0.2270  -0.2757  -0.4865  0.0568   0.1027   0.0324

THE SOLUTION TO THE EQUATIONS
 0.0514
 0.6635
 0.7149
 0.9689
-0.3054
-0.2541
```

FIGURE 6.10 Computed inverse of the coefficient matrix and the solution to the equations in Example 6.4, using matrix inversion.

6.6 ITERATIVE METHODS

In the preceding sections, we have discussed direct methods for solving a set of simultaneous linear equations. These methods are appropriate for a small number of equations, typically fewer than 20, if the coefficient matrix A is dense. The main limitation arises from the round-off error, which is incurred in each computation and affects the overall accuracy of the solution. The tridiagonal system is a special case for which the computation effort involved is much smaller than that for a general matrix. Thus, for the tridiagonal case, many more equations may be solved while the desired accuracy level is preserved. Direct methods provide the solution in a finite number of steps, and, except for the round-off error, the solution is exact. As seen earlier, MATLAB is particularly convenient for the direct solution of linear systems. However, for a large number of equations, typically on the order of several hundred, iterative methods, which start with an assumed solution and iterate to the desired solution of the system of equations, within a specified convergence criterion, are often more efficient.

Large sets of linear equations are generally *sparse*, and iterative methods, which consider only the nonzero coefficients in the computation, use this sparseness advantageously. Moreover, the round-off error after each iteration simply results in a less accurate input for the next iteration. Therefore, the resulting round-off error in the numerical solution is only what arises in the computation for the final iteration. The error does not accumulate as in direct methods. However, the solution is not exact but is obtained to an arbitrary, specified, convergence criterion.

6.6.1 BASIC APPROACH

Let us consider the set of linear equations given by Equation 6.2. These equations may be rewritten, by solving for the unknowns x_i; as follows:

$$
\begin{aligned}
x_1 &= \frac{b_1 - a_{12}x_2 - a_{13}x_3 - \cdots - a_{1n}x_n}{a_{11}} \\
x_2 &= \frac{b_2 - a_{21}x_1 - a_{23}x_3 - \cdots - a_{2n}x_n}{a_{22}} \\
&\vdots \\
x_n &= \frac{b_n - a_{n1}x_1 - a_{n2}x_2 - \cdots - a_{n,n-1}x_{n-1}}{a_{nn}}
\end{aligned}
\tag{6.58}
$$

This system may be written more concisely as

$$
x_i = \frac{b_i - \sum_{j=1,j\neq i}^{n} a_{ij}x_j}{a_{ii}} \quad \text{for } i = 1, 2, \ldots, n \tag{6.59}
$$

We need initial guesses for the unknowns to start the iterative process in the above equations. If $x_1^{(0)}, x_2^{(0)}, \ldots, x_i^{(0)}, \ldots, x_n^{(0)}$ are taken as the initial values, the value of x_1 after the first iteration, $x_1^{(1)}$, is obtained from

$$x_1^{(1)} = \frac{b_1 - a_{12}x_2^{(0)} - a_{13}x_3^{(0)} - \cdots - a_{1n}x_n^{(0)}}{a_{11}}$$

Similarly,

$$x_i^{(1)} = \frac{b_i - \sum_{j=1, j \neq i}^{n} a_{ij}x_j^{(0)}}{a_{ii}} \quad \text{for } i = 1, 2, \ldots, n \tag{6.60}$$

The values obtained after the first iteration are then used for the next iteration. Thus, this iterative process may be written as

$$x_i^{(l+1)} = \frac{b_i - \sum_{j=1, j \neq i}^{n} a_{ij}x_j^{(l)}}{a_{ii}} \quad \text{for } i = 1, 2, \ldots, n \tag{6.61}$$

The superscript indicates the number of the iteration. This equation is also often written as

$$x_i^{(l+1)} = F_i[x_1^{(l)}, x_2^{(l)}, \ldots, x_{i-1}^{(l)}, x_{i+1}^{(l)}, \ldots, x_n^{(l)}] \tag{6.62}$$

where the function F_i is obtained from Equation 6.61 and represents the relationship between an unknown x_i and the other unknowns. The value of the unknown x_i after l iterations, $x_i^{(l)}$, does not appear on the right-hand side for linear equations. However, a term containing $x_i^{(l)}$ may be added in order to alter the convergence characteristics, as discussed in Section 6.6.5.

6.6.2 Jacobi and Gauss–Seidel Methods

The formulation for iteration given in Equation 6.62 is known as the *Jacobi iterative method.* To compute the values for a given iteration step, it employs the values from the previous iteration. Therefore, all the values are computed, using previous values, before any unknown is updated. This implies that computer storage is needed for the present iteration as well as for the previous one. For single-processor, or serial, computers, a considerable improvement in the rate of convergence and in the storage requirements can be obtained by replacing the values from the previous iteration by new ones as soon as they are computed. Then only the values of the latest iteration are stored, and each iterative computation of the unknown employs the most recent values of the other unknowns. This computational scheme, known as the *Gauss–Seidel method*, is used extensively for solving large systems of equations that frequently arise in the numerical solution of differential equations.

Let us consider the use of the Gauss–Seidel method for computing the iterative values of the unknowns, starting with x_1 and then successively obtaining $x_2, x_3, \ldots, x_n$.

Then the second iteration for, say, x_3, is obtained from

$$x_3^{(2)} = \frac{b_3 - a_{31}x_1^{(2)} - a_{32}x_2^{(2)} - a_{34}x_4^{(1)} - \cdots - a_{3n}x_n^{(1)}}{a_{33}}$$

Here, the values of x_1 and x_2 are known after the second iteration, and the others are known only after the first iteration. Similarly, the $(l + 1)$th iteration for x_i may be written as

$$x_i^{(l+1)} = \frac{b_i - \sum_{j=1}^{i-1} a_{ij}x_j^{(l+1)} - \sum_{j=i+1}^{n} a_{ij}x_j^{(l)}}{a_{ii}} \quad \text{for } i = 1,2,\ldots,n \tag{6.63}$$

or

$$x_i^{(l+1)} = F_i[x_1^{(l+1)}, x_2^{(l+1)}, \ldots, x_{i-1}^{(l+1)}, x_{i+1}^{(l)}, \ldots, x_n^{(l)}] \tag{6.64}$$

This formulation, therefore, assumes that the computation of the unknowns x_i starts with x_1 and proceeds with increasing i until all the values are obtained for a given iteration. The Gauss–Seidel method repeatedly calculates the unknowns, replacing the values from the previous iteration by new ones and thus requiring only one computer storage space for each unknown. Programming is also simplified since the most recent value of each unknown is always employed in the computations. This iterative process will converge to the solution vector if the equations have certain characteristics, as discussed below. A better initial guess of the unknowns will also lead to faster convergence, if the process is convergent.

6.6.3 Convergence

The iterative computation of the unknowns is terminated when a specified convergence criterion is satisfied. Generally, if the change in the value of each unknown from one iteration to the next is less than a given small quantity ε, convergence is assumed to have been achieved. The convergence criterion ε may be applied to the physical value of the unknown or to its normalized value. Therefore, the condition for convergence may be written as

$$|x_i^{(l+1)} - x_i^{(l)}| \le \varepsilon \quad \text{for } i = 1,2,\ldots,n \tag{6.65a}$$

or

$$\left|\frac{x_i^{(l+1)} - x_i^{(l)}}{x_i^{(l)}}\right| \le \varepsilon \quad \text{for } i = 1,2,\ldots,n \tag{6.65b}$$

The second form of the convergence criterion is more appropriate if an estimate of the magnitude of the unknown x_i is not available and none of the unknowns is expected to be zero. The choice of ε is arbitrary and may be taken as around 10^{-4} for the second form of the criterion, (Equation 6.65b), which specifies the maximum fractional change in each unknown from one iteration to the next. However, as discussed in Chapter 2, the dependence of the solution on the convergence criterion must be studied by varying ε so that the numerical solution obtained is essentially independent of the value chosen. The convergence criterion may also be applied to a few important unknowns, instead of all x_i, in order to reduce the computing time. Similarly, it may be applied to the sum of the absolute values or of the squares of the changes in all the unknowns between two successive iterations. With large systems of equations, such alternative forms of the convergence criterion are often employed to save computer time.

The conditions for convergence of the iterative process have been analyzed for the Jacobi and the Gauss–Seidel methods and presented in terms of the nature of the coefficient matrix A. Both of these methods have good convergence characteristics for diagonally dominant systems, that is, for systems in which each diagonal element a_{ii} is larger, in absolute value, than the sum of the magnitudes of the other elements in the row. Thus, if

$$|a_{ii}| > \sum_{j=1, j\neq i}^{n} |a_{ij}| \tag{6.66}$$

the system is said to be *diagonally dominant*, and convergence is guaranteed for linear systems. However, convergence is often obtained with weaker diagonal dominance. These methods are particularly useful in the solution of large systems of linear equations that arise in the numerical solution of PDEs by finite difference or finite element methods. The equations obtained in these cases usually have diagonal dominance, or conditions close to it, and the above iterative methods are convergent. With some modifications, these methods may also be employed for nonlinear equations, as discussed in Section 6.8.

6.6.4 An Example

In order to illustrate the Gauss–Seidel method, let us consider the following set of linear equations:

$$\begin{aligned} 5x + y + 2z &= 17 \\ x + 3y + z &= 8 \\ 2x + y + 6z &= 23 \end{aligned} \tag{6.67}$$

This system is diagonally dominant, since the dominant coefficient in each equation is the diagonal element, which is also larger than the sum of the absolute values of the other coefficients. Therefore, the Gauss–Seidel iteration is convergent for these equations. The above equations are rewritten as

$$x = \frac{17 - y - 2z}{5}, \quad y = \frac{8 - x - z}{3}, \quad z = \frac{23 - 2x - y}{6}$$

For the Gauss–Seidel method, the most recent values of x, y, and z are to be used in the iteration. If the starting values are arbitrarily chosen as $x = 1$, $y = 1$, and $z = 1$, the values for the first iteration are computed, by rounding off to three decimal digits, as follows:

$$x^{(1)} = \frac{17 - 1 - 2}{5} = 2.8$$

$$y^{(1)} = \frac{8 - 2.8 - 1}{3} = 1.4$$

$$z^{(1)} = \frac{23 - 5.6 - 1.4}{6} = 2.667$$

Similarly, the next four iterations are obtained as

$$x^{(2)} = 2.053, \quad y^{(2)} = 1.093, \quad z^{(2)} = 2.967$$
$$x^{(3)} = 1.995, \quad y^{(3)} = 1.013, \quad z^{(3)} = 3.000$$
$$x^{(4)} = 1.997, \quad y^{(4)} = 1.001, \quad z^{(4)} = 3.001$$
$$x^{(5)} = 2.000, \quad y^{(5)} = 1.000, \quad z^{(5)} = 3.000$$

The exact solution of the above system of equations is $x = 2$, $y = 1$, and $z = 3$. Therefore, the iterative procedure converges rapidly to yield a solution, that is, within 0.2% of the exact solution in only four iterations. The process will terminate after four iterations if ε in Equation 6.65b is taken as 0.02, and after five iterations if it is chosen as 0.002. Additional iterations may be needed for a still smaller value of ε, since changes in the fourth and higher decimal places may occur from one iteration to the next. If the Jacobi method is applied to the above system, the rate of convergence is much slower. Therefore, the Jacobi method is seldom used on traditional or single-processor computers and is considered largely in order to study the convergence characteristics of other iterative methods in terms of those of the Jacobi method. However, for multiprocessor, or parallel, computers, the Jacobi method is often more convenient and has faster convergence since the previous iteration is used in the computations and all the equations may be considered independent of the others.

6.6.5 Relaxation Methods

The convergence characteristics of the Gauss–Seidel method can often be considerably improved by the use of point relaxation, which is given by

$$x_i^{(l+1)} = \omega[x_i^{(l+1)}]_{GS} + (1 - \omega)x_i^{(l)} \tag{6.68}$$

where ω is a constant in the range $0 < \omega < 2$ and $[x_i^{(l+1)}]_{GS}$ is the value of x_i obtained for the $(l + 1)$th iteration by using the Gauss–Seidel iteration equation, Equation 6.63. For $\omega > 2$, the process is divergent. If $0 < \omega < 1$, the iterative scheme is known as *successive under-relaxation* (SUR), and if $1 < \omega < 2$, the scheme is termed *successive over-relaxation* (SOR). In the former case, the value for the $(l+1)$th iteration is a weighted average of the value from the previous iteration and that obtained by the use of the Gauss–Seidel method for the present iteration. In SOR, the change in x_i from one iteration to the next, in the Gauss–Seidel scheme, is multiplied by a factor between 1.0 and 2.0 to accelerate convergence. At the optimum value of the relaxation factor, ω_{opt}, the convergence is much faster than that for Gauss–Seidel. The relaxation method may be written, using Equation 6.63, as

$$x_i^{(l+1)} = \frac{\omega\left[b_i - \sum_{j=1}^{i-1} a_{ij}x_j^{(l+1)} - \sum_{j=i+1}^{n} a_{ij}x_j^{(l)}\right]}{a_{ii}} + (1-\omega)x_i^{(l)} \quad \text{for } i = 1, 2, \ldots, n \quad (6.69)$$

It is obvious that the Gauss–Seidel method is obtained for $\omega = 1$. In this case, $x_i^{(l)}$ drops out from the right-hand side. SUR is generally used for nonlinear equations and for systems that result in a divergent Gauss–Seidel iteration. SOR is widely used for accelerating the convergence in linear systems. However, the determination of an optimum value of the relaxation factor ω is often difficult and is generally done by trial and error. For some systems, it may be available from earlier studies or from analysis. If several similar systems are to be solved, it is generally worthwhile to obtain the optimum value of ω by trying various values, over the given range, and then use it in the computations. For further details on the use of point relaxation in engineering applications, advanced books on the numerical solution of differential equations, such as Ferziger (1998) and Jaluria and Torrance (2003), may be consulted. The following example illustrates the use of Gauss–Seidel and SOR methods for solving a system of linear equations.

Example 6.5

Solve the problem discussed in Example 6.2 by means of the Gauss–Seidel iterative procedure. Then modify the computer program to solve the problem by the SOR method. Vary the relaxation factor ω to study the effect of its value on the number of iterations needed for convergence.

SOLUTION

The system of equations to be solved is

$$-T_{i+1} + [2 + G(\Delta x)^2]^2 T_i - T_{i-1} = 0 \quad \text{for } i = 1, 2, \ldots, 29$$

with

$$T_0 = T_{30} = 100$$

The given system is rewritten as

$$T_i = \frac{T_{i+1} + T_{i-1}}{2 + S} \quad \text{for } i = 1, 2, \ldots, 29$$

where

$$S = G(\Delta x)^2 = (0.071)^2\ (1.0)^2$$

Since T_0 and T_{30} are given as 100°C, the equations for T_1 and T_{29} become

$$T_1 = \frac{T_2 + 100}{2 + S}$$

$$T_{29} = \frac{100 + T_{28}}{2 + S}$$

Therefore, the resulting system of equations can be solved by the Gauss–Seidel method to obtain the required temperature distribution.

The initial guess, or starting temperature distribution, is taken as $T(I) = 0$, for $I = 1, 2, \ldots, 29$, where $T(I)$ denotes T_i. Using the preceding equations, the temperatures for the next iteration are computed and compared with the previous values to check for convergence. It is demanded that the absolute value of the difference between the two be less than the convergence criterion ε, that is, for convergence,

$$|\,(TO)_i - T_i\,| \leq \varepsilon \quad \text{for } i = 1, 2, \ldots, 29$$

where T represents the new values and TO the previous ones. If this difference for any value of i is greater than ε, the iterative process is repeated, taking the computed new values as the starting values for the next iteration. Once convergence has been achieved, we obtain the physical temperature TP by adding the ambient temperature of 20°C to the temperature difference T. Appendix B.10 gives the MATLAB program for this example, with $x(i)$ representing the unknowns, for $i = 1, 2, \ldots, 6$, *xold* (i) the previous iterative values, *tp* the physical temperatures and k the iteration number. This program can be used to solve a given system of linear equations by the Gauss–Seidel method. In the MATLAB environment, the programming is particularly simple since *x*, *xold* and *tp* are vectors representing the temperature differences, previous values and physical temperatures. Thus, algebra can be used directly on these vectors to apply the algorithm.

The resulting numerical results in terms of the physical temperatures are shown in Figure 6.11 for three values of ε, 10^{-3}, 10^{-4}, and 10^{-5}. Only small differences in the computed values are observed in going from the smallest to the largest value. The total number of iterations increases from 454 to 741. A comparison with the results obtained in Example 6.2 for this tridiagonal system also indicates a small difference in the temperatures for the smallest value of ε. Thus, a value of 10^{-3} for ε may be used for this problem.

The computer program for the Gauss–Seidel method is clearly much simpler than that for Gaussian elimination in Example 6.1. If the system is tridiagonal, Gaussian elimination is preferable, since it takes less computer time and is generally more accurate. However, the Gauss–Seidel method is advantageous to use when the coefficient matrix is sparse, although not tridiagonal. Appendix B.11 gives a general MATLAB program for the Gauss–Seidel method as a function *m*-file. The inputs needed for invoking this function file are the coefficient matrix

$\varepsilon = 10^{-3}$	$\varepsilon = 10^{-4}$	$\varepsilon = 10^{-5}$
No. of iterations = 454	No. of iterations = 598	No. of iterations = 741
The Solution is:	The Solution is:	The Solution is:
120.0000	120.0000	120.0000
114.6502	114.6567	114.6573
109.7778	109.7905	109.7918
105.3584	105.3771	105.3789
101.3696	101.3940	101.3964
97.7916	97.8213	97.8243
94.6063	94.6410	94.6445
91.7978	91.8370	91.8409
89.3520	89.3952	89.3995
87.2567	87.3034	87.3080
85.5013	85.5509	85.5558
84.0770	84.1289	84.1340
82.9767	83.0303	83.0356
82.1949	82.2495	82.2549
81.7275	81.7826	81.7881
81.5723	81.6273	81.6328
81.7285	81.7827	81.7881
82.1968	82.2497	82.2549
82.9795	83.0305	83.0356
84.0806	84.1292	84.1341
85.5055	85.5513	85.5558
87.2614	87.3038	87.3081
89.3571	89.3957	89.3996
91.8030	91.8375	91.8410
94.6115	94.6415	94.6445
97.7964	97.8218	97.8243
101.3740	101.3945	101.3965
105.3620	105.3774	105.3790
109.7805	109.7908	109.7918
114.6517	114.6568	114.6573
120.0000	120.0000	120.0000

FIGURE 6.11 Computed physical temperatures at three values, 10^{-3}, 10^{-4}, and 10^{-5}, of the convergence parameter ε.

a, the constant vector *b*, the initial guess vector *p*, the convergence parameter *ep* and the maximum number of iterations *max1*. Though this function can be used to solve a system of linear equations by applying the Gauss–Seidel method, it does not use the sparseness of the matrix *a* as effectively as the program in Appendix B.10, which specifies only the non-zero elements of the matrix.

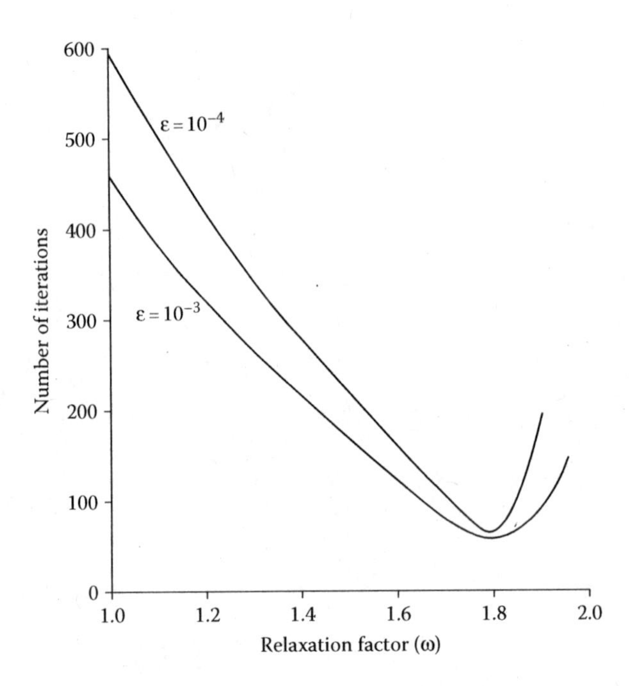

FIGURE 6.12 Variation of the number of iterations needed for convergence with the relaxation factor ω, in the solution of Example 6.5 by the SOR method.

The computer program given in Appendix B.10 can easily be modified to apply the SOR method. The relaxation factor ω must be entered as a parameter, and the recursion formula becomes

$$T_i^{(l+1)} = \omega\left[\frac{T_{i+1}^{(l)} + T_{i-1}^{(l+1)}}{2+S}\right] + (1-\omega)T_i^{(l)} \quad \text{for } i = 1, 2, \ldots, 29 \tag{6.70}$$

This equation replaces the one given earlier for the Gauss–Seidel scheme. The modified program was employed for ω varying from 1.0 to 2.0. Figure 6.12 shows the dependence of the number of iterations on the relaxation factor ω, at $\varepsilon = 10^{-3}$ and 10^{-4}. Note that the number of iterations at the optimum value, ω_{opt}, is almost one-tenth that for Gauss–Seidel iteration. Clearly, SOR is a very efficient method if the optimum value of the relaxation factor is known. Appendix C.8 gives the Fortran computer for employing the SOR method for solving the system of linear equations in this example. The similarity between the MATLAB and Fortran programs for this problem is evident, though the advantages of MATLAB in matrices and arrays, as well as in plotting, make it much easier to work in a MATLAB environment. Also, the logic presented in these programs may be employed for other high-level languages and computational environments.

6.7 HOMOGENEOUS LINEAR EQUATIONS

In many problems of engineering interest, such as those encountered in vibrating systems, stability analysis, and electrical circuits with alternating currents, the constant vector B in the system of linear equations is zero, giving rise to a set of equations of the form $AX = 0$. The system of equations is then said to be *homogeneous*. A trivial solution, $X = 0$, exists for this system. However, nontrivial solutions may be obtained only if the determinant of the coefficient matrix A is zero, that is, Det $A = 0$. This occurs when all the equations of the set are not linearly independent, and one or more equations may be obtained from a linear combination of the others.

In considering simultaneous nonhomogeneous linear equations, we noted from Cramer's rule that unique solutions may be obtained only if the determinant, Det A, is nonzero. However, in simultaneous homogeneous equations, the numerators in the solution by Cramer's rule, given in Equation 6.4, are all zero, since the constant vector B is 0. Therefore, nontrivial solutions may exist only if the denominator, which is Det A, is also zero. However, unique values of the unknowns $x_1, x_2, \ldots, x_n$ are not obtained in this case, since the solution vector X when multiplied by an arbitrary constant will also satisfy the system of homogeneous equations, $AX = 0$. Therefore, the desired solution establishes relationships between the unknowns, and the number of dependent equations in the set gives the number of unknowns that must be arbitrarily chosen to obtain the rest.

6.7.1 The Eigenvalue Problem

An important class of problems involving homogeneous equations is the eigenvalue problem, which is of considerable interest in engineering applications. Such problems occur, for instance, in the analysis of structures for critical buckling loads, in stress analysis for determining the principal normal stresses, and in the natural vibration of systems to determine the frequencies and the vibrational modes. The matrix equation for an eigenvalue problem is

$$(A - \lambda I)X = 0 \tag{6.71}$$

or

$$AX = \lambda X \tag{6.72}$$

where A is a known $n \times n$ matrix, X is the solution vector, and λ is an unknown constant. Nontrivial solutions to the above system of equations are obtained only for certain values of λ. These values are known as *eigenvalues* of the coefficient matrix A, and the solution vectors X corresponding to these eigenvalues are called the *eigenvectors*, which can be determined only to within a multiplicative constant. In a vibrating system, consisting of masses and springs, as shown in Figure 6.13, the eigenvalues are the squares of the natural frequencies of vibration, and the eigenvectors give the displacements of the masses. This problem is discussed in Example 6.6.

From Cramer's rule, it is evident that nontrivial solutions may be obtained only if

$$\text{Det}\,(A - \lambda I) = 0 \tag{6.73}$$

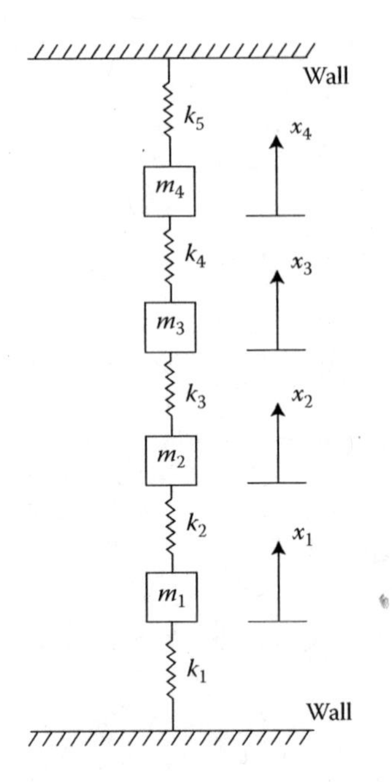

FIGURE 6.13 A vibrating system consisting of four masses, denoted by $m_1, \ldots, m_4$, and of five springs, whose spring constants are denoted by $k_1, \ldots, k_5$. The displacements are denoted by x_1, x_2, x_3, and x_4, giving a four-degrees-of-freedom system.

This may be written as

$$\mathrm{Det}\,(A-\lambda I)=\begin{bmatrix} (a_{11}-\lambda) & a_{12} & a_{13} & \cdots & a_{1n} \\ a_{21} & (a_{22}-\lambda) & a_{23} & \cdots & a_{2n} \\ \vdots & \vdots & & \vdots & a_{n-1,n} \\ a_{n1} & a_{n2} & a_{n3} & \cdots & (a_{nn}-\lambda) \end{bmatrix}=0 \qquad (6.74)$$

Expansion of this determinant results in a polynomial of order n in λ. This polynomial, which is known as the *characteristic polynomial* of matrix A, may be solved by the methods discussed in Chapter 5 to obtain the eigenvalues, see Example 5.8. Once the eigenvalues have been obtained, one can determine the eigenvector corresponding to each value by substituting the value in the given equations. If there is only one linearly dependent equation in the set of equations, one must assume the value of one unknown to obtain the corresponding values of the remaining unknowns. Similarly, if there are two dependent equations, the values of two unknowns need to be assumed, and so on. In many engineering problems, only one dependent equation arises, and, therefore, the value of only one unknown must be chosen. Textbooks on linear algebra, such as Williams (2004) and Anton (2010), may be consulted for further details on eigenvalue problems.

The preceding procedure of expanding the determinant to obtain the characteristic polynomial, which may then be solved for the eigenvalues, is computationally practical only for a small number of equations, typically up to four, and for sparse matrices. For somewhat larger systems, generally of the order of ten equations, one may obtain the characteristic polynomial by using the methods developed by Leverrier and by Faddeev; see Carnahan et al. (1969).

Since a polynomial with all real coefficients can have complex roots, complex eigenvalues may be obtained. However, in many physical problems, the coefficient matrix A is symmetric. It can be shown that all the eigenvalues of a symmetric matrix are real, which substantially simplifies the computational procedure. The solution of the eigenvalue problem for symmetric matrices is of considerable interest in engineering problems and is discussed in Section 6.7.3. It may be pointed out here that even though the characteristic polynomial may be generated, by the methods of Leverrier and Faddeev, for systems containing as many as 25 or 30 equations, the solution of the polynomial is generally very involved, and the other methods outlined here are preferred to the root solving procedures of Chapter 5.

The eigenvectors may be obtained by substitution of the eigenvalues, one at a time, into the given system of equations. The equations thus obtained may be solved by the use of the Gauss–Jordan method. The method is applied to the matrix $(A - \lambda I)$, as outlined in Section 6.3, and the process carried out until the reduced matrix is such that a further application of the method is not possible due to all possible pivot elements being zero. If the system contains only one dependent equation, then the process stops with only the last column left to be reduced. The other columns contain only zeros and one. At this stage, only the independent equations are left, and, if an arbitrary value is given to one unknown, the other unknowns may be computed from the resulting nonhomogeneous equations. Similarly, if two dependent equations are present in the given set, the Gauss–Jordan method yields two unreduced columns. This requires choosing arbitrary values for two unknowns to obtain two linearly independent eigenvectors. Column interchanges, besides row interchanges, are frequently employed in the process to avoid taking a pivot element, that is, zero. If a column interchange is employed, the components of the eigenvector corresponding to these columns must also be interchanged. Example 6.6 discusses a physical problem and the use of Gauss–Jordan elimination for determining the eigenvectors.

Example 6.6

For the natural vibration of the three masses, m, $2m$, and m, connected by the four springs shown in Figure 6.14, determine the characteristic polynomial, the eigenvalues, which correspond to the natural frequencies of vibration, and the

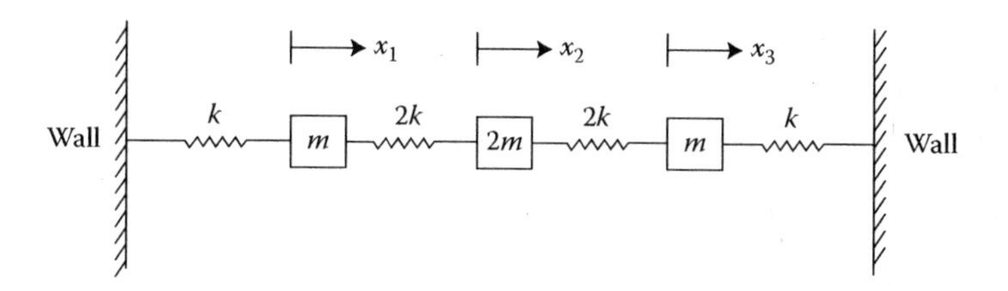

FIGURE 6.14 The vibrating mass and spring system considered in Example 6.6.

eigenvectors, which give the amplitudes of motion of the masses. The spring constants for the four springs are k, $2k$, $2k$, and k. The displacements of the three masses are defined by the coordinates x_1, x_2, and x_3, respectively, as shown. Take $k/m = 1.0$.

SOLUTION

The extension in the first spring, from the left, is x_1 and that in the second spring is $(x_2 - x_1)$. Since the inward directed force due to the extension is given by a product of the spring constant and the extension, the net force acting on the first mass in the positive x_1 direction is $[2k(x_2 - x_1) - kx_1]$. Therefore, from Newton's second law,

$$m\ddot{x}_1 = 2k(x_2 - x_1) - kx_1$$

$$m\ddot{x}_1 + kx_1 + 2k(x_1 - x_2) = 0 \tag{6.75a}$$

where $\ddot{x}_1$ is the second derivative of x_1 with respect to time t and is, thus, the acceleration of the mass. Similarly, for the other masses,

$$2m\ddot{x}_2 + 2k(x_2 - x_1) + 2k(x_2 - x_3) = 0 \tag{6.75b}$$

$$m\ddot{x}_3 + 2k(x_3 - x_2) + kx_3 = 0 \tag{6.75c}$$

From the theory of vibrations, the solution to the above equations may be taken as

$$\begin{aligned} x_1 &= X_1 \sin \omega t \\ x_2 &= X_2 \sin \omega t \\ x_3 &= X_3 \sin \omega t \end{aligned} \tag{6.76}$$

where X_1, X_2, and X_3 are the amplitudes of motion and ω is the natural frequency in radians/second. If these equations are substituted in the equations of motion, Equations 6.75, we obtain the following system of linear homogeneous equations for $k/m = 1.0$:

$$\begin{aligned} (3 - \omega^2)X_1 - 2X_2 &= 0 \\ -X_1 + (2 - \omega^2)X_2 - X_3 &= 0 \\ -2X_2 + (2 - \omega^2)X_3 &= 0 \end{aligned} \tag{6.77}$$

This system may be written in matrix form as

$$AX = \lambda X \tag{6.78}$$

where

$$\lambda = \omega^2 \tag{6.79}$$

and

$$A = \begin{bmatrix} 3 & -2 & 0 \\ -1 & 2 & -1 \\ 0 & -2 & 3 \end{bmatrix} \tag{6.80}$$

A nontrivial solution of Equation 6.77 can be obtained only if the determinant of the coefficient matrix $(A - \lambda)$ is zero. Therefore,

$$\mathrm{Det}(A - \lambda) = \begin{bmatrix} 3-\lambda & -2 & 0 \\ -1 & 2-\lambda & -1 \\ 0 & -2 & 3-\lambda \end{bmatrix} = 0 \tag{6.81}$$

We obtain the characteristic polynomial for this eigenvalue problem by expanding this determinant as

$$\lambda^3 + 8\lambda^2 - 17\lambda + 6 = 0 \tag{6.82a}$$

The roots of this polynomial equation may be determined by the root-solving methods given in Chapter 5. Using the search method to obtain the approximate location of the roots, followed by the Newton–Raphson method, we determine the eigenvalues, in s^{-2}, as follows:

$$\lambda_1 = 0.43845, \quad \lambda_2 = 3.0, \quad \lambda_3 = 4.56155 \tag{6.82b}$$

In MATLAB, the roots can be obtained simply by specifying the polynomial and invoking the *roots* command.

To determine the eigenvectors corresponding to these eigenvalues, we substitute each eigenvalue in Equation 6.77 and obtain the ratios of the amplitudes. Then, if one of the amplitudes is taken as 1.0, the others may be determined. From symmetry, $X_1 = X_3$ and both may be taken as 1.0. Then X_2 is determined for the three eigenvalues given in Equation 6.82b as, respectively,

$$X_2 = 1.28078, \quad X_2 = 0, \quad X_2 = -0.78078 \tag{6.83}$$

The problem may also be solved by applying the Gauss–Jordan method. Let us again consider the equations obtained for this problem for the first eigenvalue as

$$\begin{aligned} 2.56155X_1 - 2X_2 + 0.X_3 &= 0 \\ -X_1 + 1.56155X_2 - X_3 &= 0 \\ 0.X_1 - 2X_2 + 2.56155X_3 &= 0 \end{aligned} \tag{6.84}$$

If Gauss–Jordan elimination is applied to the coefficient matrix of these equations, we obtain, after the first two steps,

$$\begin{bmatrix} 1 & -0.78078 & 0 \\ 0 & 0.78078 & -1 \\ 0 & -2 & 2.56155 \end{bmatrix} \quad \text{and} \quad \begin{bmatrix} 1 & 0 & -1 \\ 0 & 1 & -1.28077 \\ 0 & 0 & 0 \end{bmatrix}$$

Therefore, further application of the Gauss–Jordan method is not possible. At this stage, two independent algebraic equations, containing the unknowns X_1, X_2, and X_3, are obtained from the first two rows of the reduced matrix. Therefore, if X_1 is taken as 1.0, X_3 is obtained as 1.0, and X_2 as 1.28077.

In many engineering problems, the set of n homogeneous equations contains $(n - 1)$ independent equations for determining the n components of the eigenvector. An application of Gauss–Jordan elimination, with partial and complete pivoting, if necessary, reduces the system to a set of independent equations at the stage where further reduction is not possible. If two dependent equations are present, one must assume two components in order to determine the rest, and so on. Therefore, small systems of equations may be solved by root-solving methods followed by Gauss–Jordan elimination to determine the eigenvectors. Example 6.6 also illustrates the solution of homogeneous ODEs, as considered again in Chapter 9.

6.7.2 The Power Method

The power method is a frequently employed iterative procedure for determining the eigenvalues and the corresponding eigenvectors, particularly if the largest or the smallest eigenvalue is of interest. Intermediate eigenvalues may also be determined by gradually eliminating the eigenvalues already found. However, the round-off errors accumulate, leading to lower accuracy, and the process becomes more involved as the intermediate eigenvalues are successively determined. Therefore, the method is well suited mainly for finding the largest and the smallest eigenvalues. It has the advantages of a simpler computational procedure, as compared to several other methods discussed in Section 6.7.3, and of providing the eigenvector along with the eigenvalue.

6.7.2.1 Largest Eigenvalue

Let us first consider the iterative power method for finding the largest eigenvalue of the system given by

$$AX = \lambda X \tag{6.72}$$

The method starts with an initial estimate of the eigenvector, denoted as $X^{(0)}$. Usually, all the elements of the initial vector are taken as 1 unless a better estimate is available. The vector $X^{(0)}$ is multiplied by the coefficient matrix A to obtain the vector $AX^{(0)}$. This resulting vector is normalized by dividing each of its elements by any one element, generally chosen as the largest element for accuracy or as the first element for convenience. The normalized vector is denoted as $X^{(1)}$. If the difference between the new vector $X^{(1)}$ and the old one $X^{(0)}$ is less than a chosen convergence criterion, the process is terminated. Otherwise, $X^{(1)}$ becomes the starting vector for the next iteration, and the process is repeated. When convergence has been achieved, the normalizing factor is taken as the largest component of the vector X. Then this is the largest eigenvalue λ_1 and the normalized vector is the corresponding eigenvector. The convergence of the method depends on the initial vector $X^{(0)}$ and on the ratio r of the two largest eigenvalues. Convergence is slower if the two are close to each other in magnitude, that is, if r is close to unity. The dominance ratio r is defined as

$$r = \frac{|\lambda_2|}{|\lambda_1|} \tag{6.85}$$

where λ_1 is the largest eigenvalue in magnitude and λ_2 is the next largest. Although the power method is particularly suitable for symmetric matrices, since the eigenvalues are all real in this case, it can also be used for nonsymmetric matrices. The convergence of the method to the largest eigenvalue be proved mathematically for symmetric matrices.

6.7.2.2 Smallest Eigenvalue

In several engineering problems, the smallest eigenvalue is of particular interest. For instance, designers are interested in the lowest frequency of the natural vibration of civil engineering structures, such as buildings and bridges, so that they can design the structures to avoid certain externally induced vibrations. The power method may be used to determine the smallest eigenvalue and the corresponding eigenvector by pre-multiplication of the original system, Equation 6.72, by the inverse A^{-1} of the coefficient matrix. Therefore,

$$A^{-1} AX = \lambda A^{-1} X$$

Now, $A^{-1}A = I$, and if both sides are divided by λ, the result is

$$HX = \frac{1}{\lambda} X \tag{6.86}$$

where the inverse matrix A^{-1} is denoted as H. Therefore, if the power method is applied to Equation 6.86, the largest eigenvalue obtained will be $1/\lambda$, which arises from the matrix H. This largest value of $1/\lambda$ corresponds to the smallest eigenvalue in magnitude. However, one must determine the inverse H of the matrix A in order to apply this method. Therefore, the procedure may not be practical for very large matrices.

6.7.2.3 Intermediate Eigenvalues

Several procedures are available for obtaining the intermediate eigenvalues, lying between the smallest and the largest eigenvalues. Most of these methods gradually remove the known eigenvalues from the problem so that the method converges to the next eigenvalue. However, this procedure, known as *deflation*, is suitable if only a few eigenvalues are needed, since the growth of round-off error often severely limits the accuracy of the results. If the largest eigenvalue λ_1 and the corresponding eigenvector X_1 have been found for the given matrix A, a new matrix A may be formed in terms of the transpose X_1^T of the matrix, as

$$A_1 = A - \frac{\lambda_1(X_1 X_1^T)}{X_1^T X_1} \tag{6.87}$$

It can be shown that $X_1^T X_1$ is a scalar and equal to the sum of the squares of the components of the X_1 eigenvector. The matrix $X_1 X_1^T$ is a symmetric matrix of the same dimension as A. It can also be shown that A_1 has the same eigenvalues and eigenvectors as A, except for λ_1 which is replaced by zero. Therefore, if the power method is applied to A_1, it will converge to the second largest eigenvalue λ_2 and the corresponding eigenvector X_{2}.. Similarly, the next largest eigenvalue A_3 and the associated eigenvector X_3 may be obtained by applying the power method to a matrix A_2, given as

$$A_2 = A_1 - \frac{\lambda_2(X_2 X_2^T)}{X_2^T X_2} \tag{6.88}$$

The power method can be applied quite easily on the computer, particularly if only the largest eigenvalue is desired. For other eigenvalues, the techniques outlined here may be applied, although the method is rarely used if more than a few eigenvalues are needed. The rate of convergence can sometimes be accelerated by the addition of a constant to each diagonal element of the matrix. This shifts all the eigenvalues by a constant value and may change the dominance ratio favorably to accelerate convergence. However, the suitable amount of shift must be obtained by trial and error. Over-relaxation or under-relaxation, similar to that discussed in Section 6.6, may also be used to achieve faster convergence. Again, the optimum value of ω for the fastest convergence must often be obtained by trying several values. Example 6.7 discusses the solution of an eigenvalue problem by the power method.

Example 6.7

For the vibrating system considered in Example 6.6, obtain the largest eigenvalue and the corresponding eigenvector, using the iterative power method.

SOLUTION

The given system of equations is written as

$$AX = \lambda X \tag{6.78}$$

where λ and A are given by Equations 6.79 and 6.80. To use the power method, an initial guess for the eigenvector is taken as (1, 1, 1), and the coefficient matrix A is multiplied with this vector, employing the formula for matrix multiplication. The resulting vector is normalized with the largest component, which gives the first approximation to the largest eigenvalue of the system. The new vector is compared with the starting vector. If the absolute value of the difference is larger than the convergence criterion ε, the new vector is taken as the starting vector for the next iteration. The process is continued until

$$| X_i - (XO)_i | \leq \varepsilon \quad \text{for } i = 1,2,3 \tag{6.89}$$

where X_t is the ith component of the eigenvector after the present iteration and $(XO)_i$ is that after the previous iteration. At convergence, the normalizing factor is the largest eigenvalue, and the computed vector the desired eigenvector.

Appendix B.12 gives the MATLAB *m*-file for the Power method. It is particularly simple because of the ease of defining and multiplying matrices. The initial guess is given and the iterative process is carried out till convergence is achieved, as specified in terms of a given convergence criterion. The results then yield the largest eigenvalue and the corresponding eigenvector, as shown in Figure 6.15.

The convergence parameter is taken as 10^{-3} here. Smaller values of the parameter were also considered, and a negligible difference in the solution was obtained. Similarly, other starting values were tried, and convergence was found to occur with essentially the same results for different values close to (1, 1, 1). Convergence was not achieved if values very far from these were taken, as expected. The converged results, shown in Figure 6.15 are very close to those obtained analytically in Example 6.6.

The program may easily be used for other such eigenvalue problems. For instance, a system of six equations was solved by this program as follows:

```
>>x= [1;1;1;1;1;1];
>>a= [2 3 4 2 -1 1;1 2 5 -2 2 1;2 -2 -3 4 3 1;2 5 3 1 1 2; ...
     2 1 -3 -2 3 2;1 4 2 5 2 -1];
```

THE LARGEST EIGENVALUE IS = 4.5613
THE EIGENVECTOR IS
$X(1) = 1.0000$
$X(2) = -0.7808$
$X(3) = 1.0000$

FIGURE 6.15 Computed results for the largest eigenvalue and the corresponding eigenvector for the vibrating system of Example 6.6 by the power method.

and the results obtained were

```
THE LARGEST EIGENVALUE IS=8.5963
THE EIGENVECTOR IS
   0.9498
   0.4225
   0.5418
   1.0000
   0.0994
   0.9298
```

Clearly, it is a fairly simple approach to obtain the largest or the smallest eigenvalue and the corresponding eigenvector.

6.7.3 Other Methods

There are several other methods that are available for the solution of eigenvalue problems. A brief outline is given here for completeness. Among the most important of these is *Householder's* method, used in conjunction with the *QL* algorithm. This approach is applicable only to symmetric matrices. Householder's method is used to convert an $n \times n$ symmetric matrix into a symmetric tridiagonal matrix. This form is convenient for matrix decompositions and transformations, since the number of operations for each such manipulation varies as n, rather than as n^3, which applies for the full matrix. Once the tridiagonal form has been obtained, several techniques are available for finding the eigenvalues. The *LR* algorithm of Rutishauser (1958) decomposes the original symmetric matrix into a product of lower triangular and upper triangular matrices. The *QR* algorithm of Francis (1962) decomposes the matrix into a product of an orthogonal matrix and an upper triangular matrix. Using similarity transformations, which preserve the eigenvalues of the original matrix, the first algorithm converges to a lower triangular matrix, and the second to an upper triangular matrix, with the desired eigenvalues in decreasing order of magnitude on the main diagonal. The decomposition in the *LR* algorithm may be done very efficiently by the use of *Choleski* decomposition, outlined in Section 6.4, if the matrix A is positive definite, which requires that all the eigenvalues be positive. The *QR* algorithm is generally more stable than the *LR* method and is often preferred. See Carnahan et al. (1969) and Hornbeck (1982) for details.

The *QL* method decomposes the matrix into the product of an orthogonal matrix and a lower triangular matrix. The method is particularly suited for tridiagonal matrices, such as those produced by Householder's method. The eventual result of decomposition and transformation is a diagonal matrix, with the eigenvalues on the diagonal. In some engineering problems, such as those concerned with the solution of the differential equations that govern the stresses in a structure, the matrix that arises is tridiagonal in form, and the above algorithms may be used efficiently, without the need of transformation by Householder's method.

Nonsymmetric matrices are also of interest in engineering problems, and the methods discussed in the preceding subsections may be employed for these. Also, the above approach for symmetric matrices may be modified to obtain the eigenvalues of an unsymmetric matrix. The application of Householder's method leads to an upper triangular form with an additional band of elements adjacent to the main diagonal. This form, known as the *Hessenberg form*, may also be obtained by elimination methods. The *LR* or the *QR* algorithms may then be employed to obtain the eigenvalues. As mentioned earlier, unsymmetric matrices may have complex eigenvalues. The power method can be modified to deal with complex eigenvalues. For further details, see the treatment of eigenvalue problems by Hornbeck (1982) and Wilkinson (1988).

Before leaving this section, we mention the Jacobi method, which is a classic, although inefficient, method for finding all the eigenvalues and eigenvectors of a symmetric matrix by the use of orthogonal transformations, which preserve the symmetry as well as the eigenvalues. If Q denotes an orthogonal matrix, we wish to use a transformation of the form $Q^T AQ$ to reduce the elements in the ith row and the jth column of the matrix to zero. However, the reduction of one element to zero often introduces nonzero elements at positions that have been previously converted to zero. The process is, therefore, an infinite one, and the matrix eventually tends toward a diagonal form. At convergence, the eigenvalues are obtained from the diagonal elements. Eigenvectors may also be obtained along with the eigenvalues by application of the reduction procedure to an identity matrix along with the given matrix. The columns of the resulting modified matrix are then the desired eigenvectors. Various modifications of the Jacobi method, such as the *threshold* method, which eliminates elements larger than a given threshold value, have been developed. The Jacobi method is not efficient in terms of computing time, but it is often available in computer libraries and is frequently used because it is applicable to a wide variety of eigenvalue problems.

6.8 SOLUTION OF SIMULTANEOUS NONLINEAR EQUATIONS

So far, we have discussed the solution of systems of linear equations, considering both homogeneous and nonhomogeneous equations. However, in engineering problems, we are frequently faced with nonlinear equations, for which no direct methods are available and iterative procedures must be used. The solution of single, nonlinear algebraic equations, in order to find the roots, was discussed in Chapter 5. We are concerned here with the solution of a system of nonlinear equations. Such systems arise in a wide variety of problems. Thermal radiation from a heated body, for instance, varies as T^4, where T is the surface temperature. Material properties often vary nonlinearly with temperature and pressure. The forces acting on a moving particle or in a flow often have a nonlinear relationship with velocity. The iterative methods outlined for linear equations are often modified and employed for nonlinear equations. However, the methods discussed in Chapter 5 may also be considered for solving a system of nonlinear equations. An important method, which is used extensively for solving small sets of equations, is the Newton–Raphson method.

6.8.1 Newton–Raphson Method

The Newton–Raphson method is based on Taylor series expansions of the functions $f_1, f_2, \ldots, f_n$ of Equation 6.1. The function f_1 may be expanded in a Taylor series about $(x_1, x_2, \ldots, x_n)$, which represents an approximation to the solution. If only the first-order terms are retained and if the exact solution $(\bar{x}_1, \bar{x}_2, \ldots, \bar{x}_n)$ is substituted for the unknowns, we obtain the relationship

$$f_1(\bar{x}_1, \bar{x}_2, \ldots, \bar{x}_n) \simeq f_1(x_1, x_2, \ldots, f_1\, x_n) + \left(\frac{\partial f_1}{\partial x_1}\right)_{xi} (\bar{x}_1 - x_1) + \left(\frac{\partial f_1}{\partial x_2}\right)_{xi} (\bar{x}_2 - x_2) + \cdots + \left(\frac{\partial f_1}{\partial x_n}\right)_{xi} (\bar{x}_n - x_n) \tag{6.90}$$

where the partial derivatives are evaluated at $(x_1, x_2, \ldots, x_n)$. The exact solution is not known, but Equation 6.90 provides a method for improving the approximation to the solution. If the other functions, $f_2, f_3, \ldots, f_n$ are similarly expanded about $(x_1, x_2, \ldots, x_n)$ and the exact solution is substituted for the unknowns, as seen in Equation 6.90, a set of linear equations is obtained. Since $(\bar{x}_1, \bar{x}_2, \ldots, \bar{x}_n)$ is the solution vector, the functions $f_i(\bar{x}_1, \bar{x}_2, \ldots, \bar{x}_n)$ for $i = 1, 2, \ldots, n$, are all zero. In this set of linear equations, the unknowns are the change in x_i, Δx_i, where $\Delta x_i = x_i' - x_i, x_i'$ being the next approximation. Then, Δx_i may be computed from the following:

$$\begin{bmatrix} \frac{\partial f_1}{\partial x_1} & \frac{\partial f_1}{\partial x_2} & \cdots & \frac{\partial f_1}{\partial x_n} \\ \frac{\partial f_2}{\partial x_1} & \frac{\partial f_2}{\partial x_2} & \cdots & \frac{\partial f_2}{\partial x_n} \\ \vdots & \vdots & & \vdots \\ \frac{\partial f_n}{\partial x_1} & \frac{\partial f_n}{\partial x_2} & \cdots & \frac{\partial f_n}{\partial x_n} \end{bmatrix} \begin{bmatrix} \Delta x_1 \\ \Delta x_2 \\ \vdots \\ \Delta x_n \end{bmatrix} = \begin{bmatrix} -f_1 \\ -f_2 \\ \vdots \\ -f_n \end{bmatrix} \tag{6.91}$$

The functions and the derivatives are all evaluated at the approximate solution $(x_1, x_2, \ldots, x_n)$. The matrix containing the derivatives is known as the *Jacobian* and its determination may be quite time-consuming if the number of independent variables is high. Since only linear terms were retained in the Taylor series expansion, the exact solution is generally not obtained by solving this system of equations. However, the next approximation to the solution may be obtained as

$$\begin{aligned} x_1^{(l+1)} &= x_1^{(l)} + \Delta x_1^{(l)} \\ x_2^{(l+1)} &= x_2^{(l)} + \Delta x_2^{(l)} \\ &\vdots \\ x_n^{(l+1)} &= x_n^{(l)} + \Delta x_n^{(l)} \end{aligned} \tag{6.92}$$

where the superscript (l) represents the values after a given iteration and ($l + 1$) those obtained by solving Equation 6.91 for the next iteration.

Equations 6.92 provide an iterative method for solving the system of nonlinear equations given by Equation 6.1. An initial, starting, guess of the unknowns is taken, and the functions $f_1, f_2, \ldots, f_n$ and the derivatives, needed in Equation 6.91, are evaluated at these x values, denoted as $x_1^{(0)}, x_2^{(0)}, \ldots, x_n^{(0)}$. The set of linear equations is solved to obtain Δx_i, which is then used to obtain the next iteration $x_i^{(1)}$ from Equation 6.92. The process is continued until all the f's are close to zero or the unknowns do not change from one iteration to the next, within a specified convergence criterion.

The method may diverge if the initial guess is too far off from the exact solution. In physical problems, some prior information is often available on the nature of the functions and on the expected solution. This information may be used advantageously in choosing the initial values. However, if no information is available, several trials, with different starting values, may be needed before the process converges. The partial derivatives are generally computed numerically, since the functions may be quite involved. This method is extensively employed in the numerical simulation of engineering systems. It is also used as a correction scheme for the solution of boundary-value problems in ODEs, as discussed in Chapter 9. Because of the computational effort required for the evaluation of the derivatives, the Newton–Raphson method is generally used when the system consists of only a relatively small number of nonlinear equations, typically less than ten. Other iterative methods, such as those based on the Jacobi and the Gauss–Seidel methods, are more suitable for large systems and are discussed below.

6.8.2 Modified Jacobi and Gauss–Seidel Methods

The system of nonlinear equations may be considered to be of the form given by Equation 6.1. These equations are rewritten, by solving for the unknowns $x_1, x_2, \ldots, x_n$, as follows:

$$x_i = F_i(x_1, x_2, \ldots, x_i, \ldots, x_n) \quad \text{for } i = 1, 2, \ldots, n \tag{6.93}$$

Therefore, the unknown x_i is retained on the right-hand side for nonlinear equations, since it would not, in general, be possible to solve for x_i in terms of just the other unknowns because of the nonlinearity in x_i. The nonlinearity may arise because transcendental equations are involved, because products of the unknowns are present in the equations, or because x_i, appears as x_i^n, where $n \neq 1$. Also, if $\bar{x}_1, \bar{x}_2, \ldots, \bar{x}_n$ represent the solution of the given system of equations, a rearrangement of Equation 6.1 gives, at the solution,

$$\bar{x}_i = F_i(\bar{x}_1, \bar{x}_2, \ldots, \bar{x}_i, \ldots, \bar{x}_n) \quad \text{for } i = 1, 2, \ldots, n \tag{6.94}$$

We may now develop an iterative procedure for solving the given set of nonlinear equations. Similar to the *Jacobi* method for linear equations, the recursion relation may be written from Equation 6.93 as

$$x_i^{(l+1)} = F_i[x_1^{(l)}, x_2^{(l)}, \ldots, x_i^{(l)}, \ldots, x_n^{(l)}] \quad \text{for } i = 1, 2, \ldots, n \tag{6.95}$$

where all the unknowns are computed for the $(l+1)$th iteration using the known values from the previous iteration. We may also replace the unknowns by the new values as soon as they are computed. This procedure is similar to the *Gauss–Seidel* method for linear equations and is given by

$$x_i^{(l+1)} = F_i[x_1^{(l+1)}, x_2^{(l+1)}, \ldots, x_{i-1}^{(l+1)}, x_i^{(l)}, \ldots, x_n^{(l)}] \quad \text{for } i = 1, 2, \ldots, n \tag{6.96}$$

where $x_1^{(l+1)}$ is computed first followed by $x_2^{(l+1)}$, and so on for increasing i. Therefore, the most recently computed values of the unknowns are used for evaluating the function F_i. The formulation for this method is similar to that for the *successive substitution method* discussed in Chapter 5. Therefore, this method is also sometimes known as the successive substitution method for solving a system of nonlinear equations.

6.8.3 Convergence

The convergence characteristics of nonlinear equations are not as well established as those for linear equations. A general theory for the iterative solution of nonlinear equations is not available, although the behavior of certain special sets of equations has been studied in detail. The equations may also yield multiple solutions, and one would then need information on the physical aspects of the problem in order to choose the correct solution. However, the solution of equations that characterize a physical problem usually results in only one physically realistic solution, and this solution is the one that is obtained most easily. The other solutions may be physically unacceptable and are usually not readily obtained when the system of equations is solved by the above methods.

We may use relaxation to alter the convergence characteristics of the iterative process. However, it is often difficult to predict the resulting behavior. Over-relaxation may even slow the convergence in some cases for nonlinear equations and accelerate it in others. SUR is particularly useful in obtaining convergence for nonlinear systems. From Equation 6.96, the relaxation method may be written as

$$x_i^{(l+1)} = \omega F_i[x_1^{(l+1)}, x_2^{(l+1)}, \ldots, x_{i-1}^{(l+1)}, x_i^{(l)}, \ldots, x_n^{(l)}] + (1-\omega)x_i^{(l)} \quad \text{for } i = 1, 2, \ldots, n \tag{6.97}$$

where $0 < \omega < 1$ for under-relaxation. Similarly, relaxation may be applied to the *modified Jacobi method*, Equation 6.95. For highly nonlinear equations, such as equations where the unknowns appear in powers substantially different from linear, or 1, and transcendental equations, quite small values of ω, such as 0.1 or smaller, may be needed to obtain convergence. Convergence is strongly dependent on the nature of the equations, and very often several trials, with different starting values, are needed before convergence is achieved. Also, the modified Jacobi method is particularly suitable for parallel computer systems, since each equation may be considered independently.

Example 6.8

In the ammonia production system shown in Figure 6.16, a mixture of 90 moles/s of nitrogen, 270 moles/s of hydrogen, and 0.9 moles/s of argon, which is present as an impurity, enters the chemical plant and is mixed with the residual mixture crossing a bleed valve. In the chemical reactor, a fraction of the entering mixture combines to give ammonia, which is removed by condensation. A bleed of 23.5 moles/s of the mixture is employed to avoid a buildup of argon, which adversely affects the reaction. The conversion efficiency of the reactor is 0.57exp(−0.0155 F_1), where F_1 is the argon flow rate in moles/s. This efficiency represents the fraction of the mixture that is converted to ammonia (Stoecker, 1989). When mass conservation is applied to the process, the following equations are obtained:

$$F_1 = \frac{0.9}{(1 - B)} \tag{6.98a}$$

$$P = 1 - 0.57e^{-0.0155F_1} \tag{6.98b}$$

$$F_2 = \frac{90}{(1 - B \times P)} \tag{6.98c}$$

$$B = 1 - \frac{23.5}{4F_2P + F_1} \tag{6.98d}$$

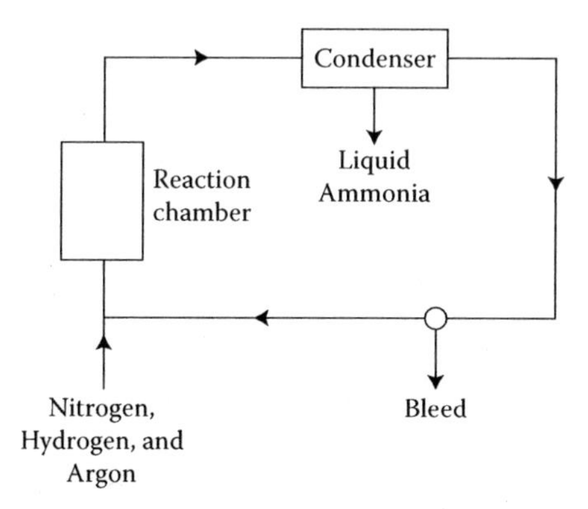

FIGURE 6.16 The ammonia production system considered in Example 6.8.

where F_1 is the flow rate of argon entering the reaction chamber, F_2 is the flow rate of nitrogen, and B and P are parameters defined above. Solve this system of nonlinear equations by the successive substitution method to obtain the flow rates and the amount of ammonia produced.

SOLUTION

This example presents a typical case of systems of nonlinear equations that often arise in the analysis of chemical reactors. The successive substitution method is often applied to solve the problem. However, the convergence is dependent on the sequence of equations solved, as well as on the initial guessed values. One particular solution process is outlined here.

The four unknowns are taken as F_1, P, F_2, and B. The starting value of B is arbitrarily chosen as 0.1, and the other quantities are computed in the sequence given by Equations 6.98. The total flow rate of the mixture entering the plant is denoted by C, where $C = F_1 + 4F_2$, since the flow rate of nitrogen is F_2 moles/s and that of hydrogen is $3F_2$ mol/s. The amount of ammonia produced is denoted by D and is given by

$$D = 2F_2 \times 0.57e^{-0.0155F_1} \tag{6.99}$$

since each mole of nitrogen gives two moles of ammonia, as seen from the chemical equation

$$N_2 + 3H_2 = 2NH_3 \tag{6.100}$$

The convergence criterion may be applied to the total flow rate C or to B as follows:

$$|C - CO| \le \varepsilon \quad \text{or} \quad |B - BO| \le \varepsilon \tag{6.101}$$

where CO and BO are the values after the previous iteration and ε is the convergence parameter.

The physical problem discussed here is evidently quite involved. However, the computer program, shown in Appendix B.13, is quite simple. It is based on the successive substitution or the modified Gauss–Seidel method outlined earlier. The convergence parameter ε is denoted by *ep* and is taken as 10^{-7}. It was ascertained that a still smaller value of ε did not significantly affect the numerical results, though larger values slightly changed the converged results. The results obtained after each iteration, for the flow rate of argon, the total flow rate, and the amount of ammonia collected, in moles/s, are indicated in Figure 6.17. The convergence is slow, partly because of the nature of the equations and partly because of the first-order convergence of this method. Convergence was not obtained for values of B very far from the chosen starting value of 0.1. Also, a change in the sequence of the computations performed resulted in divergence in some cases. If convergence is not achieved, under relaxation can be used to

ARGON	TOTAL FLOW	AMMONIA
1.0000	377.5205	105.6579
6.3653	621.9412	158.9564
11.6436	708.7917	165.8785
14.4396	749.7037	167.5278
15.8801	770.5167	168.1452
16.6299	781.3446	168.4220
17.0234	787.0322	168.5568
17.2309	790.0337	168.6252
17.3406	791.6215	168.6606
17.3987	792.4624	168.6792
17.4295	792.9081	168.6890
17.4458	793.1444	168.6942
17.4545	793.2697	168.6969
17.4591	793.3361	168.6984
17.4615	793.3713	168.6991
17.4628	793.3900	168.6995
17.4635	793.3999	168.6998
17.4638	793.4052	168.6999
17.4640	793.4080	168.6999
17.4641	793.4095	168.7000
17.4642	793.4102	168.7000
17.4642	793.4107	168.7000
17.4642	793.4109	168.7000
17.4642	793.4110	168.7000
17.4643	793.4111	168.7000

Iteration has converged
Converged results are
ARGON = 17.4643 TOTAL FLOW = 793.4111 AMMONIA = 168.7000

FIGURE 6.17 Convergence of the computed results for the solution of the system of nonlinear algebraic equations in Example 6.8 by the successive substitution method.

improve the convergence characteristics. Also, the sequence of equations and the initial guess may be varied to obtain convergence. Such variations are commonly used for solving sets of nonlinear equations. Despite these problems with convergence, the successive substitution method is a much simpler method than the Newton–Raphson method, for nonlinear equations, and is widely used in engineering applications.

Example 6.9

For the physical problem described in Example 5.7, employ the Newton–Raphson method to solve the system of nonlinear equations, given by Equations 5.32 and 5.33, to obtain the flow rate R and the pressure P.

SOLUTION

The system of equations that govern the flow through a duct due to a fan, as given in Example 5.7, are

$$R = \left(\frac{P - 80}{10.5}\right)^{3/5} \tag{6.102a}$$

$$P = \left(\frac{15 - R}{75 \times 10^{-6}}\right)^{1/2} \tag{6.102b}$$

To use the Newton–Raphson method, we take the two functions that are to be reduced to zero as

$$R_1 = \left(\frac{P - 80}{10.5}\right)^{3/5} - R = R_1(P,R) \tag{6.103a}$$

$$P_1 = \left(\frac{15 - R}{75 \times 10^{-6}}\right)^{1/2} - P = P_1(P,R) \tag{6.103b}$$

The initial guesses for R and P and the convergence criterion ε are inputs to the program, which is shown for MATLAB in Appendix B.14. In order to consider both the functions R_1 and P_1, the convergence criterion is applied to a parameter B, where

$$B = R_1^2 + P_1^2 \tag{6.104}$$

Other choices are obviously possible, including considering functions R_1 and P_1 separately.

The four partial derivatives, $\partial R_1/\partial R$, $\partial R_1/\partial P$, $\partial P_1/\partial R$, and $\partial P_1/\partial P$, are computed at the starting, guessed, values of R and P, employing analytical differentiation of the functions R_1 and P_1. These derivatives are denoted by *rr*, *rp*, *pr*, and *pp*, respectively, in the program. The increments in R and P, ΔR, and ΔP, which are denoted by d*r* and d*p*, respectively, in the program, for the next iteration are then obtained from Equation 6.91, which gives

$$\frac{\partial R_1}{\partial R}\Delta R + \frac{\partial R_1}{\partial P}\Delta P = -R_1 \tag{6.105a}$$

$$\frac{\partial P_1}{\partial R}\Delta R + \frac{\partial P_1}{\partial P}\Delta P = -P_1 \tag{6.105b}$$

The new values of R and P are determined from Equation 6.92. The iterative process is repeated until convergence is achieved, as indicated by $B < \varepsilon$. Figure 6.18 shows the results for the starting values of R and P taken as 2 and 100, respectively, and ε, or *ep*, taken as 10^{-4}. For the same convergence criterion, the flow rate was obtained as 6.7320 m³/s in Example 5.7, indicating a close agreement with the present results. For smaller values of ε, the converged results were not significantly affected. Also,

Enter the value of parameter r, $r = 2$
Enter the value of parameter p, $p = 100$
Enter the value of convergence parameter ep, ep = 0.0001
$R = 9.8736$ $P = 290.2550$
$R = 6.8644$ $P = 338.1764$
$R = 6.7326$ $P = 332.0233$
$R = 6.7321$ $P = 332.0223$
THE REQUIRED SOLUTION IS
The flow rate $R = 6.7321$ The pressure $P = 332.0223$

Enter the value of parameter r, $r = 1$
Enter the value of parameter p, $p = 200$
Enter the value of convergence parameter ep, ep = 0.0001
$R = 7.2408$ $P = 335.7523$
$R = 6.7342$ $P = 332.1454$
$R = 6.7321$ $P = 332.0223$
THE REQUIRED SOLUTION IS
The flow rate $R = 6.7321$ The pressure $P = 332.0223$

Enter the value of parameter r, $r = 0$
Enter the value of parameter p, $p = 300$
Enter the value of convergence parameter ep, ep = 0.0001
$R = 6.9444$ $P = 343.6930$
$R = 6.7337$ $P = 332.0180$
$R = 6.7321$ $P = 332.0223$
THE REQUIRED SOLUTION IS
The flow rate $R = 6.7321$ The pressure $P = 332.0223$

FIGURE 6.18 Numerical results from the solution of the system of nonlinear equations of Example 6.9 by the Newton–Raphson method.

convergence to the same results was obtained for a fairly wide range of starting values. It can be seen that convergence is very rapid, despite the considerable difference between the starting and the converged values. The convergence was much slower with the successive substitution method, as illustrated by Example 5.7. Due to its superior convergence characteristics, the Newton–Raphson method is the preferred method if the number of equations is small.

It must be noted that the two equations, Equation 6.105, for the given problem are solved directly without resorting to methods available for linear systems. These methods are needed if the number of independent variable is larger, typically 4 or higher. Then the matrix of the derivatives, Equation 6.91, is determined at each step and the increments in each of the independent variables are calculated to advance to the next iteration. For the problem considered here as well, the matrix may be formed and the increments determined by solving the matrix equation. For instance, after the derivatives of the functions are obtained, the matrix may be formulated and the backslash operator employed to obtain the increments as

```
a = [rr rp;pr pp];
b = [-r1; -p1];
```

```
dd = a\b;
dr = dd(1);
dp = dd(2);
r = r + dr;
p = p + dp;
```

This approach can thus be extended to cases where the number of independent variables is larger than two.

6.9 SUMMARY

The solution of simultaneous linear and nonlinear algebraic equations is considered in this chapter. Several methods are discussed, and their advantages over other methods and their applicability to the various types and systems of equations that arise in engineering problems are indicated. The choice of the method for the solution of a given system depends on whether the equations are linear or nonlinear and on whether they are homogeneous or nonhomogeneous. It also depends on the number of equations to be solved and the nature of the equations, particularly the sparseness of the coefficient matrix. The selection of the method for a given situation is also often influenced by the need to determine other quantities, such as the inverse or the determinant of the coefficient matrix, besides the unknown vector X. Similarly, several systems of equations with the same coefficient matrix but different constant vectors may have to be solved. This additional consideration is often an important factor in the selection of the method. In several cases, the available software in the computer library may also make a given method more attractive than the others. MATLAB is particularly well suited for solving systems of equations because of its inherent advantages in matrix algebra. Linear systems can be solved very easily by matrix inversion, decomposition and matrix manipulation in MATLAB. These approaches can also be used for simplifying the solution of nonlinear and homogeneous equations.

For linear, nonhomogeneous equations, the methods discussed here include Gaussian elimination, Gauss–Jordan elimination, Crout's method, matrix inversion, and iterative methods. Gaussian elimination is the simplest direct method, in terms of computer programming, and is appropriate for a small number of equations, typically of the order of 20 or less, because of the round-off error. However, if the system is tridiagonal, this method may be used advantageously for several hundred equations without significant loss of accuracy due to round-off error. In the finite difference solution of ODEs and PDEs, tridiagonal systems are often obtained, and Gaussian elimination (TDMA) is the preferred method. If many systems that have the same coefficient matrix A but different constant vector B are to be solved, Gauss–Jordan elimination may be used advantageously, since all systems are solved in one elimination process. Similarly, matrix inversion determines A^{-1}, which is the same for all the systems, and the solution X is obtained simply by multiplication of the inverse with the constant vector, that is, $X = A^{-1}B$. This approach is very easily applied in MATLAB to solve linear systems by using the available software.

Crout's method and other compact methods based on decomposition of the matrix A are generally more efficient than elimination methods. The round-off error is also less since the number of operations is smaller than that for Gaussian elimination.

Computer programming is somewhat more involved. However, because of their advantage over elimination methods in computing time, compact methods have become quite popular in recent years. Again, matrix decomposition can be conveniently and efficiently applied in MATLAB by using the available software.

Iterative methods, such as Jacobi, Gauss–Seidel, and relaxation methods, are particularly suitable for large systems of equations, typically of order 100 or larger, and for sparse coefficient matrices. The round-off error in the solution is due only to the error that arises in the final iteration. If the optimum value of the relaxation factor ω is known, the SOR method generally requires less computing time than most direct methods. The optimum value ω_{opt} is usually not known, and several values may have to be tried to determine it numerically. The Jacobi method is seldom used, since it requires greater computer storage and computational effort, on traditional or single-processor computers, than the Gauss–Seidel method, which remains a popular choice, along with SOR, for solving large systems of equations. These methods, with some modifications, are also suitable for nonlinear equations, which generally cannot be solved by direct methods. Therefore, the successive substitution or the modified Gauss–Seidel method is frequently used. Relaxation may also be used in this case, but the effect of relaxation on convergence is often unpredictable in nonlinear equations. SUR is commonly used to achieve convergence in nonlinear equations. For a small number of equations, typically less than 10, the Newton–Raphson method is preferable since its convergence is more rapid.

The solution of linear, homogeneous equations requires methods quite different from those for nonhomogeneous equations. The eigenvalue problem is discussed in detail. For a small number of equations, the characteristic polynomial may be solved to obtain the eigenvalues, and the Gauss–Jordan method may be applied to determine the corresponding eigenvectors. The power method, which is an iterative method for determining the largest eigenvalue and the related eigenvector, may be used for moderately sized systems. The smallest eigenvalue can also be determined easily. However, intermediate eigenvalues can be obtained accurately if only a few of them are desired, since the round-off error accumulates as these are successively determined. The Jacobi method is a classical iterative method that yields all the eigenvalues and the eigenvectors of symmetric matrices by obtaining the matrix in a diagonal form. This method is frequently employed, even though it is quite inefficient, because the software is available on many computer systems. The most efficient method for large matrices is Householder's method used in conjunction with the QL algorithm. The former converts a symmetric matrix into a tridiagonal form. Various methods, such as the QL algorithm, are available for extracting the eigenvalues from a tridiagonal matrix by the use of decompositions and transformations.

PROBLEMS

6.1. Compare the Gaussian elimination and Gauss–Jordan elimination methods for solving a system of linear equations. Which one is more accurate? Which one is more efficient? Indicate the advantages, if any, of the latter method over the former.

6.2. Draw the flow chart for solving a system of linear equations by Gaussian elimination.

6.3. Consider the electrical network shown and write down the algebraic equations for determining the four loop currents indicated. Solve this system of linear equations by Gaussian elimination. Check the results by also solving these equations employing the backslash operator in MATLAB.

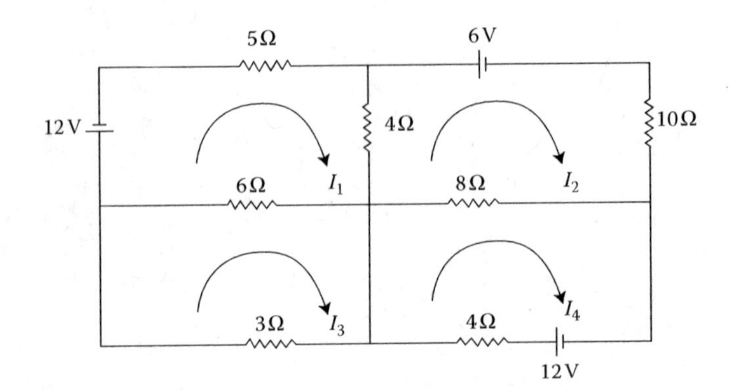

6.4. a. If the 3 Ω resistance in the network for Problem 6.3 is replaced by an open circuit, resulting in only three loops, compute the three loop currents by the Gaussian elimination method.

b. If the currents through the remaining six resistances are denoted by I_1, I_2, …, I_6, write down the six linear equations that govern these currents. Compute their values using Gaussian elimination, and compare the results with those for the loop currents in the first part of the problem.

6.5. A third-order polynomial of the form $y = Ax^3 + Bx^2 + Cx + D$ is to be fitted to four time-velocity data points. At time $x = 0$, 1, 2, and 3 s, the velocity is measured as 7, 14, 29, and 58 m/s. Using Gauss–Jordan elimination, find the curve that passes through these points. Also, solve the problem by employing the backslash operator in MATLAB and compare the results with those obtained earlier.

6.6. A fourth-order polynomial passes through the five points for which the independent and dependent variables, x and y, respectively, are given as (–2, 37), (–1, 7), (0, 5), (1, 13), and (2, 61). Find the polynomial by any suitable method. Here, x represents the spatial location and y the species concentration in a chemical reactor.

6.7. Six data points are obtained in the calibration of a velocity-measuring device. At velocities of 0, 0.2, 0.4, 0.6, 0.8, and 1.0 m/s, the voltage signals from the instrument are obtained as, respectively, 1.2, 1.74, 2.63, 3.99, 6.04, and 9.1 volts. Find the fifth-order polynomial that passes through these points. Also, solve the problem by employing the matrix inversion command in MATLAB and compare the results with those obtained earlier.

6.8. For the physical problem described in Example 6.2, take the temperature at the left face as 200°C and that at the right face as 20°C. Write down the resulting tridiagonal system of equations, and obtain the solution by Gaussian elimination.

6.9. Derive the recursion formulas for solving a tridiagonal system of equations by Gaussian elimination.

6.10. Write a MATLAB script file to compute the magnitude of an $n \times n$ determinant by Gaussian elimination, where n can be up to 10. Using this program, compute the magnitudes of the determinants of the following matrices that arise in the dynamic analysis of structures:

$$\begin{bmatrix} 1 & 2 & -1 & 0 \\ 3 & 2 & 1 & 0 \\ 2 & 1 & 1 & 3 \\ 0 & 1 & 2 & 0 \end{bmatrix} \quad \text{and} \quad \begin{bmatrix} 2 & 0 & 3 & 1 \\ 3 & 1 & 2 & 2 \\ -1 & 0 & -2 & 4 \\ 3 & 2 & 1 & -2 \end{bmatrix}$$

6.11. A system of linear equations is given as follows:

$$x_1 - 2x_2 + 2x_3 + 3x_4 = 5$$
$$x_1 + 5x_2 - 3x_3 - 5x_4 = 13$$
$$-x_1 + x_2 + 3x_3 - 4x_4 = -6$$
$$2x_1 + 3x_2 - x_3 - 2x_4 = 18$$

Determine whether all the equations in this set are independent, using a computer program as well as matrix methods in MATLAB.

6.12. Solve the following system of linear equations by the Gauss–Jordan method, to indicate the basic procedure involved:

$$2x_1 + 2x_2 + 3x_3 = 15$$
$$x_1 + 3x_2 + x_3 = 10$$
$$3x_1 - x_2 + 2x_3 = 7$$

6.13. Using the Gauss–Jordan method with partial pivoting, invert the matrices given in Problem 6.10. Also, multiply the inverse A^{-1} with the corresponding matrix A, for the two cases, and compare the values obtained with the identity matrix. Comment on the difference, if any, between the two. Also, use the *inv*(A) command in MATLAB to solve this problem and compare the results with those obtained earlier.

6.14. Find the currents in the three resistances located on the three sides of the triangle in the network shown. The voltages shown at the ends are the positive voltages imposed at these points. Use matrix inversion to solve this set of equations, using both a script file and the *inv*(A) command in MATLAB.

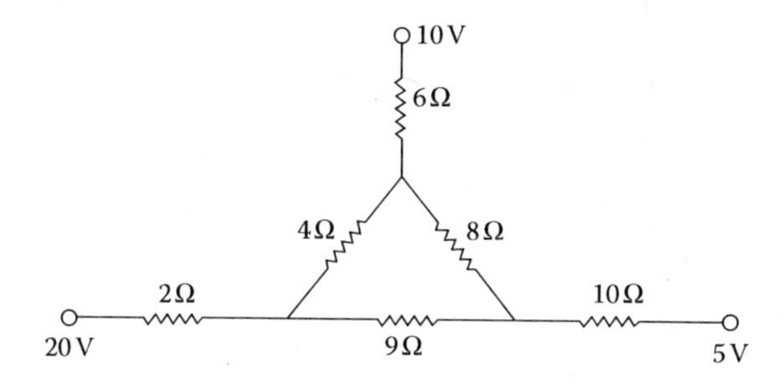

6.15. For the electrical network shown, obtain the governing set of linear equations using the voltage drops across the resistances as unknowns. Use Crout's method to solve this system of equations. Also solve the system by Gaussian elimination, and compare the accuracy obtained by the two methods.

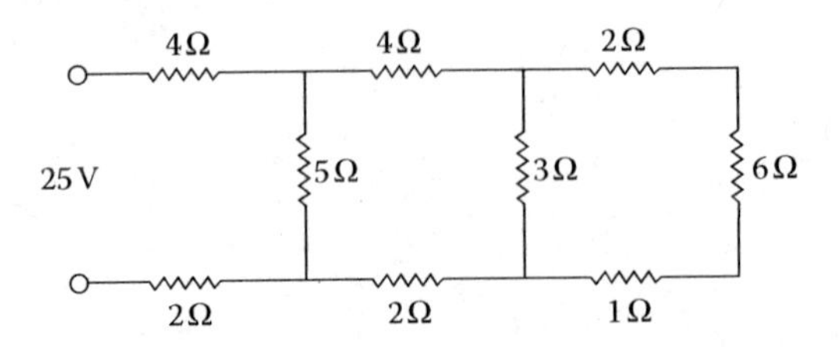

6.16. Solve the following system of linear equations using Gaussian elimination and also Gauss–Jordan elimination, both being employed with and without partial pivoting:

$$x_1 + 2x_2 + 4x_3 = 18$$
$$2x_1 + 3x_2 - 5x_3 = -18$$
$$4x_1 - x_2 - x_3 = -14$$

Compare the results obtained with each other and with the analytical solution. Does pivoting improve the accuracy of the results?

6.17. An industrial organization produces four items, x_1, x_2, x_3, and x_4. A portion of the amount produced for each is used in the manufacture of other items, and the net product is sold. The balance between the output and the production rate, resulting from the various inputs, gives rise to the following four equations, corresponding to the four items:

$$2x_1 + x_2 + 0.x_3 + 6x_4 = 64$$
$$5x_1 + 2x_2 + 0.x_3 + 0.x_4 = 37$$
$$0.x_1 + 7x_2 + 2x_3 + 2x_4 = 66$$
$$0.x_1 + 0.x_2 + 8x_3 + 9x_4 = 104$$

Using Gauss–Jordan elimination, with pivoting, solve this set of equations. Also, use Crout's method and compare the results obtained with those from the former method. Comment on the difference between the two methods. Check the results obtained by also solving the problem using the *lu* matrix decomposition command in MATLAB.

6.18. Compute the magnitude of the determinant of the coefficient matrix of the following system of equations to determine whether it is a singular matrix.

$$3x + 7y - 6z = -15$$
$$x - 3y + 2z = 27$$
$$9x + 5y - 6z = 51$$

What can you say about the given system on the basis of the computed value? How will you solve such a system?

6.19. Using the power method, determine the largest eigenvalue and the corresponding eigenvector of the following matrices.

$$\begin{bmatrix} 8 & 4 & 5 \\ 6 & 10 & 3 \\ 8 & 6 & 20 \end{bmatrix} \quad \text{and} \quad \begin{bmatrix} 10 & 5 & 8 \\ 5 & 40 & 6 \\ 8 & 10 & 20 \end{bmatrix}$$

Vary the convergence criterion ε, applied to the eigenvector, from 10^{-4} to 10^{-1}, and study the resulting effect on the number of iterations needed for convergence, starting with an initial guess of the eigenvector as (1, 1, 1).

6.20. Solve the following equations, giving at least three complete steps, by the Gauss–Seidel method:

$$2x + 8y + 3z = 27$$
$$x + 3y + 5z = 22$$
$$6x + y + 2z = 14$$

Do you expect the numerical scheme to converge? Justify your answer.

6.21. Consider the following diagonally dominant system of equations:

$$8x + y + 2z = 29$$
$$x + 9y + z = 34$$
$$2x + 3y + 7z = 48$$

Using the *Jacobi* method and also the *Gauss–Seidel* method, solve this system. Compare the number of iterations needed for convergence in the two cases, if the convergence criterion ε is taken as 10^{-3} and applied to the computed value of the unknowns.

6.22. Solve the system of equations in the preceding problem by the *LU* decomposition method, using the *lu* command in MATLAB. Print the solution from the LU decomposition method and also print the upper triangular matrix obtained. If the first and third equations were interchanged, would you expect the Gauss–Seidel iteration to converge?

6.23. Find the optimum value ω_{opt} for fastest convergence, if the SOR method is employed for the system of equations given in Problem 6.21. Plot the number of iterations to convergence against the relaxation factor ω and discuss your findings.

6.24. The temperature distribution in the square region shown is governed by the *Laplace equation*, discussed in Chapter 10. The finite difference approximation to this equation yields a system of algebraic equations given by

$$T_{i,j} = \frac{T_{i+1,j} + T_{i-1,j} + T_{i,j+1} + T_{i,j-1}}{4}$$

where i is the number of the row and j is the column in which a grid point is located. For the nine points shown, obtain nine linear equations for determining the temperatures at these points. Solve this system by the Gauss–Seidel method. Note that the temperatures at the boundaries are given and are used in the equations for all the temperatures, except for the one at position number 5. Also, using the *inv*(*A*) command in MATLAB, solve this problem and compare the results with those obtained earlier.

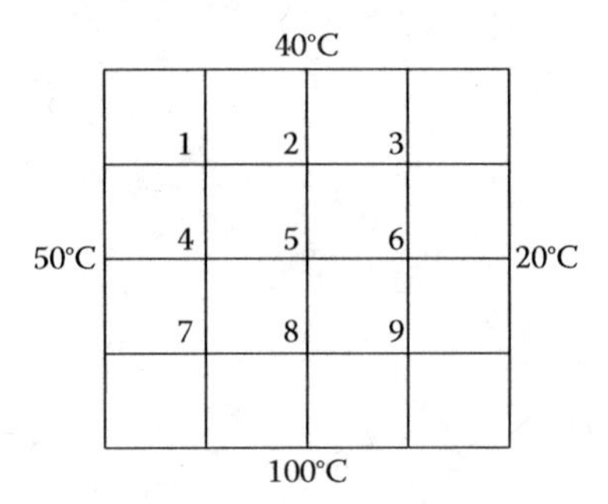

6.25. Solve the preceding problem if the bottom surface is at a temperature of 1.0, while the others are at 0.0.

6.26. Using the Gauss–Seidel iterative method, solve the following system of equations:

$$\begin{aligned} 8x_1 - x_2 - 2x_3 + 3x_4 &= 22 \\ 2x_1 + 6x_2 - x_3 + x_4 &= 12 \\ x_1 + 2x_2 + 10x_3 + 2x_4 &= 20 \\ 3x_1 - 3x_2 + 2x_3 + 9x_4 &= 32 \end{aligned}$$

Calculate the results for three tolerance values, 0.1, 0.01, and 0.001. Discuss the effect of the tolerance on the number of iterations and on the numerical solution. Also, solve the equations by using MATLAB commands for matrices. Compare the results from the two methods.

6.27. Find the optimum value ω_{opt} of the relaxation factor for solving the system of equations in preceding problem by the SOR method, and compare the number of iterations to convergence with that for the Gauss–Seidel method.

6.28. The following three linear equations describe the mass balance of an engineering system, with X, Y, and Z as design variables.

$$\begin{aligned} X - 5.7Y + 4.2Z &= 14.7 \\ -11X + 1.1Y - 7.6Z &= -81.2 \\ -2.2X - Y + 10.7Z &= 90.5 \end{aligned}$$

Find X, Y, and Z using the Gaussian elimination method. Also, set up the equations for the Gauss–Seidel method and discuss if the scheme is expected to converge (Do not actually solve the problem by this method).

6.29. A system of linear equations is given as

$$\begin{aligned} 6x + 2y + z + 2p &= 15.4 \\ 4y + 2z + p &= 12.2 \\ x + 5z + 3p &= 17.6 \\ 2y + 3z + 8p &= 26.6 \end{aligned}$$

where x, y, z, and p are the unknowns. Write a script-m file to do the following:

a. Solve the system of equations by the Gauss–Seidel iterative method to obtain x, y, z, and p. Print the solution obtained.
b. Solve the system of equations by the LU decomposition method, using the *lu* command in MATLAB.
c. Print the solution from the LU decomposition method and also print the lower and upper triangular matrices obtained.
d. Give the value of the determinant of the coefficient matrix.
e. In Gauss–Seidel, what would you expect if the first equation is solved for z, instead of x, the second equation for p, third for x and fourth for y?

6.30. The mass balance for three items x, y, and z in a chemical reactor is governed by the following linear equations:

$$\begin{aligned} 4.8x + y + 2.5z &= -1.62 \\ 2.2x + 4.5y + 1.1z &= 11.14 \\ -2.1x - 3.1y + 10.1z &= 15.57 \end{aligned}$$

Solve this system of equations by the Gauss–Seidel iteration method to obtain the values of the three items. The initial guess may be taken as $x = y = z = 0.0$ or 1.0. Also, solve the equations by using the matrix commands in MATLAB and compare the results with those obtained earlier.

6.31. Solve the following set of linear equations by the Gauss–Seidel iteration method. The initial guess may be taken as 0.0 or 1.0.

$$\begin{aligned} 5x + y + 2z &= 17 \\ x + 3y + z &= 8 \\ 2x + y + 6z &= 23 \end{aligned}$$

Vary the convergence parameter to ensure that results are independent of the value chosen.

6.32. As done for Example 6.6, obtain the system of linear homogeneous equations that govern the vibration of the three-mass system shown. Using the power method, determine the largest eigenvalue and the corresponding eigenvector.

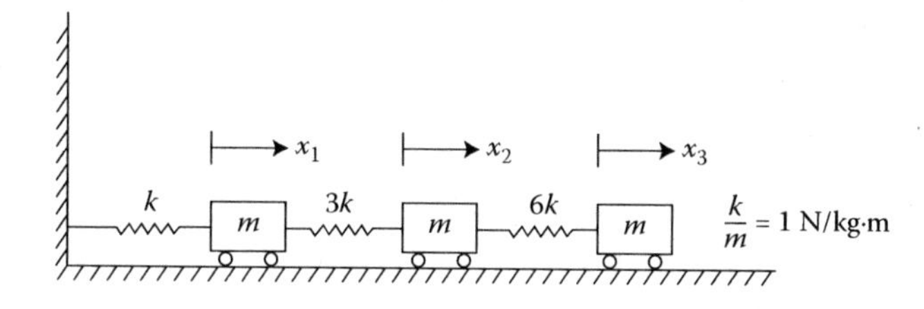

6.33. Obtain the characteristic polynomial for Problem 6.32 and find all the eigenvalues by root solving. Determine the eigenvectors by the Gauss–Jordan method, as outlined in Example 6.6.

6.34. For the vibrating system shown, use the power method to obtain the largest and the smallest eigenvalues and the corresponding eigenvectors. Neglect the effect of gravity.

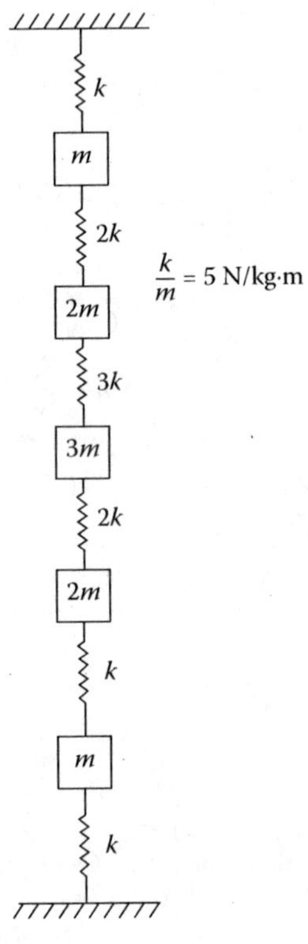

6.35. The forces acting on a body give rise to stresses in the material. At a given point in the material, the state of stress is given by the matrix

$$\begin{bmatrix} 8 & 3 & 6 \\ 5 & 10 & 2 \\ 6 & 7 & 20 \end{bmatrix} \times 10^6 \text{ N/m}^2$$

The largest principal stress, which determines the failure of the material, is the largest eigenvalue of the stress matrix. Using the iterative power method, find the largest eigenvalue and the corresponding eigenvector.

6.36. Obtain all the eigenvalues of the stress matrix

$$\begin{bmatrix} 12 & 6 & 8 \\ 8 & 40 & 7 \\ 6 & 12 & 20 \end{bmatrix} \times 10^6 \text{ N/m}^2$$

and compare the largest value obtained with the elements of the given matrix. Is the result physically expected?

6.37. For the four-mass vibrating system shown, obtain the governing algebraic equations, neglecting gravitational effects. Determine the largest eigenvalue and the corresponding eigenvector, using the power method. Also obtain the smallest eigenvalue and compare it with the computed largest eigenvalue. Comment on the physical significance of the difference.

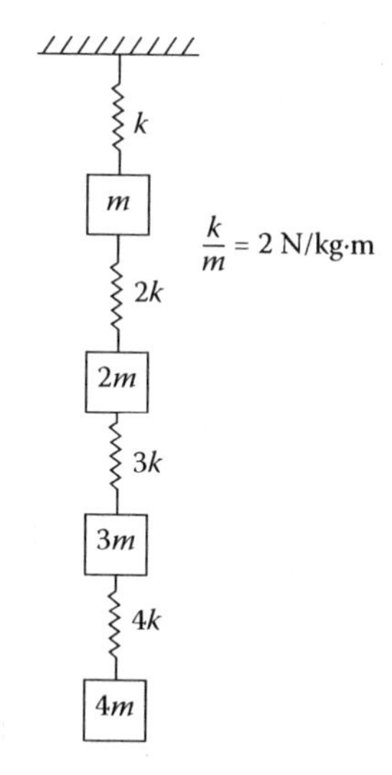

6.38. Water flows through two parallel pipe networks, each of which contains a pump to provide the necessary pressure difference Δp. The water flow rates through the two circuits are Q_1 and Q_2, the total flow rate being Q. Therefore,

$$Q = Q_1 + Q_2$$

The characteristics of two pumps are given in terms of the relationship between the pressure difference and the flow rate as

$$\Delta p = 550 - 10Q_1^2$$
$$\Delta p = 700 - 15Q_2^2$$

Also, the pressure difference may be computed from changes in elevation and friction in the pipes to give

$$\Delta p = 68 + 8Q^2$$

Using the successive substitution method, solve this system of nonlinear equations to obtain the flow rates and the pressure difference.

6.39. Solve Problem 6.38 by the Newton–Raphson method, and discuss the difference in the convergence characteristics and in the programming from the successive substitution method.

6.40. Solve the following system of nonlinear equations by using Newton's method:

$$x^3 + 3y^2 = 21$$
$$x^2 + 2y + 2 = 0$$

Show two complete cycles of iteration to locate the root for $x > 0$.

6.41. Consider the physical problem discussed in Example 5.4. The problem may be posed in terms of the single equation given earlier or in terms of the two equations, with θ and w as the two unknowns,

$$\theta = 70 - 70\exp\left[-\frac{1000}{21(5+20w)}\right]$$
$$250 = 4.2w\theta$$

Solve this system of nonlinear equations by the Newton–Raphson method, and compare the results with those obtained earlier in Example 5.4.

6.42. Solve the following set of nonlinear equations, which govern the flow rates in a network of four pipes, by the Gauss–Seidel iterative method:

$$7a + b^2 + c + d = 3.7$$
$$a^2 + 8b + 3c - d = 4.9$$
$$2a - 2b + 5c + d^2 = 8.8$$
$$a - b + c^2 + 14d = 18.2$$

Because of the physical nature of the problem considered, a, b, c, and d are all real and may be positive or negative. A negative value indicates flow in a direction opposite to that assumed in the analysis.

6.43. Solve the problem discussed in Example 6.9 by the successive substitution method, and compare the convergence characteristics with those for the Newton–Raphson method used in the given example.

6.44. Alternating current electrical circuits are generally solved by the use of complex variables, since the sinusoidal variation can be represented conveniently by complex quantities. The impedance for a resistor is simply its resistance, whereas the impedances for inductors and capacitors are functions of the frequency ω. For an inductor, the impedance is $i\omega L$, where $i = \sqrt{-1}$, and for a capacitor it is–$i\omega C$, where L is the inductance in henries and C the capacitance in farads. For the circuit shown, the ac power source is 15 V with a frequency of ω. The phase angle is arbitrarily taken as zero. Considering the two loop currents I_1 and I_2 and using Kirchhoff's law, obtain two algebraic equations for the currents. Then taking the currents, voltages, and impedances as complex, separate the real and imaginary parts to obtain four linear equations. Solve these equations by using a MATLAB script file as well as by using the backslash operator in MATLAB to obtain the loop currents.

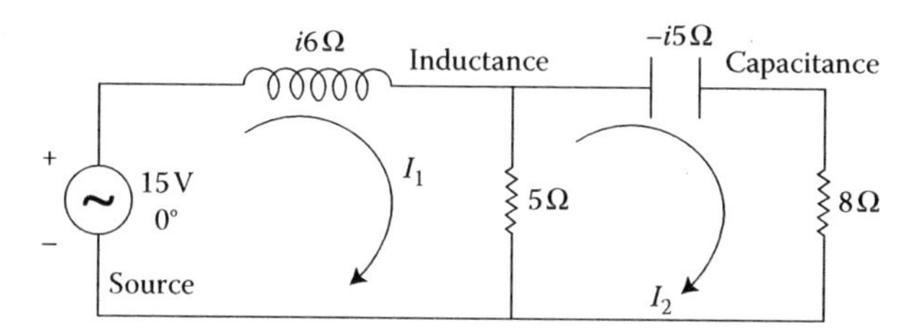

6.45. Following the procedure outlined in Problem 6.44, obtain the linear equations, with complex coefficients, for the ac electrical circuit shown. Again, obtain the corresponding linear equations with real coefficients by separating the real and imaginary parts. Solve these equations to obtain the magnitude and phase angle of the currents.

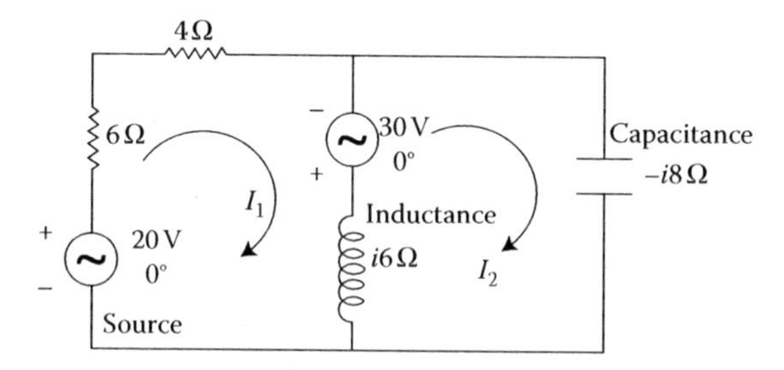

6.46. Under what conditions would the use of under-relaxation in an iterative scheme, such as Gauss–Seidel, be valuable in altering the convergence characteristics of the method? Give examples.

6.47. What would be the result of using over-relaxation with the *Jacobi* method?

6.48. Discuss some physical circumstances for which the iterative methods will be more advantageous to use than the direct methods. Justify your answer.

6.49. Several simple matrix commands in MATLAB were discussed in this chapter. Using any system of equations from the preceding problems, obtain the solution by employing the backslash, *inv*(*a*) and *lu* commands. Comment on the results and computational procedures involved.

7 Numerical Curve Fitting and Interpolation

7.1 INTRODUCTION

A problem of considerable interest in engineering applications is that of representing data at a set of discrete points by means of a smooth and continuous function. Experimental and numerical studies generally yield results at a finite number of data points. Such results are often tabulated. However, a much more useful representation of the data is by means of a smooth curve that passes through the data points or one that is as close as possible to these. This process is known as curve fitting and the equation of the curve can be employed to obtain values at intermediate points where tabulated results are not available. Also, in the numerical simulation of engineering processes and systems, it is more convenient to use a curve fit of the available data on the characteristics of the components, such as pumps and blowers, rather than tabulated results, to obtain the values needed.

Curve fitting is needed in a wide variety of engineering problems. The property data for materials are generally available at discrete values of the independent variable, such as pressure, temperature, and concentration. Curve fitting yields a function $f(x)$, where x is the independent variable and $f(x)$ is a material property, such as density, specific heat, equilibrium constant for a chemical reaction, and electrical resistance. Then this function $f(x)$ may be used to obtain the desired material property at arbitrary values of the independent variable over a given range. Curve fitting of property data is needed in the simulation and study of many diverse engineering applications, such as power plants, refrigeration systems, chemical reactors, environmental processes, electronic systems, and building structures. Similarly, experimental data on many processes of engineering interest, such as wind speed at various heights above the surface of a lake, velocity of a moving body as a function of time, the electrical current in an electronic circuit as a function of the input voltage, and the deflection of a structure under a changing load, are generally obtained in terms of a continuous function $f(x)$, which can be subsequently employed in the analysis and design of relevant engineering processes and systems. The calibration curves for measuring devices, such as pressure transducers and flow meters, are similarly obtained from data taken at discrete points.

7.1.1 EXACT AND BEST FIT

There are two basic approaches to curve fitting. The first one involves determining a curve that passes through every given data point and is known as an *exact fit*, see

Figure 7.1a. Therefore, at the given data points, the curve obtained yields values that are identical to the given data. An exact fit is appropriate if the data have a high level of accuracy, as is often the case for numerical simulations and for material property data. The number of parameters that must be determined for obtaining the approximating curve, which is often taken as a polynomial, must be equal to the number of data points. If the data set is large, the determination of the unknown parameters, which are also consequently large in number, becomes quite involved. Also, the curve obtained is not very convenient to use and is often ill-conditioned. Thus, if a large number of data points is available, the second approach, known as the *best fit*, is more appropriate. In this case, the curve does not pass through every data point. However, the difference between the values given by the approximating curve and the given data is minimized, so that the error in obtaining the values from the curve is small, see Figure 7.1b. The number of parameters in the curve is typically much smaller than the number of data points, and simple curves, such as linear and exponential distributions, are frequently used for curve fitting. This approach is also suitable if the error in the data is significant, so that the fitted curve need not pass through each data point and a best fit is more appropriate.

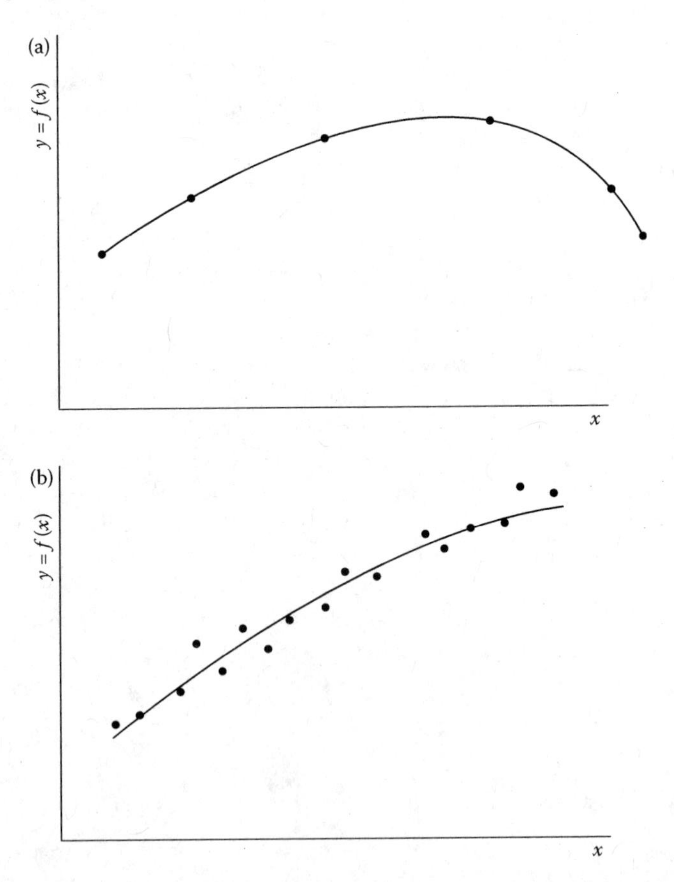

FIGURE 7.1 Curve fitting to given sets of data points: (a) an exact fit and (b) a best fit.

7.1.2 Interpolation and Extrapolation

Interpolation is employed to determine the dependent variable $y = f(x)$ at intermediate values between the given data points, and *extrapolation* is used for finding $f(x)$ outside the range of the given data. Both are extensively used in engineering applications and also in developing numerical procedures for differentiation, integration, root solving, and the solution of differential equations. The use of interpolation and extrapolation in numerical differentiation and in root solving was demonstrated in Chapters 4 and 5. The application to other problems in numerical analysis will be outlined in the following chapters. The basic approach involves fitting an exact curve to a finite number of discrete points and then applying the desired mathematical operation, such as differentiation or integration, to the smooth function obtained.

There are several numerical methods that may be employed for determining the interpolating curve from a given set of data points. Besides the direct evaluation of the parameters of an interpolating polynomial by substituting the given data and solving the resulting set of linear equations, as outlined in Chapter 6, interpolation with *Lagrange polynomials* and *Newton's divided-difference polynomials* is also discussed here. *Splines*, which fit subsets of the data with lower-order polynomials, such as a cubic, are also important in interpolation and are presented in this chapter.

Caution is needed when extrapolation is employed to compute values beyond the range of the given data, since the variation of the dependent variable beyond the given range is not known. There may be substantial changes in the variation as we move outside the given domain of data points. However, extrapolation is frequently used to predict values and trends in order to plan and to take decisions. For example, companies routinely depend on consumer spending trends and predictions of inflation, inventory, money supply and the stock market to plan for the future.

7.1.3 Basic Approach

The choice of the function $f(x)$ to obtain a best fit to a given data set is also an important consideration. Although polynomials, particularly straight lines which lead to *linear regression*, are very often employed for curve fitting, other forms, such as exponential and sinusoidal functions, are also used. As mentioned in Chapter 1, the physical nature of the given problem may often be employed to choose the appropriate form of the function $f(x)$ for a best fit. Periodic processes, such as those encountered in natural phenomena, are usually fitted with sinusoidal functions, as shown in Figure 1.5. Similarly, in chemical reactions where the rate of change of concentration is proportional to the concentration at any given time, the concentration varies exponentially. Therefore, if measurements are taken in such processes, exponential functions are employed for curve fitting. Calibration curves for devices, such as those for measuring pressure and velocity, are generally obtained as polynomials by using curve-fitting techniques for a best fit. A few examples of curve fitting are shown in Figure 7.2. The use of the physical background of the given problem or of any prior information on the variation of the dependent variable helps in the choice of the most appropriate form of the curve for a best fit. A proper choice of the function $f(x)$ for

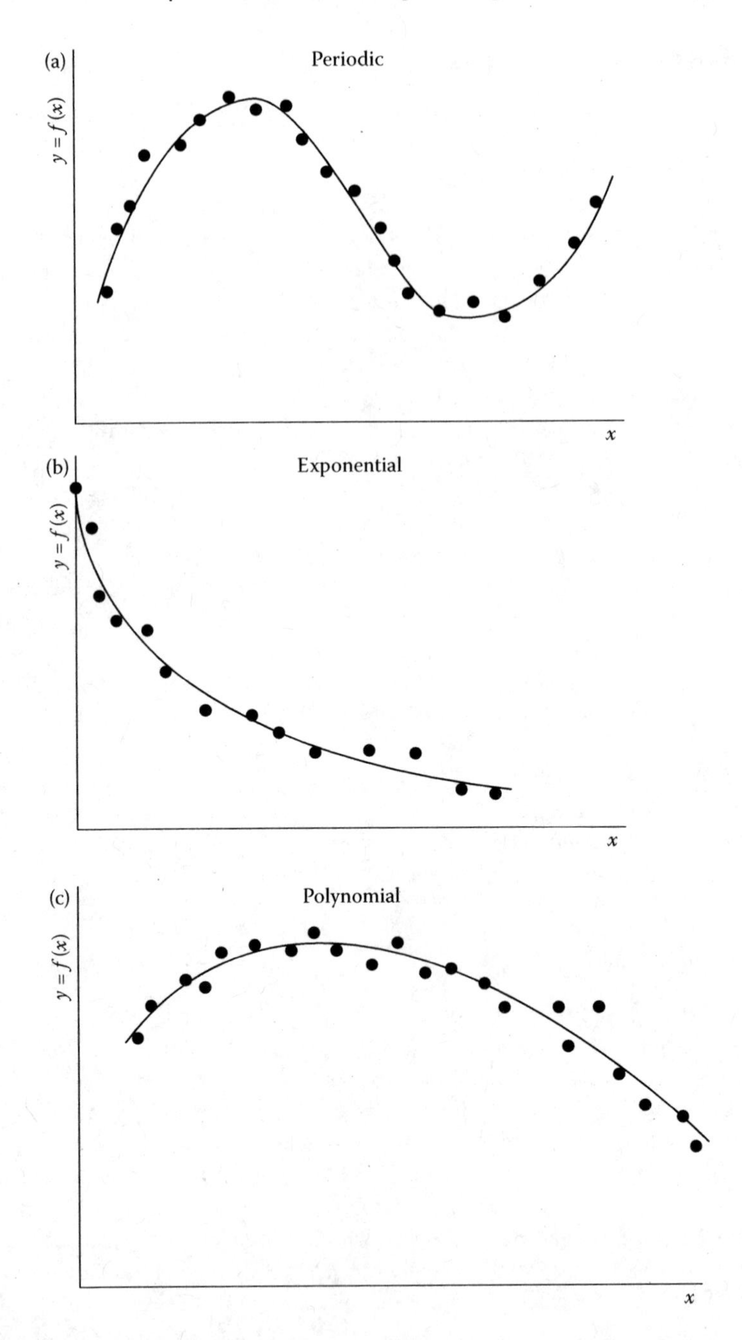

FIGURE 7.2 A few examples of curve fitting, employing different forms of the function $f(x)$ for a best fit.

curve fitting not only reduces the number of parameters to be determined but also yields the desired result in a simple and useful form.

There are several numerical methods for obtaining the best fit to a given data set. The most widely used technique is based on the minimization of the sum of the squares of the differences between the actual data and the values obtained from the best fit. This method, known as the *least-squares fit*, is discussed in detail and applied to different forms of the chosen function $f(x)$ for curve fitting. Other methods are also outlined. Also considered is the circumstance where the dependent variable is a function of more than one independent variable, say, $f(x, y)$. Such problems are of interest, for instance, in chemical reactors and power plants, where the fluid properties and system characteristics depend on two or more independent variables, for instance, pressure, temperature, and concentration.

7.1.4 Use of MATLAB® Commands

As briefly discussed in Chapter 3, MATLAB® has several commands that may be used easily and directly to obtain a best fit to given data and to obtain interpolated values. These include the *polyfit* command, which gives a best fit to the given data set using the specified order of the polynomial for curve fitting, such as 1 for linear, 2 for parabola, and 3 for cubic. Similarly, the *interp1* command is useful in obtaining interpolated values from the given data set, using a specified interpolating polynomial, such as linear, cubic, and spline, where the last one refers to a piecewise exact fit to the given data and is presented in detail later. The *interp2* command is used for two-dimensional curve fitting and *interp3* for three-dimensional. These commands are discussed in greater detail and employed for various examples here.

In this chapter, the numerical methods for obtaining an exact fit to tabulated data at discrete points are considered first. Various interpolation formulations and techniques are discussed. Interpolation with splines, particularly cubic splines, is outlined. The use of the least-squares method for obtaining a best fit with simple polynomials is then discussed in detail. Other forms of the curve for a best fit are also considered. Finally, functions of more than one independent variable are considered, and the corresponding numerical curve-fitting procedures are presented.

7.2 EXACT FIT AND INTERPOLATION

An exact fit of tabulated data is frequently obtained in engineering applications, using polynomials in most cases. One chooses a general form of the approximating polynomial and substitutes the given data into the equation for the polynomial in order to evaluate the parameters in the chosen curve. Therefore, the approximating polynomial passes through each data point and yields the exact value, as the given data, at these points. Since the polynomial is exact at the given data points, it is known as an *exact fit*. Once the approximating curve has been found, one can employ it to determine the values of the dependent variable y, which is a function $f(x)$ of the independent variable x, at arbitrary values of x not included in the tabulated data. As mentioned earlier, if the chosen value of x lies within the range covered by the given data, the function $f(x)$ is found at an intermediate value of x, and the process is known as *interpolation*. If x lies beyond the range of the data, the process is called *extrapolation*.

Both interpolation and extrapolation are widely used in engineering and in numerical analysis. Extrapolation is employed less frequently than interpolation, since there are uncertainties associated with evaluating the function at x values beyond the range for which data are available. If the function is known to be well behaved beyond the range of data, extrapolation may be used. Otherwise, substantial error may arise in the extrapolated value. However, interpolation is routinely used for a wide variety of engineering problems. Measurements and numerical simulations are carried out at a finite number of discrete data points, and curve fitting is used to obtain values at intermediate points. Several examples of engineering problems where this approach is used were given in the preceding section. As mentioned earlier, an exact fit is appropriate if the given data are very accurate and if the number of data points is relatively small, typically less than 10.

7.2.1 Exact Fit with an nth-Order Polynomial

A polynomial of degree n can be devised to exactly fit $(n + 1)$ data points. The general form of the polynomial may be taken as

$$y = f(x) = a_0 + a_1x + a_2x^2 + \cdots + a_nx^n \tag{7.1}$$

where y is the dependent variable, x is the independent variable, and the a's are constants. Two available data points are adequate to describe a first-degree, or linear, equation. Similarly, three data points are needed for a second-degree, or quadratic, equation and four points for a third-degree, or cubic, equation as shown in Figure 7.3. The available data may be denoted as (x_i, y_i) for $i = 0, 1, 2, \ldots, n$, where y_i is the value of the function y at $x = x_i$. Then these values may be substituted into the chosen general form of the polynomial, Equation 7.1, to yield

$$y_i = a_0 + a_1x_i + a_2x_i^2 + \cdots + a_nx_i^n \quad \text{for } i = 0,1,2,\ldots,n \tag{7.2}$$

Since x_i and y_i are known for the given $(n + 1)$ points, Equation 7.2 yields $(n + 1)$ linear equations for the unknown constants a_0 to a_n, as i is varied from 0 to n.

A numerical solution of this linear system, employing the methods given in the preceding chapter, will give these constants, and will thus determine the polynomial that exactly fits the given data points. Example 6.1 demonstrated the solution of such a linear system for determining a fifth-order polynomial that provides an exact fit to the given six data points on the specific volume of steam. The matrix equation that represents the system of equations yielded by Equation 7.2 is

$$\begin{pmatrix} 1 & x_0 & x_0^2 & \cdots & x_0^n \\ 1 & x_1 & x_1^2 & \cdots & x_1^n \\ \vdots & \vdots & \vdots & & \vdots \\ 1 & x_n & x_n^2 & \cdots & x_n^n \end{pmatrix} \cdot \begin{pmatrix} a_0 \\ a_1 \\ \vdots \\ a_n \end{pmatrix} = \begin{pmatrix} y_0 \\ y_1 \\ \vdots \\ y_n \end{pmatrix} \tag{7.3}$$

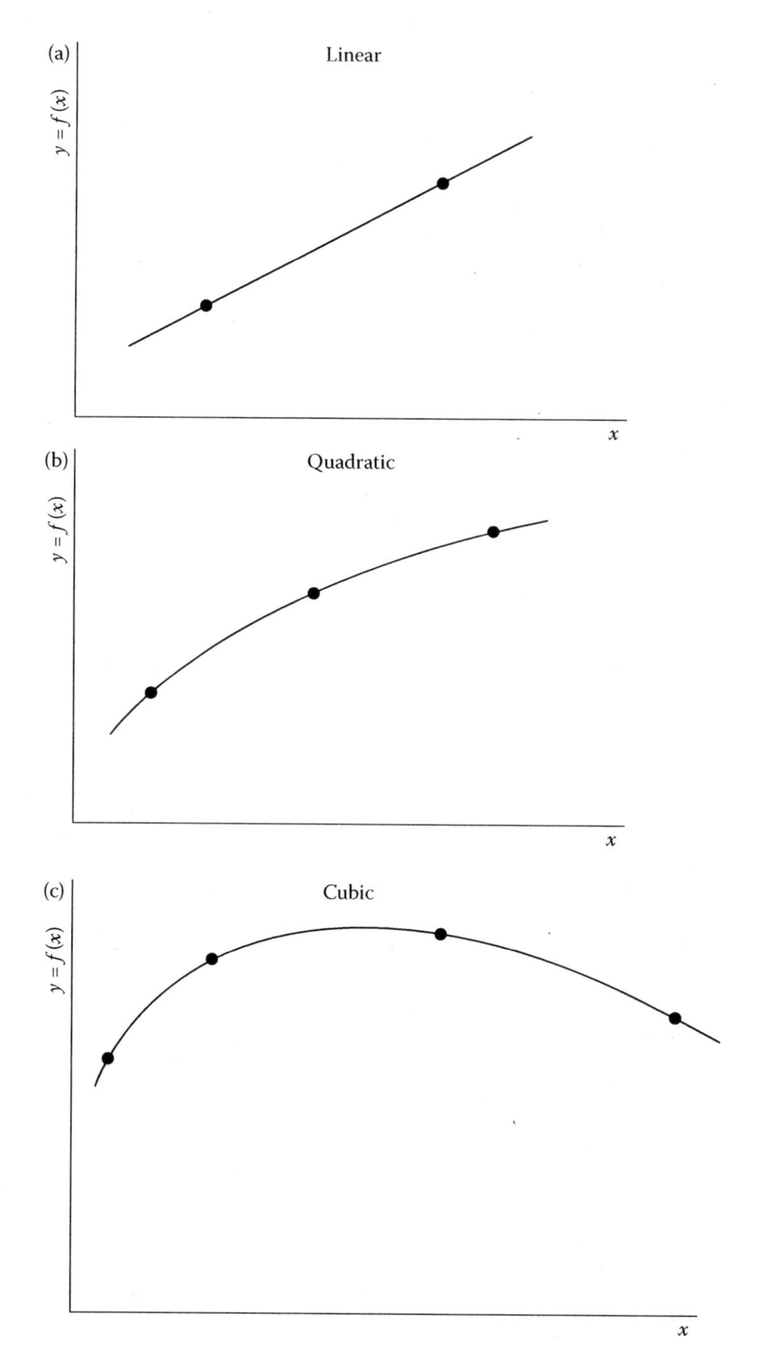

FIGURE 7.3 Exact fits to given data, using polynomials of first, second, or third order.

The determinant of the above coefficient matrix is known as the *Vandermond determinant.* It is nonzero unless a point is duplicated, that is, $x_i = x_j$ for $i \neq j$. Therefore, as shown in Chapter 6, a unique solution may be obtained for $a_0, a_1, \ldots, a_n$ from Equation 7.3, giving a unique polynomial that yields the exact value of the dependent variable at the given data points.

The approach outlined above is fairly simple and can be used for an arbitrary distribution of data points. However, as mentioned earlier, it is appropriate for relatively small sets of data and for cases where the given data are very accurate. An exact fit with a single curve is generally employed if the number of data points is typically less than 10. For a larger number of points, higher-order polynomials are needed. The coefficients of the polynomial may then be quite small, particularly for the higher-order terms and if the independent variable attains large values. An example of such a representation is the variation of a material property, such as electrical resistivity ρ with temperature T, given by the polynomial

$$\rho(T) = a_0 + a_1T + a_2T^2 + \cdots + a_{19}T^{19} + a_{20}T^{20}$$

where T could vary from, say, 20°C to 300°C. Then the value of the coefficients, particularly the higher-order ones such as a_{19} and a_{20}, will be very small, giving rise to accuracy problems in interpolation. The polynomial may also be ill-conditioned, so that small changes in T result in large changes in $\rho(T)$. Then even the round-off error is magnified to yield inaccurate results from the use of such a polynomial for interpolation.

One method of avoiding these problems is to use a polynomial of lower order, based on a corresponding smaller data set chosen from the given data. Sometimes, normalization of the independent variable, say, by defining a new variable $\tilde{T}$ where $\tilde{T} = T/20$ in the above example, reduces the range of variation of the independent variable and thus avoids very small values of the coefficients. This approach was employed in Example 6.1. The normalizing characteristic quantity, such as 20°C in the above example or 10°C in Example 6.1, may often be chosen arbitrarily or on the basis of physical reasoning to reduce the range of the independent variable to a desired level.

Considering Example 6.1 again, six data points are given and a fifth-order polynomial is to be determined for an exact fit. The matrices corresponding to Equation 7.3 are obtained from the data and the constants $a_0, a_1, \ldots, a_n$ are determined. Appendix B.15 gives a MATLAB script file for this problem, using the backslash operator to solve the system of linear equations to obtain the a's. The results obtained from this program are shown in Figure 7.4.

It is easy to see that the results obtained are close to those presented earlier in Example 6.1, in terms of the a's and the curve fit obtained. The plot of the polynomial obtained and the data points is essentially identical to that obtained earlier and shown in Figure 6.3. Once the polynomial is obtained, the values at an intermediate point may be determined. For instance, the value at $x = 3.4$ is determined here from the exact fit obtained. We may also use the command *interp1*, with specified curve fit

Coefficients of the polynomial are

a =

201.2600
−128.8210
40.6742
−7.4229
0.7408
−0.0311

Current plot held
Interpolated value from exact fit y = 26.5890
Value from linear interpolation y = 27.5480
Value from spline interpolation y = 26.5999

FIGURE 7.4 Coefficients of the polynomial for an exact fit to the data in Example 6.1 and interpolated values at $x = 3.4$ obtained from this fit, as well as from linear and spline interpolations.

for interpolation, to obtain the value at an intermediate point. The value at $x = 3.4$ is obtained here by using a linear interpolation, as well as a spline interpolation. The corresponding MATLAB commands are given in Appendix B.15. It is seen that the two values are somewhat different, with the latter being closer to that from the polynomial exact fit. Splines are discussed in Section 7.5.

7.2.2 Uniformly Spaced Independent Variable

In certain cases, the values of the function y are given at uniformly spaced values of the independent variable x. Such a circumstance arises, for instance, in numerical calculations and experimental studies where the values of x are taken as equally spaced for convenience. This is particularly true for tabulated data of material properties.

If the independent variable is uniformly spaced, as shown in Figure 7.5, the determination of the polynomial $f(x)$ which exactly fits the given data can be simplified by choosing the following alternate form of the polynomial, instead of the general form given by Equation 7.1:

$$y - y_0 = a_1\left[\frac{n}{L}(x - x_0)\right] + a_2\left[\frac{n}{L}(x - x_0)\right]^2 + \cdots + a_n\left[\frac{n}{L}(x - x_0)\right]^n \tag{7.4}$$

where L is the range, $x_n - x_0$, of x, n is the order of the polynomial, and the a's are the coefficients to be determined. Thus, the number of data points is $(n + 1)$, with (x_0, y_0), $(x_1, y_1), \ldots, (x_i, y_i), \ldots, (x_n, y_n)$ representing the given data that are to be numerically curve fitted with a polynomial. Here,

$$\frac{L}{n} = x_1 - x_0 = \frac{x_2 - x_0}{2} = \frac{x_3 - x_0}{3}, \ldots$$

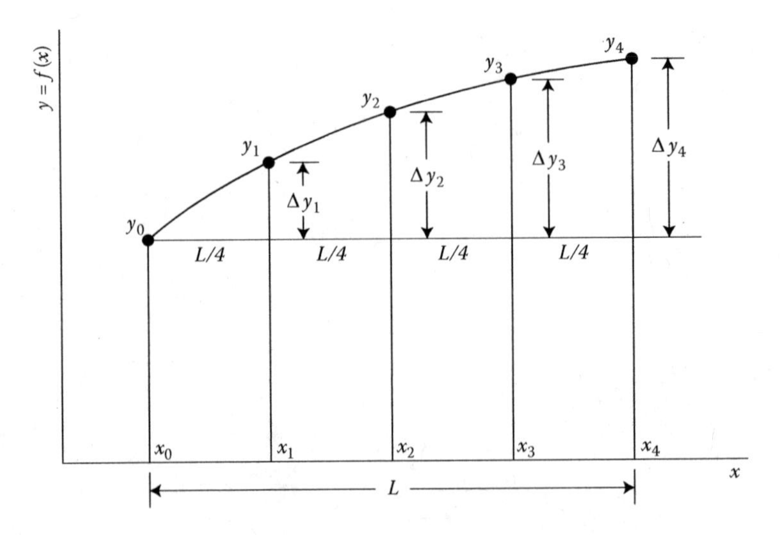

FIGURE 7.5 Exact fit to uniformly spaced data.

since the values of x are equally spaced. If the data points are successively substituted into Equation 7.4, we obtain the following in terms of the Δy's shown in Figure 7.5:

$$
\begin{aligned}
\Delta y_1 &= y_1 - y_0 = a_1 + a_2 + a_3 + \cdots + a_n \\
\Delta y_2 &= y_2 - y_0 = 2a_1 + 2^2 a_2 + 2^3 a_3 + \cdots + 2^n a_n \\
\Delta y_3 &= y_3 - y_0 = 3a_1 + 3^2 a_2 + 3^3 a_3 + \cdots + 3^n a_n \\
&\vdots \\
\Delta y_n &= y_n - y_0 = na_1 + n^2 a_2 + n^3 a_3 + \cdots + n^n a_n
\end{aligned}
\tag{7.5}
$$

Equations 7.5 can be solved for $a_1, a_2, \ldots, a_n$ in terms of $\Delta y_1, \Delta y_2, \ldots, \Delta y_n$ for chosen values of n to obtain the constants for an nth-order polynomial. Table 7.1 gives the constants for polynomials of the first four degrees, that is, $n = 1$ through $n = 4$. Then, for a given set of $(n + 1)$ data points, the degree of the polynomial is n, and the values of the constants are obtained simply by substituting the values of the differences $\Delta y_1, \Delta y_2, \ldots, \Delta y_n$ into the expressions given in the table. If several data sets are to be considered, the expressions for the coefficients of the polynomials, such as those given in Table 7.1, may be stored in the computer and the constants computed as each data set is entered. If only one or two sets of data are to be curve fitted or if higher-order polynomials are to be used, it would generally be easier to simply solve Equation 7.5 for the constants, using the methods given in Chapter 6 for systems of linear algebraic equations, see Appendix B.15. However, programs based on this approach for curve fitting are frequently available on computers, and the expressions for the coefficients may already be stored, so that the given data, at uniformly spaced values of the independent variable, are entered and the program yields the constants of an approximating polynomial of the form given by Equation 7.4. Interpolated

TABLE 7.1
Constants for Polynomials of the Form Given by Equation 7.4 for Data in Which the Independent Variable Is Uniformly Spaced

Polynomial	a_4	a_3	a_2	a_1
Fourth degree, $n = 4$	$\frac{1}{24}(\Delta y_4 - 4y_3 + 6\Delta y_2 - 4\Delta y_1)$	$\frac{1}{6}(3\Delta y_1 - \Delta y_3 - 3\Delta y_2) - 6a_4$	$\frac{1}{2}(\Delta y_2 - 2\Delta y_1) - 3a_3 - 7a_4$	$\Delta y_1 - a_2 - a_3 - a_4$
Cubic, $n = 3$		$\frac{1}{6}(3\Delta y_1 - \Delta y_3 - 3\Delta y_2)$	$\frac{1}{2}(\Delta y_2 - 2\Delta y_1) - 3a_3$	$\Delta y_1 - a_2 - a_3$
Quadratic, $n = 2$			$\frac{1}{2}(\Delta y_2 - 2\Delta y_1)$	$\Delta y_1 - a_2$
Linear, $n = 1$				Δy_1

values are then obtained from the resulting polynomial, as shown in the following example.

Example 7.1

In a fluid flow system, which experimentally simulates the flow generated in a room due to a fire, the flow rate F is measured at several values of a governing parameter R, known as the *Richardson number*, where R depends on the heat input by the fire, room dimensions, and fluid properties. The data obtained are as follows:

R	0.025	0.05	0.1	0.2	0.3	0.4	0.5
F	1.4198	2.548	4.2	5.978	6.908	7.613	7.799

Employing the last five data points, obtain an exact fit and compute the interpolated values at $R = 0.25$ and 0.35. Also, obtain the flow rates at $R = 0$, 0.025, and 0.05 by extrapolation. Compare the last two extrapolated values with the given data.

SOLUTION

Since the five data points to be considered for an exact fit, with a fourth-order polynomial, are uniformly spaced, the approach given in Section 7.2.2 may be employed. The independent variable is R, and the dependent variable is F. Denoting the data points as (R_0, F_0) (R_1, F_1), …, (R_4, F_4), we may compute the ΔF's as

$$\begin{aligned}\Delta F_1 &= F_1 - F_0 = 5.978 - 4.2\\ \Delta F_2 &= F_2 - F_0 = 6.908 - 4.2\\ \Delta F_3 &= F_3 - F_0 = 7.613 - 4.2\\ \Delta F_4 &= F_4 - F_0 = 7.799 - 4.2\end{aligned} \tag{7.6}$$

where $R_0 = 0.1$ and $F_0 = 4.2$. Also, the uniform spacing is $R_1 - R_0 = R_2 - R_1 = R_3 - R_2 = R_4 - R_3 = 0.1$. The total range L is 0.4, and the degree n of the polynomial is 4. Therefore, from Equation 7.4, the polynomial to be determined may be written as follows:

$$F - F_0 = a_1\left(\frac{R - R_0}{0.1}\right) + a_2\left(\frac{R - R_0}{0.1}\right)^2 + a_3\left(\frac{R - R_0}{0.1}\right)^3 + a_4\left(\frac{R - R_0}{0.1}\right)^4 \quad (7.7)$$

This equation gives

$$F = F_0 + A_1(R - R_0) + A_2\,(R - R_0)^2 + A_3(R - R_0)^3 + A_4(R - R_0)^4 \quad (7.8)$$

where $A_1 = a_1/0.l$, $A_2 = a_2/(0.1)^2$, and so on.

A simple computer program may be written to compute the differences from Equation 7.6 and the coefficients in Equation 7.7 from Table 7.1. These coefficients are then converted to the coefficients of Equation 7.8 for convenience in the application of the polynomial for interpolation. Figure 7.6 presents the results from such a program. The data points are entered, and the coefficients of the polynomial in Equation 7.8 are computed. This polynomial is then employed to compute the interpolated or extrapolated values of the flow rate F at several values of R, entered interactively into the program.

Note from Figure 7.6 that the polynomial obtained yields the exact values at the given data points, as expected. The extrapolated values at $R = 0.025$ and 0.05 are close to those obtained experimentally and given in the problem. The value at $R = 0$ is expected to be zero on physical grounds. However, the extrapolated value is nonzero, although it is fairly small, being equal to 0.034. Thus, over the range of R considered here, $0 < R < 0.5$, the computed polynomial yields good accuracy. For values of R larger than 0.5, extrapolation does not give satisfactory results, since F is obtained as decreasing with increasing R, which is contrary to the behavior expected for the physical problem considered. However, interpolation in the range $0.1 \le R \le 0.5$ yields physically realistic results, being exact at the data points employed for deriving the polynomial.

7.3 LAGRANGE INTERPOLATION

A method that is widely used for obtaining an exact fit to a given data set is *Lagrange interpolation*. It is based on the use of a special form of the interpolating polynomial, known as the *Lagrange polynomial*. For a quadratic function, it is written as

$$y = f(x) = a_0\,(x - x_1)(x - x_2) + a_1\,(x - x_0)(x - x_2) + a_2\,(x - x_0)(x - x_1) \quad (7.9)$$

where (x_0, y_0),(x_1, y_1), and (x_2, y_2) are the three given data points and a_0, a_1, and a_2 are the constants to be determined from these points. It can easily be seen that Equation 7.9 represents a second-order polynomial, which can also be rewritten in the form given by Equation 7.1. Substituting the given data points successively in Equation 7.9, we can easily determine the constants as follows:

$$a_0 = \frac{y_0}{(x_0 - x_1)(x_0 - x_2)}$$

```
R0 = 0.1
F0 = 4.2
Fl = 5.978
F2 = 6.908
F3 = 7.613
F4 = 7.799

THE COEFFICIENTS OF THE POLYNOMIAL ARE
Al = 26.3892              A2  = -115.5793
A3 = 333.0837             A4  = -382.0837

ENTER THE VALUE OF R AT WHICH INTERPOLATION IS DESIRED
R = 0.0
R = 0                     F = .034
ENTER THE VALUE OF R AT WHICH INTERPOLATION IS DESIRED
R = 0.025
R = .025                  F = 1.4181
ENTER THE VALUE OF R AT WHICH INTERPOLATION IS DESIRED
R = 0.05
R = .05                   F = 2.5476
ENTER THE VALUE OF R AT WHICH INTERPOLATION IS DESIRED
R = 0.25
R = .25                   F = 6.4886
ENTER THE VALUE OF R AT WHICH INTERPOLATION IS DESIRED
R = 0.35
R = .35                   F = 7.2855
ENTER. THE VALUE OF R AT WHICH INTERPOLATION IS DESIRED
R = 0.4
R = .4                    F = 7.613
ENTER THE VALUE OF R AT WHICH INTERPOLATION IS DESIRED
R = 0.5
R = .5                    F = 7.799
```

FIGURE 7.6 Coefficients of the polynomial obtained for an exact fit in Example 7.1, along with the interpolated results for the dependent variable F at several values of the independent variable R.

$$a_1 = \frac{y_1}{(x_1 - x_0)(x_1 - x_2)} \tag{7.10}$$

$$a_2 = \frac{y_2}{(x_2 - x_0)(x_2 - x_1)}$$

Similarly, if $(n + 1)$ data points are given, an nth-order Lagrange polynomial may be written by taking n factors in each term, instead of two taken for the quadratic function of Equation 7.9. Thus, an nth-order Lagrange polynomial is written as

$$y = f(x) = a_0(x-x_1)(x-x_2)\cdots(x-x_n) + a_1(x-x_0)(x-x_2)\cdots(x-x_n) + \cdots + a_n(x-x_0)(x-x_1)\cdots(x-x_{n-1}) \quad (7.11)$$

The coefficients a_i, where i varies from 0 to n, can be determined by substitution of the $(n + 1)$ data points into Equation 7.11 to obtain expressions such as those given by Equation 7.10. The resulting interpolating polynomial is

$$y = f(x) = \sum_{i=0}^{n} y_i \prod_{\substack{j=0 \\ j \neq i}}^{n} \left(\frac{x - x_j}{x_i - x_j} \right) \quad (7.12)$$

where the product sign $\prod$ denotes multiplication of the n factors obtained by varying j from 0 to n, excluding $j = i$, for the quantity within the parentheses. Thus, for example, a third-order Lagrange polynomial is obtained from Equation 7.12 as

$$y = \frac{(x-x_1)(x-x_2)(x-x_3)}{(x_0-x_1)(x_0-x_2)(x_0-x_3)} y_0 + \frac{(x-x_0)(x-x_2)(x-x_3)}{(x_1-x_0)(x_1-x_2)(x_1-x_3)} y_1 + \frac{(x-x_0)(x-x_1)(x-x_3)}{(x_2-x_0)(x_2-x_1)(x_2-x_3)} y_2 + \frac{(x-x_0)(x-x_1)(x-x_2)}{(x_3-x_0)(x_3-x_1)(x_3-x_2)} y_3$$

Lagrange interpolation is applicable to an arbitrary distribution of the independent variable x. The determination of the coefficients of the polynomial does not require the solution of a system of equations, as was the case for the methods discussed in the preceding section. The interpolating polynomial, Equation 7.12, can easily be entered and the necessary calculations performed on a computer for obtaining the desired exact fit to the given data. The programming is quite simple, as illustrated in Example 7.2. Because of the applicability of the method to arbitrary distributions of data points and the ease with which it may be applied, Lagrange interpolation is widely employed for engineering applications. Programs available on many computers for interpolation are also frequently based on this method.

Example 7.2

The deflection of a structure under loading is measured at five different values of the force applied X, in kilonewtons (kN). The deflection Y is in centimeters, and the data are given as follows:

X (kN)	0.5	1.0	1.5	2.0	2.5
Y (cm)	3.0	3.9	5.2	7.3	10.5

Employing Lagrange interpolation, compute the deflection at the intermediate load values of 0.75, 1.25, 1.8, and 2.2 kN. Also obtain the extrapolated values at 0 and 3.0 kN. Such problems are of interest in civil engineering, though many more data points are generally obtained, requiring higher-order polynomials for an exact fit or resorting to a best fit.

SOLUTION

The Lagrange polynomial to be computed is given by Equation 7.12. Since five data points are given, a fourth-order polynomial can be derived to exactly fit the given points. The coefficients a_i of the polynomial in Equation 7.11 are given by the product

$$a_i = \prod_{\substack{j=0 \\ j \neq i}}^{n} \left(\frac{Y_i}{X_i - X_j} \right) \tag{7.13}$$

where X_i and Y_i are the values of the independent and dependent variables, respectively, at the data points. The interpolated value of Y at a given X is then obtained from Equation 7.11.

We can easily write a computer program to calculate the coefficients of the polynomial and then to use these to determine the corresponding interpolated value of Y. Appendix C.9 shows the program in Fortran for Lagrange interpolation. The number of data points N is read, along with the given data. Also read is the number of intermediate points M at which interpolated or extrapolated values of the dependent variable are desired. The values of the independent variable at which interpolation/extrapolation is needed are denoted by XL and are read from the data entered. The coefficients, a_i, or $A(I)$ here, of the Lagrange polynomial are calculated from Equation 7.13, and then the interpolated/extrapolated value of the dependent variable, denoted by YL, is obtained from Equation 7.11. The values of XL are sequentially changed according to the given input, and the corresponding values of YL are computed. Finally, the calculated results are printed in tabular form, as shown in Figure 7.7, along with the coefficients $A(I)$ of the Lagrange polynomial.

```
THE VALUES FROM LAGRANGE INTERPOLATION ARE
XL = 0.0000          YL = 2.0000
XL = 0.5000          YL = 3.0000
XL = 0.7500          YL = 3.4289
XL = 1.0000          YL = 3.9000
XL = 1.2500          YL = 4.4727
XL = 1.5000          YL = 5.2000
XL = 1.8000          YL = 6.3426
XL = 2.0000          YL = 7.3000
XL = 2.2000          YL = 8.4346
XL = 2.5000          YL = 10.5000
XL = 3.0000          YL = 15.0000

COEFFICIENTS OF THE LAGRANGE POLYNOMIAL ARE
A(1) = 2.0000
A(2) = –10.4000
A(3) = 20.8000
A(4) = –19.4667
A(5) = 7.0000
```

FIGURE 7.7 Computed interpolated and extrapolated values from Lagrange interpolation and coefficients of the Lagrange polynomial for Example 7.2.

Coefficients of the polynomial in descending powers of x are
-0.0667
0.8667
-1.3833
2.4833
2.0000
Interpolated values:
xp = 0.0000 yp = 2.0000
xp = 0.5000 yp = 3.0000
xp = 0.7500 yp = 3.4289
xp = 1.0000 yp = 3.9000
xp = 1.2500 yp = 4.4727
xp = 1.5000 yp = 5.2000
xp = 1.8000 yp = 6.3426
xp = 2.0000 yp = 7.3000
xp = 2.2000 yp = 8.4346
xp = 2.5000 yp = 10.5000
xp = 3.0000 yp = 15.0000

FIGURE 7.8 Coefficients of the general polynomial in descending powers of x, as calculated by the use of the MATLAB script file in Appendix B.16 for Lagrange interpolation in Example 7.2, along with computed interpolated values at various specified values of the independent variable.

Note that, as expected, the calculated values of the dependent variable are exact at the data points employed for obtaining the Lagrange polynomial. The interpolated values at XL = 0.75, 1.25, 1.8, and 2.2 kN are found to be within the expected range. Also, a deflection of 2.0 cm is obtained at zero load, indicating the deflection due only to the weight of the structure. The extrapolated value at XL = 3.0 kN is 15.0 cm, which is qualitatively satisfactory, since the deflection increases with load. However, both values at XL = 0 and 3.0 are beyond the range of the given data and their accuracy is not known. Thus, these values must be used with caution, unless validation from further experimentation is obtained.

Appendix B.16 presents the script file in MATLAB for Lagrange interpolation. The logic is similar to that discussed above. However, the ease with which polynomials may be specified and multiplied makes the program quite simple. Also, the general polynomial, such as Equation 7.1, in descending powers of x is obtained directly. This polynomial can then be employed to calculate the interpolated values at chosen values of x and also for plotting, if needed. The results obtained from this script file are given in Figure 7.8.

7.4 NEWTON'S DIVIDED-DIFFERENCE INTERPOLATING POLYNOMIAL

An extensively used form of the polynomial for interpolation is *Newton's divided-difference polynomial.* It can be used for an arbitrary distribution of data points, although simplified formulas result for uniformly spaced points and form the basis for several interpolation schemes, such as forward, backward, and central *Newton–Gregory* formulas, as outlined later in this section.

7.4.1 General Formulas

First-order, or linear, interpolation is the simplest form of interpolation and is obtained by drawing a straight line connecting two data points, as sketched in Figure 7.8. The value of the function $f(x)$ at a given value of the independent variable x can be obtained from the interpolating straight line. Thus, from geometry,

$$\frac{f(x)-f(x_0)}{x-x_0}=\frac{f(x_1)-f(x_0)}{x_1-x_0}$$

or

$$\begin{aligned} f(x) &= f(x_0)+\frac{f(x_1)-f(x_0)}{x_1-x_0}(x-x_0) \\ &= c_0+c_1(x-x_0) \end{aligned} \tag{7.14}$$

where c_0 and c_1 are coefficients of the interpolating polynomial. Here, c_1 represents a finite divided-difference approximation of the first derivative, as given by Equation 4.17. Only two coefficients are needed here because the interpolating polynomial is a straight line.

In a similar way, a second-order, or quadratic, interpolation may be considered. On the basis of Equation 7.14, the general form of the polynomial is taken as

$$y=f(x)=c_0+c_1\,(x-x_0)+c_2\,(x-x_0)(x-x_1) \tag{7.15}$$

Three data points are needed to determine the coefficients c_0, c_1, and c_2. Employing the first point, denoted by (x_0, y_0), we obtain c_0 as

$$c_0=f(x_0) \tag{7.16}$$

Similarly, the second point, (x_1, y_1), yields

$$c_1=\frac{f(x_1)-c_0}{x_1-x_0}=\frac{f(x_1)-f(x_0)}{x_1-x_0} \tag{7.17}$$

We obtain the third coefficient c_2 by substituting the third point, (x_2, y_2), in Equation 7.15. Employing the results given by Equations 7.16 and 7.17, we obtain

$$\begin{aligned} c_2 &= \frac{f(x_2)-f(x_0)}{(x_2-x_0)(x_2-x_1)}-\frac{f(x_1)-f(x_0)}{(x_1-x_0)(x_2-x_1)} \\ &= \frac{\dfrac{f(x_2)-f(x_1)}{x_2-x_1}-\dfrac{f(x_1)-f(x_0)}{x_1-x_0}}{x_2-x_0} \end{aligned} \tag{7.18}$$

Therefore, from the preceding equations, c_0 and c_1 for quadratic interpolation are identical to those for linear interpolation. The third term on the right-hand side of

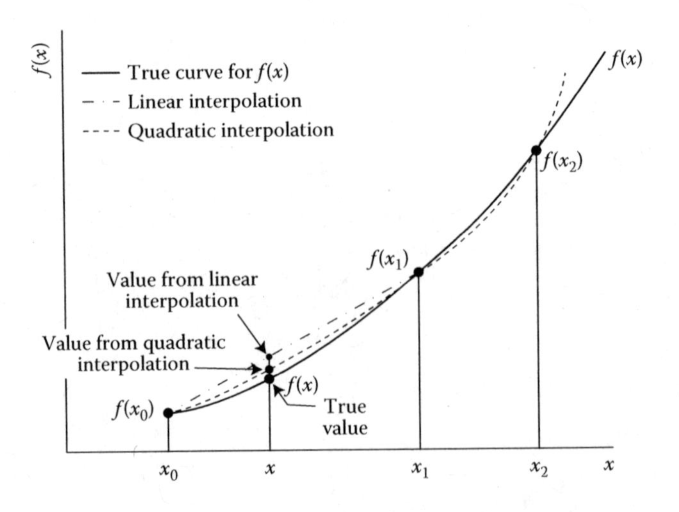

FIGURE 7.9 Interpolation with a straight line and a second-order polynomial for the derivation of Newton's divided-difference method.

Equation 7.15 improves the interpolation by introducing curvature, as shown graphically in Figure 7.9. The coefficient c_2 is similar to the finite difference representation of the second derivative; see Equation 4.21. The general form of the polynomial, Equation 7.15, is similar to the Taylor series expansion, presented in Section 4.2. Also, the first-order divided difference, Equation 7.17, can be used to determine the second-order divided difference, Equation 7.18. These features allow the development of a recursive formula for determining the coefficients of Newton's interpolating polynomials of arbitrary order.

From the above discussion, the general form for an nth-order Newton's polynomial may be written as follows:

$$\begin{aligned} y = f(x) = c_0 + c_1(x - x_0) + c_2(x - x_0)(x - x_1) + \cdots \\ + c_n(x - x_0)(x - x_1)\cdots(x - x_{n-1}) \end{aligned} \tag{7.19}$$

To determine the $(n + 1)$ coefficients, c_0, c_1, ..., c_n, in the nth-order polynomial, we need $(n + 1)$ data points. When these points, denoted by (x_0, y_0),(x_1, y_1), ..., (x_n, y_n), are substituted in the general form of the polynomial, the coefficients are given by the equations

$$\begin{aligned} c_0 &= f(x_0) \\ c_1 &= \frac{f(x_1) - f(x_0)}{x_1 - x_0} = F(x_1, x_0) \\ c_2 &= \frac{F(x_2, x_1) - F(x_1, x_0)}{x_2 - x_0} = F(x_2, x_1, x_0) \\ &\vdots \\ c_n &= F(x_n, x_{n-1}, \ldots, x_1, x_0) \end{aligned} \tag{7.20}$$

where the function F denotes finite divided differences. Therefore,

$$F(x_i, x_j) = \frac{f(x_i) - f(x_j)}{x_i - x_j}$$

$$F(x_i, x_j, x_k) = \frac{F(x_i, x_j) - F(x_j, x_k)}{x_i - x_k} \tag{7.21}$$

$$F(x_n, x_{n-1}, \ldots, x_1, x_0) = \frac{F(x_n, x_{n-1}, \ldots, x_2, x_1) - F(x_{n-1}, x_{n-2}, \ldots, x_1, x_0)}{x_n - x_0}$$

Note from the above expressions that a recursive formula may be written to determine the coefficients. Therefore, the problem is well suited for digital computation. The higher-order differences are determined from the lower-order differences. Therefore, we evaluate the coefficients by starting with c_0 and successively calculating c_1, c_2, c_3, and so on, up to c_n. Once the coefficients have been determined, the interpolating polynomial is obtained from Equation 7.19, which may also be written as

$$y = f(x) = f(x_0) + (x - x_0)F(x_1, x_0) + (x - x_0)(x - x_1)F(x_2, x_1, x_0) + \cdots$$
$$+ (x - x_0)(x - x_1)\cdots(x - x_{n-1})F(x_n, x_{n-1}, \ldots, x_1, x_0) \tag{7.22}$$

As mentioned above, the general form of Newton's interpolating polynomial is similar to the Taylor-series expansion, since terms representing higher-order derivatives are successively added to improve the accuracy of the representation. As given by Equation 4.7, the remainder term R_n in a Taylor-series expansion is

$$R_n = \frac{d^{n+1} f}{dx^{n+1}}(\xi)\frac{(x_{i+1} - x_i)^{n+1}}{(n+1)!}, \quad \text{where } x_i < \xi < x_{i+1}$$

The derivative is evaluated at a point ξ which lies in the interval from x_i to x_{i+1}. Similarly, for an nth-order Newton's interpolating polynomial, the expression for the remainder and, thus, for the error is

$$R_n = \frac{d^{n+1} f}{dx^{n+1}}(\xi)\frac{(x - x_0)(x - x_1)\ldots(x - x_n)}{(n+1)!}, \quad \text{where } x_0 < \xi < x_n \tag{7.23}$$

Since the function $f(x)$ and its derivatives are not known, in general, the $(n + 1)$th derivative may be replaced by the corresponding finite divided difference. Thus, R_n may be written as

$$R_n = [F(x, x_n, x_{n-1}, \ldots, x_1, x_0)](x - x_0)(x - x_1)\ldots(x - x_n) \tag{7.24}$$

This expression can be used to estimate the error if an additional data point (x_{n+1}, y_{n+1}) is available as follows:

$$R_n = [F(x_{n+1}, x_n, \ldots, x_1, x_0)](x - x_0)(x - x_1)\ldots(x - x_n) \tag{7.25}$$

Since the additional data point is generally not available, the interpolating polynomial itself may be used to obtain an additional point, and the error determined from Equation 7.25. Example 7.3 illustrates the use of Newton's method for interpolation.

7.4.2 Uniformly Spaced Data

Several simplified formulas can be derived from the above results if the data are given at equally spaced values of the independent variable x. If Δx is the interval between the data, the values of x are given by

$$x_i = x_0 + i\,\Delta x \quad \text{for } i = 1, 2, \ldots, n \tag{7.26}$$

Such uniformly spaced data are obtained, for instance, from numerical simulations of engineering systems, tables of material properties, and experimental studies in which the independent variable is taken at uniformly spaced intervals for convenience. Then the coefficients c_0, c_1, and c_2 are given by

$$\begin{aligned}
c_0 &= f(x_0)\\
c_1 &= \frac{f(x_1)-f(x_0)}{\Delta x} = \frac{\Delta f_0}{\Delta x}\\
c_2 &= \frac{f(x_2)-2f(x_1)+f(x_0)}{2(\Delta x)^2} = \frac{\Delta^2 f_0}{2!(\Delta x)^2}
\end{aligned} \tag{7.27}$$

where Δf_0 is known as the first forward difference and $\Delta^2 f_0$ as the second forward difference at $x = x_0$. These constitute the numerator of the forward finite difference approximations of the first and second derivatives to $0(\Delta x)$, see Figure 4.7. Therefore, in general, the coefficient c_n of the polynomial is given by

$$c_n = F(x_n, x_{n-1}, \ldots, x_1, x_0) = \frac{\Delta^n f_0}{n!(\Delta x)^n} \tag{7.28}$$

From Equation 7.22, Newton's interpolating polynomial can be written for equally spaced data as follows:

$$\begin{aligned}
y = f(x) = f(x_0) &+ \frac{\Delta f_0}{\Delta x}(x-x_0) + \frac{\Delta^2 f_0}{2!(\Delta x)^2}(x-x_0)(x-x_0-\Delta x)+\cdots\\
&+ \frac{\Delta^n f_0}{n!(\Delta x)^n}(x-x_0)(x-x_0-\Delta x)\ldots[x-x_0-(n-1)\Delta x]+R_n
\end{aligned} \tag{7.29}$$

where the remainder R_n is the same as that given by Equation 7.23. The above interpolating polynomial is known as the *Newton–Gregory forward interpolation formula.* One can generate a forward difference table by taking forward differences at each x, then taking differences of the differences, and so on. An example of such a forward difference table is shown in Table 7.2a. The general formula for computing these differences at $x = x_i$ is

$$\Delta^n f_i = \Delta^{n-1} f_{i+1} - \Delta^{n-1} f_i \tag{7.30}$$

Then these differences may be substituted into Equation 7.29 to yield the interpolating polynomial. The subscript gives the location, in x, where the difference is evaluated, and the superscript indicates the order of the difference. The lowest order differences Δf_i are given by $n = 1$. Also, $n = 0$ corresponds to the values of the function f_i.

In a similar way, a backward difference polynomial, known as the *Newton–Gregory backward interpolation polynomial*, may be derived. The backward differences at $x = x_n$ are denoted by ∇f_n, $\nabla^2 f_n$, and so on, and are obtained from Figure 4.8 or a backward difference table generated in a manner analogous to that for forward differences, see Table 7.2b. The corresponding interpolating polynomial is written as follows:

$$y = f(x) = f(x_0) + \frac{\nabla f_n}{\Delta x}(x - x_n) + \frac{\nabla^2 f_n}{2!(\Delta x)^2}(x - x_n)(x - x_n + \Delta x) + \cdots$$
$$+ \frac{\nabla^n f_n}{n!(\Delta x)^n}(x - x_n)(x - x_n - \Delta x)\ldots[x - x_n + (n-1)\Delta x] + R_n \tag{7.31}$$

TABLE 7.2
Examples of Difference Tables for Computing the Interpolation Polynomials, Using Divided Differences, for Uniformly Spaced Data

(a) Forward Differences

X	f(x)	Δf	Δ²f	Δ³f	Δ⁴f	Δ⁵f	Δ⁶f
1	−4	3	5	2	1	1	1
2	−1	8	7	3	2	2	
3	7	15	10	5	4		
4	22	25	15	9			
5	47	40	24				
6	87	64					
7	151						

(b) Backward Differences

X	f(x)	∇f	∇²f	∇³f	∇⁴f	∇⁵f	∇⁶f
1	−4						
2	−1	3					
3	7	8	5				
4	22	15	7	2			
5	47	25	10	3	1		
6	87	40	15	5	2	1	
7	151	64	24	9	4	2	1

where $\Delta x = x_0 - x_1 = x_2 - x_1$, and so on, and R_n is the remainder which can be derived in a manner similar to that given earlier for the forward difference formulation. Also, the general formula for the backward difference at $x = x_i$ is

$$\nabla^n f_i = \nabla^{n-1} f_i - \nabla^{n-1} f_{i-1} \tag{7.32}$$

where the subscripts denote the value of x at which the difference is obtained and the superscript gives the order of the difference.

Thus, if the data are given at equally spaced values of the independent variable, the above simplified formulas may be employed. The corresponding forward, or backward differences are generated, using Equation 7.30 or Equation 7.32, and the desired value of the function $f(x)$ is determined from the interpolating polynomial, for a given value of x. The choice of the formula, forward or backward, for interpolation depends on the value of x, in relation to the given data points, at which $f(x)$ is to be determined. Thus, if x is close to x_0, the forward difference formula is more appropriate than the backward formula. Similarly, if x is close to x_n, the backward difference form is used. Several central difference formulas have also been derived to accommodate interpolation in the region near the middle of the distribution of the data points. Consult Carnahan et al. (1969) and Hornbeck (1982), listed in the References, for the relevant formulas.

7.4.3 Extrapolation

The process of estimating the function $f(x)$ at a point x which lies beyond the range of the given data points is known as *extrapolation*. However, the most accurate estimation for $f(x)$ is generally obtained when x is close to the middle of the range. Also, the behavior of the function beyond the given data is not known. Thus, the estimated value of $f(x)$ could be in considerable error. Because of the element of uncertainty involved, values of the function obtained by extrapolation must be treated with caution. If any information is available on the nature of the function, and on its behavior beyond the range of the given data, one must consider the extrapolated values in terms of this information to judge their validity and accuracy.

Extrapolation is frequently needed in engineering applications. We are all familiar with predictions of weather, future trends in economic parameters, expected output from engineering systems, demand for engineering products, and so on. Extrapolation is, therefore, necessary for future planning of engineering resources and output. It is also often needed for the control and design of systems and processes. If the data are available at discrete, evenly spaced points, the *Newton–Gregory* forward or backward formula, as appropriate, may be used for extrapolation. *Lagrange interpolation* or *Newton's divided-difference polynomials* may be employed for an arbitrary distribution of data points. The procedures for extrapolation are similar to those for interpolation. However, since estimations are being made for points beyond the range of the given data, it must be reiterated that an element of uncertainty arises in the results, and extreme care must be exercised in the use of the values obtained.

Example 7.3

Solve the problem given in Example 7.1 by Newton's divided-difference interpolation, employing the data over the range $0.05 \le R \le 0.4$. Use polynomials of increasing order and compute the remainder term in each case.

SOLUTION

The data points to be considered for deriving the interpolating polynomial are as follows:

X	0.05	0.1	0.2	0.3	0.4
Y	2.548	4.2	5.978	6.908	7.613

where X is the independent variable and Y the dependent variable. We use different symbols here, as compared to those in Example 6.1, in order to derive a generalized solution procedure, based on Newton's divided-difference polynomials. Since five data points are employed for an exact fit, a fourth-order polynomial of the form given by Equation 7.19 may be derived.

Appendix B.17 gives the program in MATLAB for computing the interpolating polynomial. The number of data points and the corresponding data are entered in the program. We calculate the divided differences from the formulas given in Equation 7.21 and use them to determine the coefficients of the divided-difference polynomial. A matrix $f(i, j)$ is used to store the divided differences. The first column of this matrix consists of the given values of y at the five data points, and the first row contains the coefficients of the polynomial. The value of x at which interpolation is desired is denoted by xp and is entered interactively by the user. The interpolated value is obtained by means of Equation 7.22. The remainder term R is also calculated, employing Equation 7.25 and the computed value of the corresponding higher-order divided difference. The various symbols used are defined in the program, and the important steps in the computation are indicated.

The numerical results obtained are presented in Figure 7.10. The number of data points employed n is printed, along with the computed values of the coefficients $c(i)$. The value xp of the independent variable at which interpolation is sought is entered interactively. The program computes the interpolated value of the dependent variable y, employing zeroth, first, second, third, and fourth-order approximations. These approximations refer to the first term, the first two terms, the first three terms, and so on, in Equation 7.22. Thus, the last, or fourth-order, approximation is the most accurate one. This is also shown by the presented results since the remainder term decreases as the order of the approximation increases. The remainder term for the last approximation involves an additional point and is thus not computed here. For the other approximations, we compute the remainder term from Equation 7.25 by simply using the next-order divided difference, which is known from earlier calculations.

Note again that the interpolating polynomial yields the exact value of the dependent variable at the given data points, as expected. Also, the interpolated values at $x = 0.25$ and 0.35 are close to those obtained earlier in Example 7.1. The extrapolated values at $X = 0.025$ and 0.5 agree closely with the experimental data, and the value at $X = 0$ with that obtained earlier. Thus, Newton's method may be used as an alternative to the procedure outlined in Example 7.1. However, this method does not require uniformly spaced data points, as needed for the

```
Enter the number of data points, n = 5
Enter values of the independent variable, x = [0.05 0.1 0.2 0.3 0.4];
Enter corresponding values of the dependent variable, y = [2.548 4.2 5.978 6.908
   7.613];
Coefficients of the polynomial c0, c1, c2, ... are
c =
   2.5480
  33.0400
-101.7333
 237.3333
-381.4286

Enter x where interpolation is desired, xp = 0
xp =0.000
Interpolated value of y = 2.548
Remainder term = -1.652
Interpolated value of y = 0.896
Remainder term = -0.509
Interpolated value of y = 0.387
Remainder term = -0.237
Interpolated value of y = 0.150
Remainder term = -0.114
Interpolated value of y = 0.036

Enter x where interpolation is desired, xp = 0.025
xp = 0.025
Interpolated value of y = 2.548
Remainder term = -0.826
Interpolated value of y = 1.722
Remainder term = -0.191
Interpolated value of y = 1.531
Remainder term = -0.078
Interpolated value of y = 1.453
Remainder term = -0.034
Interpolated value of y = 1.419

Enter x where interpolation is desired, xp = 0.25
xp = 0.250
Interpolated value of y = 2.548
Remainder term = 6.608
Interpolated value of y = 9.156
Remainder term = -3.052
```

FIGURE 7.10 Numerical results obtained from Newton's divided-difference method for the problem considered in Example 7.3.

```
Interpolated value of y = 6.104
Remainder term = 0.356
Interpolated value of y = 6.460
Remainder term = 0.029
Interpolated value of y = 6.489

Enter x where interpolation is desired, xp = 0.35
xp = 0.350
Interpolated value of y = 2.548
Remainder term = 9.912
Interpolated value of y = 12.460
Remainder term = –7.630
Interpolated value of y = 4.830
Remainder term = 2.670
Interpolated value of y = 7.500
Remainder term = –0.215
Interpolated value of y = 7.285

Enter x where interpolation is desired, xp = 0.4
xp = 0.400
Interpolated value of y = 2.548
Remainder term = 11.564
Interpolated value of y = 14.112
Remainder term = –10.682
Interpolated value of y = 3.430
Remainder term = 4.984
Interpolated value of y = 8.414
Remainder term = –0.801
Interpolated value of y = 7.613

Enter x where interpolation is desired, xp = 0.5
xp = 0.500
Interpolated value of y = 2.548
Remainder term = 14.868
Interpolated value of y = 17.416
Remainder term = –18.312
Interpolated value of y = –0.896
Remainder term = 12.816
Interpolated value of y = 11.920
Remainder term = –4.119
Interpolated value of y = 7.801
```

FIGURE 7.10 Continued.

method employed in Example 7.1. Also, the method yields the remainder term which reflects the increase in the accuracy of the interpolation as the order of the approximation is increased. The program is more involved than Lagrange interpolation, as given in Example 7.2. However, this method has the important advantages of ease in employing varying orders of approximation and ease of evaluating the accuracy by means of the remainder term. Both Lagrange and Newton's divided-difference interpolation are widely used in engineering applications.

7.5 NUMERICAL INTERPOLATION WITH SPLINES

In the preceding sections, we considered several methods and forms of interpolating functions for an exact fit to a given data set. In many engineering problems, such as calibration of measuring and diagnostic instrumentation, numerical simulation of systems, and measurement of material properties, the available data points are relatively few, the function $f(x)$ is reasonably well behaved, and the accuracy level is very high, so that these techniques for an exact fit are appropriate. However, there are several cases where an alternative approach, which is based on curve fitting of small subsets of data points with lower-order polynomials, provides a better representation of the data. Such interpolating polynomials that are employed to yield a piecewise exact fit to the data are known as *spline functions*. The basic concept is based on the drafting technique of using a thin, flexible strip, known as a *spline*, to draw a smooth curve through a given distribution of points. Although the interpolating polynomial may be linear, quadratic, cubic, or of some other order, the cubic spline function is the most widely used one and is discussed here. Splines are advantageous to use when the conventional interpolation methods, such as those discussed in the previous sections, yield polynomials of higher order and the interpolating curve is of wiggly or oscillating character, as shown in Figure. 7.11. In such cases, spline interpolation often yields a better approximation. This approach is particularly valuable in the interpolation of accurate material property data over wide ranges of the independent variable such as temperature and pressure.

Let us consider two arbitrary points x_i and x_{i+1} at which the function $f(x)$ is given. The general form of the cubic that passes through these points and provides the interpolation function between the two may be taken as

$$f_i(x) = a_0 + a_1 x + a_2 x^2 + a_3 x^3 \quad \text{for } x_i \le x \le x_{i+1} \tag{7.33}$$

There are four unknown constants in this polynomial. Since the curve passes through the two points x_i and x_{i+1}, two conditions that must be used are

$$f_i(x_i) = a_0 + a_1 x_i + a_2 x_i^2 + a_3 x_i^3 \tag{7.34a}$$

$$f_i(x_{i+1}) = a_0 + a_1 x_{i+1} + a_2 x_{i+1}^2 + a_3 x_{i+1}^3 \tag{7.34b}$$

The remaining two conditions may be chosen arbitrarily to obtain a smooth transition from one cubic distribution to the adjacent ones. An effective choice is the continuity of the first and second derivatives at the two points. Thus, the slope and

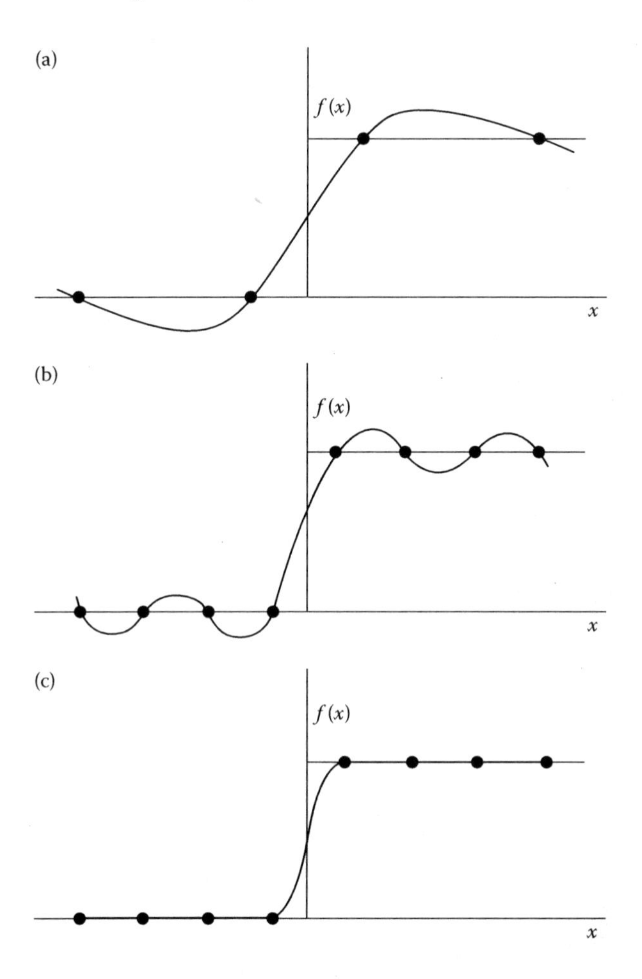

FIGURE 7.11 Interpolation with single polynomials over the entire range and with piecewise cubic splines for a step change in the dependent variable. (a) Third-order polynomial fit, (b) seventh-order polynomial fit, and (c) cubic spline interpolation.

curvature of $f_i(x)$ match those of $f_{i-1}(x)$ at $x = x_i$ and those of $f_{i+1}(x)$ at $x = x_{i+1}$. A special treatment will be needed at the end points of the given data.

Since the second derivative of a cubic is a straight line over each interval, as shown in Figure 7.12, a first-order Lagrange interpolation may be derived from Section 7.3 to represent the second derivative, over the interval $x_i \le x \le x_{i+1}$, as follows:

$$f_i''(x) = f_i''(x_i)\frac{x_{i+1} - x}{x_{i+1} - x_i} + f_i''(x_{i+1})\frac{x - x_i}{x_{i+1} - x_i} \tag{7.35}$$

Integrating this equation twice and applying Equations 7.34 to determine the constants that arise, we obtain the cubic $f_i\,(x)$ over $x_i \le x \le x_{i+1}$:

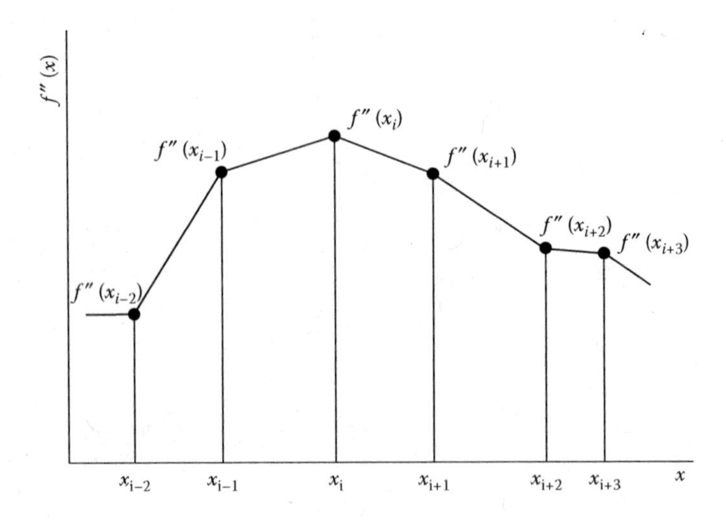

FIGURE 7.12 Variation of the second derivative $f''(x)$ over the subintervals that constitute the given range of data for spline interpolation.

$$f_i(x) = f''(x_i)\frac{(x_{i+1}-x)^3}{6\Delta x_i} + f''(x_{i+1})\frac{(x-x_i)^3}{6\Delta x_i}\left[\frac{f(x_i)}{\Delta x_i} - \frac{\Delta x_i}{6}f''(x_i)\right](x_{i+1}-x)$$
$$+\left[\frac{f(x_{i+1})}{\Delta x_i} - \frac{\Delta x_i}{6}f''(x_{i+1})\right](x-x_i), \quad \text{where } \Delta x_i = x_{i+1}-x_i \tag{7.36}$$

Equation 7.36 yields the interpolating cubic distributions over each of the subintervals in the range $x_0 \le x \le x_n$. We determine the second derivatives in Equation 7.36 by using the matching condition for the first derivative, that is,

$$f_i'(x_i) = f_{i-1}'(x_i) \tag{7.37}$$

Now, $f_i(x)$ may be differentiated and x set equal to x_i to obtain the derivative at the left-hand limit of interval i. Similarly, $f_{i-1}(x)$ is differentiated and x set equal to x_i to yield the derivative at the right-hand limit of interval (i–1). The two results obtained are equated to give a set of linear simultaneous equations of the form

$$\Delta x_{i-1} f''(x_{i-1}) + 2(x_{i+1}-x_{i-1})f''(x_i) + x_i f''(x_{i+1})$$
$$= 6\left[\frac{f(x_{i+1})-f(x_i)}{\Delta x_i} - \frac{f(x_i)-f(x_{i-1})}{\Delta x_{i-1}}\right], \quad \text{where } \Delta x_{i-1} = x_i - x_{i-1} \tag{7.38}$$

Here, the fact that $f''(x_i)$ is the same when x_i is approached from either side, as shown in Figure 7.12, has also been used. This condition may be stated as

$$f_i''(x_i) = f_{i-1}''(x_i) \tag{7.39}$$

For $(n + 1)$ data points, represented by the values of the independent variable x_i, where $i = 0, 1, 2, \ldots, n$, the number of intervals is n. Therefore, n cubics are generated by Equation 7.36 for spline interpolation. However, there are $(n + 1)$ unknown second derivatives in the equations for the n cubics. Equation 7.38, when written for all the interior points, that is, for $i = 1, 2, 3, \ldots, (n - 1)$, yields $(n - 1)$ equations for the evaluation of the second derivative. Since only $f''(x_{i-1})$, $f''(x_i)$ and $f''(x_{i+1})$ appear as unknowns in Equation 7.38, the system of equations is tridiagonal and may easily be solved by Gaussian elimination, as illustrated in Example 6.2. However, we still need two additional conditions at the end points of the data set, that is, for $i = 0$ and $i = n$, in order to determine $f''(x_0)$ and $f''(x_n)$. These conditions are usually taken as

$$f''(x_0) = 0 \quad \text{and} \quad f''(x_n) = 0 \tag{7.40}$$

Thus, the analogous elastic strip for drawing a curve through the given points is allowed to assume a natural, unconstrained straight line beyond the given range of data points. This spline, known as a *natural cubic spline*, is the one most frequently employed for an arbitrary data set.

Several other approximations for the end conditions have been employed for different types of data, see Ferziger (1998). For data that are expected to lie on a periodic curve, the end conditions are often taken as $f''(x_0) = f''(x_{n-1})$ and $f''(x_1) = f''(x_n)$, representing the repetitive nature of the curve. Another frequently used set of conditions is $f''(x_0) = f''(x_1)$ and $f''(x_{n-1}) = f''(x_n)$, which makes f'' constant in the intervals at the two ends. It also makes $f(x)$ quadratic in these intervals. Other choices for the end conditions are also possible. However, the natural spline is the most commonly employed interpolating spline. Equation 7.36 gives the cubic equation for each interval, and Equation 7.38, along with Equation 7.40 or one of the other end conditions chosen, gives the tridiagonal system for obtaining the $(n + 1)$ second derivatives. For further details on splines, see Ahlberg et al. (1967). The following example illustrates the use of spline interpolation in a problem of practical interest.

Example 7.4

Thermocouple junctions of dissimilar metals and alloys are extensively used in engineering applications for temperature measurement. A voltage difference V is generated between two junctions at different temperatures. Calibration tables, which give the voltage V in millivolts (mV) for one junction at 0°C and the other at temperature T in °C, are available in the literature for several types of thermocouple junctions. The values for a Chromel–Alumel thermocouple, which is generally known as type K thermocouple and consists of nickel–chromium and nickel–aluminum alloys, are given as follows:

T (°C)	10	20	30	40	50	60	70	80	90
V (mV)	0.397	0.798	1.203	1.611	2.022	2.436	2.85	3.266	3.681
T (°C)	100	110	120	130	140	150			
V (mV)	4.095	4.508	4.919	5.327	5.733	6.137			

Employing cubic spine interpolation, obtain the temperatures if the voltage output values are 1.0, 3.0, 4.343, 5.855, 6.0, and 6.097 mV. Compare the results obtained at 4.343, 5.855, and 6.097 mV with those given in the literature as 106°C, 143°C, and 149°C, respectively.

SOLUTION

This problem is well suited for spline interpolation, since the tabulated values are very accurate and since the large number of data points makes an exact fit with a single polynomial difficult to apply and also inaccurate, as discussed earlier. The voltage V is measured in engineering processes, and the temperature T is obtained by interpolation from the calibration data. Thus, V is taken as the independent variable and T as the independent variable.

Appendix C.10 presents the computer program in Fortran for spline interpolation. The number of data points is entered interactively by the user, and the program reads the relevant data from files V.DAT and T.DAT, for voltage and temperature, respectively, stored in the computer. Two subroutines, DERIVATIVE and SPLINE, are employed in the program. The former generates the tridiagonal matrix for the second derivative $f''(x_i)$, denoted by T2 in the program. The elements of the matrix are obtained from Equations 7.38 and 7.40. The *Thomas algorithm*, derived in Example 6.2, is employed to obtain the values of the second derivative needed for the cubic spline given by Equation 7.36. The voltage at which interpolation is desired is denoted by VP and is entered into the main program by the user. The subroutine SPLINE determines the interval in which VP lies, derives the relevant cubic spline using Equation 7.36, and computes the interpolated value TP of the temperature. The main program prints the results and inquires whether interpolation at another value of V is needed. Thus, interpolated results may be obtained at the desired values of the output voltage V.

The numerical results obtained are presented in Figure 7.13. The interpolated temperatures for $V = 1.0$, 3.0, and 6.0 mV are found to be physically realistic and to lie in the appropriate subintervals of the given data. Also, the values for $V = 4.343$, 5.855, and 6.097 are very close to those given in the literature as 106°C, 143°C, and 149°C, respectively, lending strong support to the accuracy of the interpolated results obtained. In fact, several additional values of V were considered, and the results were found to be very accurate. Thus, cubic spline interpolation may be used satisfactorily for this problem and other similar ones, such as material property data.

The algorithm presented in Appendix C.10 may be used to develop the corresponding MATLAB script *m*-file. However, the program is fairly involved and the interpolation commands discussed earlier may be used directly for spline interpolation. Thus, the command *interp*1 may be used effectively for this example to obtain the desired interpolated values. The command is given as

```
interp1(v,t,vp,'spline')
or as
yp = spline(v,t,vp)
```

where v and t are the two arrays of data for the independent and dependent variables, respectively, and vp is the array of v values where interpolated

```
      ENTER THE NUMBER OF DATA POINTS
15
      ENTER THE VALUE OF V FOR INTERPOLATION
1.0
          VOLTAGE V = 1.00000          TEMPERATURE T = 24.99982

      IF YOU WANT ADDITIONAL INTERPOLATION , TYPE 1
1
      ENTER THE VALUE OF V FOR INTERPOLATION
3.0
          VOLTAGE V = 3.00000          TEMPERATURE T = 73.60907

      IF YOU WANT ADDITIONAL INTERPOLATION , TYPE 1
1
      ENTER THE VALUE OF V FOR INTERPOLATION 4.343
          VOLTAGE V = 4.34300          TEMPERATURE T = 106.00078

      IF YOU WANT ADDITIONAL INTERPOLATION , TYPE 1
1
      ENTER THE VALUE OF V FOR INTERPOLATION 5.855
          VOLTAGE V = 5.85500          TEMPERATURE T = 143.01590

      IF YOU WANT ADDITIONAL INTERPOLATION , TYPE 1
1
      ENTER THE VALUE OF V FOR INTERPOLATION 6.0
          VOLTAGE V = 6.00000          TEMPERATURE T = 146.60565

      IF YOU WANT ADDITIONAL INTERPOLATION , TYPE 1
1
      ENTER THE VALUE OF V FOR INTERPOLATION 6.097
          VOLTAGE V = 6.09700          TEMPERATURE T = 149.00882

      IF YOU WANT ADDITIONAL INTERPOLATION , TYPE 1
2
```

FIGURE 7.13 Computed results from spline interpolation for the problem considered in Example 7.4.

results are desired, using spline interpolation. Thus, a simple script file may be written as

```
v = [0.397 0.798 1.203 1.611 2.022 2.436 2.85 3.266 3.681 ...
4.095 4.508 4.919 5.327 5.733 6.137];
t = [10 20 30 40 50 60 70 80 90 100 110 120 130 140 150];
vp = [1.0 3.0 4.343 5.855 6.0 6.097];
yp = spline(v,t,vp)
```

This yields the desired results as

```
yp = 24.9982 73.6091 106.0008 143.0141 146.6028 149.0074
```

Clearly, the results are very close to those obtained earlier by using the computer program whose algorithm was discussed above.

7.6 METHOD OF LEAST SQUARES FOR A BEST FIT

In the preceding sections, we discussed interpolation with approximating functions that pass through each given data point. Such an exact fit is appropriate if the given data are of a high level of accuracy. If the number of points is relatively small, a single polynomial approximation may be employed for interpolation. If a large number of points are given, spline interpolation, which yields lower-order polynomials, such as cubics, to fit small subsets of the data, can be used to piecewise approximate the data for obtaining values of the dependent variable at intermediate points. However, the data obtained in many engineering applications have a significant amount of associated error. Experimental data, for instance, would generally have some noise, or error, whose magnitude would depend on the instrumentation and the arrangement employed for the measurements. In such cases, a polynomial interpolation that demands that the approximating curve pass through each data point is not appropriate.

A better approach is to derive a function that provides a best fit to the given data by somehow minimizing the difference between the given values of the dependent variable and those obtained from the approximating curve. Figure 7.14 shows a few circumstances where a best fit will be much more satisfactory than an exact fit. Because of the error associated with the data, it is not necessary for the approximating curve to match each data point. A curve that adequately represents the general trend of the data, without necessarily passing through each point, will be useful in characterizing the data and deriving correlating equations for quantitatively describing the physical or chemical process under consideration. Such correlations are extremely important in engineering applications and are often the desired output from an experimental study. The measurements on the deflection of a building structure due to the flow of water, for instance, can be used to yield a best fit that then can be employed in the design of such structures for locating them in streams and in the sea. Similarly, measurements of the velocities of accelerating automobiles can be used to derive correlating equations that characterize the dependence of the acceleration on various parameters, such as shape, weight, and fuel mixture. Heat and mass transfer from surfaces are often measured for different geometries and flow conditions. The results obtained are then curve fitted to yield correlations that can be used for future analysis and design of similar processes and systems.

7.6.1 Basic Considerations

Several criteria can be used to derive the curve that best fits the data. If the approximating function is denoted by $f(x)$ and the given data points by (x_i, y_i), where y is the dependent variable, x is the independent variable, and $i = 1, 2, \ldots, n$, the error e_i at $x = x_i$ is given by

$$e_i = y_i - f(x_i) \tag{7.41a}$$

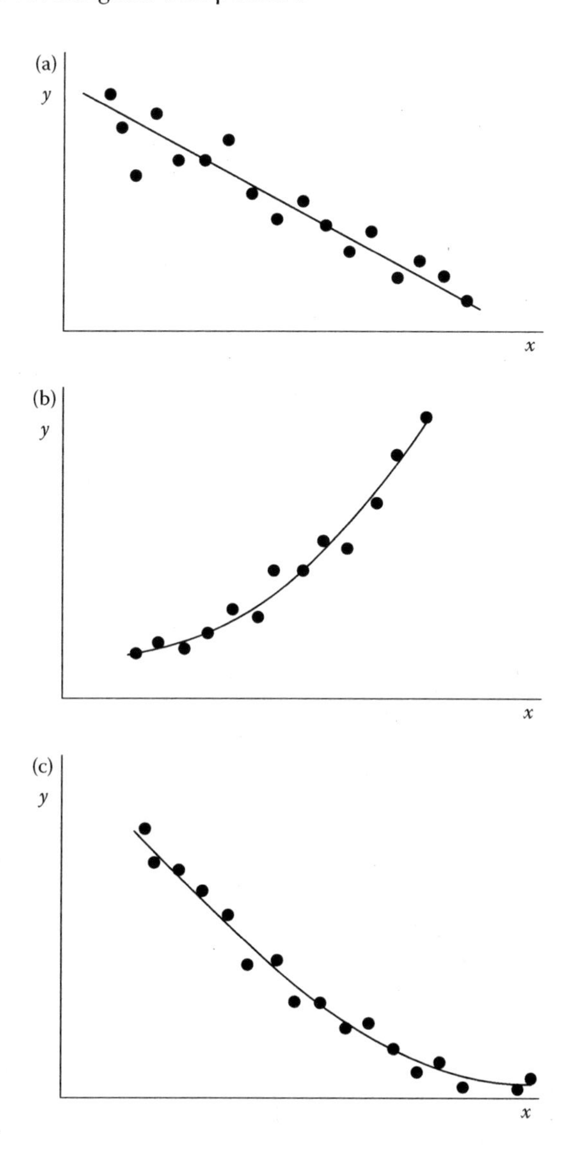

FIGURE 7.14 Data distributions for which a best fit is more appropriate than an exact fit.

One method for obtaining a best fit to the data is to minimize the sum of these individual errors, that is, minimize $\sum_{i=1}^{n} e_i$. However, this approach is not satisfactory since this criterion allows the errors to cancel out and thus does not yield a unique curve. Moreover, the curve may not represent the general trend of the data at all. If the sum of the absolute values of the errors, $\sum_{i=1}^{n} |e_i|$, is minimized, the result is better, but, again, a unique best fit is generally not obtained. Another approach that may be used is the *minimax* criterion, which minimizes the maximum error, $(e_i)_{max}$, for the data points. However, this method is heavily influenced by a single point that may have large error. Although unsuitable for obtaining a best fit in most engineering problems, this approach is often appropriate for fitting a simple function to a much more complicated one, as outlined by Carnahan et al. (1969).

The most commonly used approach for a best fit is the method of least squares. In this method, the sum S of the squares of the errors is minimized. The expression for S is

$$S = \sum_{i=1}^{n} e_i^2 = \sum_{i=1}^{n} [y_i - f(x_i)]^2 \quad (7.41b)$$

This approach generally yields a unique curve that provides a good representation of the given data, if the approximating function is properly chosen. As outlined in Section 1.4, one must employ the basic nature of the problem under consideration in choosing the form of the approximating function. Thus, for the measurements of the average daily temperature over the year, a sinusoidal function will yield a good best fit to the data, see Figure 1.5. Similarly, in most experimental studies of engineering systems and processes, the expected trends are known from the physical nature of the problem, allowing one to choose an appropriate function for curve fitting.

Let us consider, as an example, the measurement of a physical variable, which may be, say, the length, weight, or density of a given material. If n measurements are taken, the results will generally differ because of the experimental error involved. Let us denote these measurements as $l_1, l_2, \ldots, l_n$. If L denotes the desired best fit to these measurements, then

$$S = (l_1 - L)^2 + (l_2 - L)^2 + \cdots + (l_n - L)^2 \quad (7.42)$$

To minimize this sum S of the squares of the differences, we differentiate S twice to obtain

$$\frac{dS}{dL} = -2(l_1 - L) - 2(l_2 - L) - \cdots - 2(l_n - L) = -2\left[\sum_{i=1}^{n} l_i - nL\right] \quad (7.43)$$

$$\frac{d^2S}{dL^2} = 2n \quad (7.44)$$

Since n is positive, the value of L for which $dS/dL = 0$ gives a minimum value of the sum S. From Equation 7.43, this value is obtained as

$$L = \frac{\sum_{i=1}^{n} l_i}{n} \quad (7.45)$$

Therefore, if the sum S is minimized, the value of the quantity L is simply the arithmetic mean of the measurements. One will expect this value to be the best representation of the data if the measurements are all taken with equal care and are thus of comparable accuracy. This example provides a physical basis for the method of least squares and may easily be extended to a function $f(x)$, using the consideration of a single unknown variable L given above.

7.6.2 Linear Regression

The procedure of obtaining a best fit to a given data set is often known as *regression*. Let us first consider fitting a straight line to a set of data points denoted by (x_1, y_1), (x_2, y_2), . . ., (x_n, y_n), where x is the independent variable and y the dependent variable. Although engineering applications usually lead to nonlinear functions, there are several circumstances where a linear variation closely approximates the measurements. Moreover, exponential and power-law forms, which are very frequently encountered in practical problems, can often be reduced to linear variations, as illustrated later in this section. Consequently, linear regression is very important in a wide variety of engineering applications, particularly in the derivation of correlating equations from experimental data.

The equation of the straight line for curve fitting may be taken as

$$f(x) = a + bx \tag{7.46}$$

where a and b are the coefficients that must be determined from the given set of n data points. Thus, a and b are to be chosen such that the sum S of the squares of the deviations of the data points from the values obtained from the equation of the straight line, Equation 7.46, is a minimum. This implies that

$$S = \sum_{i=1}^{n} (y_i - a - bx_i)^2 \rightarrow \text{minimum} \tag{7.47}$$

The minimum occurs when the partial derivatives of S with respect to a and b are both zero. Thus,

$$\frac{\partial S}{\partial a} = \sum_{i=1}^{n} -2(y_i - a - bx_i) = 0 \tag{7.48a}$$

$$\frac{\partial S}{\partial b} = \sum_{i=1}^{n} -2(y_i - a - bx_i)x_i = 0 \tag{7.48b}$$

These equations may be simplified and expressed as

$$\begin{aligned} &\sum y_i - \sum a - \sum bx_i = 0 \\ &\sum y_i x_i - \sum ax_i - \sum bx_i^2 = 0 \end{aligned} \tag{7.49}$$

which may be written for the unknowns a and b as

$$na + b\sum x_i = \sum y_i \tag{7.50}$$

$$a\sum x_i + b\sum x_i^2 = \sum x_i y_i \tag{7.51}$$

where the summations are all from $i = 1$ to $i = n$.

Equations 7.50 and 7.51 are linear in the unknowns and may be solved simultaneously to yield the desired values of a and b. Using *Cramer's* rule, we obtain a and b in terms of the relevant determinants as follows:

$$a = \frac{\begin{vmatrix} \sum y_i & \sum x_i \\ \sum x_i y_i & \sum x_i^2 \end{vmatrix}}{\begin{vmatrix} n & \sum x_i \\ \sum x_i & \sum x_i^2 \end{vmatrix}} \tag{7.52}$$

$$b = \frac{\begin{vmatrix} n & \sum y_i \\ \sum x_i & \sum x_i y_i \end{vmatrix}}{\begin{vmatrix} n & \sum x_i \\ \sum x_i & \sum x_i^2 \end{vmatrix}} \tag{7.53}$$

where the vertical bars indicate magnitude of the determinant. We may employ the given set of n data points to compute $\sum x_i, \sum y_i, \sum x_i^2$, and $\sum x_i y_i$. Then we use these values to calculate the determinants in Equations 7.52 and 7.53. These equations then yield the coefficients a and b for the straight line, Equation 7.46, that provides a best fit to the given data.

To quantify the accuracy with which the computed straight line fits the given data, we compute the sum of the squares of the deviations of the data from the mean to represent the spread before regression is applied. Denoting this sum by S_m and the mean by $\bar{y}$, we have

$$S_m = \sum_{i=1}^{n} (y_i - \bar{y})^2 \tag{7.54}$$

The spread in the data that remains after regression is indicated by S, where

$$S = \sum_{i=1}^{n} (y_i - a - bx_i)^2 \tag{7.55}$$

Therefore, the extent of improvement due to curve fitting by a straight line is indicated by

$$r^2 = \frac{S_m - S}{S_m} \tag{7.56}$$

where r is known as the *correlation coefficient.* A good correlation for linear regression is indicated by a high value of r, the maximum of which is 1.0. However, the given data should also be plotted along with the computed curve, in order to determine, qualitatively, how good a representation of the data is provided by the fit. Equation 7.56 can also be used for higher-order polynomials and nonpolynomial forms of the function for a best fit, as outlined later in this section. See Draper and Smith (1998) for further details on the application of regression analysis.

7.6.3 Best Fit with a Polynomial

Linear regression yields a straight line that provides a best fit to a given data set. It is simple to apply, since only two unknown coefficients, a and b in Equation 7.46, are to be determined. In many cases, particularly if the range of the independent variable is relatively small, a straight line provides a fairly good representation of the data. Also, as outlined in Section 7.6.4, certain nonlinear forms, such as exponentials, may be transformed to yield linear variations. However, the data may have a definite trend that is poorly represented by a straight line. An example of such a situation is shown in Figure 7.15, which illustrates that a straight line is not a satisfactory choice for curve fitting in this case. A polynomial, such as a parabola or a cubic, will be more appropriate.

In order to obtain a best fit to the given data, let us consider an mth-order polynomial, given as

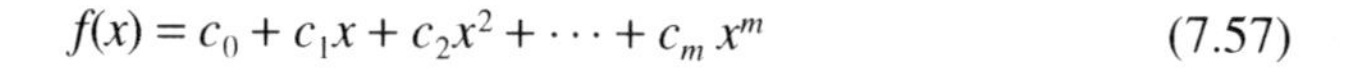

$$f(x) = c_0 + c_1 x + c_2 x^2 + \cdots + c_m x^m \tag{7.57}$$

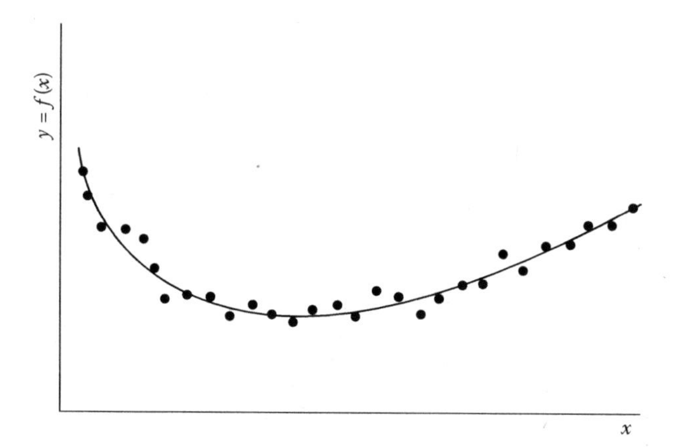

FIGURE 7.15 A polynomial best fit to given data.

Then the sum S of the squares of the deviations of the data from the curve is given by

$$S = \sum_{i=1}^{n} (y_i - c_0 - c_1 x_i - c_2 x_i^2 - \cdots - c_m x_i^m)^2 \tag{7.58}$$

We determine the coefficients $c_0, c_1, \ldots, c_m$ by extending the procedure outlined in the preceding section for linear regression. Therefore, S is differentiated with respect to each of the coefficients, and the partial derivatives are set equal to zero in order to minimize S. This gives

$$\begin{aligned}
\frac{\partial S}{\partial c_0} &= -2\sum (y_i - c_0 - c_1 x_i - c_2 x_i^2 - \cdots - c_m x_i^m) = 0 \\
\frac{\partial S}{\partial c_1} &= -2\sum x_i (y_i - c_0 - c_1 x_i - c_2 x_i^2 - \cdots - c_m x_i^m) = 0 \\
&\vdots \\
\frac{\partial S}{\partial c_m} &= -2\sum x_i^m (y_i - c_0 - c_1 x_i - c_2 x_i^2 - \cdots - c_m x_i^m) = 0
\end{aligned} \tag{7.59}$$

Equations 7.59 may be simplified and rearranged to yield the following system of $(m + 1)$ linear equations for the $(m + 1)$ unknowns $c_0, c_1, \ldots, c_m$:

$$\begin{aligned}
nc_0 + c_1 \sum x_i + c_2 \sum x_i^2 + \cdots + c_m \sum x_i^m &= \sum y_i \\
c_0 \sum x_i + c_1 \sum x_i^2 + c_2 \sum x_i^3 + \cdots + c_m \sum x_i^{m+1} &= \sum x_i y_i \\
c_0 \sum x_i^m + c_1 \sum x_i^{m+1} + c_2 \sum x_i^{m+2} + \cdots + c_m \sum x_i^{2m} &= \sum x_i^m y_i
\end{aligned} \tag{7.60}$$

where all the summations are from $i = 1$ to $i = n$. It can easily be verified that the equations for linear regression, Equations 7.50 and 7.51, are obtained for a first-order polynomial, $m = 1$. The methods given in Chapter 6 may be employed to solve the above system of equations, which are linear in the unknown coefficients $c_0, c_1, \ldots, c_m$.

Curve fitting with polynomials is generally restricted to small values of the order m of the polynomial, in order to avoid extensive calculations for the determination of the coefficients and to obtain simple correlating curves that approximate the general trends of the data. Typical values of m range from 1 to 4, the appropriate value being chosen on the basis of the accuracy and spread of the data, as well as the number of data points. For a relatively large spread of the data, a lower-order polynomial fit will generally be more appropriate. Computation is involved in evaluating the summations in Equation 7.60 and then solving this system of equations, which may be recast in matrix notation. For a second-order polynomial, for instance, we have

$$\begin{pmatrix} n & \sum x_i & \sum x_i^2 \\ \sum x_i & \sum x_i^2 & \sum x_i^3 \\ \sum x_i^2 & \sum x_i^3 & \sum x_i^4 \end{pmatrix} \begin{pmatrix} c_0 \\ c_1 \\ c_2 \end{pmatrix} = \begin{pmatrix} \sum y_i \\ \sum x_i y_i \\ \sum x_i^2 y_i \end{pmatrix} \tag{7.61}$$

Various elimination and matrix inversion or decomposition methods, given in Chapter 6, may be employed for solving Equation 7.60 or 7.61 for the coefficients. Gaussian elimination is the most popular choice because of the small number of equations to be solved in most cases. The correlation coefficient r may again be determined from Equation 7.56 to evaluate how good a fit is given by the resulting polynomial.

The *polyfit* command in MATLAB can be used conveniently for polynomial regression. Linear regression is obtained if the order of the polynomial is given as 1, as discussed in Chapter 3. Linear regression can also be used for nonpolynomial forms, as discussed in the next section. Higher order polynomials can be determined for a best fit by specifying the order of the desired polynomial. The use of the *polyfit* command is illustrated in Example 7.5.

7.6.4 Nonpolynomial Forms

The method of least squares is not restricted to polynomials for curve fitting and may easily be applied to various other forms that contain constant coefficients. An example of a physical situation where such a form is more appropriate than a polynomial is the periodic variation in ambient temperature considered in Chapter 1; see Figure 1.5. Equations 1.10 through 1.12 give some of the sinusoidal functions that may be employed for curve fitting. Considering the function given in Equation 1.11, for example, we obtain

$$f(x) = A \sin \omega x + B \cos \omega x \tag{7.62}$$

and

$$S = \sum_{i=1}^{n} (y_i - A \sin \omega x_i - B \cos \omega x_i)^2 \tag{7.63}$$

where the sum S is to be minimized for a best fit. Thus,

$$\frac{\partial S}{\partial A} = \sum_{i=1}^{n} -2(y_i - A \sin \omega x_i - B \cos \omega x_i) \sin \omega x_i = 0$$

$$\frac{\partial S}{\partial B} = \sum_{i=1}^{n} -2(y_i - A \sin \omega x_i - B \cos \omega x_i) \cos \omega x_i = 0$$

This gives the equations

$$A \sum (\sin \omega x_i)^2 + B \sum (\sin \omega x_i \cos \omega x_i) = \sum y_i \sin \omega x_i \tag{7.64}$$

$$A \sum (\sin \omega x_i \cos \omega x_i) + B \sum (\cos \omega x_i)^2 = \sum y_i \cos \omega x_i \tag{7.65}$$

which can be easily solved for A and B.

Nonpolynomial forms are important in a wide variety of engineering problems. If the function chosen for curve fitting has constant coefficients, such as A and B in Equation 7.62, the method of least squares can be easily applied. However, if the constants do not appear as coefficients, for example, the constant a in Equation 1.10, a straightforward application of the method is not possible. Therefore, the nonpolynomial forms employed for the curve fitting of various types of engineering data are chosen such that the constants to be determined appear only as coefficients.

Besides periodic processes, an example of which is considered above, several engineering applications involve power-law and exponential variations, some of which can be linearized as outlined below. The example given by Equations 1.1 and 1.2, for instance, concerns an exponential variation. Similarly, processes that approach a constant magnitude at large values of the independent variable x can often be represented by polynomials with negative exponents, for example,

$$y = c_0 + c_1 x^{-1} + c_2 x^{-2} \tag{7.66}$$

Processes where such an equation may be applicable are the charging of a capacitor in an electrical circuit, the free fall of an object under gravity to attain a terminal velocity, and the dissolution of salt in a given amount of liquid until saturation occurs. However, unless the physical or chemical nature of the given data indicates the suitability of a particular nonpolynomial form, curve fitting is first explored using a polynomial, with varying orders of the polynomial, to obtain a satisfactory representation of the data, see Example 7.5.

7.6.4.1 Linearization

In several cases, a nonlinear form chosen to curve fit the given data may be linearized by suitable transformations so that linear regression may be applied. Consider, for example, the exponential variation that is commonly encountered in engineering problems, as shown in Figure 1.2. The general form of an exponential variation may be taken as

$$f(x) = c_1 e^{c_2 x} \tag{7.67}$$

where c_1 and c_2 are constants to be determined for a best fit. In engineering applications, c_1 is generally positive, and c_2 may be positive, as in the convective heating of a metal block, or negative, as in radioactive decay and discharge of a capacitor. If the natural logarithm of Equation 7.67 is taken, we obtain

$$\log\,[f(x)] = \log c_1 + c_2 x \tag{7.68}$$

Thus, $\log[f(x)]$ is a linear function of x, and linear regression may be applied using x as the independent variable and the natural logarithm of y, where $y = f(x)$, as the dependent variable. Then y_i in Equations 7.52 and 7.53 is replaced by $\log y_i$. Also, $a = \log c_1$ and $b = c_2$, where a and b are the coefficients for linear regression, Equation 7.46. This approach is frequently employed for obtaining correlating equations for measured heat and mass transfer rates from bodies and surfaces under different physical and chemical conditions.

Similarly, the power-law variation given by the general form

$$f(x) = c_1 x^{c_2} \tag{7.69}$$

is frequently employed for the representation of certain engineering processes. Again, a natural logarithm of the equation is taken to yield

$$\log\,[f(x)] = \log c_1 + c_2 \log x \tag{7.70}$$

where the logarithm to base 10 may also be taken for convenience, instead of the natural logarithm. Again, linear regression may be applied, with $\log x$ and $\log [f(x)]$ as the independent and dependent variables, respectively, to obtain the coefficients c_1 and c_2.

Similarly, various other forms, such as

$$f(x) = \frac{c_1}{c_2 + x} \tag{7.71a}$$

$$f(x) = c_1 + c_2 x^{-1} \tag{7.71b}$$

$$f(x) = \frac{c_1 x}{c_2 + x} \tag{7.71c}$$

may be linearized by taking the reciprocal of $f(x)$, of x, or of both as the independent and dependent variables. Thus, these equations may be rewritten as

$$Y = \left(\frac{c_2}{c_1}\right) + \left(\frac{1}{c_1}\right) x \tag{7.72a}$$

$$y = c_1 + c_2 X \tag{7.72b}$$

$$Y = \left(\frac{1}{c_1}\right) + \left(\frac{c_2}{c_1}\right) X \tag{7.72c}$$

where $y = f(x)$, $Y = 1/f(x)$, and $X = 1/x$. Therefore, linear regression may be applied to these transformed equations to obtain the coefficients c_1 and c_2. These examples also indicate the importance of linear regression in the curve fitting of engineering data. Many processes of practical interest are governed by exponential, power law, and other forms given above, and linear regression is employed to determine the unknown coefficients. Example 7.5 illustrates the use of the method of least squares for obtaining a best fit to a given data set.

Example 7.5

a. In a chemical reaction, the effect of the concentration C of a catalyst on the rate R of the reaction is investigated experimentally. The measurements of C in g/m^3 and of R in g/s yield the following:

C (g/m^3)	0.1	0.2	0.5	1.0	1.2	1.8	2.0	2.6	3.5	4.0
R (g/s)	1.85	1.91	2.07	2.32	2.40	2.54	2.56	2.53	2.03	1.24

Using the method of least squares and considering polynomials up to the fifth order, obtain a best fit to these data.

b. A small, heated metal block cools in air, and its temperature T is measured as a function of time t to give the following data:

t (s)	1	2	5	10	15	20	25	30
T (°C)	109.58	99.25	73.78	45.15	26.78	17.24	9.85	6.97

From physical considerations of the problem, the temperature is expected to decay exponentially, as Ae^{-at}, where A and a are constants. Employing the program developed in Part (a), obtain a best fit to the given data and determine the constants A and a.

SOLUTION

a. From the data presented, we can see that the reaction rate R increases with concentration C of the catalyst up to a point and then decreases. Thus, we expect that linear regression would not be satisfactory, and therefore we attempt curve fitting with polynomials. However, we can also obtain the results for linear regression from the numerical scheme by choosing the order of the polynomial as 1.

Denoting the independent variable by x and the dependent variable by y, for generality, we obtain a system of linear equations, as given by Equation 7.60. Then we solve this system to obtain the coefficients c_i of the polynomial. Appendix B.18 presents the script *m*-file in MATLAB and Appendix C.11 presents the computer program in Fortran for the least-squares method for polynomial regression. In the latter, the data points are represented by $X(I)$ and $Y(I)$, and the coefficients of the polynomial by $C(I)$. The order of the polynomial is denoted by MP, which gives the number of coefficients to be determined as $N = \text{MP} + 1$. The various other symbols employed are defined in the program.

In the given program, the input data and the chosen order of the polynomial for curve fitting are read from an appended data set. The system of linear equations, given by Equation 7.60, is then generated. The corresponding augmented matrix, with the constant vector on the right-hand side of Equation 7.60 being stored as the $(N + l)$th column, is obtained. Gaussian elimination is employed for the solution of this system of equations. A subroutine called GAUSS is employed. This subroutine applies the Gaussian elimination algorithm to the system of equations. Thus, the coefficient matrix is reduced to an upper triangular matrix, and back-substitution is employed to determine the coefficients $C(I)$. The subroutine GAUSS is the same as that developed earlier for Example 6.1 and employs partial pivoting for accuracy and for avoiding a zero pivot element. The computed coefficients are printed and thus a polynomial of form given by Equation 7.57 is obtained for a best fit. The values of the dependent variable $Y(I)$ at the given data points $X(I)$ are calculated using this polynomial. These values are then compared with the given data to estimate the accuracy of the best fit obtained.

Figure 7.16 shows the computed results for polynomials of order 1, 3, and 5. A comparison of the results with the given data shows that linear regression is in considerable error, as expected. The third-order polynomial fit is fairly accurate, although the fifth-order regression is more accurate. However, because of the smaller computational effort required and the ease in application to engineering problems, a third-order polynomial best fit is very frequently employed in practice, rather than higher-order polynomials. These trends are illustrated more clearly in Figure 7.17, where the given data points are plotted, along with some of the polynomials derived from the method of least squares. Again, note that a third-order polynomial yields a fairly accurate representation of the data.

Using a somewhat similar logic, the MATLAB script file in Appendix B.18 is developed. However, this program uses the convenience of matrix specification and algebra available in the MATLAB environment, yielding a fairly simple script file. The given data are entered and the desired order of the polynomial best fit is specified. The matrices for polynomial regression are obtained by appropriate summations of the data. The backslash operator is used to solve the system of linear equations to determine the coefficients

```
THE ORDER OF THE POLYNOMIAL = 1
THE CONSTANTS OF THE POLYNOMIAL ARE

C(l) =  2.25954
C(2) = -0.06778

THE VALUES CALCULATED FROM THE BEST FIT ARE
X(1)  = 0.1000          Y(1)  = 2.2528
X(2)  = 0.2000          Y(2)  = 2.2460
X(3)  = 0.5000          Y(3)  = 2.2257
X(4)  = 1.0000          Y(4)  = 2.1918
X(5)  = 1.2000          Y(5)  = 2.1782
X(6)  = 1.8000          Y(6)  = 2.1375
X(7)  = 2.0000          Y(7)  = 2.1240
X(8)  = 2.6000          Y(8)  = 2.0833
X(9)  = 3.5000          Y(9)  = 2.0223
X(10) = 4.0000          Y(10) = 1.9884

THE ORDER OF THE POLYNOMIAL = 3
THE CONSTANTS OF THE POLYNOMIAL ARE

C(l) =  1.82437
C(2) =  0.43013
C(3) =  0.09669
C(4) = -0.05961
```

FIGURE 7.16 Calculated results for Example 7.5(a), using polynomials of order 1, 3, and 5 for a best fit.

THE VALUES CALCULATED FROM THE BEST FIT ARE

X(1) = 0.1000   Y(1) = 1.8683
X(2) = 0.2000   Y(2) = 1.9138
X(3) = 0.5000   Y(3) = 2.0562
X(4) = 1.0000   Y(4) = 2.2916
X(5) = 1.2000   Y(5) = 2.3768
X(6) = 1.8000   Y(6) = 2.5643
X(7) = 2.0000   Y(7) = 2.5945
X(8) = 2.6000   Y(8) = 2.5487
X(9) = 3.5000   Y(9) = 1.9587
X(10) = 4.0000   Y(10) = 1.2772

THE ORDER OF THE POLYNOMIAL = 5
THE CONSTANTS OF THE POLYNOMIAL ARE

C(1) = 1.79686
C(2) = 0.52743
C(3) = 0.14705
C(4) = −0.21183
C(5) = 0.06872
C(6) = −0.00884

THE VALUES CALCULATED FROM THE BEST FIT ARE

X(1) = 0.1000   Y(1) = 1.8509
X(2) = 0.2000   Y(2) = 1.9066
X(3) = 0.5000   Y(3) = 2.0749
X(4) = 1.0000   Y(4) = 2.3194
X(5) = 1.2000   Y(5) = 2.3960
X(6) = 1.8000   Y(6) = 2.5416
X(7) = 2.0000   Y(7) = 2.5618
X(8) = 2.6000   Y(8) = 2.5288
X(9) = 3.5000   Y(9) = 2.0301
X(10) = 4.0000   Y(*) = 1.2400

FIGURE 7.16 Continued.

of the polynomial. The coefficients are printed for the form represented by Equation 7.57. These are then rearranged in descending powers of the independent variable to use these with MATLAB functions, such as *polyval*, to obtain the desired values of the dependent variable from the best fit.

The results obtained from this script file for a third-order polynomial are: The constants of the polynomial are

```
c =
    1.8244
    0.4300
    0.0968
   -0.0596
```

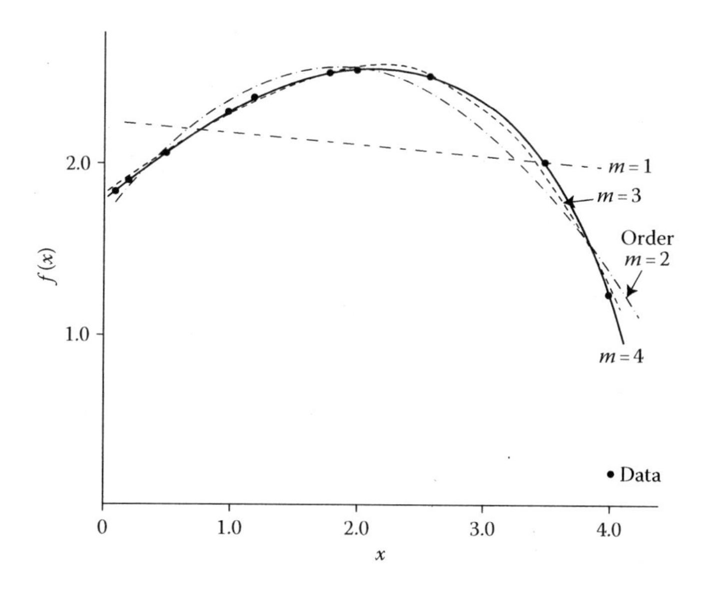

FIGURE 7.17 Comparison between given data and the best-fit obtained, using polynomials of different order.

The values calculated from the best fit are

```
y=1.8683
  1.9138
  2.0562
  2.2916
  2.3767
  2.5643
  2.5945
  2.5487
  1.9587
  1.2772
```

Thus, the results obtained are very close to those presented earlier. Similarly, the results for other orders of the polynomial, as given by the specified value of *np*, may be obtained. For linear regression, *np* is set equal to 1, as seen in the following example.

b. Consider regression with an exponential of the form

$$y = Ae^{-ax} \tag{7.73}$$

Taking natural logarithms of both sides, we obtain

$$\begin{aligned} \log y &= \log A - ax \\ &= B - ax \end{aligned} \tag{7.74}$$

where $B = \log A$ is a constant. Thus, we may apply linear regression to the given data, employing $\log y$ as the dependent variable and x as the

THE CONSTANTS OF THE POLYNOMIAL ARE

C(1) = 4.78000
C(2) = −0.09695

CONSTANT A = 119.10 EXPONENT = −0.09695

THE VALUES CALCULATED FROM THE BEST FIT ARE

X(1) = 1.0000	Y(1) = 108.0993
X(2) = 2.0000	Y(2) = 98.1109
X(3) = 5.0000	Y(3) = 73.3503
X(4) = 10.0000	Y(4) = 45.1726
X(5) = 15.0000	Y(5) = 27.8195
X(6) = 20.0000	Y(6) = 17.1326
X(7) = 25.0000	Y(7) = 10.5510
X(8) = 30.0000	Y(8) = 6.4978

THE CORRELATION COEFFICIENT = 0.9998

FIGURE 7.18 Numerical results obtained with an exponential best fit for the problem considered in Example 7.5(b).

independent variable. The results obtained are shown in Figure 7.18. The constants for the linear best fit are $C(1)$ and $C(2)$, which correspond to B and $-a$ in Equation 7.74. Therefore, $\log A = B = C(1)$, which gives $A = \exp[C(1)]$ and $a = -C(2)$. The resulting constants A and a are obtained from the program as 119.1 and 0.09695, respectively. The term exponent in this figure refers to $-a$ in Equation 7.73.

The values of the dependent variable $Y(I)$ are also calculated from the best fit and are found to be close to the given data. The correlation coefficient r for this problem is found to be 0.9998, which indicates a very good representation of the given data by the exponential function $y = 119.1 \exp(-0.09695x)$. Similarly, the program given in Appendix C.11 may be employed for other nonpolynomial forms.

Similarly, the MATLAB script file given in Appendix B.18 may be used with $np = 1$ to obtain a linear best fit to the given data, with x as the independent variable and $\log y$ as the dependent variable. The script file may be modified by employing the following after applying polynomial regression with $np = 1$:

```
disp('Constants of the linear regression are:')
c=a\b
Constant=exp(c(1))
Exponent=c(2)
plot(x1,y1,'*')
hold
x=linspace(0,40,40);
y=Constant*exp(Exponent.*x);
plot(x,y,'k-')
xlabel('x','Fontsize',14)
ylabel('y','Fontsize',14)
```

The results obtained are
Constants of the linear regression are

```
c =
    4.7800
   -0.0970
Constant = 119.1062
Exponent = -0.0970
```

The results obtained also yield the graphical representation of Figure 7.19, which shows the given data and the best fit thus obtained, indicating the close approximation of the data by the exponential best fit. We can also directly use the *polyfit* command to obtain the best fit. For instance, we could use

```
x = [1 2 5 10 15 20 25 30];
y = [109.58 99.25 73.78 45.15 26.78 17.24 9.85 6.97];
y = log(y);
p1 = polyfit(x,y,1)
```

This would then yield the two coefficients as −0.0970 and 4.7800, arranging them in descending powers of x. Thus, from Equation 7.74, $B = 4.7800$ and $a = -0.0970$ and the results are the same as those given above. Thus, the *polyfit* function may be used conveniently to obtain the best fit to the given data using a polynomial of specified order.

7.7 FUNCTION OF TWO OR MORE INDEPENDENT VARIABLES

In the preceding sections, we considered curve fitting for dependent variables that are functions of only one independent variable. However, in engineering applications,

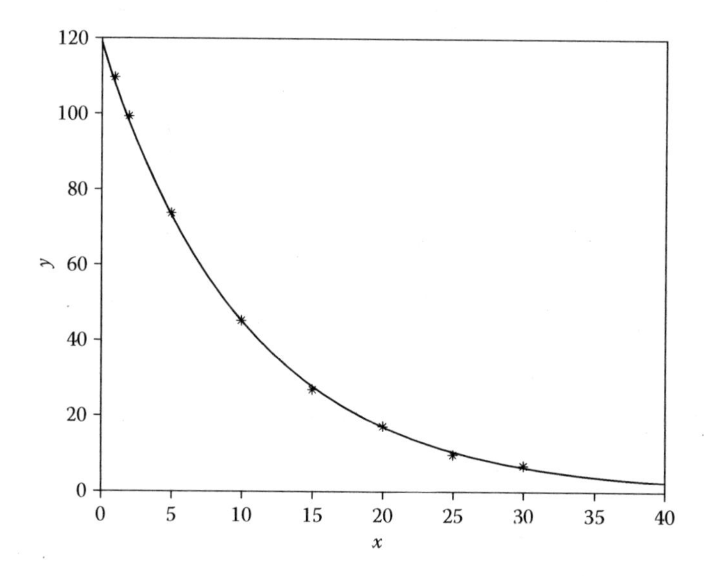

FIGURE 7.19 Given data and the exponential best-fit obtained in Example 7.5(b).

we frequently encounter functions of two or more independent variables. In many cases, interest lies in representing the dependence of such functions on only one independent variable, while the others are held constant at given values. Then, an exact or a best fit, as appropriate, may be employed, as discussed earlier, to characterize this variation. However, there are several circumstances where it is necessary to consider the variation of the dependent variable y with two or more independent variables, say, x_1, x_2, and so on. The pressure generated by a pump, for instance, depends on both the speed and the flow rate. Similarly, properties of gases, such as density, depend on the pressure as well as the temperature. Although curve fitting may be carried out with only one independent variable, taking the others at specified values and thus generating a number of curves that fit the data, it is often more convenient and desirable to seek a single function such as $f(x_1, x_2)$ that represents the dependence on all the independent variables.

7.7.1 Exact Fit

Let us consider a variable y which is a function of two independent variables x_1 and x_2. Then if an exact fit with a second-order polynomial is sought, we may employ the general equation

$$y = A + Bx_1 + Cx_1^2 \tag{7.75}$$

where the coefficients A, B, and C are functions of x_2. Again, employing second-order polynomials, we may write

$$A = a_0 + a_1 x_2 + a_2 x_2^2 \tag{7.76}$$

$$B = b_0 + b_1 x_2 + b_2 x_2^2 \tag{7.77}$$

$$C = c_0 + c_1 x_2 + c_2 x_2^2 \tag{7.78}$$

Equation 7.75 may be written at three different values of x_2 as follows:

$$y = A_1 + B_1 x_1 + C_1 x_1^2 \tag{7.79}$$

$$y = A_2 + B_2 x_1 + C_2 x_1^2 \tag{7.80}$$

$$y = A_3 + B_3 x_1 + C_3 x_1^2 \tag{7.81}$$

where (A_1, B_1, C_1) correspond to one value of x_2, (A_2, B_2, C_2) to another, and (A_3, B_3, C_3) to a third value of x_2.

The first step involves determining the coefficients in Equations 7.79 through 7.81 by employing three data points, in terms of y and x_1, at each value of x_2. Thus, as shown in Figure 7.20, we need nine data points to evaluate these nine coefficients.

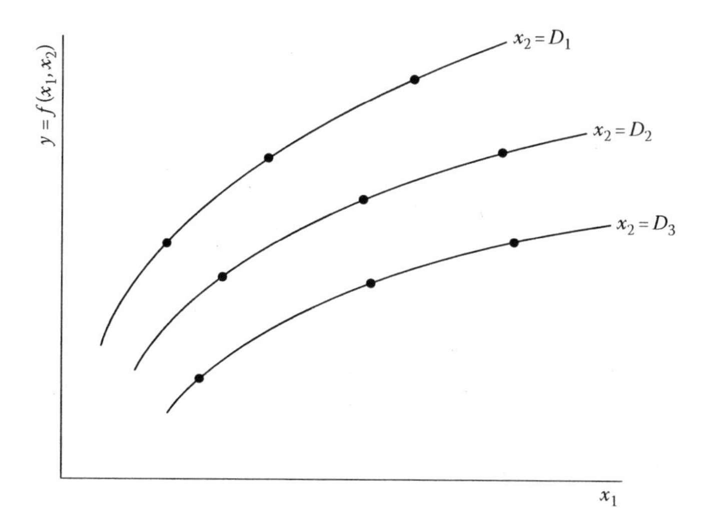

FIGURE 7.20 Sketch of a function, $f(x_1, x_2)$, of two independent variables x_1 and x_2, showing the nine data points needed for an exact fit with second-order polynomials.

Each curve in Figure 7.20 is represented by a second-order polynomial, which is determined at the given value of x_2 if three data points are available for this curve. Thus, a set of three equations is solved, as discussed in Section 7.2, to obtain the coefficients A_1, B_1, and C_1 in Equation 7.79. Similarly, the coefficients in Equations 7.80 and 7.81 are determined. Thus, we now have the values A_1, A_2, and A_3 for the variable A in Equation 7.76 at three values of x_2. Using these values, we may determine the coefficients a_0, a_1, and a_2. Similarly, we determine the coefficients in Equations 7.77 and 7.78 using the values B_1, B_2, B_3, and C_1, C_2, C_3 at the three given values of x_2. Thus, the procedure for an exact fit is applied twice to obtain all the relevant coefficients.

The coefficients obtained from the nine data points shown in Figure 7.20 yield an exact second-order polynomial fit to the given data. The resulting general equation is written as

$$y = (a_0 + a_1x_2 + a_2x_2^2) + (b_0 + b_1x_2 + b_2x_2^2)x_1 + (c_0 + c_1x_2 + c_2x_2^2)x_1^2 \quad (7.82)$$

Thus, the functional dependence of y on x_1 and x_2 is represented by this equation. This approach may easily be extended to higher-order polynomials and functions of more than two variables. However, the solution becomes more involved because of the larger number of coefficients to be determined. For instance, if third-order polynomials are employed instead of the parabolas in Equation 7.82, we must determine sixteen coefficients, employing four data points in terms of y and x_1 at four different values of x_2. Similarly, 25 data points are needed for fourth-order polynomials. Similar increase in the complexity of the solution and of the resulting polynomial fit arises if functions of more than two independent variables are considered. Other forms of the function for the exact fit, besides that given in Equation 7.75, may also be considered.

7.7.2 Best Fit

A best fit is often more appropriate than an exact fit for functions of two or more independent variables. Experimental data with a significant amount of error, for example, are better represented by a best fit than by a curve that passes through each data point. These considerations have been discussed earlier in relation to functions of one independent variable and apply equally well to multiple variables.

Let us first consider multiple linear regression, assuming the dependent variable y to be a linear function of x_1 and x_2 as

$$y = f(x_1,x_2) = c_0 + c_1x_1 + c_2x_2 \tag{7.83}$$

where c_0, c_1, and c_2 are constants to be computed. Employing the procedure given earlier for linear regression, we determine the sum S to be minimized as follows:

$$S = \sum_{i=1}^{n}(y_i - c_0 - c_1x_{1,i} - c_2x_{2,i})^2 \tag{7.84}$$

where the subscript i, which varies from 1 to n, is used to denote the n data points. Differentiating S with respect to the coefficients and setting the partial derivatives equal to zero yields the minimum value of S. Thus,

$$\frac{\partial S}{\partial c_0} = -2\sum(y_i - c_0 - c_1x_{1,i} - c_2x_{2,i}) = 0$$

$$\frac{\partial S}{\partial c_1} = -2\sum x_{1,i}(y_i - c_0 - c_1x_{1,i} - c_2x_{2,i}) = 0$$

$$\frac{\partial S}{\partial c_2} = -2\sum x_{2,i}(y_i - c_0 - c_1x_{1,i} - c_2x_{2,i}) = 0$$

where the summations are from $i = 1$ to $i = n$.

The above equations yield the following system of linear equations for the unknowns c_0, c_1, and c_2:

$$nc_0 + c_1\sum x_{1,i} + c_2\sum x_{2,i} = \sum y_i \tag{7.85}$$

$$c_0\sum x_{1,i} + c_1\sum (x_{1,i})^2 + c_2\sum x_{1,i}x_{2,i} = \sum x_{1,i}y_i \tag{7.86}$$

$$c_0\sum x_{2,i} + c_1\sum x_{1,i}x_{2,i} + c_2\sum (x_{2,i})^2 = \sum x_{2,i}y_i \tag{7.87}$$

These simultaneous equations may be solved for c_0, c_1, and c_2 to give the best fit, Equation 7.83. In this case, a regression plane is obtained instead of a line, since y varies with two independent variables x_1 and x_2. The correlation coefficient is again obtained from Equation 7.56, with appropriate change in the definition of S_m to take the dependence of y on both x_1 and x_2 into account.

By employing the following general form of the function for a best fit, we can extend the procedure outlined above for multiple regression to functions of more than two variables:

$$y = f(x_1, x_2, \ldots, x_m) = c_0 + c_1x_1 + c_2x_2 + \cdots + c_mx_m \tag{7.88}$$

The system of linear equations for evaluating the coefficients $c_0, c_1, \ldots, c_m$ may easily be derived as given above for the case of two independent variables. Similarly, multiple polynomial regression, with orders higher than linear, may be derived for introducing curvature into the best fit.

Also, linearization of nonlinear functions, such as exponentials and power-law variations, can often be carried out, as outlined earlier for functions of a single variable x. Then multiple linear regression may be applied. Thus, if y is of the general form

$$y = c_0 x_1^{c_1} x_2^{c_2} \cdots x_m^{c_m} \tag{7.89}$$

the equation may be transformed by taking its natural logarithm to give

$$\log y = \log c_0 + c_1 \log x_1 + c_2 \log x_2 + \cdots + c_m \log x_m \tag{7.90}$$

Multiple linear regression may now be applied. Example 7.6 illustrates the use of this procedure for a practical circumstance.

Example 7.6

The flow of water in an open channel with a slight downward slope is an important circumstance in civil engineering applications. The channel is specified in terms of its hydraulic radius R, which is the cross-sectional area divided by the wetted perimeter consisting of the sides and bottom of the channel, and the slope S. The slope is given as tan θ, where θ is the angle that the bottom makes with the horizontal, considered positive for downhill flow. The volume flow rate Q in m^3/s is measured as a function of R and S for certain open channels to yield the following data:

	R (m)	0.5	1.0	1.5	2.0
S					
1.5×10^{-3}		1.91	3.10	4.11	5.03
5×10^{-3}		3.48	6.66	7.51	9.19
9×10^{-3}		4.67	7.59	10.08	12.33

It is expected, from theoretical considerations, that Q varies as AS^bR^c, where A, b, and c are constants. Obtain a best fit to the given data and determine these constants.

SOLUTION

Since the dependent variable Q is a function of two independent variables R and S, multiple regression, as outlined in Section 7.7.2, may be applied. The form of the function to be employed is

$$Q = AS^b R^c \tag{7.91}$$

Taking natural logarithms of both sides, we obtain

$$\log Q = \log A + b \log S + c \log R \tag{7.92}$$

Thus, multiple linear regression may be used with log S and log R as the independent variables and log Q as the dependent variables.

The dependent variable log Q is denoted by y, and log S and log R by x_1 and x_2, respectively. Then, the system of linear equations to be solved for the constants c_0, c_1, and c_2, in the linear function $y = c_0 + c_1x_1 + c_2x_2$, for a best fit is given by Equations 7.85 through 7.87. The coefficients of this system of equations may be obtained from the twelve data points given. Thus, $n = 12$, and the summations, such as $\Sigma\, x_{1,i}$ and $\Sigma\, x_{1,i}\, y_i$, are from $i = 1$ to $i = 12$. Employing a calculator or a simple program in MATLAB, we obtain the following system of equations:

$$12c_0 - 66.045c_1 + 1.216c_2 = 20.575 \tag{7.93}$$

$$-66.045c_0 + 370.164c_1 - 6.695c_2 = -109.871 \tag{7.94}$$

$$1.216c_0 - 6.695c_1 + 3.376c_2 = 4.345 \tag{7.95}$$

The above system of linear equations can easily be solved by using matrix methods or the backslash operator in MATLAB to obtain the coefficients as

$$c_0 = 4.4235,\ c_1 = 0.505,\ c_2 = 0.6945 \tag{7.96}$$

From Equation 7.92, $c_0 = \log A$, $c_1 = b$, and $c_2 = c$. This gives $A = \exp(c_0) = \exp(4.4235) = 83.387$. Therefore, the best fit to the given data is obtained as

$$Q = 83.387S^{0.505}\, R^{0.6945} \tag{7.97}$$

A MATLAB script file may also be written to solve the preceding multiple regression problem. The given data are entered and the various summations are carried out. The matrices representing the equations, Equations 7.85 through 7.87, are formulated and the constants for multiple regression are calculated. The transformations to the original variables are then made to yield the desired results. The following script file may thus be used for this problem.

```
s1 = 1.5e-3;s2 = 5e-3;s3 = 9e-3;
s(1:4) = s1;s(5:8) = s2;s(9:12) = s3;
r1 = [0.5 1.0 1.5 2.0];
r = [r1 r1 r1];
q = [1.91 3.1 4.11 5.03 3.48 6.66 7.51 9.19 4.67 7.59 10.08 ...
    12.33];
```

```
s=log(s);r=log(r);q=log(q);
n=length(q);
a=[n sum(s) sum(r);sum(s) sum(s.*s) sum(s.*r);...
   sum(r) sum(s.*r) sum(r.*r)];
b=[sum(q) sum(s.*q) sum(r.*q)];
disp('Constants for multiple linear regression are:')
c=a\b'
A=exp(c(1));
b=c(2);
c=c(3);
fprintf('A=%.4f b=%.4f c=%.4f/n',A,b,c)
```

The results obtained from this program are
Constants for multiple linear regression are

```
c=
   4.4234
   0.5050
   0.6945

A=83.3760 b=0.5050 c=0.6945
```

In a similar way, other power-law and exponential variations may be treated for functions of two or more independent variables. Multiple polynomial regression, with polynomials of order higher than linear, may also be employed for certain circumstances, using a similar, although more complicated, approach.

7.8 SUMMARY

This chapter presents numerical methods for the curve fitting of data given at discrete points, considering both an exact fit and a best fit. In the former case, the approximating curve passes through each data point and is appropriate if the data have a high level of accuracy and a relatively small number of points are given. Various forms of the approximating function are considered, including the general equation of a polynomial, *Lagrange* polynomial, and *Newton's divided-difference* polynomials. The use of these interpolating polynomials for evaluating the function at intermediate points, where data are not available, is discussed. Lagrange interpolation is particularly useful for an arbitrary distribution of points and is widely used. If a large number of very accurate data points are given, spline interpolation, which provides a piecewise exact fit to the data, is more appropriate than a single curve, since polynomials of high order may be ill-conditioned and are also inconvenient to use in practical circumstances. The equations for cubic splines are derived. Examples are given to demonstrate the use of interpolation in engineering problems. MATLAB functions for interpolation using different forms of the interpolating curve, such as linear, cubic or spline, are presented.

A best fit, which minimizes the error between the data and the approximating curve without forcing it to pass through each given data point, is extensively employed for correlating engineering data. It is more suitable than an exact fit for data that have a significant amount of associated error. Experimental data generally do have some error, and a best fit is used for representing the observed trends. This approach is generally used with lower-order polynomials, such as straight lines, parabolas, and cubics, to obtain a best fit to a large number of data points. The method of least squares is discussed in detail, considering linear regression, polynomial regression, and nonpolynomial forms. In several important engineering applications, special forms, such as exponential and power-law variations, are of interest. These forms may often be linearized by suitable transformations, and linear regression may be applied. Finally, functions of two or more variables are considered. A few simple procedures for an exact fit, as well as for a best fit, are outlined.

The choice of the form of the approximating function for curve fitting is an important consideration. Frequently, the basic nature of the problem under consideration may be employed to determine the general nature of the variation and the function chosen appropriately. If no prior information is available on the expected trends, a rough plot of the data may be used to guide the choice of the function for curve fitting. A best fit is much more extensively used in engineering problems than an exact fit, because of the presence of significant error in most available data and also because a large number of data points are often given. One may start with simple linear regression and then proceed to parabolas and cubics, in order to check whether a better representation is obtained with a higher-order function.

Lagrange interpolation is a very popular choice for an exact fit, since a system of linear equations does not have to be solved, as is the case for the general form of an nth-order polynomial. Newton's method is particularly useful if the data points are evenly spaced. Extrapolation is also employed in some cases to compute the value of the function at a point beyond the range of the given data. However, one should exercise extreme care while using extrapolated values, since the behavior of the function beyond the given range is often not known. There are also several special interpolating functions, such as Chebyshev polynomials, that are employed in the analysis of engineering systems and processes, see Hornbeck (1982). Also, there are other methods for deriving the interpolating function. One such method is Hermite interpolation which uses both the function and its derivative at a given number of data points, as outlined by Ferziger (1998).

PROBLEMS

7.1. Consider a second-order Lagrange polynomial and show that it may be recast in the general form of a second-order polynomial given by Equation 7.1. Obtain the relationship between the coefficients of the two polynomials.

7.2. Show that Lagrange interpolation is a more efficient method for interpolation than that obtained by using the general form of an nth order polynomial, as demonstrated in Example 6.1.

7.3. Compare the Lagrange and Newton's divided-difference interpolation methods, indicating their respective advantages over the other. Which one is expected to require less computer time for interpolation with an arbitrary distribution of data points?

7.4. The specific heat C of pure copper is given at 100, 200, 400, 600, and 800 K as 252, 356, 397, 417, and 433 J/kg K. Employ Lagrange interpolation to compute the values at 300 K and 500 K. Also, compute the extrapolated value at 1000 K and compare it with the value of 451 given in the literature.

7.5. The density of air at 200, 300, 400, and 500 K is obtained as 1.7458, 1.1614, 0.8711, and 0.6964 kg/m^3, respectively, from tabulated property data in the literature. For this uniformly spaced data, obtain a third-order interpolating polynomial.

7.6. A car showroom has 100 cars at the beginning of a week, and the number left after each day is tabulated as follows:

Time (days)	0	1	2	3	4	5	6
Cars left (N)	100	75	65	52	46	39	34

We wish to extrapolate these results to predict the cars left at the end of the week. Using an exact fit, predict the number of cars left in the showroom after seven days. Comment on the result obtained.

7.7. Use a second-order and also a third-order polynomial regression for Problem 7.6. Compare the results obtained with that obtained earlier with an exact fit, and comment on the difference. Which method would you expect to yield a more dependable prediction? Discuss.

7.8. The force F on a structure due to winds is measured as a function of wind speed V. The results at speeds of 5, 10, 15, 20, and 25 m/s are obtained as 36.2, 52.5, 85.6, 150.0, and 210.9 newtons. Obtain a fourth-order interpolating polynomial that provides an exact fit to these data points.

7.9. The future worth (FW) of a given sum of money R after n years is $R(1 + x)^n$, where x is the interest rate per unit amount, say $1.00, compounded annually. Therefore, the FW ratio, FW/R gives the FW per unit deposit and may be determined at interest rates of 8%, 10%, 12%, and 15% for 15 years as 3.172, 4.177, 5.474, and 8.137. Employing Newton's divided-difference interpolation method, compute the corresponding values at 9% and 12.5% interest rates. Also give the resulting FW for a deposit of $5000 at these rates. Using the *interp*1 command in MATLAB, obtain the interpolated values and compare with those obtained earlier.

7.10. The voltage v applied across an electrical circuit is varied, and the resulting current i measured. For v values of 1, 2, 3.5, 5, and 6 V, the current is 1.5, 1.8, 2.6, 3.0, and 3.5 amperes. Use Newton's divided-difference method to obtain the electrical current at v = 4 and 5.5 V.

7.11. An important fluid property is the kinematic viscosity which determines the viscous, or frictional, forces acting in a flow. The kinematic viscosity of air multiplied by 10^6 is given at 350, 450, 500, 550, and 650 K as 20.92, 32.39, 38.79, 45.57, and 60.21 m^2/s, respectively. Using any suitable interpolation method, compute the intermediate values at 400 and 600 K. Compare the results obtained with the values given in the literature as 26.41×10^{-6} m^2/s and 52.69×10^{-6} m^2/s, respectively. Also, solve the problem using the *interp*1 command in MATLAB, and compare the results with those obtained earlier.

7.12. From the data given in Problem 7.9, determine the interest rate if the FW ratio FW/K is 6.5.

7.13. The calibration table for a copper-constantan thermocouple which is employed for temperature measurement gives the temperature T in °C for different values of the voltage output V in millivolts (mV). Using interpolation with a cubic spline for the following data, compute the temperatures corresponding to thermocouple outputs of 0.9 and 1.75 mV:

T (°C)	10	20	30	40	50	60	70	80
V (mV)	0.391	0.789	1.196	1.611	2.035	2.467	2.908	3.357

Also, solve the problem using the *interp*1 command in MATLAB for a spline exact fit and compare the results with those obtained earlier.

7.14. Using the data in Problem 7.13 with Lagrange interpolation, calculate the voltage output at $T = 65$°C. Employ a fourth-order polynomial and choose appropriate data points.

7.15. The transport rate $\dot{m}$ of a chemical species at a porous surface is measured as a function of the difference in concentration ΔC between the surface and the ambient medium. The results obtained are as follows:

ΔC (kg/m^3)	0.1	0.3	0.4	0.5	0.7	0.9	1.0
$\dot{m}$ (kg/s)	2.53	3.33	3.58	3.78	4.12	4.38	4.5

A power-law variation of the form $\dot{m} = A(\Delta C)^a$ is expected to govern this mass transfer process. Obtain a best fit to the given data by the method of least squares and determine the constants A and a. Also, use the *polyfit* command in MATLAB to solve this problem and compare the results with those obtained earlier.

7.16. Experimental runs are performed on a compressor to determine the relationship between the volume flow rate Q and the pressure difference P. It is expected that Q will be proportional to P^b, where b is a constant. The measurements yield the mass flow rate Q for different pressure differences P as

P (atm)	5.0	10.0	15.0	20.0	25.0	30.0
Q (m^3/h)	7.4	13.3	16.5	19.0	20.6	24.3

It is known that there is some error in the data. Will you use a best or an exact fit? Use the appropriate fit to these data and determine the coefficients. Is the equation obtained by you a good fit?

7.17. Tests are performed on a nuclear power system to ensure safe shutdown in case of an accident. The measurements yield the power output P versus time t in hours as

t (h)	1	3	5	9	10	12
P (MW)	13.0	7.0	5.4	4.7	4.5	4.2

From theoretical considerations, the power is expected to vary as $a + b/t$, where a and b are constants. It is also known that there is significant error in the data. What curve fitting will you use? Use an appropriate fit to these data points and determine the relevant constants. Is it a good curve fit? Briefly explain your answer.

7.18. Experiments are carried out on a plastic extrusion die to determine the relationship between the mass flow rate m and the pressure difference P. We expect the relationship to be of the form $m = AP^n$, where A and n are constants. The measurements yield the mass flow rate m for different pressure differences P as

m (kg/h)	12.8	15.5	17.5	19.8	22.0
P (atm)	10.0	15.0	20.0	25.0	30.0

Obtain a best fit to these data, using MATLAB commands, and determine the coefficients A and n. Plot the results from your best or exact fit, along with the data to see if it is a good fit.

7.19. A set of four data points is given as:

$x =$ 0.5 1.0 1.5 2.0

$y =$ 3.0 3.9 5.2 7.3

where x is the independent variable and y is the dependent variable. Write a script-*m* file to do the following:

a. Obtain the polynomial that passes through all these four points.
b. Use this polynomial to find the value at $x = 1.7$ by interpolation.
c. Obtain a linear least-squares best fit (linear regression) to these data points.
d. Use the linear regression to obtain the value at $x = 1.7$.
e. Plot the data points and the linear best fit on a x–y plot.

7.20. The concentration of salt decreases in a container because of mass transfer at the surface. The concentration C is measured as a function of time t to yield

t (s)	0.1	0.2	0.3	0.5	1.0	2.0	4.0	4.5	5.0
C (kg/m^3)	83.3	81.7	80.0	76.9	69.6	57.0	38.2	34.6	31.3

An exponential variation of the form $C = Be^{-bt}$ is expected on physical grounds. Obtain a best fit to the data using a MATLAB script file, and determine the constants B and b. Also, solve the problem using the *polyfit* function in MATLAB and compare the results with those obtained earlier.

7.21. The temperature T, pressure p (in kilopascals), and specific volume v, which is inverse of density, for saturated steam are obtained from tabulated data in the literature as follows:

T (°C)	10	20	30	40	50	60	70	80	90
P (k Pa)	1.23	2.34	4.25	7.38	12.35	19.94	31.19	47.39	70.13
v (m^3/kg)	106.4	57.79	32.90	19.52	12.03	7.67	5.04	3.41	2.36

Obtain a best fit, with a third-order polynomial, to the T–v data. Using the polynomial obtained, compute the specific volumes at 55°C and 75°C. Also calculate the value at 100°C, and compare it with the given value of 1.673 m^3/kg. Also, using the *polyfit* command in MATLAB, obtain the best fit polynomial and compare with that obtained earlier.

7.22. Using the data given in the preceding problem, obtain a best fit to the specific volume dependence on pressure. Consider both second- and third-order polynomials. Using the polynomials obtained, calculate the specific volumes at 5.0 and 25.0 kPa.

7.23. The pressure-temperature relationship for saturated steam is suggested to be of the form $\log p = C + D/T$, where C and D are constants and log represents the natural logarithm. Using linear regression with the data in Problem 7.21, determine the values of these constants. Is the given functional dependence of p on T a satisfactory representation?

7.24. The acceleration of certain objects is studied in an experimental test track for automobiles. The distance traveled by an object L is measured as a function of time t to yield the following:

t (s)	0.1	0.2	0.5	1.0	1.5	1.8	2.0	3.0
L (m)	0.26	0.55	1.56	3.90	7.41	10.28	12.6	30.9

Obtain a best fit to this data, considering first-, second, and third-order polynomials. Using these polynomials, calculate the values of the dependent variable L at the time intervals employed for the given data to evaluate the accuracy of the polynomial representations. Discuss the results obtained.

7.25. In the preceding problem, calculate the correlation coefficient to estimate the improvement in the representation of the data by means of curve fitting.

7.26. For the experimental data given in Example 7.1, obtain a best fit, using second- and third-order polynomials. Compare the interpolated values obtained by the exact fit in Example 7.1 with those obtained from the best fit. Comment on the difference.

7.27. Solve the problem given in Example 7.2 by Newton's divided-difference method, and compare the interpolated results obtained with those given earlier.

7.28. The temperature T of a small copper sphere cooling in air is measured as a function of time t to yield the following:

t (s)	0.2	0.6	1.0	1.8	2.0	3.0	5.0	6.0	8.0
T (°C)	146.0	129.5	114.8	90.3	85.1	63.0	34.6	25.6	14.1

An exponential temperature decrease is expected from theoretical considerations. Using linear regression, obtain the exponent c and the constant C, where $T = Ce^{-ct}$ represents the variation. Also, solve this problem using the *polyfit* function in MATLAB.

7.29. The temperature of a furnace wall is expected to vary sinusoidally with a time period of one day, because of the daily start-up and shut-down. The measured temperatures at several time intervals t, where t is measured from midnight, are given as follows:

t (h)	2	3	5	8	10	15	18	22	24
T (°C)	86.5	97.7	104.0	101.7	92.5	62.3	55.0	67.5	80.0

Obtain a best fit to these data, using the method of least squares and assuming a sinusoidal variation of the form $A\sin(2\pi t/24) + B\cos(2\pi t/24) + C$, where A, B, and C are constants that are to be determined.

7.30. Calculate the correlation coefficients for the various polynomials considered for least-squares best fit in Example 7.5, and discuss the trends indicated by the results obtained.

7.31. Derive an expression for the correlation coefficient corresponding to a third-order polynomial best fit to a data set represented by (x_i, y_i), from $i = 1$ to $in = n$. Discuss the physical implications of this coefficient. Can the correlation coefficient be related to the accuracy of the best fit obtained?

7.32. Consider an equation of the form $y = \sin(\pi a x) + Ax^b$, where A, a, and b are constants. Can the method of least squares be applied to this equation for a given set of data points? Discuss.

7.33. Outline a procedure for obtaining a best fit with a power-law function of the form $y = z + bx^n$, where a, b, and n are constants.

7.34. Consider a functional dependence of y on the independent variable x of the form given by Equation 7.66. Using this equation, outline a procedure for deriving a best fit to given data.

7.35. Use *polyfit* in MATLAB to get the best fit to the following data, using first, second, and third order polynomials. Then plot the data as well as the three best fit curves obtained. Which is the best fit? Discuss.

x:	0	0.1	0.2	0.3	0.4	0.5	0.6	0.8	1.0	1.2
y:	0	0.87	1.82	2.86	4.0	5.26	6.65	9.88	13.8	18.52

7.36. The flow rate F is given at various values of the pressure P as

P	0.02	0.05	0.1	0.2	0.3	0.4	0.5
F	1.7	2.9	5.6	6.6	7.8	8.7	9.3

Use the last five points to get an exact fit. Use extrapolation with this fit to obtain values at 0.025 and 0.05. Compare with given data. Comment on the results.

7.37. Obtain the first, second, and third order best fits to the data in the preceding problem. Plot all the three curves and the data to determine the best curve to use.

7.38. Six data points generated by a polynomial are given. Outline a method for finding the order of the polynomial. Also apply your method to y values of 3.61, 5.38, 11.0, 18.34, 28.63, and 35.0 corresponding to x values of 0.2, 0.5, 1.0, 1.4, 1.8, and 2.0, respectively, where x is the independent variable and y the dependent variable.

7.39. Five data points are given, with one of them in considerable error. How will you find this point, using the interpolation methods discussed in the text? Consider, as an example, the following data set:

X	0.25	0.75	1.25	2.5	3.0
Y	2.80	4.60	5.75	7.94	6.5

where x is the independent variable and y the dependent variable.

7.40. The decay of the electrical current I in an electronic circuit is measured as a function of time t, following the opening of a switch. The data obtained are given as follows:

t (s)	0.5	1.0	1.5	2.5	3.5	5.0	6.5	9.0	9.5
I (amperes)	13.2	10.1	8.7	6.9	6.3	5.1	4.7	4.2	4.0

From theoretical considerations, the current is expected to follow a variation of the form At^{-a}, where A and a are constants. Obtain a best fit to the given data, and determine the values of these constants.

7.41. The flow rate Q in circular pipes is measured as a function of the pressure difference Δp and diameter D. The resulting data for the flow rate in m^3/s are given as follows:

Δp(atm)	D (m)	0.3	0.5	1.0	1.4
0.5		0.13	0.43	2.1	4.55
0.9		0.25	0.81	4.0	8.69
1.2		0.34	1.12	5.5	11.92
1.8		0.54	1.74	8.59	18.63

Using the method of least squares, obtain a best fit for the flow rate as a function of the two independent variables D and Δp. It is expected that Q varies as $BD^a \Delta p^b$, where B, a, and b are constants to be determined.

7.42. Repeat the preceding problem if the values of the pipe diameter were given as 0.5, 0.8, 1.4, and 1.9, instead, with the remaining values unchanged. Similarly, solve the problem again if all the values were unchanged but the pressure difference values Δp were given as 0.7, 1.2, 1.5 and 2.1, instead.

8 Numerical Integration

8.1 INTRODUCTION

A problem that frequently arises in engineering applications is that of integration of a given function $f(x)$ over a specified range of the independent variable x. In many cases, the function $f(x)$ is continuous, finite, and well behaved over the range of integration $a \leq x \leq b$, where a and b are constants. Then, the integral I where

$$I = \int_a^b f(x)\mathrm{d}x \tag{8.1}$$

may often be determined by using available mathematical or analytical techniques. The results for common elementary functions such as $\sin x$, $\cos x$, e^x, x^2, $1/x$, and so on, are well known, and those for many more complicated functions are given in integral tables. Symbolic algebra available in MATLAB®, *Mathematica*, *Maple*, and other such environments may also be used in many cases to obtain the integral analytically. Analytical, or closed-form, expressions for integrals, whenever available, are of considerable value since they are exact, that is, without the errors that inevitably arise in numerical methods. Moreover, they are generally applicable over given domains without any limitations, so that the effect of varying the physical parameters, associated with the problem, on the integral may easily be investigated. In addition, analytical results can be employed in the validation of a numerical integration scheme and for estimating the accuracy of the results.

In engineering problems, the function $f(x)$ is often too complicated to be integrated analytically. One or both limits of integration may be infinite, and the function $f(x)$ itself may be discontinuous or infinite at some point. Also, the function may be available only at certain discrete points, say, from an experimental study or from the numerical solution of a differential equation. In this last circumstance, curve fitting, as discussed in the preceding chapter, may sometimes be employed to yield a function $f(x)$ that can be integrated analytically. Otherwise, numerical integration is necessary. Similarly, for the various other circumstances mentioned above, analytical methods may be unavailable, may be time consuming, or may be too difficult to apply, making it essential to use numerical integration.

Integration, which is also often called *quadrature*, basically refers to the area between the curve of $f(x)$ versus x and the x axis, from $x = a$ to $x = b$, as shown graphically in Figure 8.1. As expected, the integral I is positive if the area above the x axis is larger than that below it. This graphical representation of the integral $I = \int_a^b f(x)\mathrm{d}x$ will frequently be referred to in the development of formulas for numerical integration.

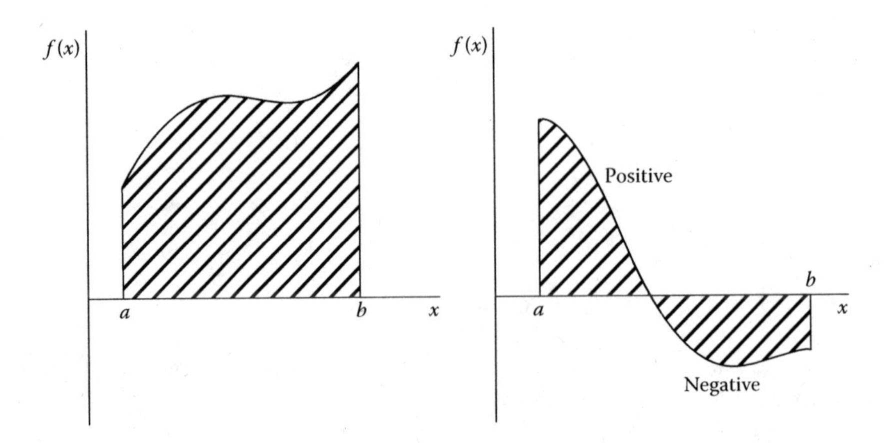

FIGURE 8.1 Graphical representation of the integral of a function $f(x)$ over x, between the limits $x = a$ and $x = b$, as the area between the curve and the x-axis.

In this chapter, various methods for the numerical integration of a continuous or discretized function $f(x)$ are presented. The most common approach is based on replacing the function $f(x)$ or the tabulated data with a simple polynomial that can be easily integrated. This approach gives rise to the *Newton–Cotes* formulas, the simplest one being obtained when the function $f(x)$ is taken as constant over the various segments into which the given range $a \le x \le b$ is divided. The most commonly used Newton–Cotes formulas are the *trapezoidal*, *Simpson's one-third*, and *Simpson's three-eighths* rules, which are based, respectively, on linear, parabolic, and cubic, or third-order, polynomial approximations. Although these formulas are derived for continuous functions, their application to evenly and unevenly spaced data is also discussed, since experimental and numerical results are generally available at such discrete values of the independent variable x.

The truncation errors (TEs) in these formulas are determined to evaluate the resulting accuracy. As the number of segments n into which the region is divided is increased, or the segment or step size Δx is reduced, the TE decreases, so that the numerical value of the integral approaches the exact value. However, as Δx is reduced to very small values, the computational effort and the round-off error increase substantially, as discussed in Section 2.3, resulting in an increase in the total error with a further reduction in Δx. Thus, even though the mathematical definition of integration demands that $\Delta x \to 0$, a lower limit on Δx is imposed by the round-off error in numerical integration. These considerations are again discussed later in this chapter.

In many engineering applications, an accuracy higher than that provided by the relatively simple trapezoidal and Simpson's rules is demanded from numerical integration. Various methods for improving the accuracy, such as *Richardson's extrapolation* and higher-order integration formulas, are discussed. *Romberg integration*, which provides very high accuracy, without an associated substantial increase in the computational effort and the round-off error, as encountered at very small segment size Δx, is of particular importance in such applications and is discussed in detail.

Also considered in this chapter is *Gauss quadrature*, which is particularly suitable for cases where the evaluation of the integrand $f(x)$ is involved and is thus time consuming. *Adaptive* methods, which increase the accuracy of the computation by focusing on intervals in which the inaccuracy is larger than that in other intervals, are also outlined. Finally, improper integrals, in which the integrand becomes infinite at some point or the limits of integration are infinite, are discussed, and some of the techniques that may be employed for computing the integral are presented.

8.1.1 Engineering Examples

Before proceeding to the various methods for numerical integration, let us consider a few examples of engineering interest in which numerical integration is needed, in order to provide a physical background for the discussion to follow. In electrical engineering, the root mean square (RMS) value of an electrical current $I(t)$, which varies periodically with time t, is given by

$$I_{\mathrm{RMS}} = \frac{1}{t_{\mathrm{c}}}\sqrt{\int_0^{t_{\mathrm{c}}} I^2(t)\,\mathrm{d}t} \tag{8.2}$$

where t_c is the time for one cycle. Numerical integration is generally needed for an arbitrary periodic variation of $I(t)$. Periodic processes are also encountered in natural phenomena, such as the daily and yearly variations of environmental temperatures, and numerical integration is employed to compute the resulting transport of mass and energy, say, from the surface of a lake. The integral of the current $I(t)$ entering a capacitor, $\int_0^t I(t)\mathrm{d}t$, gives the stored charge $Q(t)$. Thus, the voltage $V(t)$ across the capacitor, due to the current in a given electrical circuit containing the capacitor, of capacitance C, may be determined, since $V=Q/C$. A similar integral arises in civil engineering for water storage in a reservoir due to the inflow minus the outflow, both of which are time dependent. The variation with time is often very complicated, or the values are known only at certain discrete data points, making it necessary to use numerical integration.

Integration is very important in radiation heat transfer where integrals over surfaces, volumes, wavelength interval of the radiation, and total angle of the incident radiation are needed to compute the energy transport rates. In most practical cases, these integrals are too complicated to be obtained by analytical methods. The integral of the emissive power of a blackbody, Equation 4.62, over wavelength ranging from zero to infinity is one such example. The mass or energy transfer from a surface is frequently obtained from an integral of the transport rate, given as a time-dependent mass or heat transfer flux, per unit area and time. Although some simple problems may be solved analytically, most practical circumstances require numerical integration. Such problems often arise in chemical reactors and manufacturing processes.

The volume flow rate in a circular tube, Q, is obtained by an integral of the velocity distribution $V(r)$ as follows:

$$Q = \int_0^R V(r)2\pi r\,\mathrm{d}r \tag{8.3}$$

where R is the radius of the tube and r the radial distance from the axis. In most practical cases, $V(r)$ is a complicated function or is available only at discrete data points, thus requiring numerical integration for the computation of Q. In dynamic systems, the work done W is related to the force $F(x)$ and distance x as

$$W = \int_{x_1}^{x_2} F(x)\,dx \tag{8.4}$$

where x_1 and x_2 are the initial and final positions. Again, for an arbitrary functional dependence $F(x)$, numerical integration is needed.

The few examples outlined here indicate the importance of numerical integration in many diverse engineering fields. Some relatively simple integral expressions are also given. However, many more complicated forms are often encountered in engineering. For example, multiple integrals commonly arise in radiation due to integration over several independent variables. Improper integrals, due to the integrand becoming singular or the integration limits becoming infinite, are also often of interest. Many of these cases are considered in this chapter. We now proceed to the derivation of some of the commonly used formulas for numerical integration.

8.2 RECTANGULAR AND TRAPEZOIDAL RULES FOR INTEGRATION

The most commonly used schemes for numerical integration are the *Newton–Cotes* formulas, which are based on the approximation of a complicated function $f(x)$, or of tabulated data, with a simple polynomial that can be integrated easily. Thus, the integral I is written as

$$I = \int_a^b f(x)\,dx \cong \int_a^b P_m(x)\,dx \tag{8.5}$$

where $P_m(x)$ is an mth order polynomial of the form

$$P_m(x) = p_0 + p_1x + p_2x^2 + \cdots + p_mx^m \tag{8.6}$$

The p's are constants that are determined by choosing an interpolating polynomial that yields the same values of the dependent variable as the given function $f(x)$ at a finite number of points, as done in Section 7.2.1. However, the replacement of $f(x)$ by $P_m(x)$ is done piecewise over each of the n intervals into which the total range of x is subdivided. The general approach to the derivation of the Newton–Cotes formulas is based on Lagrange interpolation, which was discussed in the preceding chapter. However, the first few approximations may be derived by simple direct methods, which are based on the graphical interpretation of integration.

The first step in the numerical integration of a function $f(x)$ is the division of the integration range $a \le x \le b$ into a finite number n of intervals or strips, as shown in

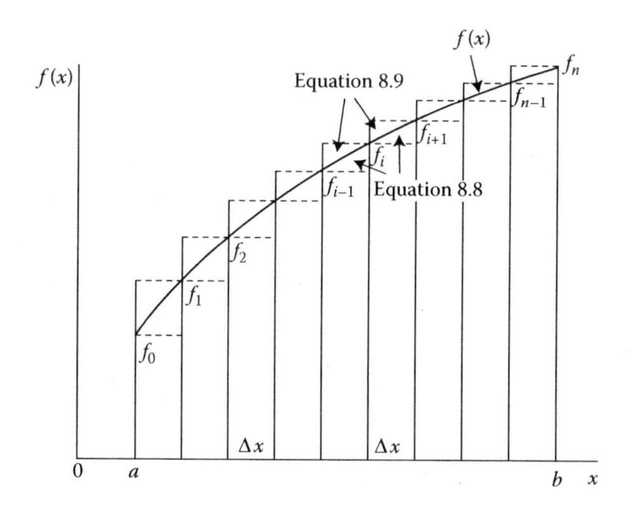

FIGURE 8.2 Approximation of an integral by a finite number of rectangular strips or segments.

Figure 8.2. If Δx is the width of each interval, then $\Delta x = (b - a)/n$. The largest value of Δx is $(b-a)$, which is obtained when the entire range of integration is taken as a single interval. The independent variable x varies from $x = a$ to $x = b$ in steps of Δx, so that x may be written as

$$x_i = a + i\Delta x, \quad \text{where } i = 0, 1, 2, \ldots, (n-1), n \tag{8.7}$$

Thus, $x_0 = a$, $x_n = b$, and x_i represents the value at an intermediate grid point, as shown in Figure 8.2. The corresponding ordinates are denoted by $f_0, f_1, f_2, \ldots, f_i \ldots, f_n$. The interpolating polynomial, Equation 8.6, is now applied piecewise to the function or data over these intervals of constant width. Each segment has two end points that can be used to determine a polynomial of order 1, that is, a straight line, $m = 1$. However, for higher order polynomials, more than one segment will be needed in order to provide the necessary number of points for the determination of all the coefficients of the interpolating polynomial.

8.2.1 The Rectangular Rule

The simplest approximation to the function $f(x)$ is a zeroth-order polynomial, that is, a constant value over each interval. Then the function $f(x)$ is approximated as a constant, at f_i or f_{i+1}, over the interval $x_i \le x \le x_{i+1}$. Thus, the area under the curve in this interval is taken as $f_i\Delta x$, or $f_{i+1}\Delta x$. For an increasing function, as sketched in Figure 8.2, the approximation of the function as f_i over the interval underestimates the actual area under the curve, and the approximation as f_{i+1} overestimates the integral. Similarly, for a decreasing function, the former approximation provides an upper bound for the integral, and the latter approximation a lower bound.

Therefore, the given integral I is approximated, in the rectangular numerical integration scheme, by

$$I = \int_a^b f(x)\mathrm{d}x \cong \sum_{i=0}^{n-1} f_i \Delta x \tag{8.8}$$

or

$$I = \int_a^b f(x)\mathrm{d}x \cong \sum_{i=0}^{n-1} f_{i+1} \Delta x \tag{8.9}$$

The first formulation sums the ordinates at the beginning of each interval and multiplies the sum with the step size Δx, to give the numerical approximation to the integral. The second formulation sums the ordinates at the end of each interval and approximates the integral by the product of this sum with Δx. As shown in Figure 8.2 and as mentioned above, the two formulations provide the upper and lower bounds for the given integral if the function $f(x)$ is a monotonically increasing or decreasing function of x. It must also be noted that the difference between the integrals from the two formulations is simply $|f_n - f_0|\Delta x$, that is, the product of Δx and the difference between the two end ordinates.

The rectangular rule yields the exact value of the integral only if $f(x)$ is a constant. For an arbitrary function, the TE is generally very large and the method is seldom used. However, this discussion serves to illustrate the basic concepts involved in numerical integration.

8.2.2 The Trapezoidal Rule

The next order approximation of the function $f(x)$ is by means of a first-order polynomial, which implies that the function is replaced by a straight line over each interval, as sketched in Figure 8.3. Then the area under the curve in each element or interval is replaced by that of a trapezoid. If the areas of these trapezoids are denoted by $I_1, I_2, \ldots, I_n$, as indicated in Figure 8.3, then

$$\begin{aligned}
I_1 &= \left(\frac{f_0 + f_1}{2}\right)\Delta x \\
I_2 &= \left(\frac{f_1 + f_2}{2}\right)\Delta x \\
&\vdots \\
I_i &= \left(\frac{f_{i-1} + f_i}{2}\right)\Delta x \\
&\vdots \\
I_n &= \left(\frac{f_{n-1} + f_n}{2}\right)\Delta x
\end{aligned} \tag{8.10}$$

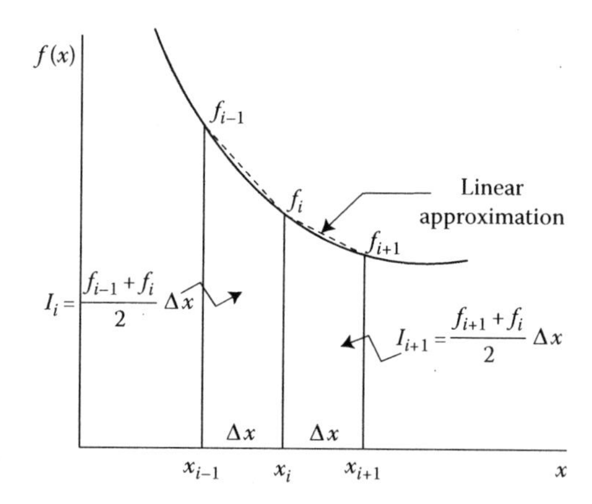

FIGURE 8.3 Approximation of the given function by straight lines over each of the segments, in which the integration domain is subdivided, for the trapezoidal rule.

Therefore, the integral I may be approximated by

$$I = \int_a^b f(x)\mathrm{d}x \cong \frac{\Delta x}{2}(f_0 + 2f_1 + 2f_2 + \cdots + 2f_{n-1} + f_n) \tag{8.11}$$

It can easily be shown that the result obtained by this method is simply the average of the results from the two formulations of the rectangular rule, given by Equations 8.8 and 8.9.

The trapezoidal rule for numerical integration is extensively used in engineering applications. It is fairly simple to program and it also imposes no constraints on the choice of the number of intervals n. Simpson's one-third rule, which is discussed later in this chapter, for instance, requires n to be even. Since each interval can be treated separately by the trapezoidal rule, as given in Equation 8.10, the method can easily be extended to numerical integration with intervals of unequal width. This is of particular importance in the integration of a function that is given at a finite number of data points, as is the case in several engineering applications.

8.2.3 Truncation Error

In order to derive the TEs associated with the rectangular and trapezoidal rules for numerical integration, let us define a function $y(x)$ as

$$y(x) = \int_a^x f(x)\mathrm{d}x \tag{8.12}$$

so that $y(x)$ is the integral of the function $f(x)$ from $x = a$ to x, as shown graphically in Figure 8.4. Also, from Equation 8.12 and from basic calculus,

$$y'(x) = f(x), \quad y''(x) = f'(x), \quad y'''(x) = f''(x)\ldots \tag{8.13}$$

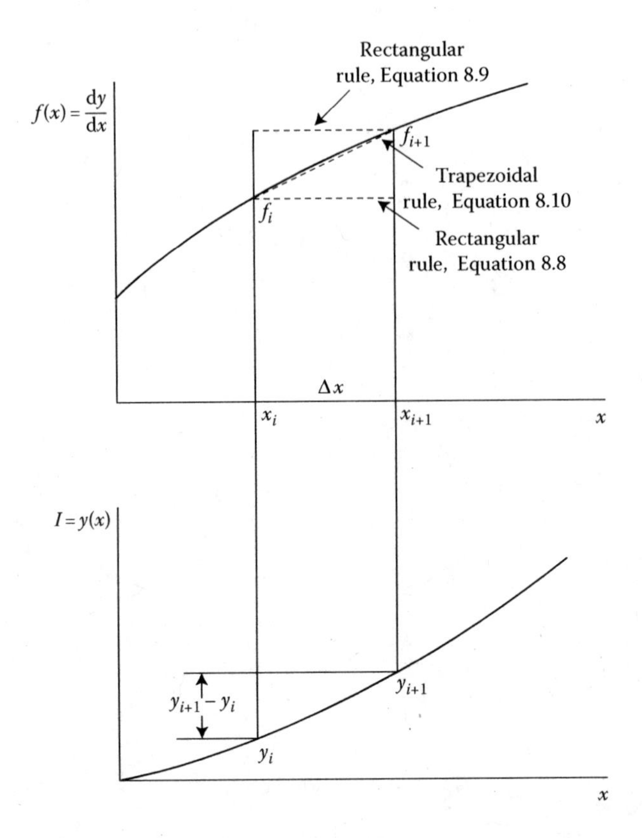

FIGURE 8.4 Sketch of a function and its integral for the estimation of TE in numerical integration by the rectangular and trapezoidal rules.

Since $y(x)$ represents the integral of the function, the exact integral over the range $x_i \leq x \leq x_{i+1}$ is $y(x_{i+1}) - y(x_i)$. With $y(x_i)$ denoted by y_i, the exact area under the curve in the given interval is, therefore, $y_{i+1} - y_i$.

Assuming both $y(x)$ and $f(x)$ to be continuous and smooth over the interval $x_i \leq x \leq x_{i+1}$, we may expand y_{i+1} and f_{i+1} in Taylor series about $x = x_i$. Thus, if the derivatives are denoted by primes, the expansion for y_{i+1} is given by

$$y_{i+1} = y_i + \Delta x y_i' + \frac{(\Delta x)^2}{2!} y_i'' + \frac{(\Delta x)^3}{3!} y_i''' + O\left[(\Delta x)^4\right]$$

Using the relations in Equation 8.13, this formula gives the exact integral over the interval $x_i \leq x \leq x_{i+1}$ as

$$\begin{aligned} y_{i+1} - y_i &= \Delta x y_i' + \frac{(\Delta x)^2}{2!} y_i'' + \frac{(\Delta x)^3}{3!} y_i''' + O\left[(\Delta x)^4\right] \\ &= \Delta x f_i + \frac{(\Delta x)^2}{2!} f_i' + \frac{(\Delta x)^3}{3!} f_i'' + O\left[(\Delta x)^4\right] \end{aligned} \quad (8.14)$$

Similarly, y_i may be expanded in a Taylor series about $x = x_{i+1}$ as follows:

$$y_i = y_{i+1} - \Delta x f_{i+1} + \frac{(\Delta x)^3}{2!} f'_{i+1} - \frac{(\Delta x)^3}{3!} f''_{i+1} + O\left[(\Delta x)^4\right]$$

Thus, the exact integral over the interval $x_i \le x \le x_{i+1}$ is given by

$$y_{i+1} - y_i = \Delta x f_{i+1} - \frac{(\Delta x)^2}{2!} f'_{i+1} + \frac{(\Delta x)^3}{3!} f''_{i+1} + O[(\Delta x)^4) \tag{8.15}$$

8.2.3.1 Rectangular Rule

In the rectangular rule, the integral over the interval $x_i \le x \le x_{i+1}$ is approximated by $f_i\,\Delta x$ in the first formulation, Equation 8.8, and by $f_{i+1}\,\Delta x$ in the second formulation, Equation 8.9. The exact integral is $y_{i+1} - y_i$. Therefore, from Equation 8.14, the TE in the first formulation of the rectangular rule is

$$\begin{aligned} \text{TE} &= \underbrace{\left(y_{i+1} - y_i\right)}_{\substack{\text{Exact}\\ \text{value}}} - \underbrace{f_i \Delta x}_{\substack{\text{Numerical}\\ \text{approximation}}} \\ &= \frac{(\Delta x)^2}{2} f_i' + O\left[(\Delta x)^3\right] \end{aligned}$$

This implies that the leading term of the TE associated with this step is $[(\Delta x)^2/2]f_i'$. Similarly, from Equation 8.15, the TE in the second formulation of the rectangular rule is obtained as follows:

$$\begin{aligned} \text{TE} &= \underbrace{\left(y_{i+1} - y_i\right)}_{\substack{\text{Exact}\\ \text{value}}} - \underbrace{f_{i+1} \Delta x}_{\substack{\text{Numerical}\\ \text{approximation}}} \\ &= -\frac{(\Delta x)^2}{2} f'_{i+1} + O\left[(\Delta x)^3\right] \end{aligned}$$

Using the remainder theorem, discussed in Chapter 4, we write the error per step in the two formulations, respectively, as follows:

$$\frac{\text{TE}}{\text{step}} = \frac{(\Delta x)^2}{2} f'(\xi) \quad \text{and} \quad -\frac{(\Delta x)^2}{2} f'(\xi) \quad \text{where } x_i < \xi < x_{i+1} \tag{8.16}$$

8.2.3.2 Trapezoidal Rule

Expanding the function $f(x)$ in a Taylor series about $x = x_i$, we obtain

$$f_{i+1} = f_i + \Delta x f_i' + \frac{(\Delta x)^2}{2!} f_i'' + \frac{(\Delta x)^3}{3!} f_i''' + O\left[(\Delta x)^4\right]$$

Therefore,

$$f_i' = \frac{f_{i+1} - f_i}{\Delta x} - \frac{\Delta x}{2} f_i'' - \frac{(\Delta x)^2}{6} f_i''' + O\left[(\Delta x)^3\right] \tag{8.17}$$

This formula is simply the forward difference approximation for f_i', along with the associated TE, as derived in Chapter 4. Substituting this expression into Equation 8.14, we obtain

$$\begin{aligned} y_{i+1} - y_i &= \Delta x f_i + \frac{(\Delta x)^2}{2}\left[\frac{f_{i+1} - f_i}{\Delta x} - \frac{\Delta x}{2} f_i'' - \frac{(\Delta x)^2}{6} f_i''' - \cdots\right] \\ &+ \frac{(\Delta x)^3}{3!} f_i'' + O\left[(\Delta x)^4\right] \\ &= \Delta x \frac{f_{i+1} + f_i}{2} - \frac{1}{12}(\Delta x)^3 f_i'' + O\left[(\Delta x)^4\right] \end{aligned} \tag{8.18}$$

Since $(y_{i+1} - y_i)$ is the exact area under the curve and $\Delta x(f_{i+1} + f_i)/2$ the trapezoidal area in the interval considered, the TE per step is the difference between the two, that is,

$$\frac{\text{TE}}{\text{step}} = -\frac{1}{12}(\Delta x)^3 f_i'' + O\left[(\Delta x)^4\right]$$

Again, using the remainder theorem, we write the TE for integration over the interval $x_i \le x \le x_{i+1}$ as

$$\frac{\text{TE}}{\text{step}} = -\frac{1}{12}(\Delta x)^3 f''(\xi) \quad \text{where } x_i < \xi < x_{i+1} \tag{8.19}$$

This expression gives the TE in the $(i+1)$th strip, or subinterval, see Figure 8.2. Therefore, the TE per step in the trapezoidal rule is $O[(\Delta x)^3]$. Since the error is zero if $f'' = 0$, the method is exact only for a linear function.

8.2.3.3 Total Error

The total error in integrating the function $f(x)$ over the entire interval $a \le x \le b$ is obtained by the summation of the errors over n subintervals. Therefore, the total truncation error E for the trapezoidal rule is

$$E = \sum_{i=0}^{n-1}\left[-\frac{1}{12}(\Delta x)^3 f''(\xi_i)\right], \quad \text{where } x_i < \xi_i < x_{i+1} \tag{8.20}$$

The maximum total error may be estimated from this expression by employing the largest value of f'' in each subinterval. However, the second derivative f'' may not be easy to evaluate in many practical circumstances. A more useful alternative expression for the total error is obtained by defining an arithmetic mean f''_{av} of the values of $f''(\xi_i)$ in the n strips. Then

$$\sum_{i=0}^{n-1} f''(\xi_i) = nf''_{av} \tag{8.21}$$

Therefore, the total truncation error E may be expressed in terms of f''_{av} as follows:

$$\begin{aligned} E &= -\frac{1}{12}(\Delta x)^3 nf''_{av} = -\frac{1}{12}(\Delta x)^3 \frac{b-a}{\Delta x} f''_{av} \\ &= -\frac{1}{12}(\Delta x)^2 (b-a) f''_{av} \end{aligned} \tag{8.22}$$

Assuming f''_{av} to remain essentially constant as the step size Δx is varied, we write the total TE as

$$E \cong S_T(\Delta x)^2 = O[(\Delta x)^2] \tag{8.23}$$

where $S_T = -(b-a)f''_{av}/12$ and is assumed to be a constant, as indicated by the approximation ($\cong$) sign. Thus, the trapezoidal rule is a second-order method.

Proceeding in a similar manner for the rectangular rule, we can show that the total TEs, for the two formulations of Equations 8.8 and 8.9, are, respectively,

$$E = \frac{\Delta x}{2}(b-a)f'_{av} \quad \text{and} \quad -\frac{\Delta x}{2}(b-a)f'_{av} \tag{8.24}$$

where f'_{av} is the average of the $f'(\xi_i)$ values in the n strips. This expression may again be written as

$$E \cong S_R(\Delta x) = O(\Delta x) \tag{8.25}$$

where $S_R = (b-a)f'_{av}/12$ and is again assumed to be essentially a constant. Therefore, the rectangular rule is a first-order method. Both the trapezoidal and the rectangular rules for numerical integration are quite simple to program. The difference between the two lies only in the incorporation of the ordinates at the ends of the total range in the summation of the ordinates for the numerical scheme. The rectangular rule uses only the ordinate at $x = a$ in the first formulation and at $x = b$ in the second formulation, whereas the trapezoidal rule uses the average of the two. The ordinates in the interior region are summed in all three cases, see Equations 8.8, 8.9, and 8.11. Since the trapezoidal rule is more accurate than the rectangular rule, there is no reason to use the latter. In fact, the trapezoidal rule is among the most widely used schemes for numerical integration in problems of engineering interest because of its simplicity.

8.2.3.4 Accuracy

As shown in the expressions for the TEs in the rectangular and trapezoidal integration methods, the error decreases as the step size Δx is decreased, that is, as the number of segments n is increased. This behavior is expected, as discussed in Section 2.3. Thus, the accuracy of the numerical results can be improved by decreasing Δx, a process generally known as *grid refinement*. However, as Δx is decreased, the number of segments increases and so does the computational effort. This results in an increase in the round-off error. Therefore, the total error, which includes the TE and round-off errors, is reduced by decreasing Δx to a certain point, beyond which the round-off error becomes substantial and the total error increases with decreasing Δx, see Figure 2.12. All of these considerations were discussed in Section 2.3 and are repeated here to emphasize the importance of numerical errors and the need to vary the grid size, Δx, keeping it larger than the constraint imposed by the round-off error, to ensure that the numerical solution is essentially independent of the value chosen. Figure 8.5 shows a typical variation of the numerical value of the integral I with the segment size Δx. Then the largest grid size at which the solution becomes essentially independent of Δx, so that a further reduction in the segment size does not significantly affect the results, is chosen, as shown in the figure.

Example 8.1

A capacitor in an electrical circuit is initially at zero charge. At time t of 1 s, a switch is closed, and a time-dependent electric current $I(t)$ charges up the capacitor. The current is given as

$$I(t) = 4(1 - e^{-0.5})e^{-0.5(t-1)}(1 - e^{-t}) \tag{8.26}$$

Using the trapezoidal rule for numerical integration, compute the charge Q and the voltage V across the capacitor as functions of time up to $t = 20$ s. The capacitance C of the capacitor is 0.025 farad.

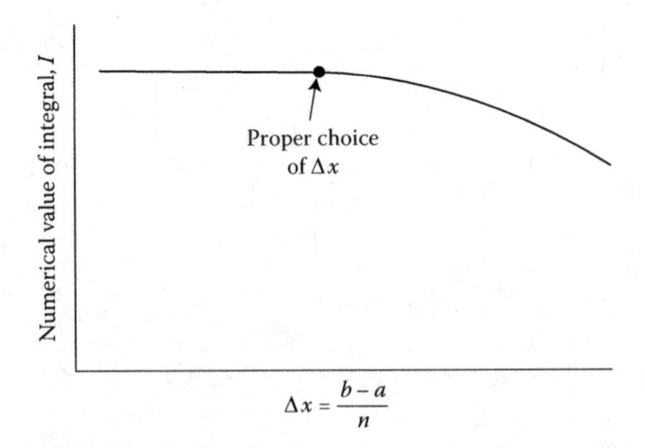

FIGURE 8.5 Dependence of the numerical value of the integral I on the segment size Δx and a suitable choice of Δx for further computations.

SOLUTION

The charge Q stored by the capacitor is given by the integral

$$Q(t) = \int_0^t I(t')\,dt' \tag{8.27}$$

where t is the time at which the charge is to be determined and t' is simply a dummy variable. The voltage across the capacitor is given by

$$V(t) = \frac{Q(t)}{C} \tag{8.28}$$

Here, it is assumed that the charge and thus the voltage across the capacitor are zero at $t = 1$ s, as given in the problem. Therefore, this problem requires the application of the trapezoidal rule for evaluating the integral

$$Q(t) = \int_1^t 4(1 - e^{-0.5})e^{-0.5(t'-1)}(1 - e^{-t'})\,dt' \tag{8.29}$$

Appendix B.19 presents a MATLAB function-*m* file for solving this problem. The current $I(t)$ is the function to be integrated and is entered as a string. Thus, a function file *f81.m*, as given below, is defined for this problem and the function f in the function *m*-file in Appendix B.19 is entered as 'f81'.

```
function z=f81(x)
z=4*(1-exp(-0.5)).*exp(-0.5*(x-1)).*(1-exp(-x));
end
```

Note that .* is used instead of * for multiplication in order to allow x to be specified as an array, if needed.

The lower limit of the integral is a and the upper limit is b. The number of subintervals m is specified, so that the segment size h is given by $h = (b - a)/m$. Therefore, the function *m*-file given in Appendix B.19 is invoked as *trap('f81', 1, 2, 8);* for 1 as the lower limit, 2 as the upper limit, and 8 as the number of segments, yielding $h = 0.125$. The upper limit is varied in increments of 2 each to study its effect on the integral. For the given problem, b is thus varied from 2 to 20 s. The sum of the ordinates in the interior region of the integration domain is computed, and the trapezoidal rule is applied to yield the numerical value of the integral in Equation 8.29. This gives the electrical charge Q at time t. From the computed value of Q, the voltage V is determined from Equation 8.28. An alternative, more compact, implementation of the algorithm for the trapezoidal rule is also shown in Appendix B.19. Here, x is employed as an array, for which the function definition given above is needed. Obviously, there are different ways of applying the formulas to obtain the integral.

Figure 8.6 presents some of the numerical results obtained, with the upper limit of the integral b being varied from 2 to 20 s. The number of segments and thus the step size h was also varied, starting with 2 s and then successively halving it. The results remain essentially unchanged as h is decreased from 0.125 to 0.0625 s, indicating the former to be adequate for this computation. The results at $h = 2$ s were found to be in considerable error. The effect of the segment size h is shown more clearly in Figure 8.7 by a plot of the charge Q at $t = 2$, 4, and 8 s versus h. Clearly, the accuracy is improved as h is decreased, over the range considered, due to the decrease in the TE. Also, the effect of h on the results is smaller at large values of t. This behavior is expected from the function being integrated, since the integrand approaches zero at large time, giving a constant value of the integral as t becomes large.

```
Step size = 2.0000
Time =  2     Charge = 1.8203     Voltage = 72.8117
Time =  4     Charge = 1.3396     Voltage = 53.5851
Time =  6     Charge = 2.2241     Voltage = 88.9637
Time =  8     Charge = 2.5659     Voltage = 102.6346
Time = 10     Charge = 2.6924     Voltage = 107.6964
Time = 12     Charge = 2.7390     Voltage = 109.5602
Time = 14     Charge = 2.7561     Voltage = 110.2460
Time = 16     Charge = 2.7625     Voltage = 110.4982
Time = 18     Charge = 2.7648     Voltage = 110.5910
Time = 20     Charge = 2.7656     Voltage = 110.6252

Step size = 1.0000
Time =  2     Charge = 0.9101     Voltage = 36.4059
Time =  4     Charge = 2.0454     Voltage = 81.8159
Time =  6     Charge = 2.4938     Voltage = 99.7509
Time =  8     Charge = 2.6603     Voltage = 106.4101
Time = 10     Charge = 2.7216     Voltage = 108.8629
Time = 12     Charge = 2.7441     Voltage = 109.7654
Time = 14     Charge = 2.7524     Voltage = 110.0974
Time = 16     Charge = 2.7555     Voltage = 110.2195
Time = 18     Charge = 2.7566     Voltage = 110.2644
Time = 20     Charge = 2.7570     Voltage = 110.2810

Step size = 0.2500
Time =  2     Charge = 0.9368     Voltage = 37.4712
Time =  4     Charge = 2.0624     Voltage = 82.4962
Time =  6     Charge = 2.5028     Voltage = 100.1135
Time =  8     Charge = 2.6662     Voltage = 106.6470
Time = 10     Charge = 2.7263     Voltage = 109.0531
Time = 12     Charge = 2.7485     Voltage = 109.9384
Time = 14     Charge = 2.7566     Voltage = 110.2641
Time = 16     Charge = 2.7596     Voltage = 110.3839
Time = 18     Charge = 2.7607     Voltage = 110.4280
Time = 20     Charge = 2.7611     Voltage = 110.4442

Step size = 0.1250
Time =  2     Charge = 0.9382     Voltage = 37.5280
Time =  4     Charge = 2.0634     Voltage = 82.5346
Time =  6     Charge = 2.5034     Voltage = 100.1359
Time =  8     Charge = 2.6666     Voltage = 106.6631
Time = 10     Charge = 2.7267     Voltage = 109.0669
Time = 12     Charge = 2.7488     Voltage = 109.9513
Time = 14     Charge = 2.7569     Voltage = 110.2767
Time = 16     Charge = 2.7599     Voltage = 110.3964
Time = 18     Charge = 2.7610     Voltage = 110.4404
Time = 20     Charge = 2.7614     Voltage = 110.4566
```

FIGURE 8.6 Numerical results obtained for the charge Q and the voltage V, as functions of time t, in Example 8.1, at several values of the step size h.

Step size = 0.0625

Time = 2	Charge = 0.9386	Voltage = 37.5422
Time = 4	Charge = 2.0636	Voltage = 82.5442
Time = 6	Charge = 2.5035	Voltage = 100.1416
Time = 8	Charge = 2.6667	Voltage = 106.6672
Time = 10	Charge = 2.7268	Voltage = 109.0704
Time = 12	Charge = 2.7489	Voltage = 109.9546
Time = 14	Charge = 2.7570	Voltage = 110.2799
Time = 16	Charge = 2.7600	Voltage = 110.3996
Time = 18	Charge = 2.7611	Voltage = 110.4436
Time = 20	Charge = 2.7615	Voltage = 110.4598

FIGURE 8.6 Continued.

The results show that the charge Q approaches a constant value of around 2.76 coulombs as time increases beyond about 18 s. This indicates that the capacitor is fully charged by this time. A constant voltage difference of 110.46 V across the capacitor is also attained. The electrical current asymptotically approaches zero, as expected from the form of the given function $I(t)$. Thus, as $t \to \infty$, the charge Q and voltage V approach constant values, indicating a finite constant value of the integral as the upper integration limit approaches infinity. The integral for $t \to \infty$ may also be evaluated analytically to yield Q as 2.76176, which agrees closely with the numerical result obtained. Appendix C.12 gives the corresponding program in Fortran for this problem and a similar logic is employed to implement the algorithm for the trapezoidal rule.

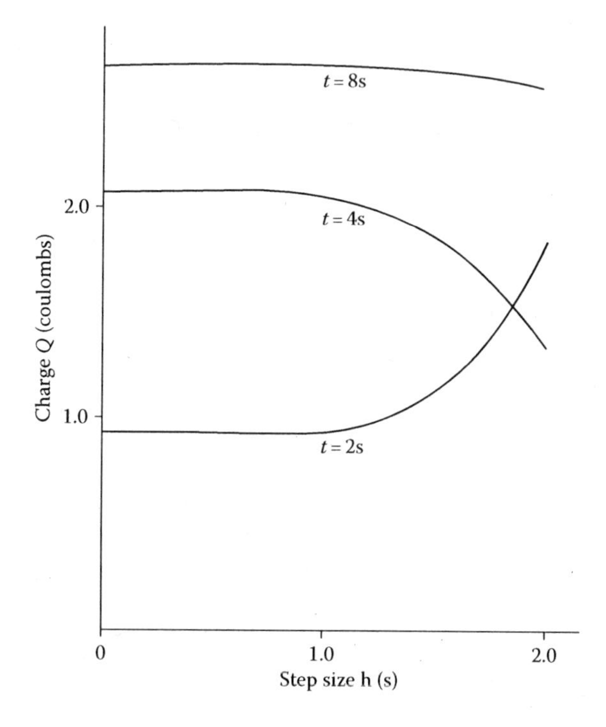

FIGURE 8.7 Variation of the computed capacitor charge Q, at $t = 2$, 4, and 8 s, with the time step h.

8.3 SIMPSON'S RULES FOR NUMERICAL INTEGRATION

One can improve the accuracy with which an integral is computed by increasing the number of segments n into which the range of integration is divided, constrained by the round-off error which becomes significant as the step size Δx is reduced to very small values, or by employing higher-order polynomials $P_m(x)$ to approximate the function $f(x)$. The trapezoidal rule uses a straight line to approximate the function over each segment. Simpson's one-third rule, usually referred to as simply *Simpson's rule*, uses second-order polynomials, that is, parabolas, to approximate the function. One connects successive groups of three points on the $f(x)$ versus x curve with parabolas to determine the area under the curve over the interval defined by these points. Similarly, a third-order polynomial, $m = 3$, requires four points on the curve for the approximation of the function and leads to what is known as *Simpson's three-eighths rule*. Since each segment of the integration domain is associated with only two points on the curve, as shown in Figure 8.2, Simpson's one-third rule requires a minimum of two segments and an even number of segments n into which the total range of integration is subdivided. Simpson's three-eighths rule requires a minimum of three segments; and, if it is used in conjunction with the one-third rule, n may be odd or even.

8.3.1 Simpson's One-Third Rule

The function $f(x)$ in the integral

$$I = \int_a^b f(x)\mathrm{d}x \tag{8.1}$$

is replaced by a second-order polynomial, or a parabola, for numerical integration by Simpson's rule. Three points on the curve of $f(x)$ versus x are needed to determine this parabola. Consider the three points (x_{i-1}, f_{i-1}), (x_i, f_i), and (x_{i+1}, f_{i+1}), as shown in Figure 8.8. A parabola that passes through these three points may be found and the area under the curve of $f(x)$ approximated by that under the parabola. Two segments, each of width Δx, are involved in this computation, since three points are needed to define the parabola.

We may employ the various methods of interpolation discussed in Chapter 7 to determine the parabola passing through the three given points. Lagrange interpolation provides the general method for deriving Newton–Cotes formulas. However, because only a second-order polynomial, $m = 2$, is under consideration here, we may simply employ the general form of the equation for a parabola and determine the coefficients by substituting the coordinates for the three points into this equation, as done in Section 7.2.1. Therefore, the second-order polynomial may be taken as

$$P_2(x) = Ax^2 + Bx + C \tag{8.30}$$

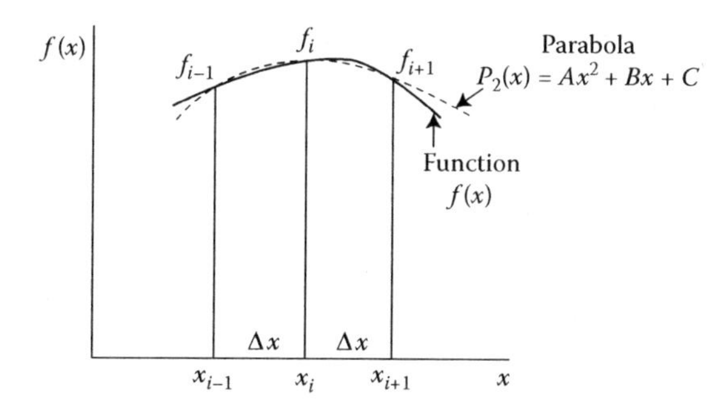

FIGURE 8.8 Replacement of the function $f(x)$ over the width of two segments by a parabola for the derivation of Simpson's one-third rule.

Since this parabola passes through the three points being considered, as shown in Figure 8.8, the constants A, B, and C can be determined from

$$\begin{aligned} f_{i-1} &= A(-\Delta x)^2 + B(-\Delta x) + C \\ f_i &= C \\ f_{i+1} &= A(\Delta x)^2 + B(\Delta x) + C \end{aligned}$$

where x_i has been taken at the origin, $x = 0$, to simplify the calculation. Such a choice does not affect the generality of the derivation. From the above equations,

$$A = \frac{f_{i-1} - 2f_i + f_{i+1}}{2(\Delta x)^2}, \quad B = \frac{f_{i+1} - f_{i-1}}{2\Delta x}, \quad C = f_i$$

The area under the polynomial of Equation 8.30 is denoted by I_p and is given by

$$I_p = \int_{-\Delta x}^{\Delta x} (Ax^2 + Bx + C)\,\mathrm{d}x = \left[A\frac{x^3}{3} + B\frac{x^2}{2} + Cx\right]_{-\Delta x}^{\Delta x} = \frac{2}{3}A(\Delta x)^3 + 2C(\Delta x)$$

Therefore, the area under the curve in the two segments is approximated by

$$I_p = \left[\frac{f_{i-1} - 2f_i + f_{i+1}}{3} + 2f_i\right]\Delta x = \left(\frac{f_{i-1} + 4f_i + f_{i+1}}{3}\right)\Delta x \tag{8.31}$$

where the expressions for A and C, given above, have been substituted. This formula is known as *Simpson's rule*, or as *Simpson's one-third rule*, because the step size Δx is divided by 3 in the formula. The use of the one-third in the terminology distinguishes this method from a similar one, derived later, in which Δx is multiplied by 3/8, instead of 1/3, and which is known as *Simpson's three-eighths rule*.

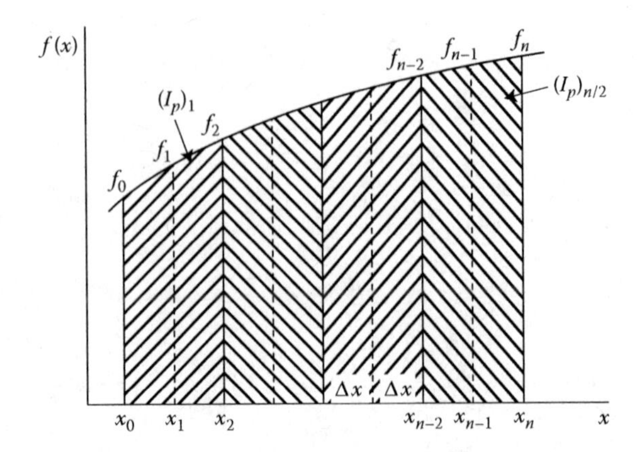

FIGURE 8.9 Application of Simpson's rule, with an even number of strips, for the numerical integration of a function $f(x)$ over the range $x = a$ to $x = b$.

We may use the expression in Equation 8.31 for the integral over two segments to determine the integral over the entire range $a \le x \le b$, which is divided into n segments of equal width Δx. Here, n must be even in order to consider groups of two segments for the application of Equation 8.31, see Figure 8.9. Then the total integral I is approximated by the following:

$$I = \int_a^b f(x)\mathrm{d}x \cong \sum_{j=1}^{n/2} (I_\mathrm{p})_j$$

where $(I_p)_j$ is the integral given by Equation 8.31 for the jth group of two segments. Thus, i in Equation 8.31 is given by $i = 2j - 1$. Therefore,

$$I \cong \frac{\Delta x}{3}\left(f_0 + 4\sum_{i=1,3,5,\ldots}^{n-1} f_i + 2\sum_{i=2,4,6,\ldots}^{n-2} f_i + f_n \right) \qquad (8.32)$$

It is shown later in this section that the TE per step in numerical integration by Simpson's rule is $O[(\Delta x)^5]$, which results in a total error of $O[(\Delta x)^4]$. Therefore, this is a fourth-order method and is much more accurate than the trapezoidal rule for an arbitrary function $f(x)$. If the function being integrated is a polynomial of order zero, one, two, or three, Simpson's rule is exact, that is, there is no TE. This is because the leading term in the TE contains only the fourth derivative f'''', all terms containing the lower derivatives having canceled out. Computer programming for Simpson's rule is more involved than that for the trapezoidal rule. However, because of its much higher accuracy level, Simpson's rule is widely used in engineering applications, where accuracy is usually important. Example 8.2 demonstrates the use of this method in a problem of practical interest.

8.3.2 Simpson's Three-Eighths Rule

If a third-order polynomial, $m = 3$, is employed to approximate the integrand $f(x)$ by requiring that it pass through four points on the curve of $f(x)$, Simpson's three-eighths

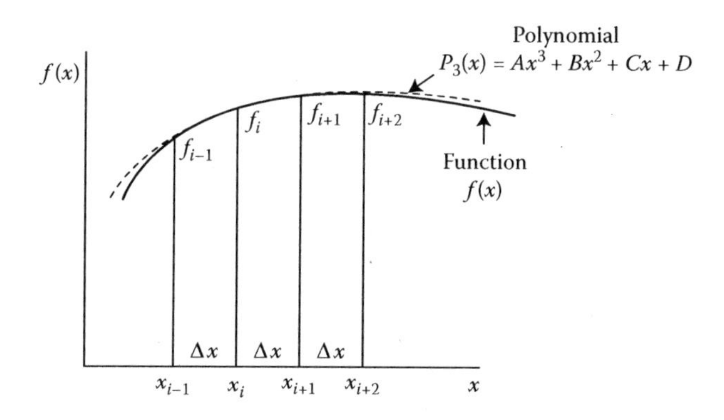

FIGURE 8.10 Replacement of the function $f(x)$ over the width of three segments by a third-order polynomial for the derivation of Simpson's three-eighths rule.

rule is obtained. A minimum of three segments are needed in order to provide the four points for the determination of the polynomial, as shown in Figure 8.10. The general equation for the polynomial is taken as

$$P_3(x) = Ax^3 + Bx^2 + Cx + D \tag{8.33}$$

In a manner similar to that given in the preceding for the derivation of the one-third rule, the polynomial is determined by substitution of the coordinates of the four points through which it passes. The integral over the three segments is approximated by the corresponding integral I_p of the polynomial $P_3(x)$. The resulting expression for I_p is

$$I_\text{p} = \frac{3}{8}\Delta x\left(f_{i-1} + 3f_i + 3f_{i+1} + f_{i+2}\right) \tag{8.34}$$

For application of this method over the entire range of integration, n must be a multiple of 3. Then the integral I is approximated by

$$I = \int_a^b f(x)\mathrm{d}x \cong \sum_{j=1}^{n/3} (I_\text{p})_j$$

where $(I_p)_j$ is the integral for the jth group of three segments, giving i in Equation 8.34 as $i = 3j - 2$. Thus,

$$I = \frac{3}{8}\Delta x\left[f_0 + 3\sum_{i=1,4,7,\ldots}^{n-2}\left(f_i + f_{i+1}\right) + 2\sum_{i=3,6,9\ldots}^{n-3} f_i + f_n\right] \tag{8.35}$$

It is shown below that the TE in numerical integration by Simpson's three-eighths rule is of the same order as that for the one-third rule. Because of this and the requirement that the number of segments n must be a multiple of 3, the three-eighths rule is seldom used by itself. Simpson's one-third rule is easier to program, and the constraint on n is only that it must be even. However, if the two methods are used together, no such constraint on n is needed. If n is even, Simpson's one-third rule is employed for numerical integration over the entire region. If n is odd, one can use Simpson's three-eighths rule, for instance, to compute the area under the curve in the first three segments and the one-third rule for the remaining even number of segments. Thus, a combination of the two methods provides fourth-order accuracy in the numerical results, without restricting the number of segments that may be employed, except that $n \geq 2$.

8.3.3 Truncation Errors

The derivation of the TEs associated with the two Simpson's rules for numerical integration follows the procedure presented for the trapezoidal rule. Thus, $y(x)$ represents the integral $\int_a^x f(x)\,dx$, and $y'(x) = f(x)$, $y''(x) = f'(x)$, and so on. The exact area under the curve of $f(x)$ over the two segments shown in Figure 8.11 is $y_{i+1} - y_{i-1}$. Expanding y_{i+1} and y_{i-1} in Taylor series about y_i, we obtain

$$y_{i+1} = y_i + \Delta x y_i' + \frac{(\Delta x)^2}{2!} y_i'' + \frac{(\Delta x)^3}{3!} y_i''' + \frac{(\Delta x)^4}{4!} y_i'''' + \frac{(\Delta x)^5}{5!} y_i''''' + \cdots$$

$$y_{i-1} = y_i - \Delta x y_i' + \frac{(\Delta x)^2}{2!} y_i'' - \frac{(\Delta x)^3}{3!} y_i''' + \frac{(\Delta x)^4}{4!} y_i'''' - \frac{(\Delta x)^5}{5!} y_i''''' + \cdots$$

Therefore,

$$y_{i+1} - y_{i-1} = 2\Delta x f_i + \frac{(\Delta x)^3}{3} f'' + \frac{(\Delta x)^5}{60} f_i'''' + O\left[(\Delta x)^7\right] \tag{8.36}$$

where the relationships between $y(x)$ and $f(x)$, from Equation 8.13, have been used. The finite difference approximation for f_i'' is needed to obtain the expression for Simpson's one-third rule on the right-hand side of Equation 8.36. From Section 4.4.2,

$$f_i'' = \frac{f_{i+1} - 2f_i + f_{i-1}}{(\Delta x)^2} - \frac{1}{12}(\Delta x)^2 f_i'''' + O\left[(\Delta x)^4\right]$$

Substituting this expression for f_i'' into Equation 8.36, we obtain the resulting equation:

$$\underbrace{y_{i+1} - y_{i-1}}_{\text{Exact Value}} = \underbrace{\frac{\Delta x}{3}(f_{i+1} + 4f_i + f_{i-1})}_{\text{Simpson's One-Third Rule}} \underbrace{- \frac{1}{90}(\Delta x)^5 f_i'''' + O\left[(\Delta x)^7\right]}_{\text{TE}}$$

From the remainder theorem, the truncation error per step (TE/step) of Simpson's one-third rule for numerical integration is

$$\frac{\text{TE}}{\text{step}} = -\frac{1}{90}(\Delta x)^5 f''''(\xi), \quad \text{where } x_{i-1} < \xi < x_{i+1} \tag{8.37}$$

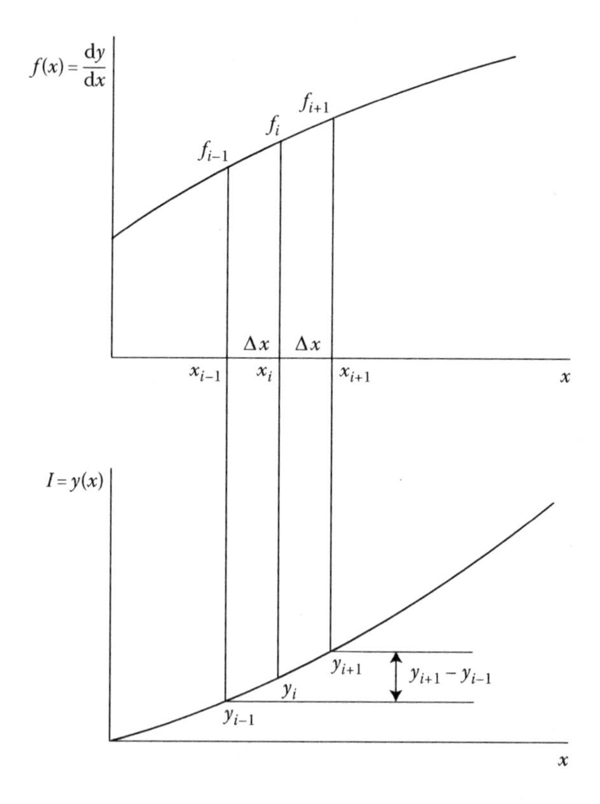

FIGURE 8.11 Sketch of a function $y'(x)$ and its integral $y(x)$ for the estimation of TE in Simpson's rule for numerical integration.

Therefore, the error will be zero if $f'''' = 0$. This implies that the method is exact for polynomials up to third-order. We obtain the total error E by summing the errors in all the steps:

$$E = \sum_{i=1}^{n/2}\left[-\frac{1}{90}(\Delta x)^5 f''''(\xi_i)\right], \quad \text{where } x_{2i-2} < \xi_i < x_{2i}$$

Again, defining an arithmetic mean f''''_{av} of the values of f'''' in the $n/2$ subintervals, each of width $2\Delta x$, we may write the total TE as

$$\begin{aligned} E &= -\frac{1}{90}(\Delta x)^5 \frac{n}{2} f''''_{\text{av}} = -\frac{1}{90}(\Delta x)^5 \frac{b-a}{2\Delta x} f''''_{\text{av}} \\ E &= -\frac{1}{180}(\Delta x)^4 (b-a) f''''_{\text{av}} \end{aligned} \tag{8.38}$$

If f_{av}'''' is assumed to remain essentially unchanged as Δx is varied, the total error E may be expressed as

$$E \cong S_s(\Delta x)^4 = O[(\Delta x)^4] \tag{8.39}$$

where $S_s = -(b-a)\,f_{av}''''/180$ and is assumed to be a constant. This indicates that Simpson's one-third rule is fourth-order accurate. On the basis of this expression for the TE, higher-order accuracy may be obtained by the use of *Richardson extrapolation*, as discussed in the next section.

The TE for Simpson's three-eighths rule may be derived in a similar way. It can be shown that the TE per step in the expression given by Equation 8.34 is

$$\frac{\text{TE}}{\text{step}} = -\frac{3}{80}(\Delta x)^5 f''''(\xi), \quad \text{where } x_{i-1} < \xi < x_{i+2}$$

Proceeding as before, we can show that the total error E is

$$E = -\frac{1}{80}(\Delta x)^4(b-a)f_{av}'''' \cong S_{ST}(\Delta x)^4 \tag{8.40}$$

where f_{av}'''' is the arithmetic mean of the values of f'''' in the $n/3$ subintervals, each of width $3\Delta x$. Therefore, the method is also fourth-order accurate, and the total error is somewhat larger than that for the one-third rule for a given step size Δx. However, if the total integration region is divided into three segments for the application of the three-eighths rule and into two segments for the one-third rule, the former yields more accurate results, because the step size is $(b-a)/3$ in the first case and $(b-a)/2$ in the second case. The smaller step size for the application of the three-eighths rule then yields a smaller total TE.

Example 8.2

The velocity profile in the turbulent flow of a fluid in a smooth circular pipe may be represented by the empirical power-law equation

$$U(x) = 5\left(1 - \frac{x}{R}\right)^{1/7} \tag{8.41}$$

where $U(x)$ is the axial velocity in the pipe, in m/s, x is the radial distance from the axis, in meters, and R is the radius of the pipe. The total volume flow rate in the pipe is then given by the integral $\int_0^R U(x)2\pi x\,dx$. Using Simpson's one-third rule, compute this integral as accurately as possible for $R = 0.1$ m. Also determine the average velocity.

SOLUTION

The integral to be evaluated numerically is

$$I = \int_0^{0.1} 5\left(1 - \frac{x}{0.1}\right)^{1/7} 2\pi x\,dx \tag{8.42}$$

Since x and R are in meters and the velocity U is in m/s, this integral will yield the volume flow rate in m^3/s. We obtain the average velocity V_{av} by dividing the flow rate by the area of cross section of the pipe. Therefore,

$$V_{av} = \frac{I}{\pi R^2} = \frac{I}{\pi (0.1)^2}\ \text{m/s} \tag{8.43}$$

Appendix B.20 presents the computer program in MATLAB as a function *m*-file for the numerical integration of a given function by Simpson's one-third rule. The function file is *simp(f,a,b,n)*, where *f* is the function to be integrated, *a* and *b* are the lower and upper limits of integration, respectively, and *n* is the total number of segments. In employing this function file, the function to be integrated is defined separately as another function file, *f82.m*, given as

```
function z=f82(x)
z=5*((1-x/0.1).^(1/7))*2*pi.*x;
end
```

Again. .* and .^ are employed so that x can be treated as an array. Then, the function *m*-file for Simpson's rule is invoked by the command *simp('f82',0,0.1,10);*, where the given function to be integrated is entered as a string, the limits of integration a and b are 0 and 0.1 here and the number of segments n is entered as 10. The integral I is computed by Simpson's one-third rule to yield the resulting flow rate. Then, the average velocity is obtained by the use of Equation 8.43. As given earlier for the trapezoidal rule, an alternative, more compact, implementation of the algorithm for the Simpson's one-third rule is also shown in Appendix B.20.

The numerical results obtained from the numerical scheme are presented in Figure 8.12. The computed flow rate and the average velocity are shown for the number of subdivisions n ranging from 10 to 5120. Note that $n = 320$ is quite adequate for this problem. Again, at much larger values of n, the round-off error is expected to become significant and to increase the total error, acting against the decrease in the TE, as n is increased. Although n is successively doubled a chosen number of times in this program, a better approach would be to continuously monitor the effect of increasing n on the numerical value of the integral. If the change in I from one value of n to the next higher value is smaller than a chosen convergence criterion, then the computation may be terminated. In the given function *m*-file, n is increased to values much larger than necessary, in order to

```
n =   10     Flow rate = 0.1230     Avg. vel. = 3.9165
n =   20     Flow rate = 0.1259     Avg. vel. = 4.0079
n =   40     Flow rate = 0.1272     Avg. vel. = 4.0492
n =   80     Flow rate = 0.1278     Avg. vel. = 4.0679
n =  160     Flow rate = 0.1281     Avg. vel. = 4.0763
n =  320     Flow rate = 0.1282     Avg. vel. = 4.0802
n =  640     Flow rate = 0.1282     Avg. vel. = 4.0819
n = 1280     Flow rate = 0.1283     Avg. vel. = 4.0827
n = 2560     Flow rate = 0.1283     Avg. vel. = 4.0830
n = 5120     Flow rate = 0.1283     Avg. vel. = 4.0832
```

FIGURE 8.12 Numerical results on the flow rate and the average velocity, obtained in Example 8.2, for various values of the number of subdivisions *n*.

determine if the effect of round-off error becomes significant at the larger values. The results shown indicate that the trends are pretty much as expected, and the round-off error is small over the range of n considered.

The ratio of the average velocity V_{av} to the velocity V_{max} at the axis can be shown analytically to be given by the expression

$$\frac{V_{av}}{V_{max}} = \frac{2n^2}{(n+1)(2n+1)} \tag{8.44}$$

where $1/n$ is the exponent in Equation 8.41. In the present case, $n = 7$ and $V_{max} =$ 5 m/s, being the velocity at the axis, $x = 0$. This gives $V_{av} = 0.8167V_{max} = 0.40835$ m/s. This analytical value agrees closely with the numerical result obtained, lending support to the accuracy of the numerical scheme.

8.3.4 Use of MATLAB® Integration Commands

We have discussed some of the commonly used methods for numerical integration. The basic formulas and the algorithms have been presented, along with MATLAB and Fortran programs. However, there are several commands for numerical integration that are available in the MATLAB environment and that can be used quite effectively in many cases. Among the most common commands is *quad*, which is invoked to obtain the integral s as

```
s=quad('f',a,b)
```

where f is a function file, which defines the function to be integrated, a is the lower limit of integration and b the upper limit. The command numerically obtains an approximation to the integral of scalar-valued function f from a to b to within an error of 1.e–6 using recursive adaptive Simpson quadrature. The adaptive scheme allows it to use finer subdivisions in regions with accuracy lower than the desired value, as discussed later. An *inline* definition of the function f can also be used. Then the command used is

```
s=quad(f,a,b)
```

As an example, if the speed v is given as a function of time t as $v = 2 + 3t + 2t^2 + t^3$. Then the integral $\int_0^t v(t)dt$ gives the total distance traveled. This integral can be obtained by using the *quad* command as

```
>> v=inline('2+3*t+2*t.^2+t.^3');
>> dist=quad(v,0,2)
```

This gives the result as

```
dist=
  19.3333
```

Similarly, the function may be defined as

```
function z=fn(t)
z=2+3*t+2*t.^2+t.^3;
end
```

Then the *quad* command is invoked as

```
>> dist = quad('fn',0,2)
```

yielding the result as

```
dist =
   19.3333
```

Other limits may be easily employed as

```
>> dist = quad('fn',0,4)
dist =
   138.6667
>> dist = quad('fn',0,1)
dist =
   4.4167
```

There are several other such commands, including *quad2d* for numerically evaluating a double integral over a planar region, *triplequad* for numerically evaluating a triple integral, and *trapz*, which computes the integral using the trapezoidal method. All these commands, along with the *quad* command discussed above, may be employed instead of the various methods for integration given here to simplify the programming or to verify the results. For instance, Examples 8.1 and 8.2 may be solved by using the *quad* command as

```
>> int = quad('f81',1,6)
int =
   2.5036
>> int = quad('f82',0,0.1)
int =
   0.1283
```

where the given function files *f81.m* and *f82.m* are used and the limits of integration are specified. Clearly, it is easier to use the available commands for integration. However, greater flexibility and control, particularly on the numerical parameters, is obtained by algorithms and programs developed by the user.

As mentioned earlier, symbolic algebra can also be used in MATLAB to obtain analytical expressions for the integral in many cases. We first need to construct symbolic numbers, variables and objects. The command x = sym('x') creates the symbolic variable with name x and stores the result in x. The symbolic integration function is given by the command *int*. Thus, we could use

```
>> t = sym('t');
>> s = int(2 + 3*t + 2*t^2 + t^3)
```

which yields the result

```
s =
   t^4/4 + (2*t^3)/3 + (3*t^2)/2 + 2*t
```

Similarly, the integral $\int e^{-x^2} dx$ is obtained by

```
>> x = sym('x');
>> s = int(exp(-x^2))
s =
   (pi^(1/2)*erf(x))/2
```

where erf is the error function.

A few other examples may thus be given as

```
>> int(x^2)
ans =
x^3/3
>> int(sin(x))
ans =
-cos(x)
>> int(exp(-x))
ans =
-1/exp(x)
>> int(log(x))
ans =
x*(log(x)-1)
>> int(1/x)
ans =
log(x)
```

8.4 HIGHER-ACCURACY METHODS

Accuracy is particularly important in engineering applications. In the dynamics of bodies, for instance, the integration of the force over distance gives the energy, an accurate determination of which is necessary to study the damping and accelerating characteristics of the body for a suitable control system. Similarly, the integral of mass transfer rate over time yields the total mass transfer from a chemical reactor. An accurate evaluation of this integral is needed for supplying the required inflow of material into the system. Because of the high level of accuracy generally needed in engineering problems, methods have been developed for improving the accuracy of the numerical results obtained from integration formulas such as those discussed in the preceding sections. Higher-order Newton–Cotes formulas may also be employed for obtaining greater accuracy in numerical integration. This section discusses several of these higher-accuracy methods.

8.4.1 Richardson Extrapolation

Richardson extrapolation, which is also called *deferred approach to the limit*, is a numerical method for improving the accuracy of the results obtained from a given numerical scheme, provided an estimate of the total discretization error is available.

Although the technique is applied to numerical integration here, it can be used for a wide variety of numerical problems, such as the solution of differential equations by finite difference methods. Let us first consider integration by the trapezoidal rule. The total TE is given by Equation 8.23 as $E \cong S_T(\Delta x)^2$. Then, if I is the exact integral, and I_1 and I_2 are the numerical values of the integral obtained with step sizes Δx_1 and Δx_2, we may write, using TE to represent the total discretization error,

$$I \cong I_1 + S_T(\Delta x_1)^2 \tag{8.45a}$$

and

$$I \cong I_2 + S_T(\Delta x_2)^2 \tag{8.45b}$$

From these equations, the constant S_T may be estimated as follows:

$$S_T \cong \frac{I_2 - I_1}{(\Delta x_1)^2 - (\Delta x_2)^2} \tag{8.46}$$

Then, from Equation 8.45b,

$$I \cong I_2 + \frac{I_2 - I_1}{(\Delta x_1/\Delta x_2)^2 - 1} \tag{8.47}$$

Equation 8.47 does not yield the exact value of the integral I, since the expression for the TE is only an approximate one and since TE is employed instead of the total discretization error. However, an improved estimate for I is obtained from Equation 8.47. The second term in this equation represents the TE for integration with a step size of Δx_2. If Δx_2 is taken as half of Δx_1, that is, $\Delta x_1/\Delta x_2 = 2$, then

$$I \cong I_2 + \frac{I_2 - I_1}{3} = \frac{4I_2 - I_1}{3} \tag{8.48}$$

It can be easily shown that this expression for the integral is identical to that obtained from Simpson's one-third rule with a step size of Δx_2. Therefore, the integral is obtained to fourth-order accuracy.

Similarly, if Simpson's one-third rule is considered, the total TE from Equation 8.39 is $S_s(\Delta x)^4$. As before, we may write

$$I \cong I_1 + S_s(\Delta x_1)^4 \tag{8.49a}$$

and

$$I \cong I_2 + S_s(\Delta x_2)^4 \tag{8.49b}$$

where, again, I is the exact integral and I_1 and I_2 are the numerical values from Simpson's rule for step sizes Δx_1 and Δx_2, respectively. Then

$$S_s \cong \frac{I_2 - I_1}{(\Delta x_1)^4 - (\Delta x_2)^4} \tag{8.50}$$

and

$$I \cong I_2 + \frac{I_2 - I_1}{(\Delta x_1/\Delta x_2)^4 - 1} \tag{8.51}$$

Equation 8.51 yields an improved estimate of the integral. Thus, TE of $O[(\Delta x)^4]$ has been eliminated. Then, this expression gives the results to a sixth-order accuracy, since the next term in the TE for Simpson's rule is $O[(\Delta x)^6]$. If $\Delta x_2 = \Delta x_1/2$, a more accurate approximation to the integral is obtained from

$$I \cong I_2 + \frac{I_2 - I_1}{15} = \frac{16I_2 - I_1}{15} \tag{8.52}$$

Therefore, the accuracy of the numerical results obtained from the trapezoidal rule or from Simpson's rule can be substantially improved by computing the integral twice, with two different step sizes, and using Equation 8.47 or 8.51. Generally, numerical integration is carried out with a chosen step size, which is then halved to yield the second estimate I_2 of the integral. Then Equation 8.48 or 8.52 yields the improved estimate. This method does not require any major change in the computer program since the numerical scheme is simply applied twice. The computational effort is essentially doubled. However, because of the considerable improvement in accuracy and the simplicity of its application, Richardson extrapolation is frequently used.

8.4.2 Romberg Integration

Richardson extrapolation substantially improves the accuracy of the numerical results by eliminating the leading term in the TE. Thus, one may obtain fourth-order accuracy by applying the trapezoidal rule twice, with different step sizes, and using Richardson extrapolation to determine the improved value of the integral, as given above. One may apply this technique in succession to eliminate still higher-order terms in the TE. This leads to an efficient method, known as *Romberg integration*, which is widely used for obtaining numerical results of high accuracy.

The TE in the trapezoidal rule for numerical integration was obtained in terms of the dominant term by the use of the remainder theorem. However, it can be shown (Ralston, 1965; Davis and Rabinowitz, 1967; Ralston and Rabinowitz, 1978) that if the higher-order terms are included, the error E in the trapezoidal rule may be written as

$$E = A_1(\Delta x)^2 + A_2(\Delta x)^4 + A_3(\Delta x)^6 + A_4(\Delta x)^8 + \cdots \tag{8.53}$$

where the A's are constants. The leading term, of order $(\Delta x)^2$, was eliminated by the application of Richardson extrapolation in the preceding section. It the integral computed by the trapezoidal rule, with n segments, is denoted as $I_{0,n}$ and the improved value of the integral by Richardson extrapolation as $I_{1,n}$, then, from Equation 8.48,

$$I_{1,n} = \frac{4I_{0,n} - I_{0,n/2}}{4^1 - 1} \tag{8.54}$$

where the two integrals by the trapezoidal rule are obtained with $n/2$ and n intervals, corresponding to step sizes Δx and $\Delta x/2$, respectively.

Similarly, we eliminate the second term in the series representing the total TE, Equation 8.53, by applying Richardson extrapolation again. Since this term is of fourth order, the next extrapolation will be of sixth-order accuracy. Denoting this second extrapolation as $I_{2,n}$, we obtain

$$I_{2,n} = \frac{4^2 I_{1,n} - I_{1,n/2}}{4^2 - 1} \tag{8.55}$$

The process may be continued indefinitely, improving the value of the integral by successively eliminating the higher-order terms in the error. The general formula for the kth-order extrapolation is obtained as follows:

$$I_{k,n} = \frac{4^k I_{k-1,n} - I_{k-1,n/2}}{4^k - 1} \tag{8.56}$$

Thus, the value of the integral may be improved to the desired level of accuracy.

Figure 8.13 shows a schematic of Romberg integration. First, $I_{0,1}$ and $I_{0,2}$ are determined from the trapezoidal rule, with one and two elements, respectively. These yield the first extrapolation $I_{0,2}$. Similarly, $I_{0,4}$ used with $I_{0,2}$ gives $I_{1,4}$. The second extrapolation, $I_{2,4}$, is computed from $I_{1,2}$ and $I_{1,4}$. This computation of the integrals by the trapezoidal rule and of the improved values by Richardson extrapolation continues until the results remain essentially unchanged from one order of extrapolation to the next. Thus, if $|I_{2,4} - I_{1,4}|$ is less than a chosen convergence criterion ε, the process is terminated there. If not, we employ eight elements for the trapezoidal rule to determine the improved values of the integral for eight elements. Then, we compare $I_{3,8}$ with $I_{2,8}$ for convergence. Thus, the criterion for convergence may be written as

$$\left| I_{\lambda,n} - I_{\lambda-1,n} \right| \le \varepsilon \tag{8.57}$$

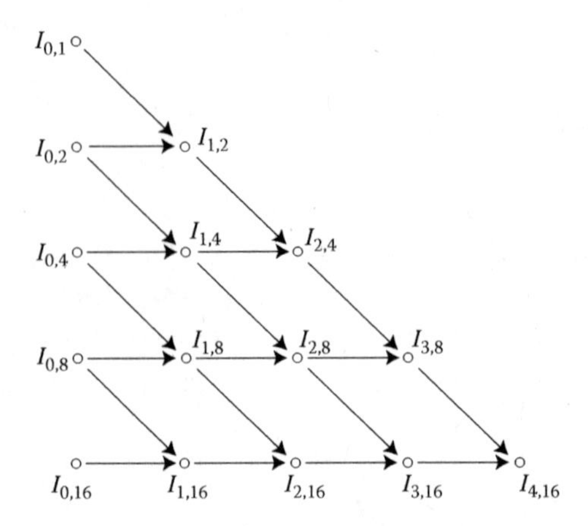

FIGURE 8.13 A schematic of Romberg integration, indicating the various levels of extrapolation.

where λ represents the highest order of extrapolation that can be obtained with n elements.

Romberg integration can, therefore, be used to obtain results of arbitrary accuracy, as far as TE is concerned. However, the round-off error, as always, imposes a limitation on the accuracy that may be achieved in practice. Example 8.3 demonstrates the use of Romberg integration in the accurate evaluation of an integral.

8.4.3 Higher-Order Newton–Cotes Formulas

So far, we have considered only the zeroth-, first-, second-, and third-order Newton–Cotes formulas. In general, these formulas are quite adequate for most engineering applications. The trapezoidal and Simpson's one-third rules are extensively used. The three-eighths rule is generally used in conjunction with the one-third rule if an odd number of segments is to be employed. We derived the formulas for these cases simply by taking the general form of the polynomial, whose coefficients were determined by making this curve pass through the required number of points on the plot of $f(x)$ versus x. However, the general approach for the derivation of Newton–Cotes formulas is based on the use of Lagrange interpolation, presented in Chapter 7. Since the segments are of uniform width, the points employed in the determination of the interpolating polynomial are uniformly distributed.

Closed Newton–Cotes formulas are those for which the data points at the two ends of the integration interval are known. Thus, the rectangular, trapezoidal, and Simpson's rules are all closed formulas. *Open Newton–Cotes formulas* are based on integration limits that lie beyond the range of available data. Although seldom used for numerical integration, open formulas are of interest in the solution of ODEs. Table 8.1 gives the formulas and the TEs per step for several Newton–Cotes closed integration schemes. Note that the accuracy improves substantially as the order of the polynomial increases from zero to two. The error for Simpson's three-eighths

TABLE 8.1
Newton–Cotes Closed Integration Formulas, along with the TE per Step and the Minimum Number of Segments Needed

Minimum No. of Segments	Points	Name	Formula	TE per Step
1	2	Rectangular rule	$f_i \Delta x$ or $f_{i+1} \Delta x$	$\frac{1}{2}(\Delta x)^2 f'(\xi)$ or $-\frac{1}{2}(\Delta x)^2 f'(\xi)$
1	2	Trapezoidal rule	$\frac{f_{i-1} + f_i}{2} \Delta x$	$-\frac{1}{12}(\Delta x)^3 f''(\xi)$
2	3	Simpson's one-third rule	$\frac{f_{i-1} + 4f_i + f_{i+1}}{3} \Delta x$	$-\frac{1}{90}(\Delta x)^5 f''''(\xi)$
3	4	Simpson's three-eighths rule	$\frac{3(f_{i-1} + 3f_i + 3f_{i+1} + f_{i+2})}{8} \Delta x$	$-\frac{3}{80}(\Delta x)^5 f''''(\xi)$
4	5	Boole's rule	$\frac{2(7f_{i-2} + 32f_{i-1} + 12f_i + 32f_{i+1} + 7f_{i+2})}{45} \Delta x$	$-\frac{8}{945}(\Delta x)^7 f^{VI}(\xi)$
5	6	—	$\frac{5(19f_{i-2} + 75f_{i-1} + 50f_i + 50f_{i+1} + 75f_{i+2} + 19f_{i+3})}{288} \Delta x$	$-\frac{275}{12,096}(\Delta x)^7 f^{VI}(\xi)$

rule is of the same order as that for the one-third rule. In fact, the former is somewhat larger in magnitude for the same segment size.

The accuracy again increases substantially as the order of the polynomial changes from two or three to four, the formula for which is often known as *Boole's rule*. The next-order polynomial is found to yield an accuracy of the same order as Boole's law, as observed for the two Simpson's rules. Because of the constraints imposed on the minimum number of points and, therefore, on the minimum and total number of segments that must be used for higher-order formulas, these are generally more difficult to program and are less versatile. As a consequence, the trapezoidal and Simpson's rules are the most extensively employed methods for numerical integration in engineering problems. However, Richardson extrapolation is frequently used, as in Romberg integration, to improve the accuracy of the numerical results.

Example 8.3

A mathematical function frequently encountered in the analysis of several engineering problems is the Gaussian error function, erf z, which is defined as

$$\text{erf } z = \frac{2}{\sqrt{\pi}} \int_0^z e^{-x^2} dx \tag{8.58}$$

Using Romberg integration, compute the value of the error function at $z = 0.5$, 1.0, 1.5, and 2.0, with a convergence criterion parameter ε in Equation 8.57 of 10^{-5}.

SOLUTION

The recursion formula for computing the various orders of extrapolation in Romberg integration is given by Equation 8.56. For convenience, the first-order approximation, or extrapolation, is taken as the trapezoidal rule, with Richardson extrapolation being the second-order extrapolation. Then $I_{1,m}$ represents the integral by the trapezoidal rule with m segments. With this change, the recursion formula for the kth-order extrapolation becomes

$$I_{k,n} = \frac{4^{k-1} I_{k-1,n} - I_{k-1,n-1}}{4^{k-1} - 1} \tag{8.59}$$

where $I_{k-1,n}$ represents the more accurate extrapolation of (k–l)th order, and $I_{k-1,n-1}$ the less accurate one, that is, at half the number of segments as the former. The recursion formula given by Equation 8.59 is more convenient to use than that given earlier by Equation 8.56, since we start with the first-order extrapolation, which is simply the trapezoidal rule, and we can successively double the number of segments, starting with 1 and keeping track of increasing accuracy by means of the subscript n.

The computer program for solving this problem is presented in Appendix C.13 for Fortran and Appendix B.21 for MATLAB. The given function $f(x) = (2/\sqrt{\pi})\exp(-x^2)$, which is to be integrated, is defined, the input variables, such as the convergence criterion ε, are entered, and the value of z at which the error function erf z is to be computed is given. Integration by the trapezoidal rule is carried out with one segment, over the range $x = 0$ to $x = z$. The segment size is successively halved, making maximum use of the segment areas already calculated. The program then computes the higher-order extrapolations, using Equation 8.59. At each iterative step, corresponding to a given number of subdivisions, the difference between the two highest possible extrapolations $I_{\lambda,n}$ and $I_{\lambda-1,n}$, as defined in Equation 8.57, is determined. Using Equation 8.57, if this difference is less than or equal to ε, the program is terminated; otherwise, the segment size is halved and the computation continued. After convergence has been achieved for a given value of z, other values of z, as given in the problem, are successively entered until all the required numerical values have been obtained.

The numerical results obtained from the given computer program are shown in Figure 8.14. The number of iterative steps needed in each case are shown, along with the computed value of the error function. The number of iterations needed are found to increase with z, over the range considered. The value of the error function is listed in most books on mathematical functions, and the values corresponding to $z = 0.5$, 1.0, 1.5, and 2.0 are given, respectively, as 0.5205, 0.8427, 0.9661, and 0.9953. Clearly, these values given in the literature are very close to those obtained from the present computation. Since the number of segments needed for four iterative steps is eight, the results indicate the rapid increase in accuracy as the segment width is halved. As mentioned earlier, Romberg integration is an extremely efficient and accurate method. Consequently, it is widely employed in engineering problems.

```
Enter the value of z = 0.5
No. of iterations = 1        Erf(z) = 0.501790
No. of iterations = 2        Erf(z) = 0.520602
No. of iterations = 3        Erf(z) = 0.520500

Enter the value of z = 1.0
No. of iterations = 1        Erf(z) = 0.771743
No. of iterations = 2        Erf(z) = 0.843103
No. of iterations = 3        Erf(z) = 0.842712
No. of iterations = 4        Erf(z) = 0.842701

Enter the value of z = 1.5
No. of iterations = 1        Erf(z) = 0.935482
No. of iterations = 2        Erf(z) = 0.954758
No. of iterations = 3        Erf(z) = 0.966707
No. of iterations = 4        Erf(z) = 0.966097

Enter the value of z = 2.0
No. of iterations = 1        Erf(z) = 1.149046
No. of iterations = 2        Erf(z) = 0.936492
No. of iterations = 3        Erf(z) = 0.998921
No. of iterations = 4        Erf(z) = 0.995266
No. of iterations = 5        Erf(z) = 0.995322
```

FIGURE 8.14 Numerical results obtained for Example 8.3, indicating the number of iterative steps needed for Romberg integration and the computed values of the error function.

8.5 INTEGRATION WITH SEGMENTS OF UNEQUAL WIDTH

All of the formulas for numerical integration presented so far were based on segments of equal width. This implies that the points, on the curve of the function $f(x)$, that were used for determining the interpolating polynomial were equally spaced. However, this procedure is not necessarily the most efficient one. In regions where the function varies very gradually or where its value is small, the number of points for function evaluation may be reduced without significantly affecting the accuracy of the results. The optimum distribution of points for the numerical integration of a given function may also be derived to obtain maximum accuracy with a given number of function evaluations. In these cases, the segments into which the range of integration is subdivided are not of equal width. Similarly, experimental or numerical data may be available at only specified points which may be unevenly distributed. Procedures for the numerical integration of such data are needed. This section considers various methods for improving the efficiency of numerical integration by using unequal segments and also the methods that may be employed for integrating a function whose value is given at unevenly spaced data points.

8.5.1 UNEQUALLY SPACED DATA

Experimental results are often obtained at unevenly spaced values of the independent variable. In an experimental study of the displacement of a moving body such as a car as a function of time, for instance, more frequent measurements are generally taken at small times, just after the onset of motion, than at large times. Similarly, the pressure loading on a building due to the wind is measured only at discrete locations, which may not be evenly spaced. Numerical solutions may also yield results at unequally spaced data points. Thus, we are faced with the problem of integrating a function $f(x)$ which is available simply as data at arbitrary, unevenly spaced values of the independent variable.

One approach for solving this problem is to employ the curve-fitting techniques given in the preceding chapter, in order to obtain a continuous function $f(x)$. Then the range of integration may be subdivided into segments of equal width, and numerical integration may be carried out by, say, the Newton–Cotes formulas, using the ordinates obtained from the function $f(x)$ derived from curve fitting. This approach is often employed in engineering problems, since experimental and numerical data are often curve fitted for using the results in other computations, as discussed in Chapter 7.

The second approach employs the data as given and simply obtains the integral in each segment. The trapezoidal rule may be applied to each segment and the results summed to yield the integral I as follows:

$$I = \Delta x_1 \frac{f_0 + f_1}{2} + \Delta x_2 \frac{f_1 + f_2}{2} + \Delta x_3 \frac{f_2 + f_3}{2} + \cdots + \Delta x_n \frac{f_{n-1} + f_n}{2} \tag{8.60}$$

where Δx_1; Δx_2, …, Δx_n are the widths of the n segments that correspond to $(n + 1)$ data points, x_i represents the value of the independent variable at a given point, and f_i

is the corresponding value of the function. Here, Δx_1 Δx_2, ..., Δx_n are not equal, as was assumed for deriving Equation 8.11. A computer program can be easily written to compute the integral I, using Equation 8.60 for unequal segment widths.

If two adjacent segments are of equal width, Simpson's one-third rule may be used to obtain the integral over these segments. Thus, if Δx_{i-1} and Δx_i are equal, the integral I_s for these two subdivisions is given by

$$I_s = \left(\frac{f_{i-1} + 4f_i + f_{i+1}}{3} \right) \Delta x \quad (8.61)$$

Similarly, if three segments are of equal width, Simpson's three-eighths rule, Equation 8.34, may be employed. Since Simpson's rules are more accurate than the trapezoidal rule, an implementation of Simpson's rule in the numerical integration, wherever possible, will increase the accuracy of the results. Thus, a program may be developed that checks the widths of adjacent segments before applying numerical integration to the data. If two consecutive segments are of equal width, Simpson's one-third rule is employed, and if three segments are of equal width, Simpson's three-eighths rule is used. If the widths of two adjacent segments are different, the trapezoidal rule is employed. Example 8.4 demonstrates an application of this procedure to unevenly spaced experimental data.

8.5.2 Adaptive Quadrature

One can employ Romberg integration to obtain numerical results to any desired accuracy, within the constraints imposed by the round-off error. However, since segments of equal width are employed, the entire range of integration is treated uniformly. This approach is not the most efficient one if the function is slowly varying or small in magnitude in certain regions, where fewer points can be taken. Adaptive quadrature enables one to increase the number of points in regions where the accuracy is not at the desired level, while keeping fewer points in regions where satisfactory accuracy has been attained.

Several methods have been developed to achieve such an uneven distribution of points. The main idea is to focus on regions where the error is larger than the desired value. Suppose the integral

$$I = \int_a^b f(x)\,\mathrm{d}x \quad (8.1)$$

is to be computed with total error less than ε. Then the error in each subinterval of width Δx must be less than $\varepsilon\ \Delta x/(b - a)$, so that the total error is less than ε. We start by dividing the range of integration into equal segments of width Δx. The integral over each segment is determined by use of, say, the trapezoidal rule. Then each segment is subdivided into two subintervals of width $\Delta x/2$, and the integral is computed. From Richardson's extrapolation, Equation 8.46, an estimate of the error in these segments of width $\Delta x/2$ is $(I_2 - I_1)/3$, where I_2 is the integral with two segments of width $\Delta x/2$ each and I_1 is the integral over the segment of width Δx. If this error is less than $\varepsilon(\Delta x/2)/(b - a)$, no further subdivision of the corresponding segments is needed, and the estimate of the integral over the segments is taken for the computation of the total integral over the entire region. If the error is larger than $\varepsilon\,(\Delta x/2)/(b - a)$, in certain segments, these segments are halved and the above procedure is repeated until the specified accuracy has been attained in these.

This method, therefore, allows one to systematically reduce the error in regions where it is too large, while keeping regions where satisfactory accuracy has been attained unaffected. Simpson's rule may also be used instead of the trapezoidal rule. However, since the final accuracy is prescribed, the trapezoidal rule is more appropriate because it is simpler to use. Adaptive quadrature is particularly useful for complicated functions that have a large variation over certain regions and a small variation over others. In the integration of $\exp(-50x^2)$ over $0 \le x \le 1$, for instance, many more subintervals are needed at small x than at large x to attain uniform accuracy over the entire region. Adaptive quadrature is a valuable method in such cases. For further details, see Forsythe et al. (1977), Ferziger (1998), and Gerald and Wheatley (2003).

Example 8.4

In an experiment on the motion of accelerating bodies, the velocity *V*, in m/s, of a body is measured at several time intervals *t*, in seconds. The data obtained are tabulated for time ranging from 0 to 2.0 s as follows:

t	0.0	0.1	0.2	0.3	0.5	0.7	0.8	1.0
V	9.50	10.00	10.57	11.24	12.97	15.38	16.93	20.9
t	1.1	1.3	1.5	1.6	1.7	1.8	2.0	
V	23.41	29.74	38.17	43.33	49.21	55.88	71.90	

Compute the distance traveled by the body *x* as a function of time *t*.

SOLUTION

The experimental data are given at unevenly distributed values of the independent variable *t*. The distance traveled *x* is given by the integral

$$x = \int_0^x V(t')\,dt' \tag{8.62}$$

where *t′* is a dummy variable. We shall employ the procedure outlined in Section 8.5.1 for numerical integration to obtain *x*(*t*). As discussed earlier, Simpson's three-eighths rule is employed when three adjacent segments are of equal width, and Simpson's one-third rule when only two adjacent segments are of equal width. If the size of a given subdivision is different from that of the next subdivision, the trapezoidal rule must be used.

The MATLAB computer program used to evaluate the integral in Equation 8.62, employing the unevenly spaced data given in the problem is given in Appendix B.22.

Simpson three-eighths rule
I = 4 Time = 0.3000 Velocity = 11.2400 Distance = 3.0919

Simpson one-third rule
I = 6 Time = 0.7000 Velocity = 15.3800 Distance = 8.3252

Trapezoidal rule
I = 7 Time = 0.8000 Velocity = 16.9300 Distance = 9.9407

Trapezoidal rule
I = 8 Time = 1.0000 Velocity = 20.9000 Distance = 13.7237

Trapezoidal rule
I = 9 Time = 1.1000 Velocity = 23.4100 Distance = 15.9392

Simpson one-third rule
I = 11 Time = 1.5000 Velocity = 38.1700 Distance = 27.9752

Simpson three-eighths rule
I = 14 Time = 1.8000 Velocity = 55.8800 Distance = 41.9128

Trapezoidal rule
I = 15 Time = 2.0000 Velocity = 71.9000 Distance = 54.6908

FIGURE 8.15 Computed distance traveled x as a function of time t from the velocity data given in Example 8.4. The various integration schemes employed over different regions of the given data are also indicated.

The given data are entered, along with an arbitrary small quantity *eps*. The widths of three adjacent subdivisions, starting with $t = 0$ and proceeding in the direction of increasing time, are determined. The small allowable difference *eps* is employed to determine whether the segment widths are equal. Here, *eps* is taken as 10^{-6}, which is an arbitrary small quantity and will not affect the numerical results for the given data set. Obviously, *eps* is used to avoid problems caused by round-off error, because of which two equal segment widths may be indicated as different due to differences in round-off error. If the first two segments are of different width, the trapezoidal rule is employed. If the widths are equal, the third segment is also considered, and if all three are equal in width, Simpson's three-eighths rule is used. Otherwise, Simpson's one-third rule is employed. Thus, it is a fairly straight forward application of the integration formulas.

The numerical results obtained are shown in Figure 8.15. The data point up to which the given integral is computed, the corresponding time, the velocity, and the total distance x traveled up to this point are given. The method used for numerical integration in the preceding subdivision(s) is also indicated. Thus, the three-eighths rule is employed first, followed by Simpson's one-third rule, then the trapezoidal rule, and so on. The velocity V is printed in order to ensure that the input data have been correctly read. At the end of 2 s, the total distance traveled is obtained as 54.69 m. This program is quite flexible and can be used for a wide variety of unevenly spaced data.

8.5.3 Gauss Quadrature

In many engineering problems, the evaluation of the integrand $f(x)$ is very involved and time-consuming. *Gauss quadrature* is based on a variety of interpolating functions and gives maximum accuracy for a given number of function evaluations.

However, the x locations where the function $f(x)$ is to be evaluated are adjustable. Thus, for a two-point formula, the integral

$$I = \int_{-1}^{1} f(x)\mathrm{d}x \tag{7.63a}$$

is approximated by

$$I \cong A_1 f(x_1) + A_2 f(x_2) \tag{8.63b}$$

where A_1, A_2, x_1, and x_2 are all unknowns. The integration limits are taken as −1 and 1. The integral between the finite limits a and b can be transformed into the limits ±1 by means of the transformation

$$\xi = \frac{2x - (a+b)}{b-a} \tag{8.64}$$

Thus, the integral may be taken over the limits – 1 to 1 without loss of generality. This simplifies the computation and generalizes the formulation. Example 8.5 illustrates how this transformation is carried out in practice.

Now, if we require that Equation 8.63b yield the integrals for constant, linear, parabolic, and cubic functions exactly, the four coefficients in the equation can be determined. Thus, employing 1, x, x^2, and x^3 as the functions, and substituting these for $f(x_1)$ and $f(x_2)$, we have

$$A_1 f(x_1) + A_2 f(x_2) = \int_{-1}^{1} 1\mathrm{d}x = 2 \quad \text{or} \quad A_1 + A_2 = 2$$

$$A_1 f(x_1) + A_2 f(x_2) = \int_{-1}^{1} x\mathrm{d}x = 0 \quad \text{or} \quad A_1 x_1 + A_2 x_2 = 0$$

$$A_1 f(x_1) + A_2 f(x_2) = \int_{-1}^{1} x^2\mathrm{d}x = \frac{2}{3} \quad \text{or} \quad A_1 x_1^2 + A_2 x_2^2 = \frac{2}{3}$$

$$A_1 f(x_1) + A_2 f(x_2) = \int_{-1}^{1} x^3\mathrm{d}x = 0 \quad \text{or} \quad A_1 x_1^3 + A_2 x_2^3 = 0$$

which gives

$$A_1 = A_2 = 1, \quad x_1 = -\frac{1}{\sqrt{3}}, \quad x_2 = -\frac{1}{\sqrt{3}}$$

Therefore,

$$I \cong f\left(-\frac{1}{\sqrt{3}}\right) + f\left(\frac{1}{\sqrt{3}}\right) \tag{8.65}$$

which is known as the *two-point Gauss–Legendre formula*. This integral estimate is exact for polynomials up to the third order, that is, cubics, and is, therefore, of the same order of accuracy as Simpson's rule. It is interesting to note that this accuracy is achieved on the basis of only two function evaluations, at $x = \pm 1/\sqrt{3}$.

The above discussion indicates the general features and the power of Gauss quadrature. The general formula for this method may now be written as follows:

$$I \cong A_1 f(x_1) + A_2 f(x_2) + A_3 f(x_3) + \cdots + A_n f(x_n) \tag{8.66}$$

where n is the number of points in the range $-1 \le x \le 1$ at which the function $f(x)$ is calculated. The integral under consideration is $\int_{-1}^{1} f(x)\mathrm{d}x$, which is obtained from the integral $\int_a^b f(x)\mathrm{d}x$ by means of the transformation given by Equation 8.64. The derivation of higher-order formulas for Gauss quadrature is quite involved. Basically, orthogonal polynomials, such as *Legendre*, *Chebyshev*, *Hermite*, and *Laguerre* polynomials, are taken to represent the function over the range $-1 \le x \le 1$. These polynomials are generally discussed in books on advanced calculus. The locations where the function is to be evaluated are actually the n zeros of an nth-degree Legendre polynomial. Table 8.2 gives the n locations, along with the corresponding weights $A_1, A_2, \ldots, A_n$, for the Gauss–Legendre formulas, considering n up to 24, which should be adequate for most practical problems. Other commonly employed quadrature formulas are the *Gauss–Chebyshev*, *Gauss–Laguerre*, and *Gauss–Hermite* integration formulas, given by Abramowitz and Stegun (1964) and Stroud and Secrest (1966). Although the derivation of these formulas is complicated, the application to numerical integration is not, as illustrated in Example 8.5.

The error E in Gauss–Legendre quadrature, which is usually referred to simply as *Gauss quadrature*, is obtained for an n-point formula as

$$E \approx \frac{2^{2n+1}(n!)^4}{(2n+1)(2n!)^3} f^{(2n)}(\xi), \quad \text{where} \quad -1 < \xi < 1 \tag{8.67}$$

Therefore, a polynomial of degree $(2n - 1)$ is integrated exactly, since the $(2n)$th derivative, $f^{(2n)}$, is zero in this case, resulting in zero error. This implies that if n points are employed in Gauss quadrature, the accuracy obtained is of the same order as that obtained with a polynomial of order $(2n - 1)$ in Newton–Cotes formulas. Therefore, Gauss quadrature is a powerful method that is frequently used in engineering applications. However, a systematic reduction in error, as achieved by Romberg integration, is not possible in Gauss quadrature, and one must repeat the entire integration scheme with higher-order formulas if a greater accuracy is needed. Gauss quadrature maximizes the accuracy for a given number of function evaluations and is particularly suitable for complicated functions. However, the method is not applicable to problems where tabulated data are given at arbitrary locations, since function evaluations at definite points are needed. In some cases, it may be possible to take the data at the points specified by the quadrature formula and, thus, use Gauss quadrature for numerical integration.

TABLE 8.2
Weighting Factors *A* and the Values of the Independent Variable *x* at Which the Function *f*(*x*) Must Be Evaluated for the *Gauss–Legendre* Formulas, Considering up to the 24-Point Approximation

n	$\pm x_i$	A_i
2	0.5773502692	1.0000000000
3	0.0000000000	0.8888888889
	0.7745966692	0.5555555556
4	0.3399810436	0.6521451549
	0.8611363116	0.3478548451
5	0.0000000000	0.5688888889
	0.5384693101	0.4786286705
	0.9061798459	0.2369268850
6	0.2386191861	0.4679139346
	0.6612093865	0.3607615730
	0.9324695142	0.1713244924
7	0.0000000000	0.4179591837
	0.4058451514	0.3818300505
	0.7415311856	0.2797053915
	0.9491079123	0.1294849662
8	0.1834346425	0.3626837834
	0.5255324099	0.3137066459
	0.7966664774	0.2223810345
	0.9602898565	0.1012285363
9	0.0000000000	0.3302393550
	0.3242534234	0.3123470770
	0.6133714327	0.2606106964
	0.8360311073	0.1806481607
	0.9681602395	0.0812743884
10	0.1488743390	0.2955242247
	0.4333953941	0.2692667193
	0.6794095683	0.2190863625
	0.8650633667	0.1494513492
	0.9739065285	0.0666713443
12	0.1252334085	0.2491470458
	0.3678314990	0.2334925365
	0.5873179543	0.2031674267
	0.7699026742	0.1600783285
	0.9041172564	0.1069393260
	0.9815606342	0.0471753364
16	0.0950125098	0.1894506105
	0.2816035508	0.1826034150
	0.4580167777	0.1691565194
	0.6178762444	0.1495959888

TABLE 8.2 (continued)
Weighting Factors *A* and the Values of the Independent Variable *x* at Which the Function *f*(*x*) Must Be Evaluated for the *Gauss–Legendre* Formulas, Considering up to the 24-Point Approximation

n	$\pm x_i$	A_i
16	0.7554044084	0.1246289713
	0.8656312024	0.0951585117
	0.9445750231	0.0622535239
	0.9894009350	0.0271524594
20	0.0765265211	0.1527533871
	0.2277858511	0.1491729865
	0.3737060887	0.1420961093
	0.5108670020	0.1316886384
	0.6360536807	0.1181945320
	0.7463319065	0.1019301198
	0.8391169718	0.0832767416
	0.9122344283	0.0626720483
	0.9639719273	0.0406014298
	0.9931285992	0.0176140071
24	0.0640568929	0.1279381953
	0.1911188675	0.1258374563
	0.3150426797	0.1216704729
	0.4337935076	0.1155056681
	0.5454214714	0.1074442701
	0.6480936519	0.0976186521
	0.7401241916	0.0861901615
	0.8200019860	0.0733464814
	0.8864155270	0.0592985849
	0.9382745520	0.0442774388
	0.9747285560	0.0285313886
	0.9951872200	0.0123412298

Example 8.5

In a civil engineering application, a vertical plate 1 m high and 1.2 m wide is positioned in a stream of flowing water in a channel. The pressure p exerted on the plate due to the flow is measured at several vertical locations x, where $x = 0$ represents the top edge of the plate. Curve fitting is employed to obtain p as a function of x. The resulting expression for pressure in Newtons/(meters squared) and x in meters is

$$p(x) = 10 + 4.6x - 16.2x^2 + 8.9x^3 - 41.3x^4 + 22.6x^3 \quad (8.68)$$

Using Gauss quadrature with two as well as four function evaluations, compute the total force exerted on the plate due to the flow.

SOLUTION

The resulting force on the plate F is given by the equation

$$F = \int_0^1 p(x)1.2dx$$
$$= 1.2\int_0^1 p(x)dx \quad \text{Newtons} \tag{8.69}$$

where 1.2dx represents the area of a differential surface element at a vertical location given by x. Since $p(x)$ is a fifth-order polynomial in x, the above integral can easily be evaluated analytically to yield 4.631667, giving the force F as 5.558 N. However, let us apply Gauss quadrature to this problem in order to demonstrate the use of this method for numerical integration.

We must first change the limits of the integration to −1 and 1 by employing Equation 8.64. Thus, since $a = 0$ and $b = 1$, ξ is given by

$$\xi = \frac{2x - (0 + 1)}{1} = 2x - 1$$

or

$$x = 0.5\xi + 0.5 \tag{8.70}$$

which gives

$$dx = 0.5d\xi \tag{8.71}$$

Therefore, the expression for the total force F on the plate becomes

$$F = 1.2\int_{-1}^{1}\Big[10 + 4.6(0.5\xi + 0.5) - 16.2(0.5\xi + 0.5)^2 + 8.9(0.5\xi + 0.5)^3$$
$$- 41.3(0.5\xi + 0.5)^4 + 22.6(0.5\xi + 0.5)^4\Big]0.5d\xi \tag{8.72}$$

or,

$$F = 1.2\int_{-1}^{1} f(\xi)d\xi \tag{8.73}$$

where $f(\xi)$ is the function to be integrated, as given above.

Using the two-point Gauss–Legendre formula, F is given by

$$F = 1.2\left[f\left(-\frac{1}{\sqrt{3}}\right) + f\left(\frac{1}{\sqrt{3}}\right)\right] \tag{8.74}$$

Similarly, for the four-point Gauss–Legendre formula,

$$F = 1.2\left[A_1 f(x_1) + A_2 f(x_2) + A_3 f(x_3) + A_4 f(x_4)\right] \tag{8.75}$$

where the required x's and A's are given in Table 8.2. Two function evaluations are involved in the former case, and four in the latter. Thus, for the two-point formula,

$$f\left(-\frac{1}{\sqrt{3}}\right) = 5.129892$$

$$f\left(\frac{1}{\sqrt{3}}\right) = -0.5826686$$

which gives

$$F = 1.2\ (4.547223) = 5.456668 \text{ Newtons}$$

Similarly, for the four-point formula,

$$F = 1.2[Af(-C) + Bf(-D) + Bf(D) + Af(C)]$$

where $A = 0.347854845$, $B = 0.652145155$, $C = 0.861136312$, and $D = 0.339981044$. This gives $F = 5.558$ N. Thus, the four-point formula gives a very high level of accuracy with only four evaluations of the function $f(\xi)$. Even for the two-point formula, with only two function evaluations, the error is only 1.82%. This error figure indicates the efficiency of this method and its considerable value for complicated functions frequently encountered in engineering applications.

8.6 NUMERICAL INTEGRATION OF IMPROPER INTEGRALS

In the preceding sections, the limits of integration a and b in the integral $\int_a^b f(x)dx$ were taken as finite, and the integrand $f(x)$ was assumed to be continuous and finite over the range $a \le x \le b$. However, in engineering computations, we are often faced with integrals in which either the limits of integration are infinite or the integrand is singular at some point in the range of integration. Such integrals are known as *improper integrals*, and special procedures are often required for their evaluation. Some of these integrals are discussed here. It is assumed that the integral exists and is finite. This assumption is often based on the nature of the physical quantity represented by the integral. For instance, if the integral represents the total energy lost by a given body, the integral is expected to be finite. Similarly, if an integral over time yields the total distance traveled by a body that is decelerating due to an applied force, the integral must approach a finite value as the upper limit of integration approaches infinity. The analytical methods for proving that an integral exists and is finite are generally given in most advanced calculus books.

8.6.1 Integrals with Infinite Limits

In problems of engineering interest, we often encounter integrals of the form $\int_a^\infty f(x)\mathrm{d}x$, $\int_{-\infty}^b f(x)\mathrm{d}x$ or $\int_{-\infty}^\infty f(x)\mathrm{d}x$, where either one or both of the limits of integration are infinite. In the example of the retarding body, outlined above, the lower limit is finite, say, time $t = 0$, and the upper limit is infinite. The mass transfer from an infinite surface, which approximates, say, the surface of a large lake, would involve an integral over the range $-\infty \le x \le \infty$. Similarly, flow rates in jets and plumes often require integration from $-\infty$ to ∞, since no walls, which limit the extent of the flow, are assumed to be present. Statistical distributions, like Gaussian and Poisson distributions, also generally involve infinite limits of integration.

There are several methods by which such integrals may be evaluated. The most common and often convenient approach is to write a given integral of the form $\int_a^\infty f(x)\mathrm{d}x$ as $\int_a^b f(x)\mathrm{d}x$ and to evaluate the integral with increasing values of b, until any further increase in b results in a negligible change in the integral. This approach was demonstrated in Example 8.1, where the charge Q in the capacitor was computed by the integral $\int_0^t I(t')\mathrm{d}t'$, $I(t)$ being the current, t the time, and t' simply a dummy variable. Thus as $t \to \infty$, the charge attains a finite value. We obtained this result by increasing the upper limit of integration until the charge Q remained unchanged as t was increased further. Similarly, for integrals of the form $\int_{-\infty}^b f(x)\mathrm{d}x$ or $\int_{-\infty}^\infty f(x)\mathrm{d}x$, this approach may be used, suitably decreasing the lower limit and/or increasing the upper limit.

The integrand $f(x)$ may approach zero in an asymptotic manner as $x \to \infty$. In some cases, the dominant terms at large x can be employed to simplify the function and integrate it analytically. Thus, the given integral is written as

$$\int_a^\infty f(x)\mathrm{d}x = \int_a^s f(x)\mathrm{d}x + \int_s^\infty \tilde{f}(x)\mathrm{d}x \tag{8.76}$$

where s is chosen to be sufficiently large so that the function $f(x)$ may be replaced by a simpler asymptotic approximation $\tilde{f}(x)$ for $x \ge s$. Then the first integral on the right-hand side of Equation 8.76 is evaluated numerically and the second integral analytically. Examples of functions for which this approach may be employed are $f(x) = 1/(e^x + e^{-2x} + 3x^{-2})$ and $f(x) = 1/(e^{5x} + 2x^3 + 1)$. At large x, the first function may be approximated as e^{-x}, and the second as e^{-5x}, both of which may be integrated analytically over $s \le x < \infty$ to yield e^{-s} and $e^{-5s}/5$, respectively. Again, s may be varied until the numerical value of the total integral $\int_a^\infty f(x)\mathrm{d}x$ shows a negligible change with a further increase in s. This procedure, wherever applicable, is more efficient than replacing the upper limit by a large number b and computing the integral for increasing values of b, as outlined earlier.

In some cases, a transformation of the independent variable may be employed to change the infinite limit of integration into a finite one. Commonly used transformations are $y = x^{-n}$ and $y = e^{-x}$, both of which give zero for the new variable y as x goes to infinity. For instance, consider the following two integrals:

$$I = \int_{1}^{\infty} \frac{x}{1+x+x^3} \, dx \tag{8.77a}$$

$$I = \int_{0}^{\infty} \frac{1}{e^x + e^{-x}} \, dx \tag{8.77b}$$

We can transform the first integral into one with finite limits by using the transformation $y = 1/x$. This gives $dx = -dy/y^2$, and the integral becomes

$$I = \int_{1}^{\infty} \frac{x}{1+x+x^3} \, dx = \int_{1}^{0} \frac{1/y}{1+(1/y)+1/y^3} \cdot -\frac{dy}{y^2} = \int_{0}^{1} \frac{dy}{1+y^2+y^3} \tag{8.78a}$$

Similarly, we transform the integral in Equation 8.77b by employing $y = e^{-x}$ as follows:

$$I = \int_{0}^{\infty} \frac{dx}{e^x + e^{-x}} = \int_{1}^{0} \frac{1}{y+(1/y)} \cdot -\frac{dy}{y} = \int_{0}^{1} \frac{dy}{1+y^2} \tag{8.78b}$$

Thus, integrals over infinite regions may be transformed into integrals over finite regions. However, in some cases, the transformed integrand may be singular at one of the limits. Then the problem with an infinite limit is replaced by one involving a singular integrand, discussed in the following subsection. In general, it is easier to handle an infinite limit of integration than an integrand that becomes singular. Therefore, one must consider whether or not a given transformation simplifies the problem.

8.6.2 Singular Integrand

Another class of improper integrals is the one in which the limits of integration are finite but the integrand is singular in the range of integration, generally at one or both limits. However, the singularity is assumed to be gentle enough for the integral to exist and be finite. Examples of such integrals are

$$\int_{0}^{2} \frac{e^x}{x} \, dx \qquad \int_{0}^{1} \frac{1}{\sqrt{x}} \, dx \qquad \int_{0}^{1} \frac{x^3+1}{\sqrt{1-x^2}} \, dx \qquad \int_{0}^{\pi/2} \frac{d\theta}{2\sqrt{\cos\theta}}$$

In all of these cases, the integrand becomes singular at the lower or the upper limit of integration, and the integral can be shown to exist.

There are several methods for dealing with such improper integrals. Among these are integrating by parts, subtracting out the singularity, using a power series to approximate the integral near the point of singularity, and transforming the variables. Obviously, the appropriate method depends on the nature of the singularity, and all of these techniques may be considered to determine if one of them would work. Let us illustrate the use of some of these strategies to eliminate the singularity by means of examples.

The integral

$$\int_0^2 \frac{e^x}{\sqrt{x}}\,dx \tag{8.79a}$$

can be integrated by parts to yield

$$\int_0^2 \frac{e^x}{\sqrt{x}}\,dx = 2\sqrt{x}\,e^x\Big|_0^2 + \int_0^2 2\sqrt{x}e^x\,dx \tag{8.79b}$$

The first term on the right-hand side can easily be evaluated, and the second term is an integral that is not singular. This remaining integral can thus be computed analytically or by means of the various numerical methods presented in this chapter. The singularity in this integral can also be subtracted out as follows:

$$\int_0^2 \frac{e^x}{\sqrt{x}}\,dx = \int_0^2 \frac{1}{\sqrt{x}}\,dx + \int_0^2 \frac{e^x - 1}{\sqrt{x}}\,dx \tag{8.80}$$

The first integral on the right-hand side is singular, but it can easily be integrated analytically. The second integral is nonsingular, since $\left(e^x - 1\right)/\sqrt{x}$ can be shown to approach zero at the lower limit of integration.

The singularity in the integral

$$\int_0^1 \frac{x^3 + 1}{\sqrt{1 - x^2}}\,dx \tag{8.81a}$$

can be eliminated by use of the transformation $x = \sin y$. Then $\sqrt{1 - x^2} = \cos y$, and $dx = \cos y\,dy$. The integral becomes

$$\int_0^1 \frac{x^3 + 1}{\sqrt{1 - x^2}}\,dx = \int_0^{\pi/2} \frac{(\sin y)^3 + 1}{\cos y}\cos y\,dy = \int_0^{\pi/2} \left[(\sin y)^3 + 1\right]dy \tag{8.81b}$$

The resulting integral is nonsingular and can be integrated by any numerical method.

Another strategy that is sometimes applicable is the expansion of the integrand in a power series about the singular point and retention of a few leading terms that can be integrated by standard methods. Gauss quadrature is also particularly suitable for certain types of singularities, since formulas are sometimes available that have already accounted for the singularity in the choice of the weighting function.

A frequently used procedure, if the above strategies do not work, is to replace the integration limit where the singularity exists by a quantity close to this limit, but not equal to it. Thus, $\int_a^b f(x)\,dx$ is replaced by $\int_{a+\varepsilon}^b f(x)\,dx$, where ε is a small quantity, if $f(x)$ is singular at $x = a$. Then numerical integration is carried out over the range $a + \varepsilon \le x \le b$, and ε is made smaller, starting with a chosen small value, until the computed integral is not significantly affected by a further reduction in ε. This method is not very efficient, particularly if equally wide intervals are used. However, the range of integration may be broken down into a region close to the point of singularity and others farther away, so that a finer mesh may be used near the singularity. For example, the following integrals may be written as

$$\int_0^1 \frac{1}{\sqrt{1-x}}\,dx = \int_0^{0.99} \frac{1}{\sqrt{1-x}}\,dx + \int_{0.99}^{1-\varepsilon} \frac{1}{\sqrt{1-x}}\,dx \tag{8.82a}$$

$$\int_0^\infty \frac{1}{\sqrt{x}(1+x)}\,dx = \int_\varepsilon^{0.01} \frac{1}{\sqrt{x}(1+x)}\,dx + \int_{0.01}^\infty \frac{1}{\sqrt{x}(1+x)}\,dx \tag{8.82b}$$

with ε being reduced toward zero till the integrals do not vary significantly with a further reduction in ε. The upper limit in the second integral in Equation 8.82b is infinity and can be treated by the methods given in Section 8.6.1. The integration region $\varepsilon \le x \le 0.01$ may be further subdivided as $\varepsilon \le x \le 0.001$ and $0.001 \le x \le 0.01$, if necessary. Adaptive quadrature can also be used advantageously for this problem.

Example 8.6

Many physical measurements follow the symmetrical, bell-shaped curve of the Gaussian, or normal, frequency distribution, sketched in Figure 8.16. Repeated measurements of the fluid velocity in a hydraulic control system are found to closely approximate the Gaussian distribution

$$f(x) = \frac{1}{\sqrt{2\pi}\sigma} \exp\left[-\frac{1}{2}\left(\frac{x-\mu}{\sigma}\right)^2\right] \tag{8.83}$$

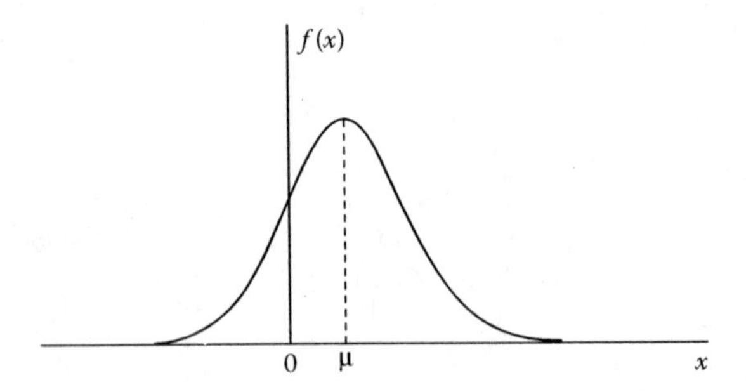

FIGURE 8.16 Sketch of the Gaussian, or normal, frequency distribution, with μ as the mean value.

where $f(x)$ is the height of the frequency curve corresponding to a given velocity x, μ is the mean value, and σ is known as the *standard deviation*. For the given circumstance, $\mu = 10$ m/s and $\sigma = 5$ m/s. Compute the fraction of the measurements for which the velocity is larger than or equal to 0, 5.0, 10.0, and 15.0 m/s, respectively.

SOLUTION

The normal distribution extends from $x = -\infty$ to $x = +\infty$ and is symmetrical about the mean μ. The area under the curve from, say, $x = x_1$ to $x = x_2$, gives the fraction of the total measurements for which the velocity lies between x_1 and x_2. Therefore, the integral to be computed is

$$I = \int_{x_{min}}^{\infty} f(x)dx \tag{8.84}$$

where x_{min} is the minimum value of velocity considered. For the given problem, $x_{min} = 0$, 5.0, 10.0, and 15.0 m/s, respectively. Since $\int_{-\infty}^{\infty} f(x)\,dx$ covers the entire range of measurements, this integral equals 1.0. Similarly, $\int_{\mu}^{\infty} f(x)dx = 0.5$, since it represents half of the measurements taken.

The given problem involves the evaluation of an improper integral, since the upper limit is infinity. Thus, following the approach given in the preceding section for such problems, we may compute the integral

$$\hat{I} = \int_{x_{min}}^{x_{max}} f(x)dx \tag{8.85}$$

by any standard method for numerical integration, such as Simpson's rule, and increase x_{max} until the value of the integral remains essentially unchanged as x_{max} is increased further. This approach is followed in the computer program for this problem. As given in Appendix B.23, the function *m*-file for Simpson's method, given in Appendix B.20, is easily modified for this problem. The segment width h is chosen, and the four given values for x_{min} are successively entered. The upper limit of integration x_{max} is varied until the integral varies by less than a convergence parameter ε, taken as 10^{-5} here, with a further increase in x_{max}. The segment size h and the convergence parameter must be varied to ensure that the results obtained are not significantly dependent on the values chosen.

The numerical results obtained are shown in Figure 8.17, in terms of the integral at various values of x_{max}, for each of the four given values of x_{min}. It is found that an x_{max} of 40.0 m/s is adequate for the approximation of infinity, which is the upper limit of integration in Equation 8.84. The convergence parameter ε and the segment width h were also varied. The values chosen were found to be quite satisfactory. Thus, from the results obtained, 97.725% of the measurements, or 0.97725 in fractional notation, yield a positive fluid velocity, that is, $x \geq 0$. Similarly, 84.135% of the measurements give a velocity larger than or equal to 5 m/s, and so on. Tabulated results for the integration of the normal distribution curve are available in the literature. For the four cases considered here, the values given in the literature are 97.72, 84.13, 50.0, and 15.87%, respectively. Clearly, these values are very close to those obtained from the given computer program. Further details on this problem and tabulated results on the area under the frequency distribution curve may be obtained from any statistics textbook.

$x_{min} = 0.0000$
Integral = 0.47725 $x_{max} = 10.0000$
Integral = 0.818595 $x_{max} = 15.0000$
Integral = 0.9545 $x_{max} = 20.0000$
Integral = 0.9759 $x_{max} = 25.0000$
Integral = 0.977218 $x_{max} = 30.0000$
Integral = 0.97725 $x_{max} = 35.0000$
Integral = 0.97725 $x_{max} = 40.0000$

$x_{min} = 5.0000$
Integral = 0.682689 $x_{max} = 15.0000$
Integral = 0.818595 $x_{max} = 20.0000$
Integral = 0.839995 $x_{max} = 25.0000$
Integral = 0.841313 $x_{max} = 30.0000$
Integral = 0.841344 $x_{max} = 35.0000$
Integral = 0.841345 $x_{max} = 40.0000$

$x_{min} = 10.0000$
Integral = 0.47725 $x_{max} = 20.0000$
Integral = 0.49865 $x_{max} = 25.0000$
Integral = 0.499968 $x_{max} = 30.0000$
Integral = 0.5 $x_{max} = 35.0000$
Integral = 0.5 $x_{max} = 40.0000$

$x_{min} = 15.0000$
Integral = 0.157305 $x_{max} = 25.0000$
Integral = 0.158624 $x_{max} = 30.0000$
Integral = 0.158655 $x_{max} = 35.0000$
Integral = 0.158655 $x_{max} = 40.0000$

FIGURE 8.17 Numerical results obtained from the integration of the Gaussian distribution from $x = x_{min}$ to $x = \infty$, for Example 8.6, at several values of x_{min}.

8.6.3 Multiple Integrals

There are a few other forms of integrals that have not been discussed thus far. Among these are multiple integrals, which arise for functions that depend on more than one independent variable. Integrals over the surface area, for example, to compute the evaporation from a pond, or over the volume of a body involve multiple integrals. Similarly, the work done in the two-dimensional motion of a particle on a flat surface and the total force acting on a vertical surface, such as a building, require double integrals of the form

$$I = \int_{a}^{b} \int_{g(x)}^{h(x)} f(x, y)\mathrm{d}y\,\mathrm{d}x \tag{8.86}$$

We evaluate such multiple integrals by twice applying the numerical methods discussed in this chapter, first for the inner integral over y and then for the outer integral over x. If the range of integration $a \le x \le b$ is divided into n segments, so that $x_i = a + i\Delta x$, then we may define a function $F(x)$ as

$$F(x_i) = \int_{g(x_i)}^{h(x_i)} f(x_i, y)\,\mathrm{d}y \tag{8.87}$$

so that x is held constant at x_i. This integral may be computed as a one-dimensional integral on y. Thus, $(n + 1)$ ordinates, corresponding to $F(x_i)$, with $i = 0, 1, \ldots, n$, are generated. Since these ordinates are at evenly spaced points, we can employ Simpson's one-third rule to evaluate the integral I, provided n is even. Similarly, other numerical methods may be employed.

8.7 SUMMARY

This chapter presents several available methods for the numerical integration of a given continuous function $f(x)$ over a finite range of the independent variable x. These methods, which include the rectangular, trapezoidal, Simpson's one-third, and Simpson's three-eighths rules for numerical integration, form the first four orders of the Newton–Cotes formulas. They are discussed in detail in this chapter, particularly the trapezoidal and Simpson's one-third rules, because of their wide usage. The TEs associated with these formulas are also derived. Simpson's one-third rule is a very popular choice in engineering problems, since it is fourth-order accurate, as compared to the trapezoidal rule which is second-order accurate. Also, when it is used in conjunction with the three-eighths rule, it imposes no constraints on the choice of the number n of the segments, or subintervals, of the integration region, except that n be two or larger. The trapezoidal rule is also widely used because of its simplicity. Higher-order Newton–Cotes formulas are also presented, although they are used only if a very high level of accuracy is needed. The accuracy of the numerical results can also be improved by a reduction in the step size Δx. However, at very small values of Δx, the round-off error may become significant. Higher-order formulas can then be employed more advantageously. MATLAB integration commands that may be used directly are also presented.

Various methods for the successive improvement in accuracy are also discussed, including Richardson's extrapolation, which is applicable to several other numerical procedures as well, and Romberg integration, which is a very efficient method for achieving any desired accuracy level. Romberg integration is based on the trapezoidal rule and uses a procedure similar to Richardson's extrapolation to successively eliminate the higher-order terms in the TE. It is presently one of the most widely used methods for the numerical integration of well-behaved functions.

Gauss quadrature uses the minimum number of function evaluations for computing the integral and is therefore particularly suitable for very complicated functions. Different formulas can be derived for a wide variety of functions and integration limits so that the results are very accurate and the number of function evaluations is minimized. Singularities can also be effectively dealt with in several cases. However, the tables of the weight factors and the zeros must be stored or computed. The programming is more involved and less versatile than that for, say, Simpson's rule. Also, this method is generally not applicable for data available at arbitrarily spaced values of the independent variable.

For unevenly spaced data points, one can use the trapezoidal and Simpson's rules, using the latter if two or three adjacent segments are of the same width and the trapezoidal rule if the widths of adjacent segments are unequal. It is also possible to use curve fitting to obtain a continuous function to represent the data. Then the standard methods for numerical integration may be used. By focusing on regions where the accuracy is less than that in others and retaining fewer points in regions where the desired accuracy level has been attained, one can effectively use the method of adaptive quadrature for functions that are small in magnitude or slowly varying in certain regions.

This chapter also discusses the various methods for treating improper integrals, which exist and are finite although either the integrand blows up within the range of integration or the integration limits are infinite. Analytical procedures, such as transformation of the independent variable and integration by parts, can often be employed to eliminate the singularity. However, the most common approach is to replace the integration limit that is infinite or where the integrand is singular by a quantity that is large or close to, but not equal to, the limit. Numerical integration is then carried out and this quantity is varied until the numerical results are not significantly affected by a further variation. Although inefficient, this approach is applicable to most engineering problems involving improper integrals. Gauss quadrature can also be used advantageously for certain types of functions.

PROBLEMS

8.1. Consider the integral $\int_0^3 f(x)\mathrm{d}x$ for the linear function $f(x) = 3 + 5x$. Show that the numerical results obtained by use of the trapezoidal and Simpson's one-third rules for this integral are exactly equal to the analytical value, except for the round-off error. Employ two and then four subdivisions of the integration domain.

8.2. For the parabola $f(x) = 3 + 2x + 3x^2$, show that the numerical integration $\int_1^3 f(x)dx$ by Simpson's rule yields the exact analytical value, except for the round-off error. What effect would you expect an increase in the number of subdivisions to have on the accuracy of the numerical results? Explain.

8.3. Consider the integral $\int_0^2 f(x)dx$, where $f(x) = 3 + 5x + 2x^2 + x^3$. Estimate the total TE and the maximum TE per step for evaluating this integral by the rectangular and trapezoidal rules. Consider the three segment sizes $\Delta x = 0.1, 0.2$, and 0.5.

8.4. Calculate the TE per step at $x = 0.5$ and $x = 1.0$ for the integral $\int_0^{1.5}(x^5 + 2x^4 + x^3 + 4x^2 + 2x + 6)$, taking $\Delta x = 0.1$, 0.25, and 0.5. Consider the trapezoidal rule and both Simpson's rules for numerical integration. Discuss the effect, on the error, of a reduction in segment size and also of the numerical method employed.

8.5. The TE in the numerical evaluation of the integral $I = \int_a^b f(x)dx$ by the trapezoidal rule is given by Equation 8.22. If f''_{av} is approximated as $[f'(b) - f'(a)/(b - a)$, obtain the resulting estimate of the error, and add it to the formula for numerical integration by the trapezoidal rule to obtain a more accurate scheme known as the *trapezoidal rule with end correction*.

8.6. Apply the procedure outlined in Problem 8.5 to the rectangular rule, and compare the resulting formula with that for the trapezoidal rule.

8.7. Show that if Richardson's extrapolation is applied to the trapezoidal rule, the formula obtained is the same as that for Simpson's one-third rule. Also apply Richardson's extrapolation to the rectangular rule, and discuss the resulting formula for numerical integration.

8.8. The temperature T at the wall of a furnace varies periodically over the day as

$$T(t) = 125 + 50\sin\frac{2\pi}{24}(t - 6)$$

where t is the time in hours measured from midnight and T is in °C. The ambient temperature T_a is 25°C, and the surface area A of the wall is 10 m². If the heat transfer coefficient h is given as 20 W/m²°C, the heat transfer from the wall is given by $\int[T(t) - T_a]hA\,dt$. Using the trapezoidal rule, compute this integral as accurately as possible for the time interval $t = 6$ to $t = 12$. Also evaluate the integral analytically and compare the result with the computed value. Use the *quad* function in MATLAB to verify the results obtained.

8.9. In chemical engineering, we frequently need to evaluate the amount of heat required to raise the temperature of a given material from a value T_1 to T_2. If $C(T)$ is the specific heat of the material, the amount of energy needed is $\int_{T_1}^{T_2} mC(T)dT$, where m is the mass of the material, since the specific heat is the energy required to raise the temperature of unit mass of the material by unit temperature. The average specific heat C_{av} is given by $\left[\int_{T_1}^{T_2} C(T)dT\right]/(T_2 - T_1)$. The specific heat of a material is given, in J/kg K, as follows:

$$C(T) = 200 + 7.5\frac{T}{T_0} + 2.8\left(\frac{T}{T_0}\right)^2 + 0.42\left(\frac{T}{T_0}\right)^3$$

where T_0 is the reference temperature of 100 K. For 1 kg of the material, compute the total energy, in joules, needed to raise the temperature from 100 to 1000 K. Also determine the average specific heat over this temperature range. Use the trapezoidal rule, and reduce the segment size, starting with 100 K, until the results remain essentially unchanged with further reduction.

8.10. The pressure p on a 10 m high structure due to the wind is given by the expression

$$p(x) = \frac{150x}{1+e^x}$$

where x is measured in meters from the bottom of the structure and the pressure is in N/m^2. If the structure is 2 m wide, the total force due to wind is given by the integral

$$\int_0^{10} 2 \cdot \left(\frac{150x}{1+e^x} \right) dx$$

Compute this integral as accurately as possible by the trapezoidal rule. Also, use the *quad* function in MATLAB and compare the result with that obtained earlier.

8.11. Consider the Gaussian error function *erf* z defined in Example 8.3. Write a script-m file to do the following:

a. Calculate the integral using the trapezoidal rule with 20 subdivisions of the integration domain. You may use the available function-*m* file for the trapezoidal rule.
b. Vary z from 1 to 2. Use a *For ... End* loop to calculate the integral with $z = 1, 1.1, 1.2, \ldots, 2.0$.
c. Output the values of the integral for $z = 1.0$ and 2.0.
d. Using the values of the integral calculated in Part b, plot the integral versus z.

8.12. Consider the expression for blackbody radiation given by Equation 4.62. The integral of this expression over all wavelengths, that is, $\int_0^\infty E_{b\lambda} d\lambda$, gives the total energy radiated by a blackbody per unit area and time. Using Simpson's rule, compute this integral at $T = 1000$ K as accurately as possible. The analytical result is given in the literature as σT^4, where σ is known as the *Stefan-Boltzmann constant* and has a value of 5.67×10^{-8} W/m^2 K^4. Compare your numerical result with the analytical value at 1000 K.

8.13. Using Simpson's rule, repeat the problem given in Example 8.1, and compare the results obtained with those given in the example. Discuss the observed differences between the trapezoidal and Simpson's rules for this problem.

8.14. Using Simpson's three-eighths rule, compute the total momentum flow in the problem outlined in Example 8.2. The momentum flow is given by the integral $\int_0^R \rho[U(x)]^2 2\pi x \, dx$, where ρ is the fluid density, given as 1 kg/m^3 for the fluid considered. Also, calculate the integral using the *quad* function and compare the result with that obtained earlier.

8.15. The RMS value of an electric current $I(t)$, where I varies periodically with time t, is given by the expression

$$I_{\text{RMS}} = \frac{1}{t_c}\sqrt{\int_0^{t_c} I^2 \, dt} \, .$$

where t_c is the time period for one cycle in the variation of $I(t)$. If $I(t)$ is given as $5e^{-t} \sin 4\pi t$, with $t_c = 0.5$ s, compute the RMS value, using Simpson's rule.

8.16. The force $F(x)$ exerted per centimeter on a vertical plate immersed in flowing water is given by the expression

$$F(x) = 1.5x^3e^{-x}$$

where x is measured from the top of the plate and $F(x)$ is in N/cm. If the plate is 10 cm high, the total force F_T, in Newtons, on the plate is given by

$$F_T = \int_0^{10} F(x)\mathrm{d}x$$

Employing Romberg integration, compute F_T to a convergence criterion of 10^{-4}.

8.17. For the preceding problem, write a script-*m* file to do the following:

a. Calculate the integral using the trapezoidal rule with 1, 2, 4, 8, 16, 32, 64, 128, and 256 subdivisions of the integration domain, in sequence. You may use the available function-*m* file for the trapezoidal rule.

b. Print the results for the nine cases.

c. Richardson extrapolation gives

$$S(h) = \frac{4T(h) - T(2h)}{3}$$

where $T(h)$ is the integral from the trapezoidal rule for step size h and $T(2h)$ that for step size $2h$ and S is the result from Simpson's rule. Using this equation and results from Part a, calculate the Simpson's rule results for the different subdivisions considered.

d. Print the results obtained.

e. From the results, how many subdivisions are satisfactory for the trapezoidal and Simpson's rules?

8.18. The meniscus of a liquid film supported by surface tension can often be represented as

$$h(x) = Ae^{-a^2x^2}$$

where $h(x)$ is the height as a function of horizontal distance x and A and a are constants. The total volume of liquid supported by surface tension is then given by the integral $W\int_0^L h(x)\,\mathrm{d}x$, where W is the width of the meniscus and L is its length. If W, L, h, and x are all in centimeters, compute this volume for $A = 0.8$, $a = 2.0$, $W = 1$, and $L = 1$ cm, using Romberg integration.

8.19. Calculate the integral in the preceding problem using Simpson's rule. Also use the *quad* command in MATLAB to get the integral and compare the results. Do you expect the two-point Gauss–Legendre method to give accurate results for this problem? Explain your answer.

8.20. The velocity $v(t)$ of a moving particle is given as $v(t) = 5(1 - e^{-t/10})$. Using Romberg integration, compute the total distance S traveled by the particle from $t = 10$ s to $t = 20$ s. Note that the distance traveled between $t = t_1$ and $t = t_2$ is simply given by the integral $\int_{t_1}^{t_2} v(t)\mathrm{d}t$.

8.21. We wish to evaluate the integral $\int_0^{6\pi} \sin^2 \mathrm{d}x$ by means of Romberg integration. Are any difficulties encountered in the application of Romberg integration to this problem? If so, suggest methods to overcome them.

8.22. Using Romberg integration, repeat the problem given in Example 8.2. Compare the results obtained with those given in the example, and comment on the numerical accuracy and the computational effort involved in the two methods used for this problem.

8.23. The velocity v of a moving body is measured at several time intervals t and is tabulated as follows:

t (s)	0	1	2	3	5	7	8	10
v (m/s)	10	11.5	14.8	21.1	47.5	100.3	139.6	250.0

Using this uneven distribution of data points, compute the distance traveled x as a function of time t, and find the total distance traveled by the body in 10 s.

8.24. The pressure p in a gas being compressed by a moving piston is measured at various positions x of the piston, where x is measured from the starting position of the piston. The data are tabulated as follows:

x (m)	0	0.1	0.2	0.4	0.5	0.8	0.9	1.0
p (N/m^2)	4.0	4.23	4.53	5.34	5.88	8.03	8.96	10.0

The total work done over this distance of 1 m by the piston is given by the integral $A\int_0^1 p(x)\mathrm{d}x$, where A is the cross-sectional area of the piston. Compute this integral, taking $A = 0.5$ m^2.

8.25. The fluid velocity V is measured at several radial locations r for flow in a circular pipe of radius 1 cm. The velocities in cm/s are tabulated as follows:

r (cm)	0	0.2	0.5	0.6	0.8	0.9	1.0
V (cm/s)	1.0	0.96	0.75	0.64	0.36	0.19	0.0

The volume flow in the pipe is given by the integral $\int_0^R V(r)2\pi r\,\mathrm{d}r$, where R is the radius of the pipe. Using the data given, compute this integral.

8.26. The turbulent flow in a pipe of diameter 0.2 m is given by the velocity distribution

$$V = U\left(1 - \frac{r}{R}\right)^{1/6}$$

where U is the velocity at the axis, r is the radial distance from the axis, and R is the radius of the pipe. Using the four-point Gauss–Legendre integration formula, compute the volume flow rate in the pipe. Take $U = 2$ m/s. Verify the result obtained by using the *quad* function. Also, see if an analytical result can be obtained by using symbolic algebra in MATLAB.

8.27. Using the two-point, as well as the four-point, Gauss–Legendre integration scheme, solve Problem 8.18. Then compare the results obtained and the computational effort involved for the two methods, Romberg integration and Gauss quadrature, considered for this problem.

8.28. Using the trapezoidal rule, compute the value of the improper integral

$$\int_0^\infty \frac{7.5\,\mathrm{d}x}{\sqrt{x}(1+x)}$$

as accurately as possible.

8.29. Using Simpson's rule, compute the improper integral

$$\int_0^\infty \frac{2\,\mathrm{d}x}{1+\mathrm{e}^{-x}+x^2}$$

as accurately as possible.

8.30. Using any convenient integration scheme, determine the integral

$$\int_0^\infty \frac{\mathrm{d}x}{\mathrm{e}^{x}+\mathrm{e}^{-x}}$$

and compare the numerical value obtained with the analytical result of $\pi/4$.

8.31. An integral commonly encountered in the estimation of the moisture lost at the surface of a wet body, such as paper or cloth, is of the form $\int_0^1 x^{-1/4}\,\mathrm{d}x$, which is singular at $x = 0$. Compute this integral by varying the lower limit, starting with 0.1, and then reducing it to values as small as needed to make the neglected area under the curve negligible. Compare the numerical result obtained with the analytical value obtained by using mathematics as well as symbolic algebra in MATLAB.

8.32. Determine the integral

$$\int_0^1 \frac{5(1+x^2)}{x^{0.2}}\,\mathrm{d}x$$

which arises in a mass transfer calculation for a chemical engineering process. Employ any suitable method.

8.33. The work done W by a force $F(x, y)$ in a two-dimensional dynamics problem is given by the integral

$$W = \int_0^1\int_0^2 \left(x^2 - 2xy + 3xy^2 + y^3\right)\mathrm{d}y\,\mathrm{d}x$$

Using the trapezoidal rule as outlined in Equations 8.86 and 8.87, calculate this integral.

8.34. The area A enclosed by a curve on a plane is given by the integral

$$A = \int_0^2 \int_y^{2y} \left(x^2 + y^2\right) \mathrm{d}x\, \mathrm{d}y$$

Using the trapezoidal rule, compute this integral.

8.35. Calculate the integral $\int_0^1 \mathrm{e}^x\, \mathrm{d}x$ by using four important methods for numerical integration, namely, the trapezoidal rule, Simpson's one-third rule, Romberg integration, and Gauss quadrature, as well as the *quad* function. Compare the results obtained and the computational efforts involved in all of these methods. Comment on the conclusions that may be drawn from such a comparison.

9 Numerical Solution of Ordinary Differential Equations

9.1 INTRODUCTION

Ordinary differential equations, or ODEs, which are equations that consist of functions of a single independent variable and their derivatives, arise in many diverse engineering problems. Several physical laws, such as those concerned with the transport of mass, momentum, and energy, are expressed in terms of differential equations. In many cases, only one independent variable, such as time or distance, exists in the problem, because of the nature of the problem or because of simplifications and approximations made with respect to the other variables. Also, in a few circumstances, the functional dependence on two or more independent variables can be expressed, by suitable transformations, in terms of a single variable. Therefore, many problems of engineering interest are described by ODEs. These equations arise, for instance, in heat and mass transfer, dynamics of particles, vibrations of systems, electrical circuitry, and chemical kinetics. Although analytical methods may be employed for the solution of some ODEs, numerical techniques are generally needed for most of the equations that arise in engineering applications.

A general ODE may be written in terms of the independent variable x and the dependent function $y(x)$ as

$$f\left(x, y, \frac{dy}{dx}, \frac{d^2y}{dx^2}, \ldots, \frac{d^ny}{dx^n}\right) = 0 \tag{9.1}$$

where the highest derivative is of order n. Then the equation is known as an *nth-order ODE*. The highest-order derivative is often separated, to obtain the preceding equation as

$$\frac{d^ny}{dx^n} = F\left(x, y, \frac{dy}{dx}, \ldots, \frac{d^{n-1}y}{dx^{n-1}}\right) \tag{9.2}$$

A function $y(x)$ that satisfies this equation is said to be a *solution* of the equation. There may be many functions $y(x)$ that satisfy a given differential equation. To obtain a unique solution, which is obviously needed in a physical problem, n conditions on

$y(x)$ and/or on its derivatives must be given at specified values of x. These conditions must be independent, that is, one condition may not be derived from a combination of the others.

9.1.1 Initial and Boundary Value Problems

If all of the n conditions are specified at the same value of x, say, $x = x_1$, then the problem represented by the ODE and the given conditions is termed an *initial-value problem*. If the conditions are specified at more than one value of the independent variable x, the problem is termed as *boundary-value problem* (*BVP*). In the former case, there is a definite starting point, and one can obtain the solution by varying x in order to move outward from this starting point. An example of an initial-value problem was given by Equation 1.1, which is of the form $dy/dx = Ay + B$, where A and B are constants, and by the initial condition $y = y_0$ at $x = x_0$. This equation governs, for instance, the temperature of a small, heated metal sphere being cooled by a stream of cold air and the charge in the capacitor which forms part of the electrical circuit shown in Figure 1.2. Boundary value problems (BVPs) are more involved, since conditions specified at different values of x are to be satisfied. Because at least two conditions are needed for specification at different x values, the ODE must be at least of second order for a BVP. Several of the methods employed for BVPs are based on those for initial-value problems, often employing the root-solving procedures of Chapter 5 to satisfy the given boundary conditions.

9.1.2 Reduction of Higher-Order Equations to First-Order Equations

The nth-order equation, given by Equation 9.2, can be reduced to a system of n first-order equations by defining $(n–1)$ new variables Y_i, where $i = 1, 2, \ldots, (n–1)$, as follows:

$$
\begin{aligned}
Y_1 &= \frac{dy}{dx} \\
Y_2 &= \frac{d^2y}{dx^2} \\
&\vdots \\
Y_{n-1} &= \frac{d^{n-1}y}{dx^{n-1}}
\end{aligned}
\tag{9.3}
$$

Then, the given ODE may be written as the following n first-order equations:

$$
\begin{aligned}
\frac{dy}{dx} &= Y_1 \\
\frac{dY_{i-1}}{dx} &= Y_i : \quad \text{where } i = 2,3,\ldots,(n-1) \\
\frac{dY_{n-1}}{dx} &= F\left(x, y, Y_1, Y_2, \ldots, Y_{n-1}\right)
\end{aligned}
\tag{9.4}
$$

As an example of the above procedure, consider the following third-order, nonlinear, equation that governs the flow over a two-dimensional wedge:

$$\frac{d^3 f}{dx^3} + f\frac{d^2 f}{dx^2} + \beta\left[1 - \left(\frac{df}{dx}\right)^2\right] = 0 \tag{9.5}$$

where x is the dimensionless distance from the wedge surface, f is the nondimensional stream function, which is related to the velocity field, and β is a constant that gives the wedge angle $\beta\pi$ in radians. Figure 9.1 gives a graphical representation of this physical problem. Equation 9.5, which is typical of several fluid flow circumstances encountered in aeronautical, chemical, and mechanical engineering, may be written as three first-order equations by defining two variables, F_1 and F_2, as

$$F_1 = \frac{df}{dx}, \quad F_2 = \frac{d^2 f}{dx^2} \tag{9.6}$$

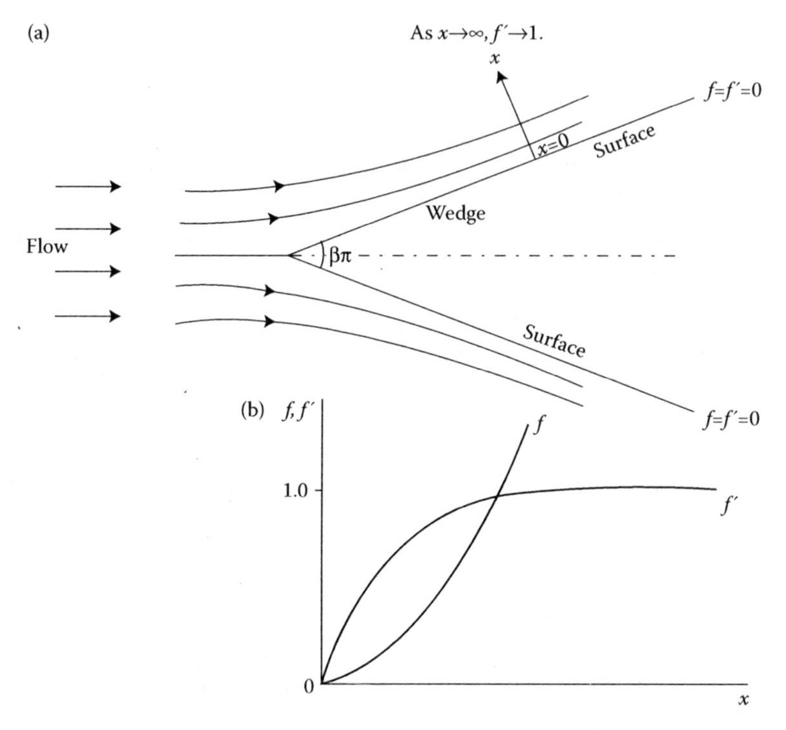

FIGURE 9.1 Graphical representation of the physical problem governed by Equation 9.5. (a) Sketch of the flow over a two-dimensional wedge; (b) qualitative sketch of the variation of the functions f and f' with x.

to yield

$$\frac{df}{dx} = F_1$$
$$\frac{dF_1}{dx} = F_2 \tag{9.7}$$
$$\frac{dF_2}{dx} = -f\ F_2 - \beta\left(1 - F_1^2\right)$$

The velocity component parallel to the wedge surfaces is given by $F_1 = df/dx$. Figure 9.1b shows, qualitatively, the distributions of f and F_1. This problem is considered again in greater detail later. Therefore, a given nth-order ODE can generally be reduced to a system of n first-order equations and we can focus our attention on the solution of first-order equations.

A first-order ODE may be written in the form

$$\frac{dy}{dx} = F\left(x, y\right) \tag{9.8}$$

The methods used for solving a system of first-order equations are based on those for a single equation, and, since most higher-order equations can generally be reduced to a system of first-order equations, most of the available methods are directed at solving a single first-order equation, given by Equation 9.8. In many problems of engineering interest, the differential equations obtained are nonlinear, with the dependent variable and its derivatives appearing as nonlinear functions in the equation. Equation 9.5 is an example of a nonlinear ODE. However, if the equation can be written in the form given by Equation 9.2, the solution procedure is essentially the same for linear and nonlinear equations. Still, nonlinear problems generally involve a greater computational effort and, in BVPs, may lead to convergence difficulties. For linear equations, one can often use the superposition of solutions to simplify the computational scheme. Linear, homogeneous, BVPs arise in some engineering applications, such as the natural vibration of systems. These situations lead to eigenvalue problems which often require special solution techniques.

In view of the above discussion, ODEs may be classified as first-order or higher-order, single equation or system of equations, initial-value or boundary-value, linear or nonlinear, and homogeneous or inhomogeneous. Although there are often large differences in the analytical solution of these different types of equations, the applicable numerical procedures are quite similar. However, the classification of the problem as initial-value or boundary-value is important, since different techniques for solving the equation or for satisfying the boundary conditions are generally needed. The solution of a single first-order equation is particularly important since it forms the basis for solving other types of equations.

Analytical solutions of ODEs may be obtained in a few simple cases, particularly for linear equations. As discussed in Chapter 1, analytical results, whenever available, are useful in the validation and testing of the numerical scheme, which may be first employed for the simple problem whose analytical solution is known. Once the procedure has been tested for correctness and accuracy, one may proceed to the solution of more involved problems that cannot be solved analytically. If no relevant analytical results are available, one considers the numerical results obtained in terms of the physical or basic nature of the problem to determine whether the results follow expected trends. In several engineering problems, some experimental data may be available on the problem being solved numerically and may be employed for the validation of the method and the numerical results.

9.1.3 Solution Methods

Several methods are available for the solution of ODEs. Although each method has its particular advantages over other methods, and also certain disadvantages, many numerical methods are generally applicable to a given problem, and the choice of the method frequently becomes a matter of personal preference. Generally, one solves higher-order equations by reducing them to a system of first-order equations, as outlined above. BVPs are often solved by *shooting methods*, which are based on the methods applicable for initial-value problems. Consider, for example, the problem shown in Figure 9.1 and governed by Equation 9.5. This is a BVP, with the boundary conditions given as follows:

$$\begin{aligned} &\text{At } x = 0: \quad f = 0 \text{ and } F_1 = 0 \\ &\text{As } x \to \infty: \quad F_1 \to 1 \end{aligned} \tag{9.9}$$

Therefore, the conditions are specified at two values of the independent variable x. If F_2, or d^2f/dx^2, is given at $x = 0$ instead of the third condition, an initial-value problem will be obtained. Therefore, F_2 at $x = 0$ may be guessed, the equation solved as an initial-value problem, and a correction scheme employed to iteratively vary the guessed value of F_2 until the third condition in Equation 9.9 is satisfied to a desired tolerance level. Such an approach is known as a shooting method and is frequently employed. Therefore, much of the discussion in this chapter is directed at the first-order initial-value problem, followed by a consideration of other problems and techniques, particularly the finite difference methods that can be used to solve BVPs directly.

There are mainly two types of methods available for solving the first-order initial-value problem given by Equation 9.8. In the first case, the desired solution at a given value of x is obtained in terms of the function $F(x, y)$ evaluated at various x values between x and $x-\Delta x$, where Δx is the chosen increment in x. The values for $x < x-\Delta x$ are not needed, and the methods are, therefore, *self-starting*, since only the initial condition is needed to obtain the solution at the next step, $x = \Delta x$. *Euler's* method and the *Runge–Kutta* methods fall in this category. The methods that constitute the second category require information at values of x less than $x-\Delta x$ and are not self-starting. These are known as *multistep methods* and require other methods to yield the solution for the first few steps beyond the initial condition. Included in this category are *Adams multistep formulas* and the *predictor–corrector methods*, such as *Hamming's* and *Milne's* methods. The multistep methods are among the most efficient numerical techniques available for solving ODEs.

This chapter discusses various methods that may be employed for solving first-order initial-value problems, outlining the important advantages and limitations of each method. The solution of a system of ODEs is considered next. The solution of BVPs is considered in terms of shooting methods, using the techniques for initial-value problems, and in terms of finite difference methods. The techniques applicable for eigenvalue problems are also discussed.

9.2 EULER'S METHOD

Let us consider the solution of the first-order ODE

$$\frac{dy}{dx} = F(x, y) \tag{9.8}$$

with the initial condition

$$y(x_0) = y_0 \tag{9.10}$$

where y_0 is the value of $y(x)$ at a given value of the independent variable, $x = x_0$. A numerical solution of this differential equation involves obtaining the numerical values of the function $y(x)$ at discrete values of x, termed *node points*, for $x > x_0$. If Δx represents a uniform step size, that is, a constant difference between successive values of x at which the numerical solution is to be obtained, the node points x_i are defined by

$$x_i = x_0 + i\Delta x, \quad \text{where } i = 0,1,2,\ldots \tag{9.11}$$

The numerical values of the solution at these points may be denoted by $y_0, y_1, \ldots, y_n, \ldots$. Therefore, the numerical scheme must provide a means of evaluating y_{i+1} from the given or computed solution at the preceding grid points. If interest lies in determining the solution for $x < x_0$, instead of $x > x_0$, x_i may be taken as $x_i = x_0 - i\Delta x$, or a simple transformation of the independent variable may be employed to yield a new variable that increases as x decreases. Therefore, we shall consider only the case of increasing x here.

9.2.1 Computational Formula and Physical Interpretation of the Method

Euler's method is one of the simplest methods available for solving ODEs. However, it is very seldom used since several more efficient methods are available. The main reason for studying this method is that it is simple and allows a consideration of many of the basic features of the numerical solution of ODEs without the additional complexity of other methods. The computational formula for solving Equation 9.8 by Euler's method is

$$y_{i+1} = y_i + \Delta x\, F(x_i, y_i) \quad \text{with } i = 0,1,2,\ldots \tag{9.12}$$

Therefore, the solution can be obtained for increasing x, starting with $x = x_0$. This implies that the method is self-starting. Figure 9.2 shows the geometric interpretation of Euler's method. The exact solution is denoted by $y(x)$, and a qualitative comparison between the computed results and the exact solution is shown. The tangent to the curve at $x = x_i$ has a slope of $F(x_i, y_i)$ and approximates the true curve for $x_i \le x \le x_{i+1}$. At the initial point, $x = x_0$, the line tangent to the graph of $y(x)$, as shown, approximates the numerical solution for $0 < x < \Delta x$. As x increases, the numerical results increasingly deviate from the exact solution, due to accumulation of error.

There are several other ways of interpreting Euler's method. If the function $y(x)$ is assumed to be analytic near x_i, it may be expanded in a Taylor series, using Equation 4.2, as follows:

$$y_{i+1} = y_i + \Delta x \left.\frac{dy}{dx}\right|_{x_i} + \frac{(\Delta x)^2}{2} \left.\frac{d^2 y}{dx^2}\right|_{\xi_i}, \quad \text{where } x_i < \xi_i < x_{i+1} \tag{9.13}$$

(a)

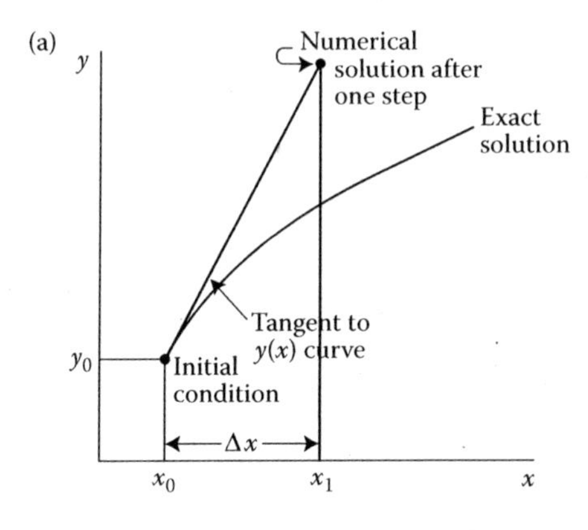

(b)

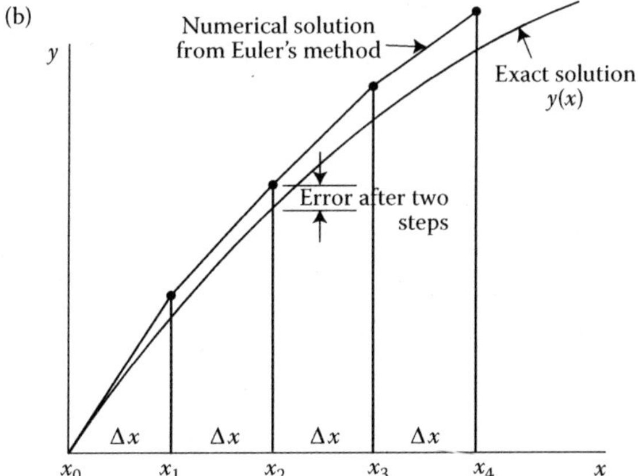

FIGURE 9.2 Graphical interpretation of Euler's method. (a) Numerical solution and error after the first step; (b) accumulation of error with increasing independent variable x.

Euler's method is obtained if the last term, which then becomes the TE for this computational step, is dropped. Therefore, this formulation for deriving Euler's method allows a determination of the error, as discussed in detail later in this section.

We may also use numerical differentiation, with a forward difference approximation for the derivative, to represent Equation 9.8 as

$$F(x_i, y_i) = \left.\frac{dy}{dx}\right|_{x_i} \cong \frac{y_{i+1} - y_i}{\Delta x} \tag{9.14}$$

This equation also gives the formula for Euler's method. Similarly, numerical integration may be applied to the given differential equation to give

$$y_{i+1} = y_i + \int_{x_i}^{x_{i+1}} F(x, y)\, dx \tag{9.15}$$

Therefore, the change in y is represented by the area under the $F(x, y)$ curve. One can obtain an approximation to the integral by taking $F(x, y)$ as constant over the interval. This is the rectangular rule for numerical integration, as discussed in the preceding chapter. Thus,

$$\int_{x_i}^{x_{i+1}} F(x, y)\, dx \cong (x_{i+1} - x_i) F(x_i, y_i) = \Delta x F(x_i, y_i)$$

which gives

$$y_{i+1} = y_i + \Delta x\, F(x_i, y_i) \tag{9.12}$$

Figure 9.3 shows a few steps of this numerical integration to obtain the function $y(x)$.

Both the Taylor-series formulation and the numerical integration procedure can be employed to generate more accurate methods. The former leads to single-step methods such as the Runge–Kutta formulas, and the latter to multistep methods, particularly the predictor–corrector methods. Euler's method does not yield a high level of accuracy in the solution and is, therefore, rarely used. However, because the method is so simple, it is used in some engineering applications to obtain an initial estimate of the physical variables, by solving the governing differential equations with a relatively small step size Δx. The method also serves to illustrate the basic considerations that arise in the numerical solution of ODEs.

9.2.2 Solution of a System of Equations

Euler's method may easily be extended to yield a solution of a system of first-order equations. Consider the following system of three equations:

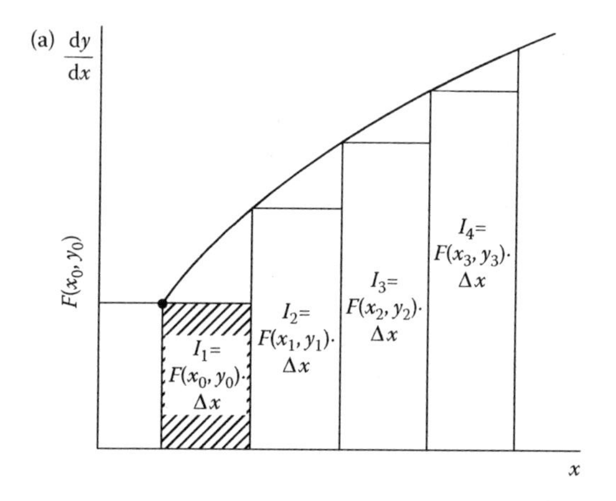

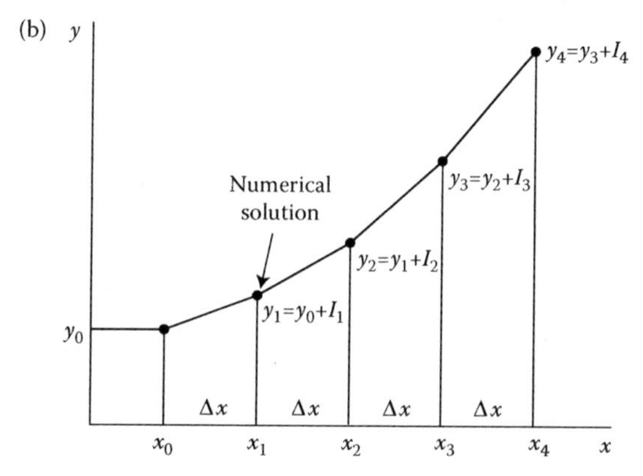

FIGURE 9.3 Sketch of a few steps in the numerical integration of the differential equation by Euler's method to yield the numerical solution $y(x)$.

$$
\begin{aligned}
\frac{dY_1}{dx} &= F_1(x, Y_1, Y_2, Y_3) \text{ with } Y_1(x_0) = Y_{1,0} \\
\frac{dY_2}{dx} &= F_2(x, Y_1, Y_2, Y_3) \text{ with } Y_2(x_0) = Y_{2,0} \\
\frac{dY_3}{dx} &= F_3(x, Y_1, Y_2, Y_3) \text{ with } Y_3(x_0) = Y_{3,0}
\end{aligned}
\tag{9.16}
$$

where Y_1, Y_2, and Y_3 are three dependent variables whose values are given at $x = x_0$. We obtain the numerical solution of these equations from Euler's method by employing the computational formulas

$$
\begin{aligned}
Y_{1,i+1} &= Y_{1,i} + \Delta x \; F_1(x_i, Y_{1,i}, Y_{2,i}, Y_{3,i}) \\
Y_{2,i+1} &= Y_{1,i} + \Delta x \; F_2(x_i, Y_{1,i}, Y_{2,i}, Y_{3,i}) \\
Y_{3,i+1} &= Y_{3,i} + \Delta x \; F_3(x_i, Y_{1,i}, Y_{2,i}, Y_{3,i})
\end{aligned}
\tag{9.17}
$$

Thus, we obtain the numerical solution by proceeding in the direction of increasing x, starting with the initial conditions at $x = x_0$, and successively calculating the three independent variables Y_1, Y_2, and Y_3 at each step.

9.2.3 Errors, Convergence, and Stability

It is important to examine the errors associated with the numerical solution of a differential equation in order to determine the accuracy of the results obtained. As discussed in Chapter 2, several types of errors arise in numerical computation. Among the most important of these are the round-off and TEs. The round-off error arises due to the retention of a finite number of significant figures by the computer. The round-off error is, therefore, a function of the computer and may be reduced by the use of double precision in the computation. The TE arises due to the approximation of a function by a finite number of terms in the infinite series that represents the function. The series is generally truncated after a few terms to develop the scheme for the numerical solution of the differential equation. Therefore, the dropping of the remaining terms leads to the TE. A very important aspect in the error analysis of numerical methods is the growth or accumulation of errors as computation progresses, since this consideration is related to the stability of the scheme, as discussed below.

To find the TE in Euler's method, let us assume that the exact solution to the differential equation, $y(x_i)$, is known at x_i and is employed in Equation 9.12 to compute the solution at x_{i+1}. Then

$$y_{i+1} = y(x_i) + \Delta x\, F\left[x_i, y(x_i)\right] \tag{9.18}$$

If the exact solution is analytic near x_i, we may represent it by a Taylor series as follows:

$$y(x_i + \Delta x) = y(x_i) + \Delta x \left.\frac{\mathrm{d}y}{\mathrm{d}x}\right|_{x_i} + \frac{(\Delta x)^2}{2}\left.\frac{\mathrm{d}^2 y}{\mathrm{d}x^2}\right|_{x_i} + \frac{(\Delta x)^3}{3!}\left.\frac{\mathrm{d}^3 y}{\mathrm{d}x^3}\right|_{x_i} + \cdots \tag{9.19}$$

From Equations 9.18 and 9.19,

$$y(x_i + \Delta x) - y_{i+1} = \frac{(\Delta x)^2}{2}\left.\frac{\mathrm{d}^2 y}{\mathrm{d}x^2}\right|_{x_i} + O\left[(\Delta x)^3\right] \tag{9.20}$$

since $(\mathrm{d}y/\mathrm{d}x)_{x_i} = F[x_i, y(x_i)]$. Here, $y(x_i + \Delta x) - y_{i+i}$ is the *TE* from x_i to x_{i+1}, starting with the exact solution at x_i. The leading term of the error is of the order of $(\Delta x)^2$ and may be denoted as $O[(\Delta x)^2]$.

Equation 9.20 gives the TE per step in Euler's formula. However, this is not the total error in the numerical solution at x_{i+1}, since the exact solution $y(x_i)$ is not known at x_i, except for the first step where the initial condition is given as exact. The value of y_i obtained by Euler's method contains the error accumulated in previous steps. The total error at a given value of x_i will be the product of the error per step and the number of steps. Since the number of steps is $x_{i+1}/\Delta x$, the total error is proportional to Δx and may, therefore, be denoted as $O(\Delta x)$ for Euler's method. For a detailed derivation of the total error, see Hornbeck (1982). Because the total error is of the first order in Δx, Euler's method is a *first-order method.*

In the above discussion, we have not considered the round-off error, which is inevitably present in any numerical solution. We can reduce the TE by making the step size Δx smaller. As illustrated above, this error decreases linearly with Δx. However, a reduction in Δx also results in an increase in the number of steps to obtain the solution over a given range in x. This results in an increase in the computing time and the round-off error, as shown qualitatively in Figure 9.4. An optimum value of Δx at which the error is minimum is, therefore, expected. Because of the round-off error, the numerical solution will always differ from the exact solution of the differential equation. However, neglecting the round-off error, if the numerical solution approaches the exact solution, as the step size Δx approaches zero, the numerical method applied to a given differential equation is said to be *convergent.* The numerical techniques discussed in this chapter are convergent when applied to most differential equations, and, therefore, the convergence of the scheme is generally assumed. This definition of convergence is different from that employed in earlier chapters to indicate a negligible change in the solution from one step to the next during an iterative computational scheme. To distinguish between these two definitions, we shall refer to an iterative process as being *iteratively convergent.*

A very important consideration in the numerical solution of differential equations is that of stability of the numerical method. Although several definitions of stability are used in the literature, the most commonly employed definition simply considers a numerical method to be unstable if it yields an unbounded solution when the exact solution is bounded. Instability arises due to the amplification of the error, and, under

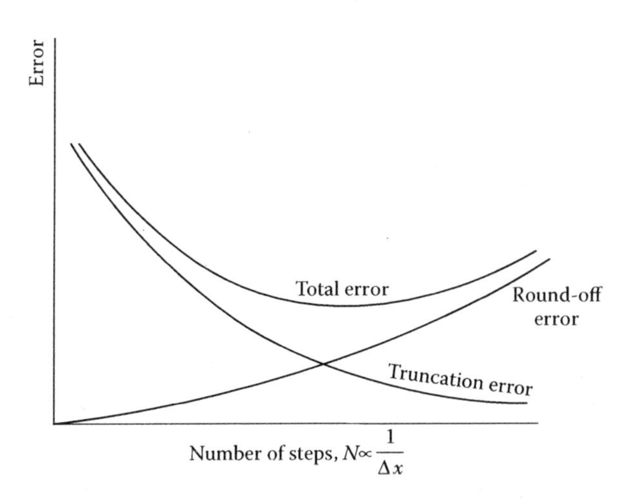

FIGURE 9.4 Qualitative representation of the variation of the truncation, round-off, and total errors, in the numerical solution of an ODE, with the total number of steps N, which varies inversely as the step size Δx.

certain conditions, an unbounded growth may arise. The stability of a numerical scheme depends on both the method and the differential equation. For instance, as discussed in detail by Ferziger (1998), if Euler's method is applied to the differential equation $dy/dx = -\alpha y$, where α is a positive constant, the scheme is conditionally stable. If $|1-\alpha\Delta x| > 1$, Euler's method gives rise to an increasing solution, whereas a decaying solution is given by analysis. Therefore, the scheme is stable only for a certain range of values of $\alpha\,\Delta x$. Oscillations that increase in amplitude with increasing x are observed, indicating the presence of instability (see Section 2.3.4). Numerical schemes, which are stable for any value of the step size and other governing parameters, are said to be *unconditionally stable*. Similarly, there are unconditionally unstable schemes that are unstable for all values. However, the computational scheme can be analyzed in only a few cases to determine its stability characteristics. A common approach employed in practice is to obtain the numerical solutions with two significantly different step sizes. If the two results are substantially different, numerical instability may be assumed to be present. If the two solutions are close to each other, then the scheme is probably stable.

In the numerical solution of a differential equation, it is important to consider the questions of accuracy, convergence, and stability, as outlined above. The exact solution is generally available only for a few simple cases. However, a comparison between the numerical solution for these cases and the exact solution will yield important information on the accuracy and correctness of the numerical results. It is also important to vary the step size Δx after the corresponding numerical solution has been obtained. By varying the step size, one can often determine whether the scheme is convergent and stable. The numerical results should be essentially independent of the step size. This process of *grid refinement* to ensure that the results do not depend on the grid and on other numerical parameters chosen by the user is often known as *verification* (Roache, 2010). Similarly, comparison of the numerical results obtained with experimental data, analytical results and other available results, as well as consideration of the basic nature of the problem, to ensure that the mathematical and numerical model satisfactorily represents the process or system is known as *validation*. Several of these considerations were also discussed earlier in Chapter 2. The following example illustrates the use of Euler's method in solving a first-order initial-value problem.

Example 9.1

An electrical circuit consists of an inductance L, a resistance R, and an emf source E, as shown in Figure 9.5. Initially, the switch is open and there is no current in the circuit. At time $t = 0$, the switch is closed and the current builds up. After 0.5 s, the switch is again opened and the current decreases with time to zero. Using Euler's method, solve this problem to obtain the variation of the current with time for (a) $E = 20$ V, $L = 5$ henries, and $R = 10\ \Omega$, and (b) $E = 20$ V, $L = 10$ henries, and $R = 5\ \Omega$.

SOLUTION

We obtain the differential equation that governs the current I for the first part of the problem, when the switch is closed, by adding the voltage changes around

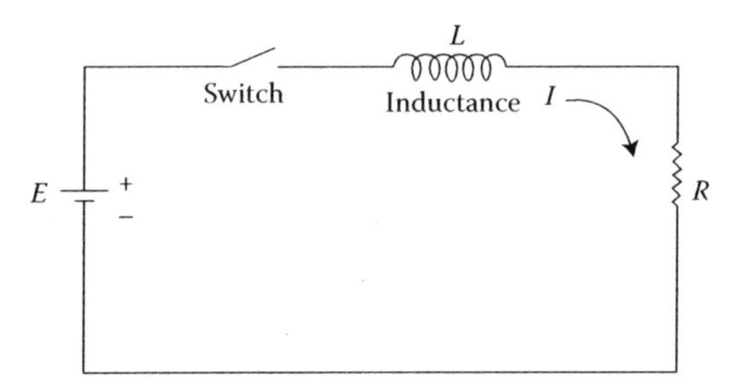

FIGURE 9.5 Electrical circuit considered in Example 9.1.

the circuit and setting the sum equal to zero. The voltage across the inductance is $L(dI/dt)$, and that across the resistance is RI. For the analysis of such electrical circuits, see, for instance, Halliday et al. (2010) and Ogata (2003). Thus,

$$L\frac{dI}{dt} + RI - E = 0 \tag{9.21}$$

or

$$\frac{dI}{dt} = \frac{E}{L} - \frac{R}{L}I \text{ with } I = 0 \text{ at } t = 0 \tag{9.22}$$

where t is the time, in seconds, elapsed following the closing of the switch. We obtain the equation that applies for the second phase when the switch is reopened by setting $E = 0$:

$$\frac{dI}{dt} = -\frac{R}{L}I \quad \text{with } I = I_1 \text{ at } t = 0.5 \tag{9.23}$$

The initial condition for this equation is the current I_1 at $t = 0.5$, where I_1 is obtained from the numerical solution of Equation 9.22.

Therefore, the problem involves the solution of two first-order ODEs. We must first solve Equation 9.22 to obtain the current I from $t = 0$ to $t = 0.5$ s. Then we solve Equation 9.23 to obtain the current until it becomes essentially zero. The problem is a simple one and can be solved analytically. Here, we will consider its solution by the simple one-step, self-starting, Euler's method and compare the numerical results with the analytical solution.

For the two sets of data given for this problem, the equations are obtained as

$$\frac{dI}{dt} = 4 - 2I \quad \text{for } 0 \le t \le 0.5 \tag{9.24a}$$

$$\frac{dI}{dt} = -2I \quad \text{for } t > 0.5 \tag{9.24b}$$

and

$$\frac{dI}{dt} = 2 - 0.5I \quad \text{for } 0 \le t \le 0.5 \tag{9.25a}$$

$$\frac{dI}{dt} = -0.5I \quad \text{for } t > 0.5 \tag{9.25b}$$

The computational formula for Euler's method is given by

$$I_{i+1} = I_i + \Delta t F(t_i, I_i) \quad \text{with } i = 0,1,2,... \tag{9.26}$$

where the subscript i represents the computed values after the ith step, and subscript $(i + 1)$ those after the $(i + 1)$th step. Here, F represents the function on the right-hand side of the equations. In the present case, F depends only on I, which in turn is a function of time t. Also, $t = i\,\Delta t$, where Δt is the time step and I_0 is the current at $t = 0$.

Appendix B.24 gives the computer programs in MATLAB® for this problem. Two algorithms are shown, the first as a function *m*-file and the other as a script *m*-file. In the former case, the function F, taken as f in the program, is given as a string, along with the beginning and end points, a and b, of the time range to be considered. The number of steps, n, is also given, so that the step size can be computed. The initial condition $y0$ is also specified. Then, Euler's method is used to yield the values of the dependent variable y at various values of the independent variable t. In order to use this function file for the given problem, the two functions $F(t, I)$ are defined as

```
function z = fe1(x,y)
z = 4-2 * y;
end

function z = fe2(x,y)
z = -2 * y;
end
```

Then, the function file *euler.m* is employed as

```
s1 = euler('fe1',0,0.5,0,50);
s2 = euler('fe2',0.5,8,s1(51,2),750);
plot(s1(:,1),s1(:,2))
hold on
plot(s2(:,1),s2(:,2))
```

This solves the two ODEs by the Euler's method and plots the results, discussed later.

The second MATLAB program solves the first ODE, followed by the second one, as given in the problem and as discussed in the preceding. The various symbols employed are defined in the program. The given parameters, E, L, and R, the time step, dt, and the total time for the computation are entered. The initial condition $I(0) = 0$ is employed to start the computational scheme. Equation 9.24a or 9.25a is solved until $t = 0.5$. At this point, the computed value of the current is employed as the initial condition for Equation 9.24b or 9.25b, and the computation is continued until the total time is reached. A convergence criterion may also be employed to terminate the computation when I does not change significantly with time, as given by a convergence parameter. The time step dt and the convergence parameter must be varied to ensure that the results obtained are essentially independent of the values chosen. The time step was varied from 10^{-1} to 10^{-4} and a value of 10^{-2} was found to be adequate, since decreasing dt further did not significantly affect the results.

Figure 9.6 shows the numerical results obtained for the two sets of data given in the problem. In both cases, the current I rises sharply from zero as the switch is closed. The maximum value of the current is obtained at $t = 0.5$ s, beyond which the current decreases because of the reopening of the switch. A larger maximum value of the current will be obtained if the switch is kept closed for a longer period of time. The second set of parameters results in a slower increase in the current, following the closing of the switch, and also a slower decrease after the switch has been reopened, as compared to the results for the first set. This behavior is expected since the derivative dI/dt, in the governing equations, is smaller in the second case.

Equation 9.21 can be solved analytically to give

$$I = \frac{E}{R}\left[1 - \exp\left(-\frac{Rt}{L}\right)\right] \tag{9.27}$$

Similarly, Equation 9.23 may be solved to give

$$I = \frac{E}{R}\left[1 - \exp\left(-\frac{Rt_1}{L}\right)\right]\exp\left[-\frac{R(t - t_1)}{L}\right] \tag{9.28}$$

where t_1 is the time at which the switch is reopened. The numerical results are found to agree quite well with the analytical solution. The current at $t = 0.5$ is

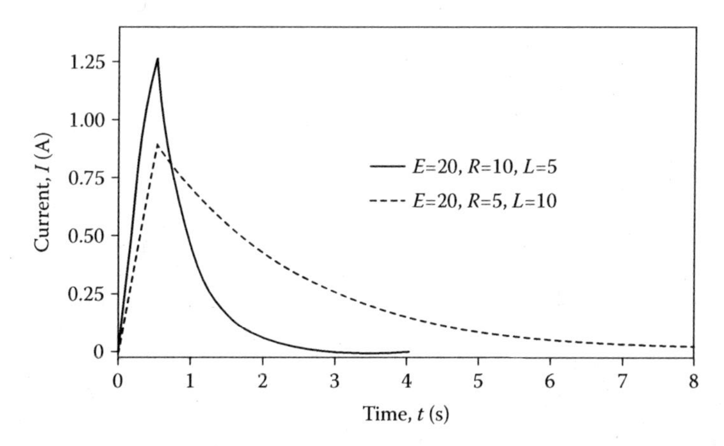

FIGURE 9.6 The computed variation of the current I with time t for the two cases considered in Example 9.1. The time step Δt is taken as 0.01 s for these results.

obtained from Equation 9.27 for the two cases as 1.264 and 0.885 A, respectively. These values are close to those obtained numerically, see Figure 9.6.

Appendix C.14 shows the program in Fortran for solving the same problem, as that given in Appendix B.24(b). As illustrated in the examples given in earlier chapters, the two programs are very similar in form. The basic logic employed is the same. The program in MATLAB is simpler because of the several features that allow considerable ease in input/output, control and plotting. As mentioned earlier, Fortran is often more appropriate for complicated circumstances. The two programs also illustrate the ease with which one may switch from one programming language to the other.

9.3 IMPROVEMENTS IN EULER'S METHOD

9.3.1 HEUN'S METHOD

There are several numerical schemes that are based on modifications of Euler's method and that can be used for solving ODEs. One of the most important is the *improved Euler's method* or *Heun's method*. In Euler's method, the function $F(x, y)$, which represents the derivative dy/dx of the dependent variable y, is taken as constant over the interval Δx at the value computed at the start of the interval. However, the derivative usually changes as x increases over the interval, and the accuracy of the solution can be improved if a better approximation is used for the derivative. Using Heun's method, one achieves this improvement by first calculating y_{i+1} from Euler's method, denoting this intermediate value as y^*_{i+1} and then using this value to obtain a better approximation to the derivative. Considering the first-order initial-value problem of Equation 9.8, Heun's method is given by

$$y^*_{i+1} = y_i + \Delta x\ F(x_i, y_i) \tag{9.29a}$$

$$y_{i+1} = y_i + \frac{\Delta x}{2}\left[F(x_i, y_i) + F(x_{i+1}, y^*_{i+1})\right] \tag{9.29b}$$

Therefore, Euler's method is employed twice in succession, and the average of the approximations to the derivative at the two ends of the interval is used to give a more accurate value of y_{i+1}. Since y_{i+1} is unknown, the derivative at x_{i+1} is approximated by $F(x_{i+1}, y^*_{i+1})$. The method is self-starting since only the conditions at x_i are needed to obtain y_{i+1}.

The preceding procedure gives one of the simplest forms of the predictor–corrector methods, which are discussed in detail later in this chapter. Equation 9.29a is a predictor equation for the first approximation to y_{i+1}, and Equation 9.29b is a corrector equation to yield an improved estimate of y_{i+1}. Figure 9.7 shows a graphical representation of this method. The improvement in the estimate of y_{i+1} is seen in terms of a better approximation to the slope of the graph of $y(x)$. Also shown is the integration of $F(x, y)$, as given by Equation 9.15, using the trapezoidal rule to obtain y_{i+1}.

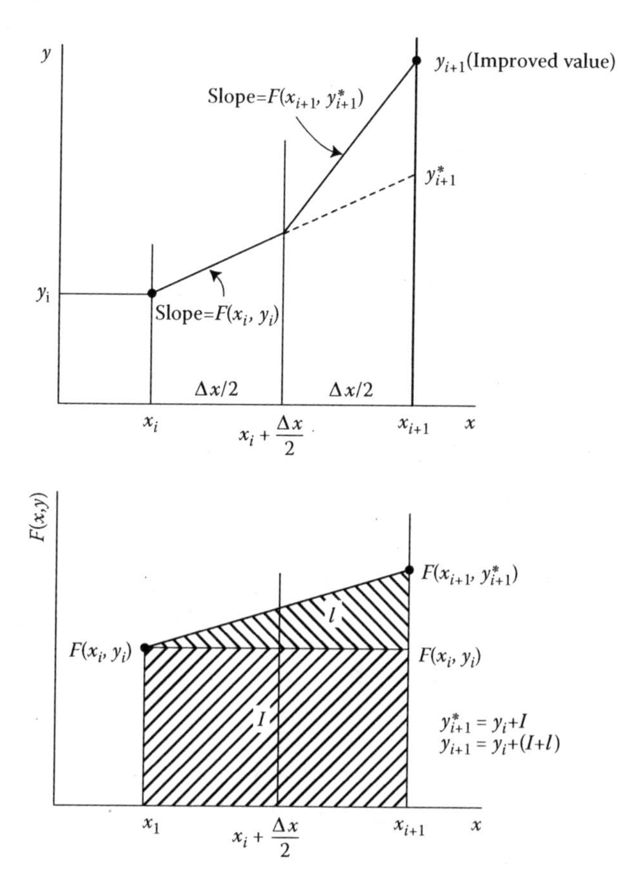

FIGURE 9.7 Graphical representation of the improved Euler's method, also known as *Heun's method.*

One can also use Equation 9.29b iteratively by substituting the value of y_{i+1}, obtained after the first computation in place of y^*_{i+1}, to obtain the next, improved, approximation to y_{i+l}. A sequence of corrected values of y_{i+1} may thus be generated. The iterative process is terminated when the change in y_{i+1} from one iteration to the next is smaller than a prescribed convergence parameter. However, greater accuracy, as compared to that obtained with Euler's method, is expected in the solution even if Equations 9.29a and 9.29b are used only once, without iteration.

From the Taylor-series expansion for $y(x)$, as given by Equation 9.19, greater accuracy is expected in the evaluation of y_{i+1}, if the terms of order $(\Delta x)^2$ are also retained. However, since the second derivative d^2y/dx^2 is not known, it may be approximated by

$$\frac{d^2y}{dx^2} \cong \frac{y'(x_{i+1}) - y'(x_i)}{\Delta x} = \frac{F(x_{i+1}, y_{i+1}) - F(x_i, y_i)}{\Delta x}$$

where the prime denotes the first derivative with respect to x. When this expression is substituted in the truncated Taylor series, using y_{i+1}^* for y_{i+1}, which is not known, we obtain

$$y_{i+1} = y_i + \Delta x\ F(x_i, y_i) + \frac{(\Delta x)^2}{2} \frac{F(x_{i+1}, y_{i+1}^*) - F(x_i, y_i)}{\Delta x}$$

Therefore,

$$y_{i+1} = y_i + \frac{\Delta x}{2}\left[F(x_i, y_i) + F(x_{i+1}, y_{i+1}^*)\right] \tag{9.29b}$$

where the derivative at x_{i+1} has been approximated by $F(x_{i+1}, y_{i+1}^*)$ and the intermediate value y_{i+1}^* is obtained from Equation 9.29a. Therefore, the method retains the second-order terms. The TE at each step is $O[(\Delta x)^3]$, and the total, or global, error is $O[(\Delta x)^2]$. The method is, therefore, a second-order method. An improved accuracy is obtained if iteration is employed with Equation 9.29b. Also, higher-order formulas may be derived by retaining additional terms in the Taylor-series expansion, as done for the Runge–Kutta methods, which are discussed in the next section. However, to achieve this increased accuracy, we need a larger computational effort for the determination of the intermediate approximations to the derivative.

The improved Euler's method may also be obtained by the application of the trapezoidal integration formula to the differential equation, Equation 9.8, instead of the rectangular rule which yielded Euler's method, as shown in Figure 9.7. Integrating the differential equation, we obtain

$$y_{i+1} - y_i = \frac{\Delta x}{2}\left[F(x_i, y_i) + F(x_{i+1}, y_{i+1})\right] \tag{9.30}$$

Again, y_{i+1} in the parentheses on the right is replaced by y_{i+1}^* which is obtained from Equation 9.29a. Then, Equation 9.29b is obtained. The TE per step for the trapezoidal rule was given in Chapter 8 as

$$-\frac{(\Delta x)^3}{12} y'''(\xi)$$

where $y'''(\xi)$ is the third derivative of y with respect to x at a point ξ which lies between x_i and x_{i+1}. If y''' is assumed to be constant over this step size, it may be evaluated at x_i to yield an estimate of the TE per step as

$$TE = -\frac{1}{12} \cdot \left.\frac{d^3y}{dx^3}\right|_{x_i} \cdot (\Delta x)^3 \tag{9.31}$$

This expression for the error applies if Equation 9.30 is solved by iteration for y_{i+1} and not if the approximation given by Equation 9.29b is employed.

If the unknown y_{i+1} appearing on the right-hand side of Equation 9.30, is not approximated as y^*_{i+1}, we cannot solve for y_{i+1} directly, except for extremely simple cases. A nonlinear algebraic equation would generally be involved, and the iterative methods of Chapter 5 may be employed to determine y_{i+1} from this equation. Such methods, which require the solution of a nonlinear algebraic equation to obtain the new value of the function y, are known as *implicit*. Another implicit method that can be derived by the application of the backward difference formula to the differential equation, Equation 9.8, gives

$$y_{i+1} - y_i = \Delta x \, F\left(x_{i+1}, y_{i+1}\right) \tag{9.32}$$

This method, known as the *implicit Euler* or the *backward Euler method*, is first-order accurate.

Implicit methods involve more computation per step, as compared to the explicit methods discussed earlier. However, these methods generally have better numerical stability and are often unconditionally stable, as is the backward Euler method given above. In explicit methods, stability considerations may sometimes restrict the step size to a small value. In such cases, implicit methods may be preferable because of weaker restrictions on step size. However, it must be noted that the stability of a numerical scheme does not indicate the accuracy of the results. In fact, the TE in the implicit Euler method is equal in magnitude, but opposite in sign, to that in Euler's method.

9.3.2 Modified Euler's Method

We can obtain another modification of Euler's method by employing the midpoint, $x_i + (\Delta x/2)$ in a given step for the evaluation of the derivative $F(x, y)$. This method, often known as the *modified Euler's method* or the *improved polygon method*, is given by

$$y_{i+1/2} = y_i + \frac{\Delta x}{2} F\left(x_i, y_i\right) \tag{9.33a}$$

$$y_{i+1} = y_i + \Delta x \, F\left(x_i + \frac{\Delta x}{2}, y_{i+1/2}\right) \tag{9.33b}$$

Euler's method is employed twice, first to obtain an approximation $y_{i+1/2}$ at the midpoint and second to evaluate y_{i+1} from the derivative approximated at the midpoint. This method is self-starting, since the computation is based on the known conditions at $x = x_i$. It is second-order accurate and is also known as the *second-order Runge–Kutta method*, as discussed in the next section. The geometrical interpretation of the method is shown in Figure 9.8, indicating the use of the midway point in determining y_{i+1}. Also, compare this figure with Figure 9.2 to see the difference between this scheme and Euler's method.

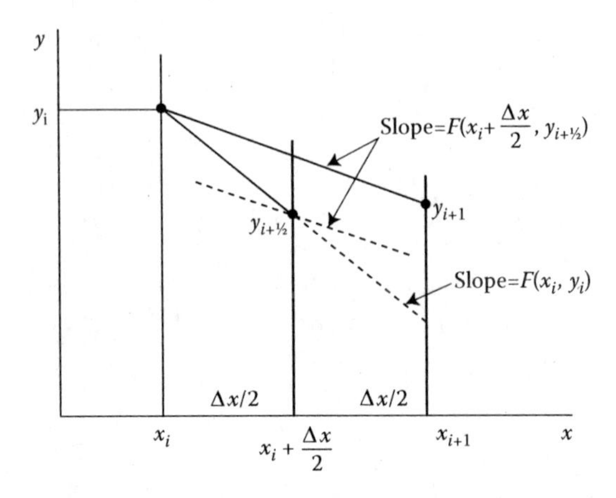

FIGURE 9.8 The modified Euler's method, also known as the *improved polygon method* or as the *second-order Runge–Kutta method.*

Both Heun's method and the modified Euler's method are employed for engineering applications where a very high level of accuracy is not necessary and a simple computational scheme is desired. The former method is a simple form of the predictor–corrector methods, which are discussed in greater detail later in this chapter. The latter method is a second-order Runge–Kutta scheme and is thus a relatively less accurate version of a class of methods extensively used for practical problems. More accurate formulas, particularly the fourth-order methods, are much more important for engineering problems. The flow charts for both of these methods are shown in Figure 9.9. A MATLAB program for Heun's method, without iteration, is given in Appendix B.25 as a function *m*-file. It can be seen that the algorithm is quite similar to that for Euler's method and can easily be used to solve ODEs. This Appendix also gives the function definitions for solving the equations given in Example 9.1 and a simple script file to use the function file to obtain the desired solution, as presented earlier for Euler's method.

Several other modifications of Euler's method are available in the literature. We have considered only self-starting methods here. Some of the modifications of Euler's method are not self-starting, and one of the self-starting methods is needed to obtain the first few steps. Similarly, other implicit and explicit modifications of Euler's method may be derived. The accuracy of these methods may be improved by *Richardson*'s deferred approach to the limit, discussed in Section 8.4.1 and also later in this chapter.

9.4 RUNGE–KUTTA METHODS

An important class of self-starting methods for the numerical solution of ODEs is based on retaining higher-order terms in the Taylor-series expansion of the dependent variable $y(x)$ and employing the computed values of the function $F(x, y)$, which represents the derivative of $y(x)$, at several values of x in the interval $x_i \leq x \leq x_{i+1}$. The resulting schemes, known as the *Runge–Kutta methods*, are widely used for the solution of various types of ODEs. In recent years, there has been an increase in the use

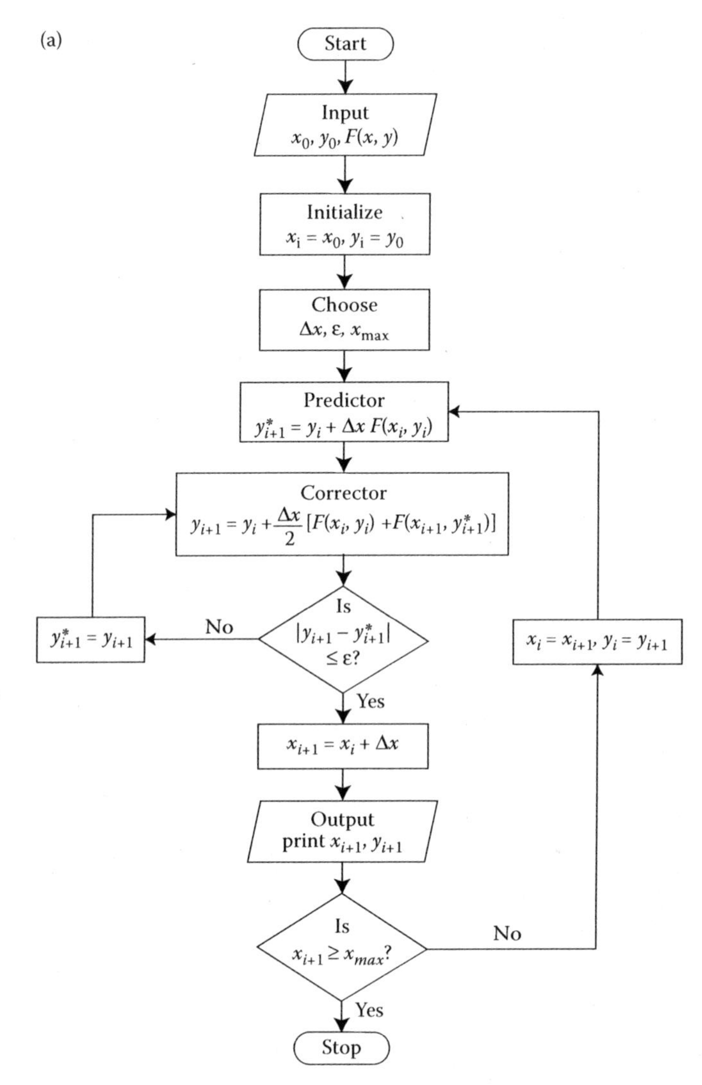

FIGURE 9.9 Flow charts for the solution of a first-order ODE by (a) Heun's method with iteration and (b) modified Euler's, or impoved polygon, method.

of other methods, such as the predictor–corrector methods, which are often more efficient. However, the Runge–Kutta methods are usually employed to start these latter methods, which are not self-starting. The Runge–Kutta methods have several important advantages over other methods, besides being self-starting. They are easy to program, they have good numerical stability, and the step size can be changed easily to improve accuracy. However, for comparable accuracy, they often require more computer time than the more efficient methods. Also, the local error is not estimated easily. Nevertheless, the Runge–Kutta methods are probably the most widely used technique for solving ODEs that arise in engineering applications.

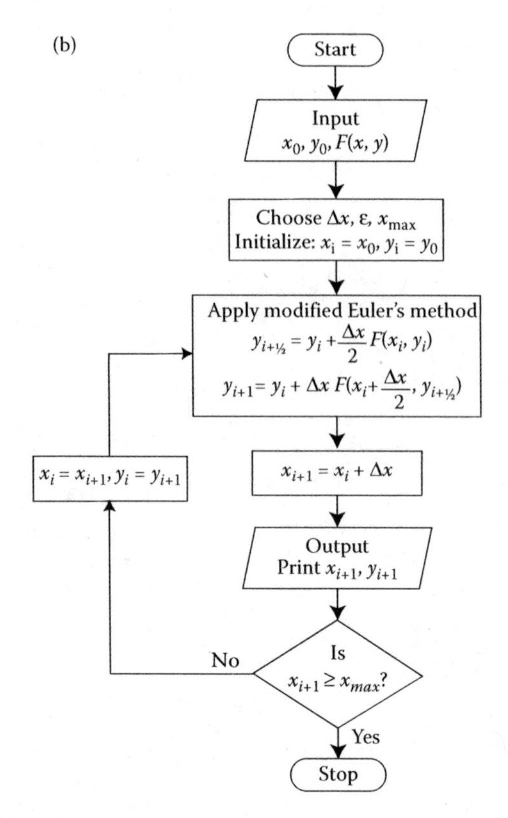

FIGURE 9.9 Continued.

9.4.1 Computational Formulas

Runge–Kutta formulas involve a weighted average of the derivative calculated at various locations within a step size Δx. Let us consider the first-order ODE given by

$$\frac{dy}{dx} = F(x, y) \tag{9.8}$$

with

$$y(x_0) = y_0 \tag{9.10}$$

If $F(x, y)$ is evaluated at two values of x within the interval $x_i \leq x \leq x_{i+1}$, we obtain the second-order Runge–Kutta method, which has the same accuracy as that obtained by retaining terms up to order $(\Delta x)^2$ in a Taylor-series expansion for $y(x)$. If $F(x, y)$ is evaluated at only one point, $x = x_i$, we obtain the first-order Euler's method. Similarly, $F(x, y)$ is evaluated at three locations in the interval for the third-order method and so on. Since the locations at which the function $F(x, y)$ is computed and the weighting factors to be used can be chosen in several ways, a family of formulas can be obtained for solving the given differential equation. The derivation of these formulas is quite involved and is based on a comparison between the terms of the Taylor-series expansion for $y(x)$, about x_i, and those of the expansion for the approximations to the change in the dependent variable over the given step. For instance, the second-order method may be expressed as follows:

$$y_{i+1} = y_i + C_1 K_1 + C_2 K_2 \tag{9.34a}$$

where

$$K_1 = \Delta x \; F(x_i, y_i) \tag{9.34b}$$

$$K_2 = \Delta x \; F(x_i + p_1 \Delta x, y_i + p_2 K_1) \tag{9.34c}$$

Thus, the K's approximate the change in y over the computational step.

The constants C_1, C_2, p_1, and p_2 are to be determined for an accurate evaluation of y_{i+1}. As discussed in detail by Ralston (1965), Ralston and Rabinowitz (1978), and Carnahan et al. (1969), one can find the relationships among these constants by comparing the expansions for $y(x)$ and K_2. One of the constants may be chosen arbitrarily. If C_2 is taken as 1/2, then it is found that $C_1 = 1/2$ and $p_1 = p_2 = 1$. This choice, therefore, leads to Heun's method, given in the preceding section. If C_2 is chosen as 1, then $C_1 = 0$ and $p_1 = p_2 = 1/2$, resulting in the modified Euler's method. Similarly, other second-order formulas may be obtained. The first-order Runge–Kutta method is simply Euler's method. One derives higher-order methods similarly by employing Taylor's expansions of $y(x)$ and retaining terms up to order $(\Delta x)^3$ for the third-order formulas and up to $(\Delta x)^4$ for the fourth-order. Still higher-order formulas have also been developed but are rarely used because of the large amount of computation involved at each step, see Butcher (1964) and Shanks (1966) for details.

The third-order method obtained by Kutta is given by

$$y_{i+1} = y_i + \frac{K_1 + 4K_2 + K_3}{6} \tag{9.35a}$$

where

$$K_1 = \Delta x \; F(x_i, y_i) \tag{9.35b}$$

$$K_2 = \Delta x \; F\left(x_i + \frac{\Delta x}{2}, y_i + \frac{K_1}{2}\right) \tag{9.35c}$$

$$K_3 = \Delta x \; F(x_i + \Delta x, y_i + 2K_2 - K_1) \tag{9.35d}$$

The fourth-order Runge–Kutta formulas are of the general form

$$y_{i+1} = y_i + (C_1 K_1 + C_2 K_2 + C_3 K_3 + C_4 K_4) \tag{9.36}$$

where the C's are constants. Depending on the choice of the locations where the K's are determined and of the appropriate parameters that arise, several fourth-order formulas have been developed. The most widely employed formula is the classical Runge–Kutta method given by

$$y_{i+1} = y_i + \frac{K_1 + 2K_2 + 2K_3 + K_4}{6} \tag{9.37a}$$

where

$$K_1 = \Delta x\ F(x_i, y_i) \tag{9.37b}$$

$$K_2 = \Delta x\ F\left(x_i + \frac{\Delta x}{2}, y_i + \frac{K_1}{2}\right) \tag{9.37c)]}$$

$$K_3 = \Delta x\ F\left(x_i + \frac{\Delta x}{2}, y_i + \frac{K_2}{2}\right) \tag{9.37d}$$

$$K_4 = \Delta x\ F(x_i + \Delta x, y_i + K_3) \tag{9.37e}$$

Therefore, four evaluations of the derivative function are made within the interval $x_i \leq x \leq x_{i+1}$ in order to obtain approximations to the change in y over this step, and a suitable weighted average is employed. The TE per step is of order $(\Delta x)^5$, since terms up to order $(\Delta x)^4$ are retained in the expansions.

Several other fourth-order formulas are available, see Carnahan et al. (1969). A formula that was quite widely used in the past is the one developed by Gill (1951). This method minimizes the round-off error and also reduces the computer storage requirements in the solution of a system of equations. The formula was, therefore, particularly useful for small computers. It is given by

$$y_{i+1} = y_i + \frac{K_1 + (2-\sqrt{2})K_2 + (2+\sqrt{2})K_3 + K_4}{6} \tag{9.38a}$$

where

$$K_1 = \Delta x\ F(x_i, y_i) \tag{9.38b}$$

$$K_2 = \Delta x\ F\left(x_i + \frac{\Delta x}{2}, y_i + \frac{K_1}{2}\right) \tag{9.38c}$$

$$K_3 = \Delta x\ F\left(x_i + \frac{\Delta x}{2}, y_i + \frac{\sqrt{2}-1}{2}K_1 + \frac{\sqrt{2}-1}{2}K_2\right) \tag{9.38d}$$

$$K_4 = \Delta x \; F\left(x_i + \Delta x, y_i - \frac{K_2}{\sqrt{2}} + \frac{\sqrt{2}+1}{\sqrt{2}} K_3 \right) \tag{9.38e}$$

The fourth-order Runge–Kutta method, given by Equations 9.37, continues to be a popular choice for the solution of ODEs. It is also frequently employed for BVPs by incorporating a root-solving scheme, such as the *secant* or the *Newton–Raphson* method, for satisfying the boundary conditions. The main attractive features of this method are, as mentioned earlier, high accuracy level, good stability characteristics, ease in programming, applicability to a wide variety of problems, self-starting computational scheme, and ease with which the step size may be changed to improve accuracy. All of these features are common to all fourth-order Runge–Kutta formulas. The classical method is the most widely used one mainly because it has been the standard method for many years and because it is available in most computer libraries. In engineering applications, where a simple method that yields reasonably accurate results is particularly attractive, the Runge–Kutta methods are commonly employed, despite the availability of more efficient methods. However, for applications that involve a substantial computational effort, the Runge–Kutta methods are generally used only to start the scheme, and other methods, such as the predictor–corrector methods, are employed after the first few steps to obtain the solution.

As briefly discussed in Chapter 3, MATLAB functions *ode*23 and *ode*45 can easily be used to solve ODEs. Both are based on Runge–Kutta methods. The function *ode*23 is a lower order method and is employed as

```
[tout, yout] =ode23('f',trange,y0)
```

where f is the function $F(x, y)$ representing the right-hand of a first-order ODE, Equation 9.8, and entered as a string, *trange* is range of the independent variable t from the initial value $t0$ to the end point *tfinal* and $y0$ is the initial condition $y = y0$ at $t = t0$. The output is given in terms of t values *tout* and corresponding values of the dependent variable *yout*. Similarly, *ode*45 is a higher order method for solving a first-order ODE and the command to use it is similar to that for *ode*23, given above. Both these methods can be used for higher order ODEs by converting these into a system of first-order equations, as discussed earlier. The application to higher order ODEs is demonstrated in an example later in the chapter. Considerable flexibility is available in employing these functions and other such methods are available in the MATLAB environment.

9.4.2 Truncation Error and Accuracy

One of the major problems with the Runge–Kutta methods is that a quantitative estimate of the local TE is usually difficult to obtain. Therefore, it is often difficult to determine whether the step size is small enough to yield numerical results of desired accuracy. A common procedure is to run the computational scheme for different step sizes and to compare the results obtained. The step size is then chosen such that a further reduction in size does not significantly affect the results. However, this process is very time-consuming, and other methods for estimating the error and for improving the accuracy of the results have been developed.

The truncation error per step, *TE*, for a *n*th-order Runge–Kutta method may be written as

$$TE = A(\Delta x)^{n+1} + O\left[(\Delta x)^{n+2}\right] \tag{9.39}$$

where *A* is a constant that depends on the function *F*(*x*, *y*) and its higher-order partial derivatives. If Δx is small, the error is determined largely by the first term, and the bounds for *A* may be determined, as discussed by Ralston and Rabinowitz (1978). One method of estimating the TE is to obtain the two solutions y_{i+l} and $\tilde{y}_{i+1}$ corresponding to step sizes Δx and $\Delta x/2$, respectively, by integrating between the two *x* values x_i and x_{i+1}. If Y_{i+1} *is* the exact solution, an estimate of the TE and an improvement in the accuracy of the numerical solution may be obtained from the above expression for TE, considering only the dominant term. This approach, known as *Richardson's extrapolation* and discussed earlier in Chapter 8, gives

$$Y_{i+1} - y_{i+1} = A(\Delta x)^{n+1} \frac{x_{i+1} - x_i}{\Delta x}$$

$$Y_{i+1} - \tilde{y}_{i+1} = A\left(\frac{\Delta x}{2}\right)^{n+1} \frac{x_{i+1} - x_i}{\Delta x/2}$$

Therefore,

$$Y_{i+1} = \frac{2^n \tilde{y}_{i+1} - y_{i+1}}{2^n - 1} \tag{9.40}$$

which gives a more accurate approximation to the solution. Also,

$$TE = Y_{i+1} - y_{i+1} = A(x)^{n+1} = \frac{2^n(\tilde{y}_{i+1} - y_{i+1})}{2^n - 1} \tag{9.41}$$

Therefore, for the fourth-order Runge–Kutta method, *TE* is obtained as

$$TE = \frac{16}{15}(\tilde{y}_{i+1} - y_{i+1}) \tag{9.42}$$

One can use this estimate of the TE to adjust the step size and thus maintain the desired accuracy level. However, if this estimation were done at each step, the number of calculations performed would rise to approximately three times that for the Runge–Kutta method. Therefore, the error may be computed once every *m* steps, where *m* is chosen arbitrarily. Collatz (1966) gave another procedure for controlling the error, based on the calculation of the absolute value $|(K_3 - K_2)/(K_2 - K_1)|$

from Equation 9.37 after each step. If this quantity becomes larger than a few hundredths, then the error is too large and the step size should be reduced. Several other similar procedures have been given in the literature for limiting the error in Runge–Kutta methods.

Note from the preceding discussion that we can decrease the TE by reducing the step size Δx. However, a reduction in Δx also results in an increase in the number of steps needed for the integration of the equation over a given interval. This, in turn, increases the round-off error, which depends on the number of arithmetic operations performed. The total TE may be estimated as a product of the error per step and the number of steps. Since the latter varies inversely with the step size, the accumulated TE for the nth-order Runge–Kutta method is of order $(\Delta x)^n$. Therefore, as Δx is reduced, the TE is decreased while the accumulated round-off error increases. An estimate for the total error due to truncation and round-off is usually difficult to obtain. However, if analytical results are available for some simple cases, a comparison between the analytical and numerical results allows one to determine the overall accuracy of the numerical scheme.

The stability characteristics of the Runge–Kutta methods are generally good. Of particular interest in any numerical scheme is its partial instability, which may arise even when the equation being solved is not inherently unstable and which is dependent largely on the step size. If $\partial F(x, y)/\partial y$ is positive, then the error may increase without bound as x increases. On the other hand, if it is negative, then the error remains bounded for small Δx, see Fox (1962) and Ferziger (1998). An important parameter is the *step factor* $\Delta x \; \partial F/\partial y$, which affects the propagation of the error. The stability of various formulas has been considered, and the criteria for choosing the step factor have been given. A practical approach in most cases is to solve the problem for different step sizes, and, if close agreement is obtained between the corresponding results, the scheme may be assumed to be stable.

9.4.3 System of Equations

The Runge–Kutta methods can also be employed for solving a system of first-order ODEs and, therefore, for solving an nth-order differential equation, since it can usually be reduced to n first-order equations, as outlined earlier.

Let us consider the two simultaneous first-order equations

$$\frac{dy}{dx} = F(x, y, z) \tag{9.43}$$

$$\frac{dz}{dx} = G(x, y, z) \tag{9.44}$$

with

$$y = y_0 \quad \text{and} \quad z = z_0 \quad \text{at } x = x_0 \tag{9.45}$$

Then the classical fourth-order Runge–Kutta method for this problem is given by

$$y_{i+1} = y_i + \frac{K_1 + 2K_2 + 2K_3 + K_4}{6} \tag{9.46a}$$

and

$$z_{i+1} = z_i + \frac{K_1' + 2K_2' + 2K_3' + K_4'}{6} \tag{9.46b}$$

where

$$\begin{aligned}
K_1 &= \Delta x\, F(x_i, y_i, z_i) & K_1' &= \Delta x\, G(x_i, y_i, z_i) \\
K_2 &= \Delta x\, F\left(x_i + \frac{\Delta x}{2}, y_i + \frac{K_1}{2}, z_i + \frac{K_1'}{2}\right) & K_2' &= \Delta x\, G\left(x_i + \frac{\Delta x}{2}, y_i + \frac{K_1}{2}, z_i + \frac{K_1'}{2}\right) \\
K_3 &= \Delta x\, F\left(x_i + \frac{\Delta x}{2}, y_i + \frac{K_2}{2}, z_i + \frac{K_2'}{2}\right) & K_3' &= \Delta x\, G\left(x_i + \frac{\Delta x}{2}, y_i + \frac{K_2}{2}, z_i + \frac{K_2'}{2}\right) \\
K_4 &= \Delta x\, F(x_i + \Delta x, y_i + K_3, z_i + K_3') & K_4' &= \Delta x\, G(x_i + \Delta x, y_i + K_3, z_i + K_3')
\end{aligned} \tag{9.46c}$$

The computation must be carried out in the sequence given above, since K_1' is needed for calculating K_2, K_2 for calculating K_3', and so on.

To illustrate the application of the above formulas to a higher-order ODE, let us consider the following equation, which governs the vibration of a mass connected to two boundaries through a spring and a damper:

$$m\frac{\mathrm{d}^2 x}{\mathrm{d}t^2} + B\frac{\mathrm{d}x}{\mathrm{d}t} + kx = P \tag{9.47}$$

Here, m is the mass of the vibrating body, B the viscous friction coefficient of the damper, k the spring constant, P an external force, x the displacement of the mass, and t the time. The system is illustrated in Figure 9.10. We can reduce it to two first-order equations by defining a new variable y as

$$\frac{\mathrm{d}x}{\mathrm{d}t} = y = F(x, y, t) \tag{9.48}$$

Therefore,

$$m\frac{\mathrm{d}y}{\mathrm{d}t} + By + kx = P$$

or

$$\frac{\mathrm{d}y}{\mathrm{d}t} = \frac{P - By - kx}{m} = G(x, y, t) \tag{9.49}$$

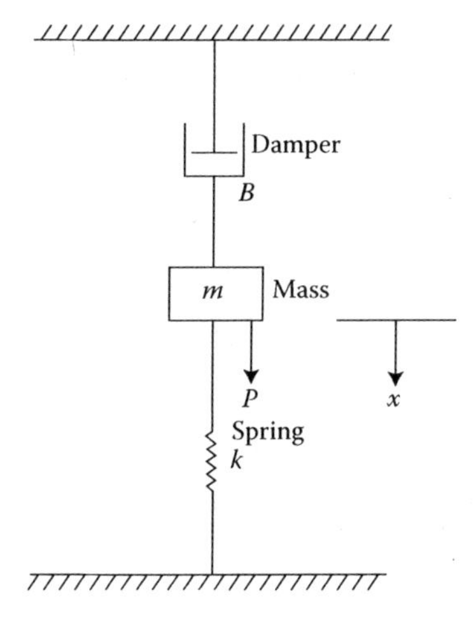

FIGURE 9.10 A vibrating system consisting of a vibrating mass m, a spring of stiffness k, a damper of friction coefficient B, and an external force P.

The two first-order equations, Equations 9.48 and 9.49, may now be solved by the Runge–Kutta method given above.

The recursion formulas for advancing from t_i to t_{i+1} are obtained as follows:

$$x_{i+1} = x_i + \frac{K_1 + 2K_2 + 2K_3 + K_4}{6} \tag{9.50a}$$

$$y_{i+1} = y_i + \frac{K_1' + 2K_2' + 2K_3' + K_4'}{6} \tag{9.50b}$$

where

$$\begin{aligned}
K_1 &= \Delta t y_i & K_1' &= \frac{\Delta t}{m}(P - By_i - kx_i) \\
K_2 &= \Delta t\left(y_i + \frac{K_1'}{2}\right) & K_2' &= \frac{\Delta t}{m}\left(P - B\left(y_i + \frac{K_1'}{2}\right) - k\left(x_i + \frac{K_1}{2}\right)\right) \\
K_3 &= \Delta t\left(y_i + \frac{K_2'}{2}\right) & K_3' &= \frac{\Delta t}{m}\left(P - B\left(y_i + \frac{K_2'}{2}\right) - k\left(x_i + \frac{K_2}{2}\right)\right) \\
K_4 &= \Delta t\left(y_i + K_3'\right) & K_4' &= \frac{\Delta t}{m}\left[P - B(y_i + K_3') - k(x_i + K_3)\right]
\end{aligned} \tag{9.50c}$$

Here, Δt is the step size in time t, and y represents the rate of change of the displacement x with time, that is, the velocity of the mass m. Note that the initial values of x and y are needed for starting the computation. These may be specified as follows:

$$\text{At } t = 0: \; x = x_0 \quad \text{and} \quad y = \frac{dx}{dt} = y_0 \tag{9.51}$$

Similarly, we can use the method to solve third-, fourth-, or still higher-order equations by reducing them to a system of first-order equations. Considerations of error, accuracy, and stability are similar to those discussed for first-order equations, although these concerns become more involved for a system of equations. The following example illustrates the use of Runge–Kutta methods for solving ODEs.

Example 9.2

A projectile of mass m is shot vertically upward at a velocity of 100 m/s. The frictional force acting on the projectile due to its motion in air is given as $m(AV + BV^2)$, where A and B are constants and V is the velocity at any given time t. Using the fourth-order Runge–Kutta method, compute the vertical position x and the velocity V of the projectile as functions of time for (a) $A = 0.01\ s^{-1}$, $B = 0.001 m^{-1}$, and (b) $A = 0.1\ s^{-1}$, $B = 0.01\ m^{-1}$. Solve for the vertical motion until the velocity becomes zero.

SOLUTION

The projectile is subjected to retardation due to the gravitational force of magnitude, mg, where g is the magnitude of the gravitational acceleration, and the frictional force due to the motion in air. We obtain the governing differential equation for the vertical displacement x by writing the force balance, from Newton's Second Law, as

$$m\frac{d^2x}{dt^2} = -mg - m(AV + BV^2), \quad \text{where } V = \frac{dx}{dt} \tag{9.52}$$

Here, d^2x/dt^2 is the acceleration of the projectile, taken as positive in the vertically upward direction. The gravitational force is negative since it acts downward. The frictional force is also negative since it acts in a direction opposite to that of the motion. Therefore, the equation to be solved is

$$\frac{d^2x}{dt^2} = -g - \left[A\frac{dx}{dt} + B\left(\frac{dx}{dt}\right)^2\right] = F(x,t) \tag{9.53}$$

This equation is solved until the velocity drops to zero and the projectile reaches its maximum height, before starting the downward motion. The initial conditions for Equation 9.53 are as follows:

$$\text{At } t = 0 : x = 0 \quad \text{and} \quad \frac{dx}{dt} = 100 \tag{9.54}$$

Since SI units are being used, $g = 9.8\ m/s^2$. Also, x, t, and V are in m, s, and m/s, respectively.

The second-order equation, given above, may be broken down into two first-order equations as

$$\frac{dx}{dt} = V \tag{9.55a}$$

$$\frac{dV}{dt} = -g - (AV + BV^2) \tag{9.55b}$$

Then the initial conditions for starting the Runge–Kutta scheme are as follows:

$$\text{At } t = 0 : x = 0 \quad \text{and} \quad V = 100 \tag{9.55c}$$

Appendix B.26 presents the MATLAB script file and Appendix C.15 the corresponding Fortran program for this problem. The various symbols employed are defined in the program. The function $F(x,t) = dV/dt = -g - (AV + BV^2)$ is defined and input parameters are entered. The fourth-order Runge–Kutta scheme is written, using the formulas given in Equations 9.46. Then, in terms of the nomenclature used here, the recursion formulas are as follows:

$$K_1 = \Delta t V_i \qquad K_1' = \Delta t\left(-g - AV_i - BV_i^2\right)$$

$$K_2 = \Delta t\left(V_i + \frac{K_1'}{2}\right) \qquad K_2' = \Delta t\left(-g - A\left(V_i + \frac{K_1'}{2}\right) - B\left(V_i + \frac{K_1'}{2}\right)^2\right)$$

$$K_3 = \Delta t\left(V_i + \frac{K_2'}{2}\right) \qquad K_3' = \Delta t\left(-g - A\left(V_i + \frac{K_2'}{2}\right) - B\left(V_i + \frac{K_2'}{2}\right)^2\right)$$

$$K_4 = \Delta t\left(V_i + K_3'\right) \qquad K_4' = \Delta t\left[-g - A(V_i + K_3') - B(V_i + K_3')^2\right]$$

$$x_{i+1} = x_i + \frac{K_1 + 2K_2 + 2K_3 + K_4}{6} \qquad V_{i+1} = V_i + \frac{K_1' + 2K_2' + 2K_3' + K_4'}{6} \tag{9.56}$$

Therefore, the displacement x and the velocity V at the next time step, denoted by the subscript $(i + 1)$, are obtained in terms of the values at the present time step, denoted by subscript i. The computation is carried out, starting with the initial conditions, until the velocity becomes less than or equal to zero. The exact point where it becomes zero may be obtained by interpolation.

Figures 9.11 and 9.12 show the numerical results obtained for the two sets of input data given. The time step Δt can be taken as larger in the first case, since the variation with time is slower due to the smaller frictional force. However, Δt must be varied to ensure that the numerical results are not significantly altered if Δt is made smaller. For the two cases shown, Δt values of 0.05 s and 0.01 s are taken,

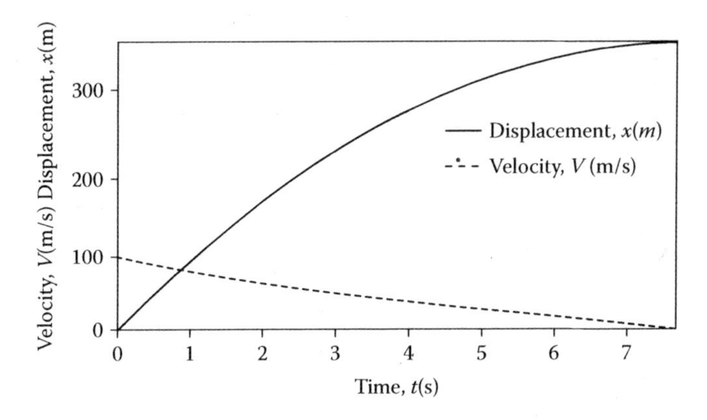

FIGURE 9.11 Computed variation of the velocity V and the displacement x with time t for Example 9.2, for $A = 0.01\ \text{s}^{-1}$, $B = 0.001\ \text{m}^{-1}$, and $\Delta t = 0.05$ s.

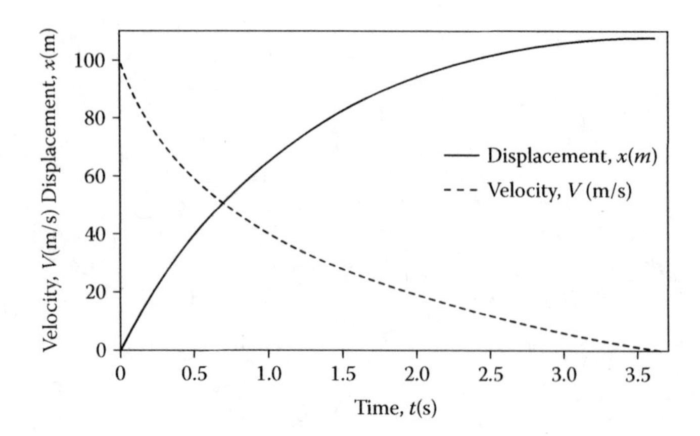

FIGURE 9.12 Computed results for $A = 0.1\ s^{-1}$ and $B = 0.01\ m^{-1}$ in Example 9.2, with $\Delta t = 0.01s$.

respectively, and found to be satisfactory, since a further reduction in Δt resulted in a negligible change in the results. As expected, the maximum height, or displacement, is found to be larger when A and B are smaller. Also, the time taken to reach this height is larger. The problem may easily be solved analytically for $A = B = 0$ (that is, no frictional drag) to obtain the maximum height as 510.2 m, and the time taken to reach it as 10.2 s. These values compare well with the results shown.

The program is quite simple because of the self-starting feature of the method. The program, written for the present initial-value problem, may be used to obtain results for arbitrary values of the input parameters. It may be modified to solve other second-order initial-value problems. The use of this method for BVPs requires a correction scheme, as discussed later in this chapter.

As discussed earlier, MATLAB functions *ode*23 and *ode*45 can be easily employed to solve this problem. The two dependent variables are x and V. A vector y is used with these two as components to specify the two ODEs and the initial conditions. The commands used are

```
y0 = [0;100];
[t,v] = ode45('rhs',10,y0);
n = length(t);
for i = 1:n
    if v(i,2) < 0
       break
    else
       t1(i) = t(i);v1(i,1) = v(i,1);v1(i,2) = v(i,2);
end
end
plot(t1,v1(:,1),'-',t1,v1(:,2))
```

where y0 represents the initial value of the dependent vector y, t is the independent variable and v is the solution matrix. Then, the first column of v represents x and the second column represents V. The results are plotted for x and V versus t. The function *rhs.m* is given as

```
function dydt = rhs(t,y)
```

```
a = 0.01;b = 0.001;
dydt = [y(2); -9.8 - (a*y(2) + b*y(2)^2)];
end
```

where the constants in the problem A and B are given as a and b and the right-hand sides of Equation 9.55 are given. The constants a and b may be changed for the second part of the problem. Additional commands are needed to stop the computation when V becomes negative. The results obtained are essentially identical to those obtained earlier.

9.5 MULTISTEP METHODS

In the methods considered so far for the solution of Equation 9.8, we computed the value of y_{i+l} by using the known conditions at x_i and the approximations to the derivative at various other points in the interval $x_i \leq x \leq x_{i+1}$. However, a large number of function evaluations are needed for each step, where this number is equal to the order of the method. Therefore, interest lies in methods that give comparable accuracy with a smaller number of function evaluations per step. The multistep methods form an important class of efficient methods that use the information at mesh points preceding x_i, along with that at x_i, to yield y_{i+1}. Several multistep formulas have been derived. Formulas in which y_{i+i} is given explicitly in terms of known values of the dependent variable y and of the function $F(x, y)$ at x_i, x_{i-1}, and so on, are termed *open formulas*. Similarly, finite difference formulas that include unknown values of y and F usually require iteration to solve for y_{i+1}. Such formulas are termed *implicit* or *closed*.

If the formula requires the values of F at mesh points preceding x_i, the method is not self-starting, since at the initial condition, $x = x_0$, the only known condition is the given value y_0 of the variable y, from which the function $F(x, y)$ may be evaluated at this point. The conditions prior to x_i are not known. Therefore, a self-starting method, such as a Runge–Kutta formula with the same order of accuracy as that of the multistep formula under consideration, must be employed in the first few steps, until the information needed to proceed with the given multistep method is obtained.

9.5.1 ADAMS MULTISTEP METHODS

The multistep formulas may easily be derived from the Taylor-series expansion of the dependent variable y, written as

$$y_{i+1} = y_i + \Delta x\ F_i + \frac{(\Delta x)^2}{2}\ F_i' + \frac{(\Delta x)^3}{3!}\ F_i'' + \cdots \tag{9.57}$$

where the primes denote differentiation with respect to x. We obtain this expansion from Equation 9.19 by noting that

$$\left(\frac{dy}{dx}\right)_{x_i} = F_i \quad \left(\frac{d^2y}{dx^2}\right)_{x_i} = F_i'$$

and so on. If the series in Equation 9.57 is truncated after the second term, Euler's method is obtained. This may be considered as the first open multistep formula. We can generate a series of higher-order formulas from Equation 9.57 by replacing the derivatives by their finite difference approximations from Chapter 4. If backward differences are used, the formulas obtained are known as *Adams–Bashforth* or *Adams open formulas*. Therefore, if F_i' is approximated as

$$F_i' = \frac{F_i - F_{i-1}}{\Delta x} + O(\Delta x) \tag{9.58}$$

the equation for y_{i+l} becomes

$$y_{i+1} = y_i + \Delta x\ F_i + \frac{(\Delta x)^2}{2}\frac{F_i - F_{i-1}}{\Delta x} + O\left[(\Delta x)^3\right]$$

or,

$$y_{i+1} = y_i + \Delta x\left[\frac{3}{2}F_i - \frac{1}{2}F_{i-1}\right] + O\left[(\Delta x)^3\right] \tag{9.59}$$

Therefore, a second-order formula, known as the *second open Adams formula*, is obtained. If F'' is taken as constant over the interval $x_i \le x \le x_{i+1}$, the TE per step in Equation 9.58 may be approximated as $\Delta x F_i''/2$, which leads to a TE of $5(\Delta x)^3 F_i''/12$ in Equation 9.59. The total error is $O[(\Delta x)^2]$, as discussed later. The method is thus a second-order method because of the order of the total TE.

Similarly, we can derive the third-order formula by employing the following backward difference approximations for F_i'' and F_i':

$$F_i'' = \frac{F_i - 2F_{i-1} + F_{i-2}}{(\Delta x)^2} + O(\Delta x) \tag{9.60a}$$

$$F_i' = \frac{3F_i - 4F_{i-1} + F_{i-2}}{2\Delta x} + O\left[(\Delta x)^2\right] \tag{9.60b}$$

which leads to the equation

$$y_{i+1} = y_i + \Delta x\left[\frac{23}{12}F_i - \frac{16}{12}F_{i-1} + \frac{5}{12}F_{i-2}\right] + O(\Delta x)^4 \tag{9.61}$$

Again, if F''' is assumed to be constant over the given step Δx, the error incurred in Equation 9.60a may be written as $\Delta x F_i'''$, and that in Equation 9.60b as $(\Delta x)^2 F_i'''/3$; see Chapter 4. Then these expressions for F_i' and F_i'' may be used in Equation 9.61 to yield a TE of $3(\Delta x)^4 F_i'''/8$.

The general formula for the Adams–Bashforth method may be written as

$$y_{i+1} = y_i + \Delta x \sum_{m=1}^{n} \beta_{nm} F_{i+1-m} + O\left[(\Delta x)^{n+1}\right] \tag{9.62}$$

Thus, the TE per step is $O[(\Delta x)^{n+1}]$. The order of the method is n, since the number of steps needed to solve Equation 9.8 up to a given value of x varies as $1/\Delta x$, leading to a total TE of order $(\Delta x)^n$. In Equation 9.62, n may be varied to obtain Adams–Bashforth methods of different order. Table 9.1 gives the corresponding values of the β's, which are the coefficients for these explicit methods, for n up to 6. One can determine the TE for higher-order formulas by retaining the error terms in the finite difference approximations for the derivatives, as outlined above. Note that these methods are explicit and, thus, no iteration is needed for computing y_{i+1}.

We derive the *Adams-Moulton or Adams closed formulas* by employing a backward Taylor-series expansion of $y(x)$, obtained as

$$y_i = y_{i+1} - \Delta x\, F_{i+1} + \frac{(\Delta x)^2}{2} F'_{i+1} - \frac{(\Delta x)^3}{3!} F''_{i+1} + \cdots$$

which gives

$$y_{i+1} = y_i + \Delta x\, F_{i+1} - \frac{(\Delta x)^2}{2} F'_{i+1} + \frac{(\Delta x)^3}{3!} F''_{i+1} + \cdots \tag{9.63}$$

If the series is truncated after the second term, an implicit formula for y_{i+1} is obtained as follows:

$$y_{i+1} = y_i + \Delta x\, F_{i+1} + O\left[(\Delta x)^2\right] \tag{9.64}$$

The TE in this equation can easily be shown to be $-(\Delta x)^2 F'/2$. Again, if the derivatives of F in the preceding series are replaced by backward difference approximations,

TABLE 9.1
The Values of the Coefficient β_{nm} for the Adams–Bashforth Method, for n up to 6

n	β	$m = 1$	2	3	4	5	6
1	β_{1m}	1					
2	$2\,\beta_{2m}$	3	−1				
3	$12\,\beta_{3m}$	23	−16	5			
4	$24\,\beta_{4m}$	55	−59	37	−9		
5	$720\,\beta_{5m}$	1901	−2774	2616	−1274	251	
6	$1440\,\beta_{6m}$	4277	−7923	9982	−7298	2877	−475

higher-order formulas are obtained. Therefore, we derive the second-order formula by employing

$$F'_{i+1} = \frac{F_{i+1} - F_i}{\Delta x} + O(\Delta x)$$

which gives

$$y_{i+1} = y_i + \Delta x \left[\frac{1}{2} F_i + \frac{1}{2} F_{i+1} \right] + O\left[(\Delta x)^3 \right] \quad (9.65)$$

This formula is the same as the trapezoidal rule for integration discussed in Chapter 8, and, therefore, as shown before, the TE per step is$-(\Delta x)^3 F''/12$. Note that since F_{i+1} is involved in Equations 9.64 and 9.65, these formulas are implicit and iteration is needed to solve for y_{i+1}.

Both the first-order and the second-order Adams-Moulton formulas are self-starting, since F_{i-1}, F_{i-2}, and so on, are not involved. However, the third- and higher-order formulas are not self-starting and need a self-starting method, such as Runge–Kutta, to solve for the first few steps. The general expression for the Adams closed formulas is

$$y_{i+1} = y_i + \Delta x \sum_{m=0}^{n-1} \beta^*_{nm} F_{i+1-m} + O\left[(\Delta x)^{n+1} \right] \quad (9.66)$$

where n is the order of the method. Table 9.2 gives the coefficients β^*_{nm} for various values of n up to 6.

The Adams open and closed formulas are an important class of multistep formulas. Although they are rarely used separately, combinations of the two sets of formulas

TABLE 9.2

The Values of the Coefficient β^*_{nm} for the Adams–Moulton Formulas, for n up to 6

n	β^*	$m = 0$	1	2	3	4	5
1	β^*_{1m}	1					
2	$2\beta^*_{2m}$	1	1				
3	$12\beta^*_{3m}$	5	8	−1			
4	$24\beta^*_{4m}$	9	19	−5	1		
5	$720\beta^*_{5m}$	251	646	−264	106	−19	
6	$1440\beta^*_{6m}$	475	1427	−798	482	−173	27

yield some of the most efficient predictor–corrector methods for solving ODEs, as discussed in the next section. One can also derive the Adams formulas by applying numerical integration to the differential equation, see Carnahan et al. (1969). The TE can be obtained quite easily for these methods. A smaller number of computations of the function F are needed in these formulas, as compared to those in the corresponding Runge–Kutta method, since the values of F at the preceding points are obtained from calculations performed for the earlier steps. The closed formulas are solved by iteration and the resulting error can be shown to be much less than that in an open formula of the same order. Also, the implicit methods generally possess better stability characteristics; see Ferziger (1998).

9.5.2 Additional Considerations

Several other multistep methods have been developed. A simple method that is often considered in studying the stability of multistep methods is the midpoint method, discussed in detail by Atkinson (1989) and given by

$$y_{i+1} = y_{i-1} + 2\Delta x\, F(x_i, y_i) \tag{9.67}$$

This method is second-order accurate and is not self-starting, since y_{i-1} is involved. A feature common to the various multistep methods is the existence of multiple solutions to the difference equation. For a convergent method, one of the solutions closely approximates the exact solution and is known as the *fundamental solution*. The other solutions, known as *parasitic solutions*, are not related to the exact solution of the differential equation. If these parasitic solutions grow with each computational step, instability arises in the numerical scheme. The growth of the parasitic solution is often exponential and oscillatory, leading to a rapid overpowering of the fundamental solution. In a stable scheme, the parasitic solution remains small compared to the fundamental solution. In practice, the numerical solution is obtained at two significantly different step sizes, and stability is assumed if the results are in reasonably close agreement. A reduction in step size also often leads to stability in a previously unstable solution.

The Runge–Kutta methods are often more stable than the corresponding multistep methods. Also, the starting method for computing the first few steps in multistep methods can substantially affect the final results. Therefore, it is important to vary the starting method to ensure that the results are essentially independent of the method used. Despite these disadvantages, multistep methods are frequently employed, particularly as predictor–corrector methods, and have, in many cases, replaced the Runge–Kutta scheme, which was the standard solution procedure for many years.

9.6 PREDICTOR–CORRECTOR METHODS

The predictor–corrector methods combine the advantages of accuracy and stability of the implicit formulas with the simplicity of the explicit formulas. The main problem with the implicit equations is the time-consuming iterative procedure necessary for obtaining the solution. However, if a first estimate y^*_{i+1} of the new value of the dependent variable is provided by the application of an explicit method at each step, the number of iterations needed for convergence to the solution by the use of the implicit method may be minimized. Therefore, an explicit formula is taken as the "predictor" to give a first estimate of the solution, followed by the use of an implicit formula as the "corrector" to obtain a better approximation to the solution.

9.6.1 Basic Features

The predictor–corrector methods are generally more efficient than the Runge–Kutta methods of the same order. The fourth-order Runge–Kutta method requires four function evaluations for advancing the solution by one step. The corresponding predictor–corrector method requires only one function evaluation for the predictor, since the other values are available from earlier computations, and, if fewer than three iterations are needed for the corrector, this method will require less computer time than the Runge–Kutta formulas of the same order. A detailed comparison between these two families of methods for solving ODEs is given by Hall et al. (1972). Another important advantage of the predictor–corrector methods is the ability to estimate the TE at each step. This ability allows one to choose a suitable step size for achieving the desired accuracy level and also to estimate the accuracy of the converged solution. Because of these advantages, the predictor–corrector methods are among the most popular ones at the present time. However, the problem of starting the scheme is usually present, since most schemes are based on multistep formulas such as those discussed in Section 9.5 and are not self-starting.

The improved Euler's method, given by Equations 9.29, is one of the simplest predictor–corrector methods. Euler's method is used as the predictor to give the first estimate y^*_{i+1} of the dependent variable as

$$y^*_{i+1} = y_i + \Delta x\ F(x_i, y_i) \tag{9.29a}$$

The trapezoidal rule for numerical integration is then used as the corrector to give

$$y_{i+1} = y_i + \frac{\Delta x}{2}\left[F(x_i, y_i) + F(x_{i+1}, y^*_{i+1})\right] \tag{9.29b}$$

This method is self-starting, unlike most other predictor–corrector methods, and is commonly employed in problems of engineering interest. As mentioned before, the method is also known as *Heun's predictor–corrector method*. The TE per step for the predictor is

$$\frac{(\Delta x)^2}{2}\frac{d^2y}{dx^2}$$

For the corrector, the error is

$$-\frac{(\Delta x)^3}{12}\frac{d^3y}{dx^3}$$

when the corrector is taken as Equation 9.30 and is solved by iteration. The error is larger if Equation 9.29b is employed, without iteration.

We apply iteration to Equation 9.30 by substituting the computed value obtained into the right-hand side of the equation to obtain an improved value of y_{i+l}. The process is repeated until a specified convergence criterion is satisfied. The above corrector equation may also be employed with other multistep formulas as the predictor. A commonly employed nonself-starting scheme is obtained by the use of the midpoint method, given in Equation 9.67, with Equation 9.29b as the corrector. Note from the above expressions for the TEs that the use of the corrector considerably improves the accuracy of the solution.

Several other predictor–corrector methods are employed in the solution of ODEs. The basic characteristics of these methods are the same as those outlined above. A predictor is employed to yield a first estimate of the dependent variable y_{i+l} at the next step, and this value is used to start the iterative solution of the corrector for an improved value of y_{i+l}. Among the most important and widely used predictor–corrector methods are the *Adams method*, *Milne's method*, and *Hamming's method*. None of these methods is self-starting, and generally a Runge–Kutta scheme, of the same order as the given predictor–corrector method, is employed for the first few steps. One can also use a Taylor-series expansion of the variable $y(x)$ to obtain the values of y_1, y_2, y_3, and so on, needed to start the method. For instance, y_i may be written as

$$y_1 = y_0 + \Delta x\ F(x_0, y_0) + \frac{(\Delta x)^2}{2}F'(x_0, y_0) + \frac{(\Delta x)^3}{3!}F''(x_0, y_0) + \cdots \tag{9.68}$$

The values of the function F and its derivatives are obtained from the given differential equation, Equation 9.8. The number of terms retained in the series should be chosen to yield an error of the same order as that obtained by the given predictor–corrector method. However, the higher-order derivatives of the given differential equation may be quite involved. In such cases, the Runge–Kutta method is preferable for obtaining the starting values.

9.6.2 Adams Method

We can obtain a family of predictor–corrector methods by employing the Adams–Bashforth formulas as the predictor and the Adams-Moulton methods of same order as the corrector (see Tables 9.1 and 9.2). If the fourth-order formulas are chosen, the predictor equation is

$$y_{i+1}^{(0)} = y_i + \frac{\Delta x}{24}\left[55F_i - 59F_{i-1} + 37F_{i-2} - 9F_{i-3}\right] \tag{9.69a}$$

and the corresponding corrector formula is

$$y_{i+1} = y_i + \frac{\Delta x}{24}\left[9F_{i+1} + 19F_i - 5F_{i-1} + F_{i-2}\right] \tag{9.69b}$$

The superscript (0) indicates the predicted value that forms the first estimate for the corrector. The values of y and F for the first three steps of Δx, beyond the initial condition, are needed for starting this method. A fourth-order Runge–Kutta method may be employed to generate these values. Then the predictor gives an estimate for the next value of y, and, using this value as the first estimate, one iterates the corrector until convergence, in terms of a specified criterion, has been achieved. Similarly, the Adams formulas of different orders may be used to generate other predictor–corrector methods. The TEs associated with this method are discussed later.

9.6.3 Milne's Method

This method is based on integrating the differential equation, Equation 9.8, to obtain

$$y_{i+1} = y_{i-3} + \int_{x_{i-3}}^{x_{i+1}} F(x,y)\,\mathrm{d}x \tag{9.70}$$

The integral may be viewed as area under the curve from x_{i-3} to x_{i+1}, as shown in Figure 9.13. If $F[x, y(x)]$ is approximated by the quadratic expression $ax^2 + bx + c$, where a, b, and c are constants, we can determine the constants by employing the

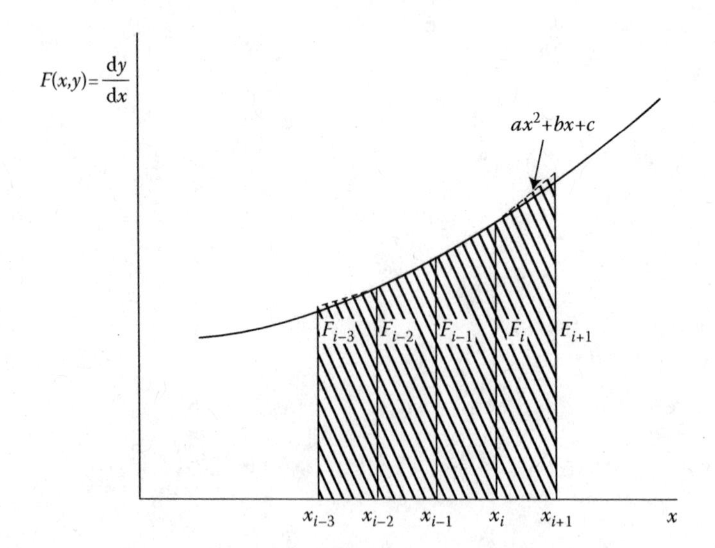

FIGURE 9.13 Sketch of the numerical integration used for the derivation of the predictor equation in Milne's method.

value of the function F_i at x_i, the value F_{i-1} at x_{i-1}, and so on. If the above integral is then carried out, we obtain

$$y_{i+1}^{(0)} = y_{i-3} + \frac{4}{3}\Delta x\left[2F_i - F_{i-1} + 2F_{i-2}\right] \quad (9.71)$$

This is the predictor equation for Milne's method. The corrector is simply obtained from Simpson's integration rule, given by Equation 8.31, as

$$y_{i+1} = y_{i-1} + \frac{\Delta x}{3}\left[F_{i-1} + 4F_i + F_{i+1}\right] \quad (9.72)$$

This is a fourth-order method since the total TE is $O[(\Delta x)^4]$, as shown later. Higher-order schemes have also been developed; see Carnahan et al. (1969) for the sixth-order Milne's method.

9.6.4 Hamming's Method

Hamming's method is based on the use of a general class of corrector equations represented by

$$\begin{aligned} y_{i+1} &= a_i y_i + a_{i-1}y_{i-1} + a_{i-2}y_{i-2} \\ &\quad + \Delta x\left(b_{i+1}F_{i+1} + b_iF_i + b_{i-1}F_{i-1} + b_{i-2}F_{i-2}\right) \end{aligned} \quad (9.73)$$

where the a's and b's are constants. This equation includes the correctors employed for the fourth-order Adams and Milne's predictor–corrector methods. We can determine the constants by employing Taylor-series expansions for all the variables and functions that appear in the above equation. Terms are retained up to $O[(\Delta x)^4]$, and the coefficients of similar terms on both sides of the equation are set equal. This results in the number of unknowns being larger than the relationships between them so that a few constants must be chosen arbitrarily. Hamming (1959) studied the stability of this corrector and chose the parameters to obtain better stability characteristics than those of the corrector used in Milne's method. The resulting equation is

$$y_{i+1} = \frac{1}{8}\left(9y_i - y_{i-2}\right) + \frac{3}{8}\Delta x\left(F_{i+1} + 2F_i - F_{i-1}\right) \quad (9.74)$$

The method employs the same predictor as that used in the fourth-order Milne's predictor–corrector method. A modifier equation is also used in order to reduce the error in the predicted value of y_{i+i}. The computational formulas for Hamming's method are, therefore, given in the order in which they are used as follows:

$$\text{Predictor:}\quad y_{i+1}^{(0)} = y_{i-3} + \frac{4}{3}\Delta x\left(2F_i - F_{i-1} + 2F_{i-2}\right) \quad (9.75a)$$

$$\text{Modifier:}\quad \bar{y}_{i+1}^{(0)} = y_{i+1}^{(0)} - \frac{112}{121}\left(y_i^{(0)} - y_i\right) \tag{9.75b}$$

$$\text{Corrector:}\quad y_{i+1} = \frac{1}{8}\left(9y_i - y_{i-2}\right) + \frac{3}{8}\Delta x\left(F_{i+1} + 2F_i - F_{i-1}\right) \tag{9.75c}$$

The superscript (0) refers to the initial estimate that may be employed for starting the iteration process in the corrector equation. The estimate from the predictor is employed in the modifier, which provides the first estimate $\bar{y}_{i+1}^{(0)}$ for computing F_{i+1} on the right-hand side of the corrector for the first iteration. In practice, the step size is chosen so that only one or two iterations are needed for convergence. In fact, the method is generally used without iteration.

9.6.5 Accuracy and Stability of Predictor–Corrector Methods

We have given the general formulas employed in several important predictor–corrector methods. The basic characteristics of all the methods are quite similar, and any one of these can generally be employed for a given ODE. The choice of a particular method is frequently made on the basis of personal preference, since the difference in the computational procedure and in the numerical results is generally small. However, the TEs associated with each formula are different from those that arise in other formulas. Similarly, the stability and convergence characteristics are different. These differences are sometimes important in the choice of the method for solving a given problem and are discussed here.

9.6.5.1 Truncation Errors

Let us first consider the TE at each step in the application of the formulas discussed above. Proceeding as outlined earlier for the second- and third-order Adams–Bashforth methods, we obtain the error that arises in the fourth-order formula as

$$\left(E_{i+1}\right)_p = \frac{251}{720}\left(\Delta x\right)^5 F''''(\xi),\quad x_{i-3}\ \xi < x_{i+1} \tag{9.76}$$

where $(E_{i+1})_p$ is the estimate of the TE in the predictor for the $(i + 1)$th step. The corresponding error $(E_{i+1})_c$ in the corrector, which is the fourth-order Adams–Moulton formula, is

$$\left(E_{i+1}\right)_c = -\frac{19}{720}\left(\Delta x\right)^5 F''''(\xi),\quad x_{i-2} < \xi < x_{i+1} \tag{9.77}$$

Note that the error in the corrector is much smaller than that in the predictor.

If we assume the value at x_i to be exact and if the round-off errors in the calculations for the $(i + 1)$th step are taken as negligible, as done before for estimating the TE per step, the exact solution at x_{i+1} may be written as

$$y\left(x_{i+1}\right) = y_{i+1}^{(0)} + \frac{251}{720}\left(\Delta x\right)^5 F_i'''' \tag{9.78}$$

or as

$$y\left(x_{i+1}\right)=y_{i+1}-\frac{19}{720}\left(\Delta x\right)^5 F_i'''' \tag{9.79}$$

where $y_{i+1}^{(0)}$ is the predicted value from the predictor, and y_{i+1} is the converged value from the corrector. From these equations, we obtain the estimate of the truncation error per step, after the application of the corrector equation, by determining $(\Delta x)^5 F_i''''$ and then using Equation 9.77 as follows:

$$E_{i+1} \cong -\frac{19}{270}\left[y_{i+1}-y_{i+1}^{(0)}\right] \tag{9.80}$$

In addition to the assumptions given above, this estimate is based on the assumptions that F'''' is essentially constant over the interval $x_{i-3} \le x \le x_{i+1}$ and that the TEs per step in the predictor and the corrector are given by Equations 9.76 and 9.77, respectively. This estimate of the error may be used for determining whether the desired accuracy level is being maintained in the computation.

Similar estimates may be obtained for Milne's method and for Hamming's method. We can obtain the TE in the predictor of Milne's method by considering the approximation employed for the function F in the integral of Equation 9.70. The corrector is based on Simpson's rule, the error for which was obtained in Chapter 8. The resulting TEs per step are thus obtained as

$$\left(E_{i+1}\right)_P=\frac{14}{45}\left(\Delta x\right)^5 F''''(\xi), \quad x_{i-3}<\xi<x_{i+1} \tag{9.81}$$

and

$$\left(E_{i+1}\right)_c=-\frac{1}{90}\left(\Delta x\right)^5 F''''(\xi), \quad x_{i-1}<\xi<x_{i+1} \tag{9.82}$$

Following the above procedure, the estimate for the TE per step, after convergence of the corrector equation, is given by

$$E_{i+1} \cong -\frac{1}{29}\left[y_{i+1}-y_{i+1}^{(0)}\right] \tag{9.83}$$

The same approach may be applied to Hamming's method. Then the TEs are obtained (Carnahan et al., 1969) as follows:

$$\left(E_{i+1}\right)_P=\frac{14}{45}\left(\Delta x\right)^5 F''''(\xi), \quad x_{i-3}<\xi<x_{i+1} \tag{9.84}$$

and

$$\left(E_{i+1}\right)_c = -\frac{1}{40}\left(\Delta x\right)^5 F''''\left(\xi\right), \quad x_{i-2} < \xi < x_{i+1} \tag{9.85}$$

This results in the estimate for the TE per step as

$$E_{i+1} \cong -\frac{9}{121}\left[y_{i+1} - y_{i+1}^{(0)}\right] \tag{9.86}$$

Note from the above expressions that the TE per step in the corrector is the smallest for Milne's method, followed by that in Hamming's method and that in the fourth-order Adams predictor–corrector method. The errors in the last two methods are close to each other and more than twice that in Milne's method. However, these methods have better stability characteristics than Milne's method, as discussed below. The TE may be estimated at each step to ensure that the numerical results have the desired accuracy. The step size can, therefore, be reduced if the error is too large, or it can be increased, to save computer time, if the error is too small. This ability to estimate the TE at each step is one of the important advantages of predictor–corrector methods over Runge–Kutta methods. However, a change in the step size is much more involved in predictor–corrector methods, since values of the derivative function F are needed at evenly spaced x values preceding x_i.

9.6.5.2 Step Size

An approach frequently adopted in changing the step size in predictor–corrector methods is simply to restart the computation scheme, with the last computed value of y as the initial condition for the new step size. Therefore, the starting method will again be needed for generating the values for the first few steps, and then the predictor–corrector method may be employed. Another approach is to use interpolation to obtain F values at the new spacing for $x \leq x_i$, using the computed values for the previous step size. A polynomial fit, as discussed in Chapter 7, may be employed to obtain a curve that passes through the available F values at the previous spacing. The required F values at the new spacing may then simply be obtained by interpolation, as shown qualitatively in Figure 9.14. The estimate of the TE per step may be employed in the subsequent calculations to determine whether a change in the step size is again needed.

In predictor–corrector methods, the step size should be small enough to ensure convergence of the corrector equation in only one or two iterations. This is necessary in order to maintain the advantage of these methods, over one-step methods of comparable accuracy. However, a larger step size is desirable for reducing the computation for a given range of the independent variable x and, therefore, also for reducing the round-off error. The step size may be changed on the basis of the TE, as outlined above, or if more than two iterations are needed for the convergence of the corrector.

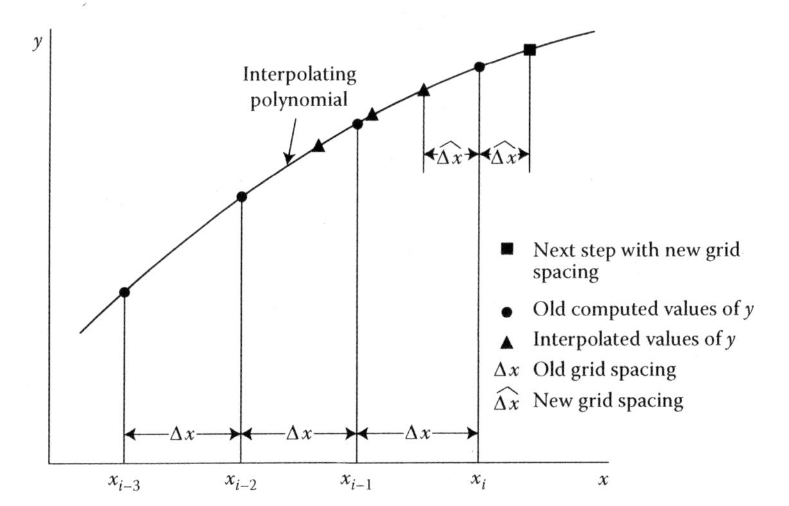

FIGURE 9.14 Interpolation of the preceding numerical results in order to vary the step size in predictor–corrector methods.

9.6.5.3 Stability

The stability characteristics of the various predictor–corrector methods discussed here have been studied in the literature, as considered in detail by Carnahan et al. (1969), Ralston and Rabinowitz (1978), and Atkinson (1989). If the corrector is iterated to convergence, then the stability of the predictor is not of much concern since it simply provides a first estimate. In fact, most predictors do not possess good stability characteristics. The stability of the corrector is, however, important, and the growth of error is studied to determine the overall stability of the scheme. If the corrector is not iterated but is employed only once, the stability of both the predictor and the corrector equations must be studied.

The corrector equations for the Adams method and Hamming's method have good stability characteristics. These methods are, therefore, often preferred over Milne's method, whose corrector equation is unstable for some differential equations. The stability analysis of multistep methods involves a consideration of the growth of the parasitic solutions, mentioned earlier. Milne's method is stable if the exact solution of the differential equation decays with increasing x and is generally known as a *marginally stable method.*

Most of the stability analyses consider simple linear equations, and the results obtained are often extended to more involved equations, particularly nonlinear equations. However, such an extension of the conclusions of the simple stability analyses may not be applicable in many cases. A practical approach employed in most engineering applications is to obtain the results for different step sizes. If the computed values do not differ significantly from each other, then the computational scheme is assumed to be stable. Also, analytical solutions may be available for a few simple circumstances. Then one can use a comparison of the numerical results with these solutions to study the accuracy of the results obtained and the stability characteristics of the method. Also, a reduction in step size usually results in an improvement in the stability of the numerical scheme.

9.6.6 Simultaneous Equations

It must be pointed out that, although the multistep and the predictor–corrector methods have been discussed for the first-order initial-value problem given by Equations 9.8 and 9.10, these methods may easily be extended to a system of simultaneous first-order equations. Since higher-order equations may be reduced to a system of first-order equations, as outlined in Section 9.1.2, these methods may be used for solving higher-order equations as well. A starting method, such as the Runge–Kutta method, is used to generate the required values for the first few steps, for each of the n dependent variables Y_j, where $j = 1, 2, \ldots, n$. The appropriate formulas are then applied to each equation in sequence at each step to obtain the values $(Y_j)_{i+1}$ at x_{i+1}. At each step, the stored function values from the preceding steps are employed to compute the new values, which are then stored for use in the next step. The procedure is a simple extension of the method for solving a single first-order equation.

There are several engineering problems that involve a wide range of scale, say, in length or time. In fluid flow and heat transfer, for instance, a small length scale may often be important in a given region, while a much larger length scale characterizes the remaining region. Similarly, in process control and chemical kinetics, a wide range of time or rate constants may arise. A system of equations that is associated with widely different time constants or eigenvalues is known as *stiff*, and special techniques are often needed to solve such a system. The step size must be small enough to treat the smallest scale or the fastest component of the process. An nth-order differential equation will, in general, have n scales. Employing an extremely small step size to take into account the smallest scale, although the other components can be treated with much larger step sizes, is obviously inefficient. A very small step size will result in large computer time and also large round-off errors. A major problem lies in maintaining a smooth behavior of the solution at large values of the independent variable, since this can lead to instability in this region. The problem of stiffness is similar to that of ill-conditioning encountered in matrices, discussed in Chapter 6. Several special techniques have been developed to solve stiff problems. The most popular among these is Gear's method. For further details, see Gear (1971), Hall and Watt (1976), and Ferziger (1998).

9.6.7 Concluding Remarks on Predictor–Corrector Methods

The preceding discussion indicates that the predictor–corrector methods are among the most efficient methods available for the solution of ODEs. In addition, the TE at each step is determined during the computation and may be employed for maintaining the desired accuracy level by changing the step size whenever the error is excessive. However, these methods are more involved than the self-starting methods, such as Runge–Kutta formulas, which continue to be a very popular choice for the solution of the ODEs that arise in engineering problems.

The choice of a predictor–corrector method, from among those considered here, is often not an easy one, since the Adams method, Milne's method, and Hamming's method are all quite comparable in terms of efficiency and accuracy. Hamming's method avoids the instability problems of Milne's method and is, therefore, often preferred. Also, it very seldom requires iteration and is generally used without iteration, making it a relatively more efficient method to use. However, if stability problems do not arise, Milne's method is superior because of its higher accuracy level. In general, personal preference and prior experience with the method are strong criteria for choosing it. Otherwise, Hamming's method may be chosen, despite the slight additional complexity in programming. The second-order predictor–corrector methods, although somewhat simpler to program, are rarely used because of the resulting lower accuracy in the results. MATLAB has several functions based on multistep methods available for the solution of ODEs. These include *ode*113, which is a multistep Adams–Bashforth–Moulton solver of varying order, and *ode*15s, which is an implicit multistep numerical solver of varying order. The latter one is often employed if *ode*45 is too inefficient or fails to yield the solution.

Example 9.3

A metal block of volume V and surface area a is initially at temperature T_i. At the surface, a constant energy input q, per unit area and time, is imposed at time $t = 0$ by thermal radiation, while the surface also loses energy by convection to air at temperature T_a surrounding the block. If the temperature in the block is assumed to be uniform at any given time and is denoted by $T(t)$, energy balance leads to the following governing equation for the temperature:

$$\rho CV \frac{dT}{dt} = qa - ha(T - T_a) \tag{9.87a}$$

where ρ and C are the density and specific heat, respectively, of the metal. The parameter h is termed the *convective heat transfer coefficient*, and its value depends on the flow of air around the block and the temperatures involved. If the temperature difference $(T–T_a)$ is taken as the dependent variable θ, the above equation may be written as

$$\frac{d\theta}{dt} = \frac{qa}{\rho CV} - \frac{ha}{\rho CV}\theta = A - B\theta \tag{9.87b}$$

where A and B are parameters defined as $A = qa/\rho CV$ and $B = ha/\rho CV$. Using the fourth-order Adams predictor–corrector method, solve this problem to obtain $\theta(t)$ if θ at $t = 0$ is given as 100°C. Consider two circumstances, given as (a) $A = 10$°C/s, $B = 0.05\ s^{-1}$ and (b) $A = 2$°C/s, $B = 0.03\ s^{-1}$.

SOLUTION

The given problem involves solving the following one-dimensional ODEs:

$$\frac{d\theta}{dt} = 10 - 0.05\theta \tag{9.88}$$

$$\frac{d\theta}{dt} = 2 - 0.03\theta \tag{9.89}$$

with the initial condition

$$\theta = 100^\circ\text{C} \quad \text{for } t = 0 \tag{9.90}$$

The given equations are quite simple and may be solved analytically. However, they may be used to demonstrate the application of the Adams predictor–corrector method to ODEs. Then we can compare the numerical results with the analytical solution to evaluate the accuracy of the numerical scheme. However, in actual practice, A and B are usually not constants but vary with θ and t, resulting in much more complicated problems for which the analytical solution may not be available.

The formulas for the fourth-order Adams predictor–corrector method are given by Equations 9.69. A starting method is needed to generate the θ values at the three time steps, Δt, 2 Δt, and 3 Δt, so that Equation 9.69a can be employed to obtain the predicted value at $t = 4\ \Delta t$, where Δt is the time step. The fourth-order Runge–Kutta scheme, given by Equations 9.37, is employed to obtain these starting values. The value at $t = 0$ is, of course, the initial condition $\theta(0) = 100$. The predicted value is corrected iteratively, using Equation 9.69b, until a specified convergence criterion has been satisfied. Thus, the value of θ at the next time step, denoted by $i + 1$, is obtained from the known values at the previous four time steps. Using this new computed value, the computation proceeds to the next time step. Thus, $\theta(t)$ is obtained with increasing time, starting with the initial condition. The computation is carried out until the temperature θ does not change significantly from one time step to the next, indicating the attainment of steady-state conditions.

Appendix B.27 shows the MATLAB script file for solving the first-order ODEs in this problem by the Adams predictor–corrector method. The fourth-order Runge–Kutta method is used for the first three steps. Then these three values, along with the initial condition, are employed to compute the intermediate value from the predictor. This value is then used to start the iteration of the corrector. This iteration is terminated when the dependent variable y changes less than a specified convergence parameter *ep1* from one iteration to the next. A time step Δt of 0.05 s is taken. For this value of Δt, only one or two iterations were needed for the convergence of the corrector. Another convergence parameter *ep* is used to terminate the overall computation, for which the criterion used is

$$\left| \frac{T_{i+1} - T_i}{T_i \Delta t} \right| \le \varepsilon \tag{9.91}$$

where ε is a chosen convergence criterion for determining steady state. A value of 10^{-4} was employed for ε, or *ep*. A value of 10^{-5} was used for *ep1*. These values, as well as Δt, were varied to ensure a negligible effect of the chosen values, on the numerical results.

Figure 9.15 shows the computed variation of temperature θ with time t for the two cases. The initial temperature difference $(T - T_a)$ is 100°C. In the first case, for which $A = 10$°C/s and $B = 0.05\ \text{s}^{-1}$, the energy input is larger than the convective

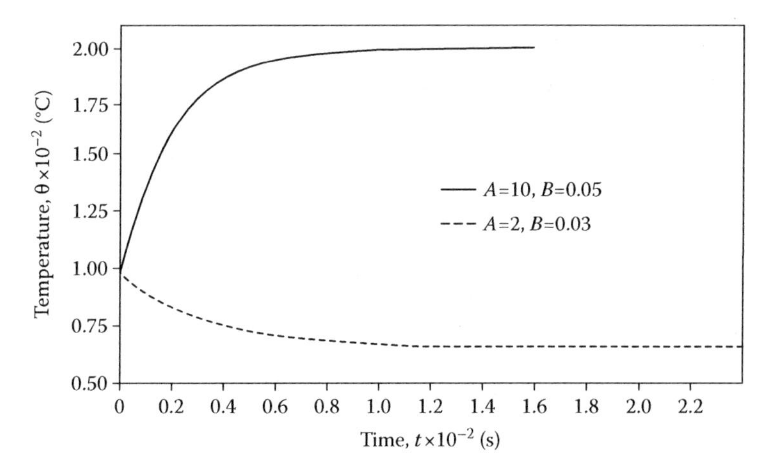

FIGURE 9.15 Variation of temperature θ with time t for the two cases of Example 9.3, computed using the fourth-order Adam's predictor–corrector method. The results are obtained with $\varepsilon = 10^{-4}$ and $\Delta t = 0.05$ s.

energy loss at $t = 0$. Thus, $d\theta/dt = 5$ at $t = 0$ from Equation 9.88. This positive initial value of the slope results in a temperature increase with time. Finally, a constant value of $\theta = 200°C$ is attained at steady state. In the second case, the energy input is less than the loss, and the metal block cools down to a temperature of $\theta = 67°C$ at steady state. The steady-state temperatures can easily be obtained analytically from Equations 9.88 and 9.89. At steady state, θ stops changing with time, and, therefore, $d\theta/dt = 0$. If $d\theta/dt$ is set equal to zero in these equations, we obtain $\theta = 200°C$ and 66.67°C in the two cases. These values agree closely with the numerical results obtained. The computed variation of θ with t was also compared with the analytical solution, and a close agreement between the two was obtained. For further details on the physical aspects of this problem and other similar ones, see Incropera et al. (2006).

Example 9.4

A metal piece of mass m is released at zero velocity in a liquid and allowed to fall freely under gravity. The frictional force, or drag, acting on the piece is $m(AV + BV^2)$, where A and B are constants and V is the downward velocity. Briefly discuss the methods that can be used to solve this problem. Then, using any appropriate method, compute the velocity V as a function of time t for $B = 0.1$ m^{-1} at two values of A given as $A = 2$ and 4 s^{-1}.

SOLUTION

The governing ODE for this problem may be derived on the basis of the discussion in Example 9.2 as

$$\frac{dV}{dt} = g - (AV + BV^2) \tag{9.92}$$

with the following initial condition:

$$\text{At } t = 0: V = 0 \tag{9.93}$$

Here, g is the magnitude of acceleration due to gravity and is equal to 9.8 m/s^2 in SI units. Therefore, the equations to be solved for the two cases are

$$\frac{dV}{dt} = 9.8 - (2V + 0.1V^2) = F(t,V) \tag{9.94}$$

and

$$\frac{dV}{dt} = 9.8 - (4V + 0.1V^2) = F(t,V) \tag{9.95}$$

Therefore, these are first-order ODEs, with the initial condition given by Equation 9.93. They can be solved by all the methods discussed so far. The fourth-order Runge–Kutta method is particularly convenient since it is accurate and self starting. The MATLAB functions *ode23* and *ode45* can also be used conveniently. The Adam's predictor–corrector method, demonstrated in the previous example, can also be employed. Details on these methods and examples have been given earlier. Let us use Hamming's method, employing the predictor, modifier and corrector in sequence only once for each step, to demonstrate the use of the predictor–corrector approach without iteration.

The numerical scheme is similar to that for the Adams predictor–corrector method, outlined in Example 9.3. We use the fourth-order Runge–Kutta method to obtain the values of the velocity at the first three time steps, the value at $t = 0$ being given by the initial condition. Once the values at time steps 0, Δt, 2 Δt, and 3 Δt have been computed, the predicted value at $t = 4\ \Delta t$ is obtained from Equation 9.75a, written for the present problem as

$$V_{i+1}^{(0)} = V_{i-3} + \frac{4}{3}\Delta t(2F_i - F_{i-1} + 2F_{i-2}) \tag{9.96a}$$

As this equation shows, to compute the predicted value of V at a given time step, we must know the values of the function F at the previous four steps. This predicted value is used in the modifier, to obtain an improved estimate $\bar{V}_{i+1}^{(0)}$ as

$$\bar{V}_{i+1}^{(0)} = V_{i+1}^{(0)} + \frac{112}{121}\left(V_i - V_i^{(0)}\right) \tag{9.96b}$$

The step size Δt was chosen as 0.05 s, and iteration is not used. The corrector is given by

$$V_{i+1} = \frac{1}{8}(9V_i - V_{i-2}) + \frac{3}{8}\Delta t(\bar{F}_{i+1}^{(0)} + 2F_i - F_{i-1}) \tag{9.96c}$$

where $\bar{F}_{i+1}^{(0)}$ is $F(x_{i+1}, \bar{V}_{i+1}^{(0)})$. The two values $\bar{V}_{i+1}^{(0)}$ and V_{i+1} were found to be very close for this value of Δt. If a larger time step is chosen, iteration may be needed to satisfy a specified convergence criterion.

Appendix B.28 gives the MATLAB script file for the solution of this problem. The computation is terminated when the velocity V stops changing, indicating the attainment of the terminal velocity. This condition is determined by the convergence criterion

$$\left|V_{i+1} - V_i\right| \leq \varepsilon \tag{9.97}$$

where ε is a chosen small quantity. It was taken as 10^{-4} and reduced to smaller values to ensure that the numerical results remain essentially unchanged. Figure 9.16 shows the computed velocity variation with time for the two cases. The velocity starts at zero, as given by the initial condition, rises sharply, and then gradually approaches the terminal velocity. The terminal velocity is attained when the net force on the body is zero, resulting in dV/dt becoming zero. Therefore, from Equation 9.94, the terminal velocity for the first case is given by the root of the equation

$$9.8 - \left(2V + 0.1V^2\right) = 0 \tag{9.98}$$

which gives V as 4.07 m/s. Similarly, the value for the second case is obtained as 2.31 m/s. The numerical results agree closely with these values.

A comparison of the programs for the Adams method and Hamming's method shows that the two are fairly similar in the general approach. Both need a starting method. However, Hamming's method is generally used without iteration, as presented here. The error may be monitored to ensure that the step size is not too large. Hamming's method is popular for problems that require a high level of accuracy. As mentioned earlier, it is generally more efficient than a Runge–Kutta scheme of the same order.

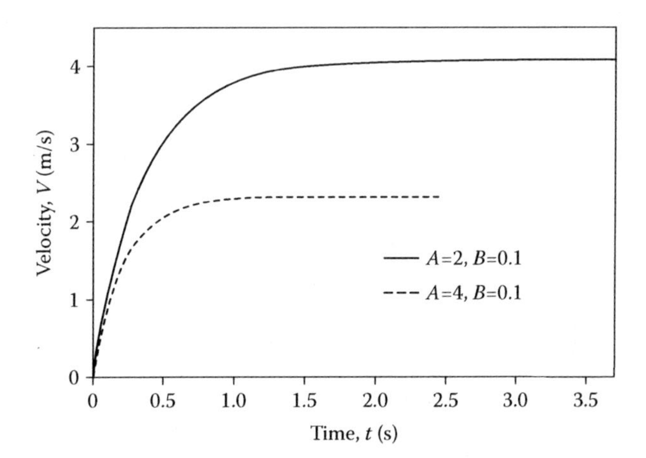

FIGURE 9.16 Variation of velocity V with time t computed using Hamming's method without iteration of the corrector for $B = 0.1$ m^{-1} and $A = 2$ and 4 s^{-1}, as given in Example 9.4.

9.7 BOUNDARY-VALUE PROBLEMS

So far, we have considered the solution of initial-value problems, in which all the conditions to be satisfied by the solution are specified at one value of the independent variable x. Integration of the ODE is started at this point, which is often specified as $x = 0$. Also, if the conditions are all specified at a given nonzero value of x, say, $x = a$, a change of variable x to $\bar{x}$, where $\bar{x} = x - a$ can be employed to impose the conditions at zero value of the transformed independent variable $\bar{x}$. However, in engineering applications, we are frequently concerned with problems in which the conditions are imposed at two, or more, different values of the independent variable. Such problems, known as boundary value problems, arise, for instance, in mass diffusion through a porous plate, conduction heat transfer in extended surfaces, vibration of strings, fluid flow over a surface, and deflection of a beam under a given loading. Since at least two conditions, specified at two different values of the independent variable, are necessary for a BVP, we are concerned with differential equations of second or higher order.

A simple example of a two-point BVP is the second-order equation

$$\frac{d^2y}{dx^2} = F(x, y, y') \tag{9.99}$$

with the boundary conditions

$$y(a) = A \quad \text{and} \quad y(b) = B \tag{9.100}$$

Here, y is the dependent variable, and y' the first derivative. Two conditions on y are specified at two values, a and b, of the independent variable x. Therefore, the solution must satisfy the given boundary conditions at $x = a$ and $x = b$. We cannot start at an initial point and march, with increasing x, to obtain the desired solution, since the derivative y' is not known at $x = a$ or $x = b$.

There are two main approaches to the numerical solution of such BVPs. The first approach reduces the problem to an initial-value problem and uses trial and error to satisfy the boundary conditions. Methods based on this approach are known as *shooting methods*, since the adjustment of initial conditions to satisfy the conditions at the other location is similar to shooting at a target. In this case, the previously discussed methods for solving initial-value problems are employed, with a root-solving method from Chapter 5, to satisfy the given boundary conditions. The second approach is based on obtaining a finite difference approximation to the differential equations and the boundary conditions and then solving the resulting algebraic equations by the methods discussed in Chapter 6. Both approaches have their advantages and disadvantages, and the choice is often made on the basis of accuracy needed, available software and personal preference. MATLAB also has built-in functions for the solution of BVPs. A particularly useful one is the function *bvp4c*, which is a BVP solver and can be used with *bvpinit* to form the initial solution guess and *bvpval* to evaluate/interpolate the solution obtained. Let us first consider shooting methods, employing the discussion given in the preceding sections of this chapter.

9.7.1 Shooting Methods

The basic approach is to convert a BVP into an equivalent initial-value problem, by applying all the conditions at one value of the independent variable x and using guessed values for those that are unknown. For instance, the problem given in Equations 9.99 and 9.100 can be reduced to the following initial-value problem:

$$\frac{d^2y}{dx^2} = F(x, y, y') \tag{9.101}$$

$$y(a) = A \quad \text{and} \quad y'(a) = P \tag{9.102}$$

where P is an unknown that must be determined so that the condition $y(b) = B$ is satisfied in order to yield the solution to the given BVP. Once a value of P is chosen, the initial-value problem, given by Equations 9.101 and 9.102, may be solved by any of the methods discussed earlier in this chapter. However, the value of $y(b)$ will not, in general, be equal to the value B demanded by the boundary condition at $x = b$. An iterative adjustment of the initial slope P is, therefore, needed to satisfy the boundary condition $y(b) = \text{B}$ within a specified convergence criterion Figure 9.17 illustrates this process of correcting the initial slope until the solution of the equivalent initial-value problem satisfies the given boundary condition at $x = b$.

From the above treatment, the BVP reduces to the solution of an equivalent initial-value problem, with the initial slope P being obtained by iteratively solving the equation

$$y_b(P) = B \quad \text{or} \quad f(P) = y_b(P) - B = 0 \tag{9.103}$$

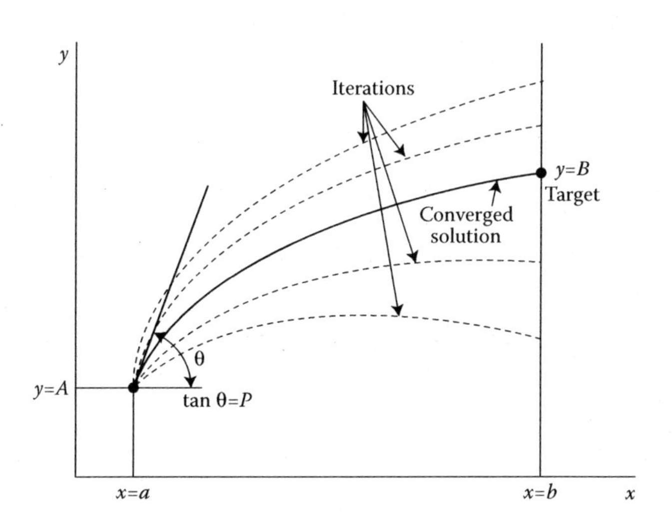

FIGURE 9.17 Sketch of the iterations to the converged solution, employing a shooting method for solving a BVP.

where y_b is the value of the dependent variable at $x = b$, and the parentheses indicate its dependence on P. This is a problem in root solving and a suitable method may be obtained from the various methods discussed in Chapter 5. The secant method, which was found to converge very rapidly in most cases, and the bisection method, which always converges if the interval containing the root is known, are both quite suitable for this application. Considering the secant method, if two solutions of the initial-value problem are obtained using P_{i-1} and P_i as two estimates of the initial slope, yielding the corresponding values of y at $x = b$ as $y_b(P_{i-1})$ and $y_b(P_i)$, then the next approximation to the root of Equation 9.103 is obtained by recognizing that P replaces x in Equation 5.11 as

$$P_{i+1} = \frac{P_{i-1} y_b(P_i) - P_i y_b(P_{i-1}) + (P_i - P_{i-1}) B}{y_b(P_i) - y_b(P_{i-1})} \tag{9.104}$$

The initial-value problem is then solved with P_{i+1} as the initial slope, and $y_b(P_{i+1})$ is obtained. If the convergence criterion, specified as, say, $|y_b(P_{i+1}) - B| \le \varepsilon$, where ε is a specified small quantity, is not satisfied, a new approximation to the root is obtained from Equation 9.104. The iterative process is continued until the specified convergence criterion is met.

The Newton–Raphson method can also be used if one numerically determines the derivative $\mathrm{d}[y_b(P)]/\mathrm{d}P$ by solving the initial-value problem for two estimates, P_i and $P_i + \Delta P$, of the initial slope, where ΔP is a small change in P. This procedure was demonstrated in Example 5.4. Once the derivative has been determined, the next estimate P_{i+1} of the root is given by

$$P_{i+1} = P_i - \frac{y_b(P_i) - B}{\left[\mathrm{d}y_b(P)/\mathrm{d}P\right]_{P_i}} \tag{9.105}$$

This technique is frequently employed in shooting methods because of the good convergence characteristics of the Newton–Raphson method. Also, this method applies for complex solutions and also if $f(P)$ is tangent to the P-axis, resulting in multiple roots, as discussed in Section 5.5.

Shooting methods may employ efficient methods, such as the predictor–corrector methods, for solving the initial-value problem, since several iterations may be needed before convergence is achieved. However, Runge–Kutta methods are often preferred because of their self-starting feature. Automatic step changes, on the basis of error estimates, are generally not used since the solution is needed at exactly $x = b$. Therefore, a fixed step size Δx is preferred. However, the step size may be changed from one solution of the initial-value problem to the next, ensuring that $x = b$ is exactly attained, in order to improve the accuracy of the results or to reduce the computer time. Although we have considered a simple second-order BVP here, the technique is applicable to higher-order equations and to more involved problems. Both linear and nonlinear differential equations can be solved by this approach.

However, superposition can be used for linear equations, making shooting particularly simple in this case.

9.7.1.1 Linear Equations

Consider a linear second-order differential equation of the form

$$\frac{d^2y}{dx^2} = g_1(x)\frac{dy}{dx} + g_2(x)y + g_3(x) \tag{9.106}$$

with boundary conditions

$$y(a) = A \text{ and } y(b) = B \tag{9.107}$$

Again, we can recast this problem as an initial-value problem by taking $y'(a) = P$, instead of the condition at $y = b$. We obtain two solutions to this initial-value problem by taking the initial slope as P_1 and P_2. With these solutions denoted as $Y_1(x)$ and $Y_2(x)$, respectively, a linear combination of these solutions is also a solution to the differential equation. Therefore,

$$y(x) = c_1Y_1(x) + c_2Y_2(x) \tag{9.108}$$

is also a solution, which satisfies the initial condition $y(a) = A$. The relationship between c_1 and c_2 may be found by substituting Equation 9.108 into the differential equation, Equation 9.106, to give

$$c_1\left[\frac{d^2Y_1}{dx^2} - g_1(x)\frac{dY_1}{dx} - g_2(x)Y_1\right] + c_2\left[\frac{d^2Y_2}{dx^2} - g_1(x)\frac{dY_2}{dx} - g_2(x)Y_2\right] = g_3(x) \tag{9.109}$$

Since both $Y_1(x)$ and $Y_2(x)$ independently satisfy the differential equation, the quantities within the parentheses are both $g_3(x)$. Therefore,

$$c_1 + c_2 = 1 \tag{9.110}$$

The value of $y(x)$ at $x = b$ is obtained as

$$y(b) = c_1Y_1(b) + c_2Y_2(b)$$

This equation may be set equal to B, to satisfy the given boundary condition. Then c_1 and c_2 may be obtained from Equation 9.110 and the following equation:

$$c_1Y_1(b) + c_2Y_2(b) = B \tag{9.111}$$

We obtain the desired solution $y(x)$ by substituting the computed values of c_1 and c_2 into Equation 9.108. Thus,

$$y(x) = \frac{1}{Y_1(b) - Y_2(b)}\left\{\left[B - Y_2(b)\right]Y_1(x) + \left[Y_1(b) - B\right]Y_2(x)\right\} \tag{9.112}$$

Therefore, this solution satisfies the given differential equation and both boundary conditions. Iteration is not needed. This technique can also be used for higher-order, linear BVPs. If the number of unknown conditions at the initial point, $x = a$, were n, we would need to obtain $(n + 1)$ solutions and to take a linear combination of these numerical solutions to obtain the required solution to the ODE.

9.7.2 Finite Difference Methods

In this approach, one reduces the solution of an ODE to the solution of a system of algebraic equations by obtaining a finite difference approximation to the differential equation at a number of mesh points. The interval $a \leq x \leq b$, over which the numerical solution is to be obtained, is divided into n equally spaced subintervals of length Δx, as shown in Figure 9.18. Then the values of x at the mesh, or node, points are denoted by x_i where

$$x_i = a + i\Delta x \quad \text{for} \quad i = 0, 1, 2, \ldots, n \tag{9.113}$$

Also,

$$x_n = a + n\Delta x = b \quad \text{and} \quad x_0 = a \tag{9.114}$$

Therefore,

$$\Delta x = \frac{b - a}{n} \tag{9.115}$$

We wish to obtain the solution y_i at these node points. The given boundary conditions are employed to compute y_0 and y_n or to obtain algebraic equations from which these may be determined. The differential equation is replaced by its finite difference

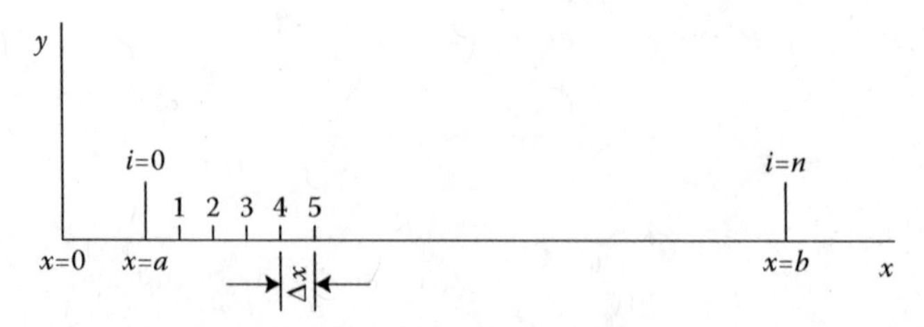

FIGURE 9.18 Subdivision of a given interval into a finite number of subintervals for employing finite difference methods to solve an ODE.

approximation at the interior mesh points, resulting in $(n-1)$ algebraic equations. One can solve this resulting set of simultaneous algebraic equations by employing the methods discussed in Chapter 6 to obtain the dependent variable y at the mesh points. The finite difference methods for solving ODEs are often of lower accuracy than the shooting methods, discussed earlier, though the programming is simpler for linear equations. The solution is more involved for nonlinear equations. Consequently, shooting methods are more frequently employed, and finite difference methods are used as an alternative technique in case problems are encountered with shooting.

As discussed in Chapter 4, there are several finite difference approximations to the first- and higher-order derivatives of the dependent variable y. If central differences are employed, the first and second derivatives of y at the ith mesh point are approximated by

$$\frac{dy}{dx} = \frac{y_{i+1} - y_{i-1}}{2\Delta x} + O\left[(\Delta x)^2\right] \tag{9.116}$$

and

$$\frac{d^2 y}{dx^2} = \frac{y_{i+1} - 2y_i + y_{i-1}}{(\Delta x)^2} + O\left[(\Delta x)^2\right] \tag{9.117}$$

Therefore, the second-order differential equation, given by Equation 9.99, is approximated at the ith node by

$$\frac{y_{i+1} - 2y_i + y_{i-1}}{(\Delta x)^2} = F\left[x_i, y_i, \frac{y_{i+1} - y_{i-1}}{2\Delta x}\right] \tag{9.118}$$

When this approximation is applied at all the interior points, $(n-1)$ equations in $(n-1)$ unknowns are obtained. The boundary conditions yield

$$y_0 = A \text{ and } y_n = B \tag{9.119}$$

These relationships are used in the system of equations generated by Equation 9.118, wherever y_0 and y_n appear.

If the differential equation is linear, the algebraic equations obtained are also linear. Similarly, a nonlinear differential equation results in nonlinear algebraic equations and a homogeneous differential equation in a homogeneous system. These different types of systems were considered in Chapter 6, and the corresponding direct and iterative methods were discussed. The same may be used for solving the resulting simultaneous algebraic equations.

If the given ODE is linear, the finite difference equations obtained from Equation 9.118 are linear and tridiagonal, since the ith equation contains only y_{i-1}, y_i, and y_{i+i}. Such a system can easily be solved by a direct method such as Gaussian elimination, and several efficient algorithms have been developed for the purpose, see Example 6.2. Because the resulting system of equations is tridiagonal, the computation of the unknowns can be carried out efficiently and with a small round-off error. The computer program is also very simple. As a consequence of these advantages, the finite difference methods are frequently employed for solving linear BVP in ODEs. Nonlinear algebraic equations, arising from a nonlinear ODE, are generally linearized. This is done by using the values from the previous iteration for the nonlinear terms, as shown in Section 6.8.2. The result is a linearized tridiagonal system of equations, which is solved with iteration, to yield the solution y_i at the nodal points. The Newton–Raphson method may also be used if the number of nodes is small. However, shooting methods are generally easier to use for nonlinear ODEs than finite difference methods.

In many cases, the boundary conditions are more complicated than the simple ones considered above. Frequently, a relationship between the derivative y' and the function y is given as, say,

$$a_1 y' + a_2 y = a_3 \text{ at } x = a \quad \text{or} \quad x = b \tag{9.120}$$

Then, one approach is to use one-sided forward or one-sided backward finite difference approximations at the two boundaries for the derivative. At $x = a$, y' may be approximated by

$$y' \cong \frac{y_1 - y_0}{\Delta x} \tag{9.121}$$

where y_0 is the value of y at $x = a$, or x_0, and y_1 that at the adjacent nodal point. When substituted in Equation 9.120, this equation gives the relationship between y_0 and y_1 as

$$a_1 \frac{y_1 - y_0}{\Delta x} + a_2 y_0 = a_3 \tag{9.122}$$

However, the TE in the approximation of Equation 9.121 is $O(\Delta x)$, whereas the error in the finite difference equations for the interior points is $O[(\Delta x)^2]$. To improve the accuracy of the finite difference equation for the boundary point, a fictitious point x_{-1} can be taken outside the computational region, as shown in Figure 9.19. Then y' is approximated by central differences to an accuracy of $O[(\Delta x)^2]$ as follows:

$$y' \cong \frac{y_1 - y_{-1}}{2\Delta x} \tag{9.123}$$

Therefore, the finite difference equation for the boundary point at $x = a$, or x_0, is

$$a_1 \frac{y_1 - y_{-1}}{2\Delta x} + a_2 y_0 = a_3 \tag{9.124}$$

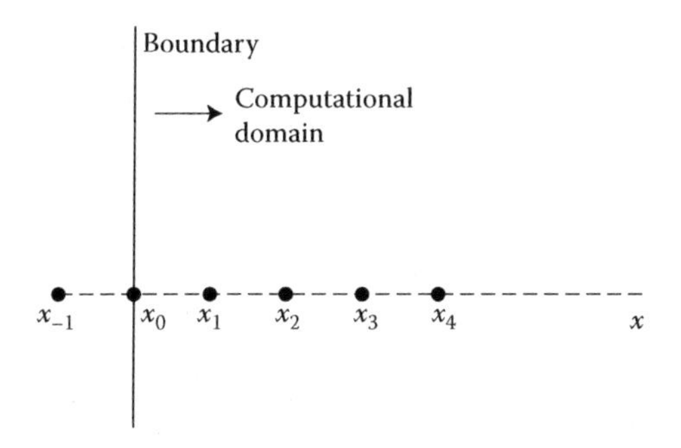

FIGURE 9.19 Use of a fictitious point x_{-1} outside the computational region for approximating a gradient condition at the boundary.

The finite difference equation, Equation 9.118, is applied at $x = a$, and the unknown y_{-1} is eliminated between the equation thus obtained and the above equation, resulting in a relationship of accuracy $O[(\Delta x)^2]$ between y_0 and y_1.

The above finite difference formulation has a TE of order $(\Delta x)^2$. Higher accuracy in the numerical results can be achieved by the use of smaller step size or higher-order methods. Richardson extrapolation, Equation 9.40, may also be used to improve the accuracy. However, higher-order methods involve more than one point on either side of the ith grid point. Therefore, a tridiagonal system is not obtained, and the treatment of the boundary conditions also becomes more involved, see Ferziger (1998) and Jaluria and Torrance (2003) for details.

The finite difference approximation to the differential equation can be obtained in many ways, leading to different sets of algebraic equations. The given equation may also be reduced to a system of first-order equations and the finite difference approximations applied to these (Keller, 1968). Equations of order higher than two can be reduced to an equivalent system of first- or second-order equations. A tridiagonal system is obtained if a system of second-order equations is employed with finite difference approximations of $O[(\Delta x)^2]$. In circumstances where the tridiagonal system is not obtained, the Gauss–Seidel and SOR iterative methods may be used. Similarly, iterative methods are used for nonlinear equations, as discussed in Section 6.8.

9.7.3 Eigenvalue Problems

Homogeneous ODEs, which give rise to eigenvalue problems, are frequently encountered in elasticity theory, vibrations, stability analysis, and mechanics of materials. In this case, the solution can be obtained only for certain values of the parameters of the system. These values, examples of which are the natural frequencies of vibration of a given structure, are related to the characteristic quantities, known as *eigenvalues*, of the problem. The eigenvalue problem for homogeneous algebraic equations was discussed in Chapter 6. One generally solves the differential equation by reducing it to an equivalent system of algebraic equations, using finite difference approximations.

Consider, for example, the following equation, which arises in the natural vibration of beams:

$$\frac{d^2y}{dx^2} + a^2 y = 0 \quad \text{with } y(0) = 0 \quad \text{and} \quad y(L) = 0 \tag{9.125}$$

where a is a constant and L is the length of the beam. If the second-order central difference approximation, Equation 9.117, is used for the second derivative, the finite difference equation is

$$y_{i+1} + \left[a^2 (\Delta x)^2 - 2\right] y_i + y_{i-1} = 0 \quad \text{for } i = 1, 2, \ldots, n-1 \tag{9.126}$$

where n is the number of subintervals and $y_0 = y_n = 0$. This system of equations may be written in matrix form as follows:

$$\left(A - \lambda I\right) y = 0 \tag{9.127}$$

where A is a tridiagonal coefficient matrix, λ is the eigenvalue, I is an identity matrix, and y is the unknown vector of the values at the nodal points.

The above system of homogeneous equations may be solved to obtain the eigenvalues and the corresponding eigenvectors. The analytical solution gives the eigenvalues as $\lambda_n = -(n\pi/L)^2$, where $n = 1, 2, \ldots$, and the eigenvectors as $\sin n\pi x$. A good approximation to the lowest eigenvalue is generally obtained by taking only a few grid points, typically around 10. However, a much larger number of subintervals is needed to accurately determine the higher eigenvalues. The power method discussed in Chapter 6 is particularly suitable for determining the smallest eigenvalue and the corresponding eigenvector. Larger eigenvalues may also be obtained in ascending order, as outlined earlier. Other methods, such as the QL algorithm, may also be used very efficiently, since the finite difference approximation leads to symmetric tridiagonal matrices in many cases.

Example 9.5

The flow of a fluid over a two-dimensional wedge, as shown in Figure 9.1, is governed by the third-order ODE

$$\frac{d^3f}{dx^3} + f\frac{d^2f}{dx^2} + \beta\left[1 - \left(\frac{df}{dx}\right)^2\right] = 0 \tag{9.128}$$

where x is the dimensionless distance away from the wedge in a direction normal to either edge as shown, f is known as the *dimensionless stream function*, so that df/dx is the dimensionless velocity in the direction parallel to either side of the wedge, and β is a constant related to the wedge angle. Thus, $\beta = 0$ gives the flow over a flat plate. Also, the stream function f lies between 0 and 1.0. The boundary conditions for this problem are as follows:

$$\text{At } x = 0: \quad f = 0 \quad \text{and} \quad \frac{df}{dx} = 0 \tag{9.129}$$

$$\text{As } x \to \infty: \quad \frac{df}{dx} \to 1 \tag{9.130}$$

Note that the boundary conditions are given in terms of both f and its derivative.

Despite the complexity of the physical phenomenon involved, this problem is chosen because it presents a third-order, nonlinear, boundary-value, ODE system. Thus, it permits the illustration of several important concepts in the solution of BVPs. Furthermore, such equations are frequently encountered in fluid flow phenomena of interest to several engineering disciplines. Using the fourth-order Runge–Kutta method, with the shooting technique, solve this problem for $\beta = 0$ and $\beta = 0.5$.

SOLUTION

The given equation may be broken down into a system of three first-order equations as outlined in Equation 9.7. Therefore, the three equations are written as

$$\begin{aligned} \frac{df}{dx} &= u \\ \frac{du}{dx} &= v \\ \frac{dv}{dx} &= -fv - \beta(1 - u^2) \end{aligned} \tag{9.131}$$

where u and v are the first and second derivatives of f, respectively. The two conditions in Equation 9.129 are given at $x = 0$. However, the third condition, given by Equation 9.130, is $u \to 1$ as $x \to \infty$, making the system a BVP. An initial-value problem is obtained if the value of $v = d^2f/dx^2$ is guessed at $x = 0$. The Runge–Kutta method may then be applied to the first-order equations in Equation 9.131 with the following boundary conditions:

$$\text{At } x = 0: \quad f = 0 \quad u = 0 \quad v = s \tag{9.132}$$

where s is a guessed value which must be adjusted by means of a correction scheme until the condition given by Equation 9.130 is satisfied.

Appendix B.29 presents the computer program in MATLAB for this problem. The *ode*45 function is used for solving the three coupled first-order ODEs. The three dependent variables are f, u, and v. A vector y is used to represent these three scalars as components, so that the right-hand sides of Equation 9.131 and the initial conditions are specified in terms of y. The initial-value problem is first solved with a chosen value of $v(0) = s$ and then with an incremented value $s + \Delta s$, with Δs taken as 0.001 here. A quantity *edge* is chosen as 6.0 to represent $x \to \infty$.

The value of $u = df/dx$ at *edge* is then determined for s and $s + \Delta s$. This allows us to compute the derivative $du_\infty(A)/dA$, where u_∞ represents the value of u for $x \to \infty$. The desired value of u_∞ is 1.0 and the Newton–Raphson correction scheme, given by Equation 9.105 is employed. This gives an improved value s_{im} of the guessed boundary condition $v(0)$ as

$$s_{im} = s - \frac{u_\infty(s) - 1}{du_\infty(s)/ds} \tag{9.133a}$$

where

$$\frac{du_\infty(s)}{ds} = \frac{u_\infty(s + \Delta s) - u_\infty(s)}{\Delta s} \tag{9.133b}$$

The improved value of s is now taken and the above procedure repeated. This process is carried out until the value of u_∞ becomes 1, within a specified convergence parameter *ep*, that is,

$$|u_\infty - 1| \le ep \tag{9.134}$$

The various symbols used are defined in the program. The right-hand sides of the three equations are defined by the function *rhs1.m*, which is also given in Appendix B.29. Thus, $b = \beta$ and the initial conditions are given by the vector $y0$. As mentioned above, the three dependent variables f, u, and v are represented by the vector y. The parameters such as *ep*, *edge*, and Δs, chosen by the user, must be varied to ensure that the results are independent of the values chosen. It was found that an *ep* value of 10^{-3}, *edge* larger than 6.0 and Δs of 10^{-3} were quite adequate.

The numerical results obtained are shown in Figures 9.20 and 9.21, in terms of the variation of f, u, and v with the independent variable x. Physically, u versus x represents the velocity distribution, which goes from zero at the wedge surface to 1.0 far away from the wedge. The dimensionless velocity gradient v is related

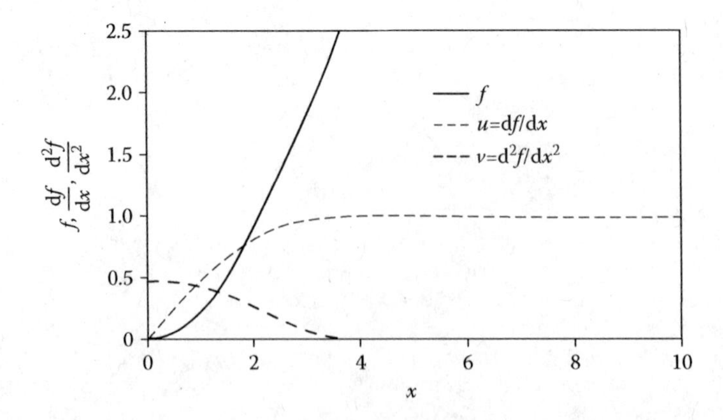

FIGURE 9.20 Computed variation of the functions f, $u = (df/dx)$, and $v = (du/dx)$, in Example 9.5, with the independent variable x, at $\beta = 0$. Here, *edge* is taken as 10 and *ep* as 10^{-3}.

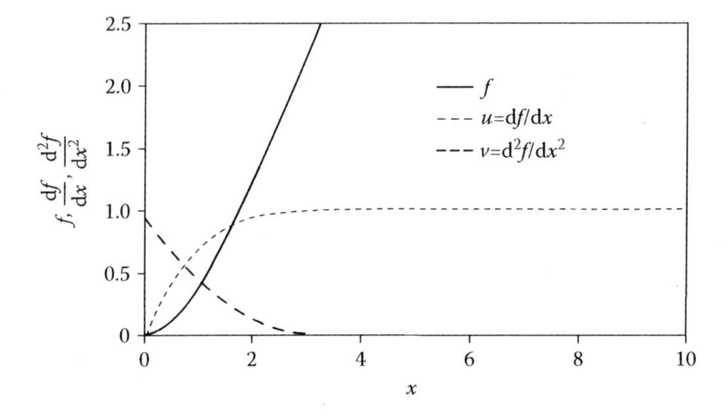

FIGURE 9.21 Computed results at $\beta = 0.5$ for Example 9.5.

to the frictional force due to the fluid. At $\beta = 0$, from Equation 9.128, $d^3f/dx^3 = 0$ at $x = 0$, since $f = 0$ at $x = 0$. This is reflected in the zero gradient, at $x = 0$, of the curve of v versus x. The results are shown for *edge* = 10 and $ep = 10^{-3}$. The choice of the value of *edge* is an important consideration. Since the boundary condition is given for $x \to \infty$, a large value of x is needed for satisfying this condition. The initial choice of *edge* may be based on results available from earlier computations. Otherwise, an arbitrary value is taken and gradually varied until the results are not affected by a further variation. It is also important to vary the initial guessed value of s to ensure that the results do not depend on the starting guess. Figure 9.22 shows the variation of s with the number of iterations, as the correction scheme proceeds. It is seen that the convergence is quite rapid and that the result is independent of the starting value of s over a fairly wide range. If

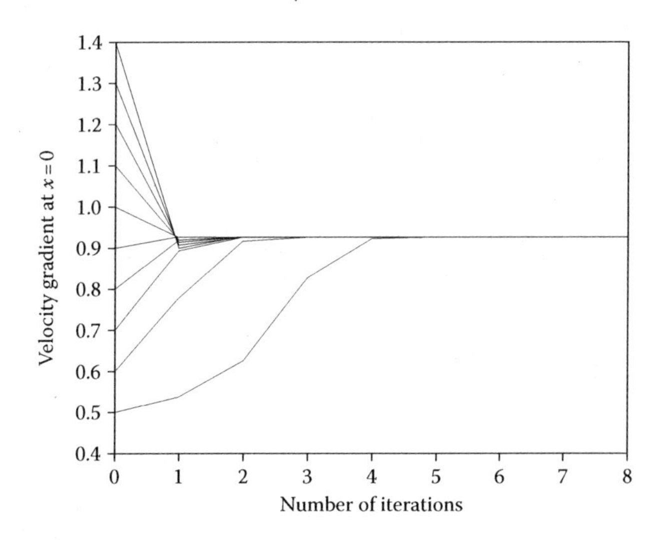

FIGURE 9.22 Convergence of the correction scheme in Example 9.5, shown in terms of the velocity gradient $v(0) = s$ versus the number of iterations.

the scheme does not converge after a large number of iterations, the initial guess should be changed and the calculation repeated.

This problem can also be solved by using the various methods and algorithms discussed earlier. The function *ode*45 is chosen here in order to demonstrate the ease and versatility of using this function for higher-order ODEs. The problem considered here is a complicated one, but it can be seen that root solving can easily be applied with methods for solving initial-value problems to obtain the desired solution. Fortran may also be employed for the solution of this problem using a similar approach. However, the programming is more complicated, as given by Jaluria (1996).

Example 9.6

A rod of length L has its two ends at temperatures T_1 and T_2. It loses energy by convection at the lateral surface, as shown in Figure 9.23. If the temperature is assumed to be uniform over any given cross section, the temperature T in the rod is a function only of x, the distance from one end. Then the governing equation for $T(x)$ is

$$kA\frac{d^2T}{dx^2} = hp(T - T_a) \tag{9.135}$$

where k is a property known as *thermal conductivity* of the rod material, A is the area of cross section of the rod, p is its circumferential perimeter, h is the convective heat transfer coefficient and T_a is the ambient air temperature. The equation may be nondimensionalized by the use of dimensionless temperature θ and distance X, defined as follows:

$$\theta = \frac{T - T_a}{T_1 - T_a} \quad X = \frac{x}{L} \tag{9.136}$$

The governing equation then becomes

$$\frac{d^2\theta}{dX^2} = \frac{hpL^2}{kA}\theta = P^2\theta \quad \text{where } P = \sqrt{\frac{hp}{kA}}\cdot L \tag{9.137}$$

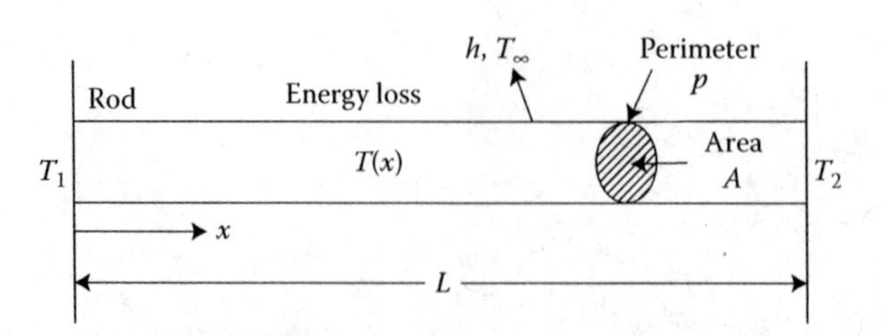

FIGURE 9.23 Conduction heat transfer in a rod, as considered in Example 9.6.

Here P is a dimensionless parameter that characterizes the problem. Solve this equation by the finite difference method for $(T_2-T_a)/(T_1-T_a) = 0.5$ and $P = 0, 0.5, 1.0, 5.0$, and 25.0.

SOLUTION

By nondimensionalizing the governing equation, we can apply the numerical results obtained to a wide range of physical parameters, given here as k, A, L, h, p, and T_a. The equation to be solved is

$$\frac{d^2\theta}{dX^2} = P^2\theta \tag{9.138}$$

with the following boundary conditions:

$$\begin{aligned} \text{At } X &= 0: \quad \theta = 1.0 \\ \text{At } X &= 1: \quad \theta = 0.5 \end{aligned} \tag{9.139}$$

These conditions follow from the definitions of θ and X and from the given temperature ratio $(T_2-T_a)/(T_1-T_a)$.

In order to use the finite difference approach for this problem, we take N nodal points over the interval, as shown in Figure 9.24. Therefore, the interval $0 \le X \le 1$ is divided into $(N-1)$ equal subintervals of length ΔX. Then the value of X at the node points is denoted by X_i, where $X_i = (i-1)\Delta X$, for $i = 1, 2, \ldots, N$. Also, $\Delta X = 1/(N-1)$, since the total dimensionless length is 1.0. The finite difference equation is

$$\frac{\theta_{i+1} - 2\theta_i + \theta_{i-1}}{(\Delta X)^2} = P^2\theta_i \quad \text{for} \quad i = 2,3,\ldots,N-2 \tag{9.140}$$

or

$$\theta_{i+1} - \left[2 + P^2(\Delta X)^2\right]\theta_i + \theta_{i-1} = 0 \tag{9.141}$$

This equation is applied at each interior nodal point, to give rise to $(N-2)$ equations which form a tridiagonal system. Also, $\theta = 1.0$ for $i = 1$, and $\theta = 0.5$ for $i = N$.

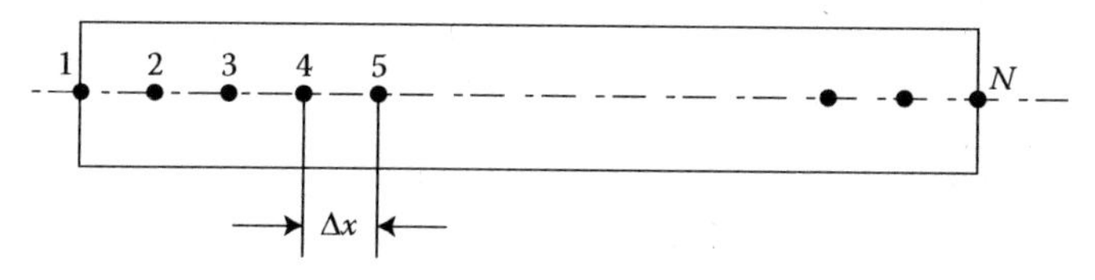

FIGURE 9.24 Subdivision of the rod in Example 9.6 for employing the finite difference approach.

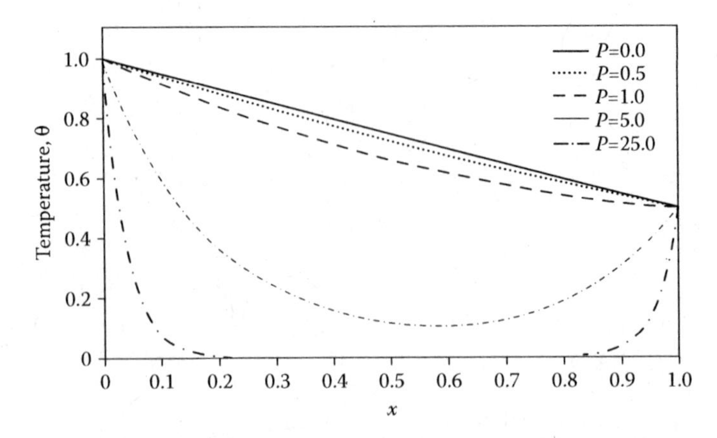

FIGURE 9.25 Computed temperature distributions for Example 9.6, with the number of grid points N taken as 51.

Appendix B.30 gives the computer program in MATLAB and Appendix C.16 the corresponding one in Fortran for solving this second-order ODE. The symbols employed in the programs are indicated at the beginning. The parameter P and the number of grid points N are chosen by the user. The boundary conditions are employed to set the values at the two extreme grid points. Equation 9.141 is used to generate the tridiagonal matrix, whose three terms in each row are denoted by $A(I)$, $B(I)$, and $C(I)$, respectively. The boundary conditions give $\theta(1)$ and $\theta(N)$, which yield the constants on the right-hand side of the first and last equations of the set. In the Fortran program, subroutine TRIDIAG is used to solve this system of equations and give the temperatures at the nodal points. In MATLAB, the script file given earlier for Example 6.2 is used. The algorithm and the computer program were discussed in detail earlier and are not repeated here.

Figure 9.25 shows the computed temperature distributions. The number of grid points N is taken as 51 for these results. At $P = 0$, the distribution is linear. This result is expected, as seen from Equation 9.137, which gives $d^2\theta/dX^2 = 0$ for $P = 0$ and, thus a linear variation for this case. As P becomes larger, the distribution deviates from the linear variation. At $P = 25.0$, the temperature θ is found to be zero, or $T = T_a$, over a substantial portion of the rod. This indicates that the heat inputs at the two ends of the rod are lost in short distances near these ends, resulting in uniform temperature at T_a over much of the rod. These results were also compared with the analytical solution of Equation 9.137, and a good agreement between the two was obtained. For further details on the physical aspects of this problem and other similar ones, see Incropera et al. (2006). Such problems commonly arise in heat and mass transfer processes of interest in chemical and mechanical engineering systems.

9.8 SUMMARY

The solution of ODEs is discussed in this chapter. The methods for solving the first-order equation are presented in detail since higher-order equations and boundary value problems are generally solved by reducing them to an equivalent system of first-order equations, which are then solved by these methods. The methods considered include Euler's method and its modifications, Runge–Kutta methods, multistep methods, and predictor–corrector methods. Euler's method is an inaccurate, although simple, method and is rarely used. It is considered here mainly because of its simplicity which allows the presentation of the basic concepts involved in solving ODEs. The modified Euler's method and the improved Euler's method, which is also known as *Heun's method*, are second-order accurate and are often used in engineering applications if a very high level of accuracy is not needed.

The Runge–Kutta methods are among the most popular numerical schemes for solving ODEs. Although usually less efficient than the corresponding predictor–corrector methods, they have the advantages of being self-starting and simpler to program. The Runge–Kutta methods are very widely used in engineering problems, particularly if the problem is to be solved only a few times with different parametric values. MATLAB functions *ode*23 and *ode*45 use these methods to generate lower and higher accuracy schemes, respectively, for solving initial-value problems.

If a substantial computational effort is involved, the predictor–corrector methods will be more appropriate, since these methods are more efficient and allow a simpler estimation of the TE, per step, which can be employed for a better control on the accuracy of the numerical results. The multistep methods, such as the Adams–Bashforth and Adams–Moulton methods, are seldom used by themselves, since a combination of the open and closed formulas leads to the predictor–corrector methods which have the advantages of both formulas. However, multistep and predictor–corrector methods are generally not self-starting, and a Taylor-series expansion of the dependent variable or a Runge–Kutta formula of the same order must be employed to compute the first few steps. Among the predictor–corrector methods discussed here are Adams method, Milne's method, and Hamming's method. All three methods are comparable in accuracy, although Milne's method has the smallest TE. However, it is unstable for certain equations. Hamming's method uses a modifier based on the TE to improve the estimate of the dependent variable obtained from the predictor and, therefore, generally requires no iteration or only one iteration for convergence of the corrector. It also has very good stability characteristics. The choice of a predictor–corrector formula for a given application is often a matter of personal preference, since all three methods are quite comparable in accuracy and efficiency.

BVPs are frequently solved by converting them into equivalent initial-value problems. Some of the conditions needed at the initial point are guessed, the differential equation is solved, and the guessed values are adjusted until the boundary conditions specified at other values of the independent variable are satisfied. Methods based on this approach are termed *shooting methods*, and the predictor–corrector methods are particularly suitable for such a solution, since several iterations may be involved and an efficient scheme is desirable. Finite difference methods may also be used for solving BVPs. These methods reduce the differential equation to a system of algebraic equations, which can be solved by the standard methods discussed in Chapter 6. Generally, finite differencing results in a tridiagonal system which can easily be solved by Gaussian elimination. This approach is particularly suitable for linear differential equations which give rise to linear algebraic equations. Nonlinear differential equations lead to nonlinear systems which must be solved iteratively. Homogeneous equations arise if the differential equation is homogeneous, and the corresponding eigenvalue problem may be solved by employing the methods given in Chapter 6.

Higher-order initial-value problems are solved by reducing them to a system of first-order equations, which are then solved by the various methods discussed here. The solution of a system of simultaneous differential equations is a simple extension of the procedure for a single equation. Accuracy and stability of the numerical scheme are important considerations in the solution of a given problem. The step size may initially be chosen on the basis of an estimate of the TE, if available. However, it is important to vary the step size in order to ensure a negligible dependence of the results obtained on the chosen value. If the results for two significantly different step sizes are close, the numerical scheme is probably stable. In some cases, the available estimate of the error and the constraints for stability may be employed for the choice of the method and the step size.

Whenever possible, the numerical results must be compared with the analytical results available for simpler cases in order to check the accuracy and correctness of the computational procedure. Richardson's extrapolation may also be used for improving the accuracy of the results. It is usually better to go to a higher-order formula for greater accuracy than continue to reduce the step size, since the round-off error and the computer time increase as the step size is reduced. Several methods have been developed in recent years to attain a high level of accuracy while retaining the efficiency of predictor–corrector methods, see Gear (1971), Lambert (1973), Shampine and Gordon (1975), and Ferziger (1998) for details.

PROBLEMS

9.1. The differential equation $dy/dx = F(x, y)$, with $y = 0$ at $x = 0$, is to be solved by Euler's method. If the function $F(x, y)$ is zero or infinite at $x = 0$, how would you start the computation? Is this approach also applicable to the fourth-order Runge–Kutta scheme?

9.2. A stone is thrown vertically upward in air at a velocity of 50 m/s. A frictional drag AV^2, where A is a constant and V is the velocity, acts on the stone in the direction opposite to that of the motion. The stone rises until the velocity becomes zero and then accelerates downward to the ground. Its velocity is governed by the following equations, while going upward and while coming down, respectively:

$$\frac{dV}{dt} = -g - AV^2$$

$$\frac{dV}{dt} = g - AV^2$$

where t is time and g is the gravitational acceleration, given as 9.8 m/s^2. If A is given as 10^{-3} m^{-1}, solve the first equation to obtain the time when velocity becomes zero. Using this as the initial condition for the second equation, find the velocity attained by the falling stone in the time taken for the upward motion. Use Euler's method and vary the time step Δt to ensure that the results are not significantly affected by the step size chosen.

9.3. Problem 9.2 may also be formulated in terms of the vertical distance x by noting that $V = \mathrm{d}x/\mathrm{d}t$. The resulting equation for the upward motion is

$$\frac{\mathrm{d}^2x}{\mathrm{d}t^2} = -g - A\left(\frac{\mathrm{d}x}{\mathrm{d}t}\right)^2$$

The following initial conditions are given

$$\text{At } t = 0: \quad x = 0 \quad \text{and} \quad \frac{\mathrm{d}x}{\mathrm{d}t} = 20 \text{ m/s}$$

Using Euler's method, solve this problem to find the maximum height to which the stone rises. Also, solve the problem by using the *ode*45 function in MATLAB and compare the results with those obtained earlier.

9.4. Repeat Problem 9.2 with $A = 0$, and compare the numerical results obtained with the corresponding values from the exact, analytical solution of the differential equation and with those obtained earlier for $A = 10^{-3}$ m^{-1}.

9.5. A stone is dropped at zero velocity from the top of a building at time $t = 0$. The differential equation that yields the displacement x from the top of the building is (with $x = 0$ at $t = 0$)

$$\frac{\mathrm{d}^2x}{\mathrm{d}t^2} = g - 5V$$

where g is the magnitude of gravitational acceleration, given as 9.8 m/s^2, and V is the downward velocity $\mathrm{d}x/\mathrm{d}t$. Using Euler's method and also the *ode*23 function, calculate the displacement x and velocity V as functions of time, taking the time step as 0.5 s.

9.6. A stone is thrown vertically upward in air at a velocity of 30 m/s. Due to gravity and air friction, the governing ODE, for the velocity V as a function of time t, is obtained as

$$\frac{\mathrm{d}V}{\mathrm{d}t} = -g - AV^{1.7}$$

where $g = 9.8$ m/s^2 and $A = 0.002$ m^{-1}. Solve this equation by Heun's method till V becomes zero or negative, using two step sizes $\Delta x = 0.01$ and 0.1 and compare the results. Plot the results obtained. Also use the *ode*23 command in MATLAB to get the solution and compare with your earlier results.

9.7. A copper sphere of diameter 5 cm is initially at temperature 200°C. It cools in air by convection and radiation. The temperature T of the sphere is governed by the equation

$$\rho C V_0 \frac{\mathrm{d}T}{\mathrm{d}t} = -\left[\varepsilon\sigma(T^4 - T_\infty^4) + h(T - T_\infty)\right]A$$

where ρ is the density of copper, C its specific heat, V_0 the volume of the sphere, t the time, ε a property of the surface known as *emissivity*,

σ a constant known as the *Stefan–Boltzmann* constant, T_∞ the ambient temperature, A the surface area of the sphere, and h the convective heat transfer coefficient. The initial condition is as follows:

$$\text{At } t = 0 : T = 200°\text{C}$$

Using Heun's method, without iteration, solve this differential equation to find the temperature variation with time, until the temperature drops below 50°C. Use the following values:

$$\rho = 9000 \text{ kg/m}^3 \quad C = 400 \text{ J/kg K} \quad \varepsilon = 0.5$$
$$\sigma = 5.67 \times 10^{-8} \text{W/m}^2 \text{ K}^4 \quad T_\infty = 25°\text{C} \quad h = 15 \text{ W/m}^2 \text{ K}$$

Employ time steps of 0.5 and 1.0 min, and compare the results obtained in the two cases.

9.8. In the preceding problem, if the surface emissivity ε is low and the convective heat transfer coefficient h is high, radiation may be neglected to obtain the governing equation as

$$\rho C V_0 \frac{dT}{dt} = -hA(T - T_\infty)$$

Using the modified Euler's (second-order Runge–Kutta) method with a time step of 1 s, solve this problem for $h = 100$ W/m^2 K. Also solve this equation mathematically, and compare the numerical results with the analytical solution. Comment on the error in the numerical solution.

9.9. A first-order ODE is given as

$$\frac{dy}{dt} = -0.5y$$

The initial condition is given as $y(0) = 5.0$. Write a script-*m* file to do the following:

a. Solve the differential equation by Heun's method to obtain y values from $t = 0$ to $t = 10$.
b. Use three step sizes: $h = 2.0$, $h = 1.0$, and $h = 0.1$.
c. Plot the results obtained for the three h values on one figure, using a different color or symbol for each case.
d. The exact (mathematical) solution to the problem is $y = 5\exp(-0.5t)$. Plot the exact solution also on the same figure as the numerical results.
e. From these results, what is the most appropriate step size?

9.10. Apply Richardson's extrapolation to Euler's method, and obtain expressions for the error and for an improved numerical solution.

9.11. Solve the nonlinear ODE

$$\frac{dy}{dx} = y^2 + x^2 \quad \text{with } y(0) = 1$$

by Euler's method, using step size Δx values of 0.1, 10^{-2}, 10^{-3}, and 10^{-4}. Compare the results obtained, and discuss the effect of decreasing the step size on the resulting error.

9.12.

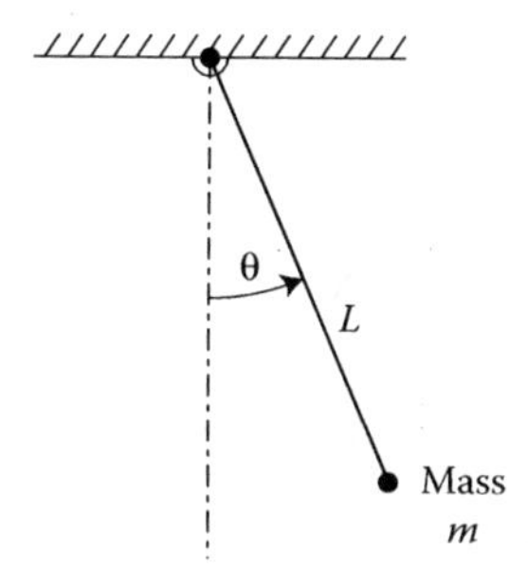

A rod of length L is attached at one end to a horizontal support and swings about this support, as shown in the figure. The motion of this pendulum is governed by

$$\frac{d^2\phi}{dt^2} + \frac{g}{L}\sin\phi = 0$$

where ϕ is the angle that the rod makes with the vertical at any given time t and g is the gravitational acceleration. The initial conditions are given as follows:

$$\text{At } t = 0: \quad \phi = \frac{\pi}{6} \quad \text{and} \quad \frac{d\phi}{dt} = 0$$

If L is given as 0.2 m and g is 9.8 m/s², obtain the numerical solution over twice the period of oscillation, using Heun's method, with iteration, to solve the corrector equation. The frequency f of the rod is given by

$$f = \sqrt{\frac{g}{4L\pi^2}}$$

9.13. A fourth-order predictor–corrector method is used to solve a differential equation from $x = 0$ to $x = B$, with a step size Δx. However, at $x = B$, the error per step is found to be too large, and it is decided to reduce the step size to $\Delta x/2$. Using extrapolation of the computed values, outline how this change may be carried out.

9.14. The height H of water in a tank, whose cross-sectional area is A, is a function of time t due to an inflow q_{in} and an outflow q_{out}. The governing differential equation arises from a mass balance as

$$A\frac{dH}{dt} = q_{in} - q_{out}$$

where q_{in} and q_{out} are the volume flow rates, and the density of water is taken as constant. The initial height, at $t = 0$, is zero. Compute the time taken for the height to rise to 2 m. The area A is given as 0.03 m^2, and $q_{in} = 6 \times 10^{-4}$ m^3/s. The outflow is given by $q_{out} = 3\times10^{-4}\sqrt{H}\,\text{m}^3/\text{s}$. Solve this problem by the fourth-order Runge–Kutta method. What is the height attained at steady state, and how long does it take to reach this value?

9.15. Solve the preceding problem numerically with an initial height of 4 m and $q_{in} = 0$. Determine the time taken for the height to drop to 2 m. Also solve the equation mathematically, and compare this computed value of the time with the analytical result.

9.16. The temperature of a metal block being heated in an oven is governed by the equation

$$\frac{dT}{dt} = 10.5 - 0.06T$$

Solve this equation by Euler's and Heun's methods to get T as a function of time t. Take the initial temperature as 100°C at $t = 0$.

9.17. A projectile of mass 0.2 kg is initially at rest. It is accelerated by the application of a constant vertical thrust of 10 N for a period of 5 s. The frictional drag on the projectile is given as $AV^{2.5}$, where $A = 10^{-2}$ m^{-1} and V is the velocity. The gravitational acceleration is 9.8 m/s^2. Following the discussion in Problem 9.2, obtain the differential equations for the upward motion. Solve these equations by the use of a fourth-order predictor–corrector method to obtain the velocity and height as functions of time t, till the projectile returns to the ground.

9.18.

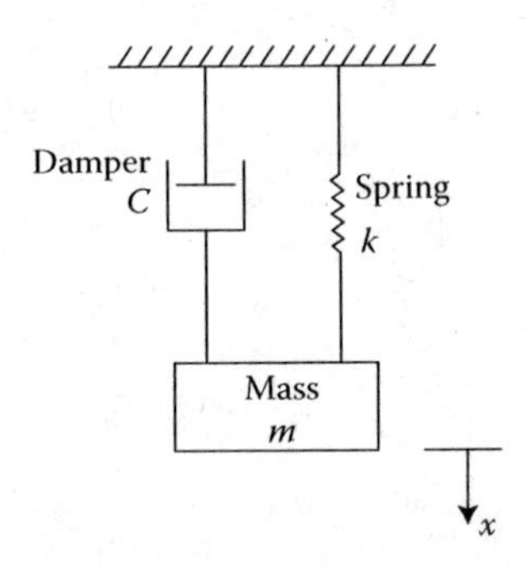

A vibrating system consists of a body of mass m attached to a wall through a spring, of spring constant k, and a damper, of damping coefficient C, as shown. The displacement x of the mass from its static-equilibrium position is governed by the equation

$$m\ddot{x} + C\dot{x} + kx = 0$$

where $\ddot{x}$ is the second derivative of x with respect to time t, and $\dot{x}$ is the first derivative. If the initial displacement $x(0)$ is given as 0.1 m and the initial velocity $\dot{x}(0)$ as zero, use Milne's method to compute the displacement x as a function of time for $0 \le t \le 2$ s. Take $m = 5$ g, $C = 20$ Ns/m, and $k = 500$ N/m.

9.19. Solve the preceding problem with the damper absent, that is, $C = 0$, and compare the results with those obtained earlier, with the damper present.

9.20. The flow of a fluid over a flat plate, aligned with the flow, is governed by the equation

$$2f''' + ff'' = 0$$

where the primes represent differentiation with respect to an independent variable η, that is, $f'' = d^2f/d\eta^2$. The dimensionless velocity in the direction along the plate is given by f', which is to be computed. The dimensionless quantity f is termed the *stream function*. The boundary conditions for this problem are as follows:

$$\text{At } \eta = 0: \quad f = 0 \quad f' = 0$$
$$\text{At } \eta = \infty: \quad f' = 1$$

Employing the fourth-order Runge–Kutta method, solve this BVP. For the application of the second boundary condition, take $\eta = 8$ as being large enough to represent infinity (see also Example 9.5).

9.21. If the flat plate in the preceding problem is heated, the temperature in the flow adjacent to the plate is governed by

$$\theta'' + \frac{Pr}{2} f\theta' = 0$$

where θ is the dimensionless temperature, f is the stream function from Problem 9.20, and Pr is a parameter, known as *Prandtl number*, which is a characteristic of the fluid. The boundary conditions are as follows:

$$\text{At } \eta = 0: \quad \theta = 1$$
$$\text{At } \eta = \infty: \quad \theta = 0$$

Using the fourth-order Runge–Kutta method, solve this problem to obtain θ as a function of η for $Pr = 1.0$.

9.22. Use the finite difference approach to solve Problem 9.20. Would you expect this method to be more efficient, in computer time, than the shooting methods? Also, which method is expected to be more accurate? Discuss.

9.23.

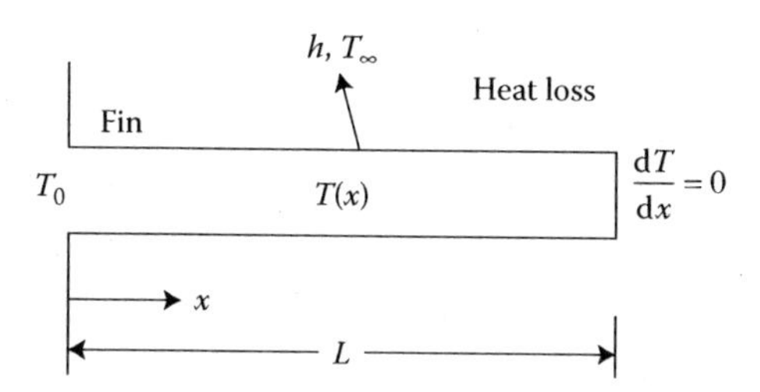

The conduction heat transfer in an extended surface, known as a *fin*, yields the following equation for the temperature T, if the temperature distribution is assumed to be one-dimensional in x, where x is the distance from the base of the fin, as shown in the figure:

$$\frac{\mathrm{d}^2T}{\mathrm{d}x^2} - \frac{hp}{kA}\left(T - T_\infty\right) = 0$$

Here, p is the perimeter of the fin, being $2\pi R$ for a cylindrical fin of radius R; A is the cross-sectional area, being πR^2 for a cylindrical fin; k is the thermal conductivity of the material; T_∞ is the ambient fluid temperature; and h is the convective heat transfer coefficient. The boundary conditions are as follows:

$$\text{At } x = 0: \quad T = T_0$$

$$\text{At } x = L: \quad \frac{\mathrm{d}T}{\mathrm{d}x} = 0$$

where L is the length of the fin. Solve this equation to obtain $T(x)$ for $R = 1$ cm, $h = 20$ W/m² K, $k = 15$ W/m K, $L = 25$ cm, $T_0 = 80$°C, and $T_\infty = 20$°C. Use a predictor–corrector method.

9.24. Solve the preceding problem with the following conditions, which make it an initial-value problem:

$$\text{At } x = 0: \quad T = T_0, \quad \frac{\mathrm{d}T}{\mathrm{d}x} = -10^4\,\text{K/m}$$

9.25.

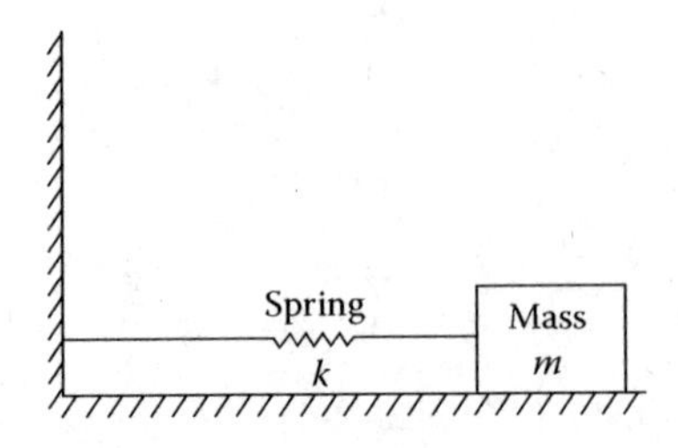

A block of wood, resting on a horizontal surface, is attached to a wall through a spring, as shown. The displacement x of the block from its equilibrium position is governed by

$$\frac{\mathrm{d}^2x}{\mathrm{d}t^2} = -\frac{k}{m}x \pm \mu g$$

where t is time, k the spring constant, m the mass of the block, μ the coefficient of friction, and g the gravitational acceleration. The frictional force is positive if the velocity $\mathrm{d}x/\mathrm{d}t$ is negative, and vice-versa. Taking $m = 2$ kg, $k = 200$ N/m, $g = 9.8$ m/s², and $\mu = 0.4$, compute x as a function of time, for time t up to 3 s. The initial conditions are as follows:

$$\text{At } t = 0: \quad x = 0.1m \quad \text{and} \quad \frac{\mathrm{d}x}{\mathrm{d}t} = 0$$

Employ the fourth-order Runge–Kutta method. Also, solve it by using the *ode*45 function and compare the results with those obtained earlier.

9.26. Solve the following initial-value problem by Hamming's method, and compare the predicted and corrected values for x up to 4.0:

$$\frac{d^2y}{dx^2} + 3\frac{dy}{dx} + 2y = 0$$

$$\text{At } x = 0: \quad y = 1 \quad \frac{dy}{dx} = 0$$

Take $\Delta x = 0.1$ and study the effect of the chosen convergence criterion on the number of iterations needed for the corrector. Compare the converged results with those obtained from the corrector without iteration. Discuss the implications of these comparisons.

9.27. Solve the following linear differential equation by the method of superposition:

$$\frac{d^2y}{dx^2} + 2y = 6$$

$$\text{At } x = 0: \quad y = 1$$

$$\text{At } x = 10: \quad y = 5$$

9.28.

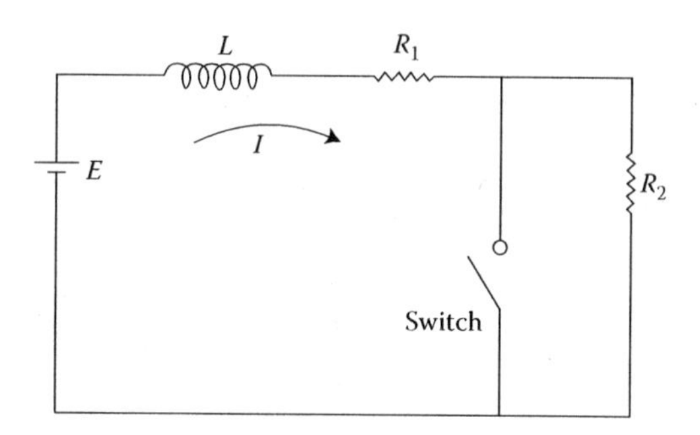

In the electrical circuit shown, the switch is open and a current I exists, where $I = E/(R_1 + R_2)$. At time $t = 0$, the switch is closed. The current $I(t)$ is then governed by the equation

$$L\frac{dI}{dt} + R_1 I = E \quad \text{with} \quad I(0) = \frac{E}{R_1 + R_2}$$

where L is the inductance in the circuit. Solve this equation by a Runge–Kutta formula to obtain I as a function of time, until the current is close to steady state. Compare the numerical results obtained with the analytical solution:

$$I(t) = \frac{E}{R_1}\left[1 - \frac{R_2}{R_1 + R_2} e^{-(R_1/L)t}\right].$$

Take $E = 10$ V, $R_1 = R_2 = 5\ \Omega$, and $L = 1$ henry.

9.29.

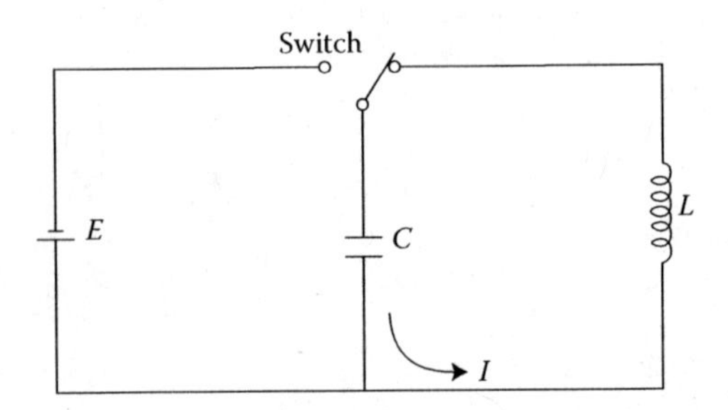

The capacitor C, in the electrical circuit shown, has an initial charge of q_0, and at time $t = 0$, the switch is closed. The governing equation for charge $q(t)$ is

$$L\frac{\mathrm{d}^2 q}{\mathrm{d}t^2} + \frac{1}{C}q = 0 \text{ with } q(0) = q_0 \text{ and } \frac{\mathrm{d}q}{\mathrm{d}t}(0) = 0$$

Using a Runge–Kutta formula, solve this equation for $q_0 = 0.1$ coulomb, $C = 10^{-4}$ farad, and $L = 0.01$ henry. The frequency of oscillation of this circuit is given by $\sqrt{1/LC}$. Obtain $q(t)$ over a few cycles.

9.30.

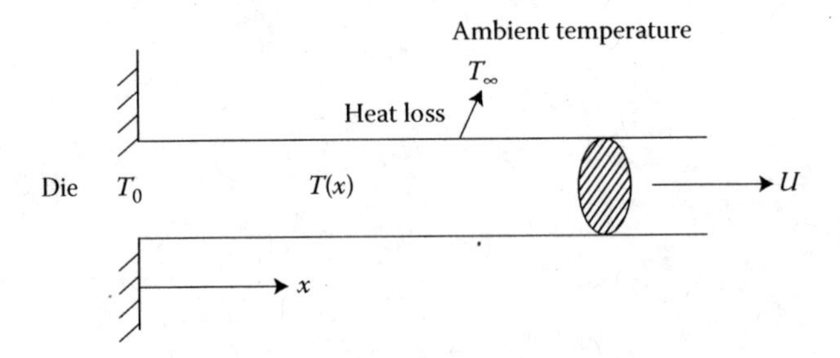

The temperature $T(x)$ in a moving rod, shown in the figure, which loses energy to the environment, at temperature T_∞, is given by the equation

$$\frac{\mathrm{d}^2 T}{\mathrm{d}x^2} - \frac{1}{\alpha}U\frac{\mathrm{d}T}{\mathrm{d}x} - \frac{2h}{kR}\left(T - T_\infty\right) = 0$$

where x is the distance from a die out of which the material emerges at temperature T_0, U is the velocity of the material, h is the convective heat transfer coefficient, R is the radius of the material, and α and k are material properties known as *thermal diffusivity* and *thermal conductivity*, respectively. The boundary conditions are as follows:

$$\text{At } x = 0: \quad T = T_0$$
$$\text{At } x = \infty: \quad T = T_\infty$$

Employing any shooting method, compute $T(x)$. Take $U = 1$ mm/s, $h = 20 \text{W/m}^2$ K, $\alpha = 10^{-4}$ m²/s, $k = 100$W/m K, $T_0 = 600$K, $T_\infty = 300$ K, and $R = 0.02$ m. For the second boundary condition, start with a large value of x, say, 1 m, to represent ∞, and then vary this value until the results are not significantly affected by a further increase.

9.31. Solve the following BVP by a shooting method:

$$x^2 \frac{d^2y}{dx^2} + 2x \frac{dy}{dx} + xy = 0$$

$$\text{At } x = 1: \quad y = 1$$

$$\text{At } x = 5: \quad y = 6$$

9.32. Formulate the preceding problem for solution by the finite difference method, and outline the numerical scheme that may be adopted to solve the resulting algebraic equations.

9.33.

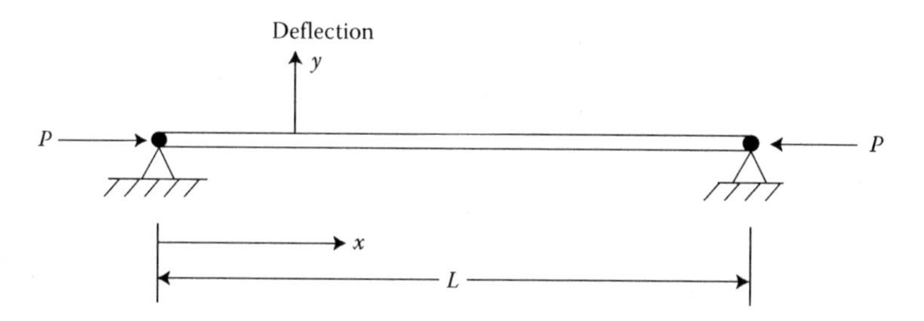

The deflection y of a beam, shown in the figure and loaded axially with force P, is governed by

$$\frac{d^2y}{dx^2} + \frac{P}{EI} y = 0 \quad \text{with } y(0) = y(L) = 0$$

where E is the modulus of elasticity of the material, I is known as its area moment of inertia, L is the length of the rod, and x is the distance from one end. We are interested in finding the smallest value of P for this eigenvalue problem, since this gives the first failure mode of the rod. Taking $P/(EI)$ as λ, solve this problem by the power method, taking five subdivisions of the rod, to obtain the smallest eigenvalue. Find the corresponding critical load P, if $EI = 1.5 \times 10^6$ N m² and $L = 1$ m.

9.34. If a heated vertical plate is placed in a quiescent ambient medium, a flow is generated adjacent to the plate because the fluid becomes buoyant due to heating. The dimensionless stream function f, which gives the velocity as f', and temperature θ in this flow are governed by the coupled system of equations

$$f''' + 3ff'' - 2(f')^2 + \theta = 0$$

$$\theta'' + 3Prf\theta' = 0$$

with the boundary conditions

$$f(0) = f'(0) = 1 - \theta(0) = f(\infty) = \theta(\infty) = 0$$

where the quantity within the parentheses represents the location, in the independent variable η, where the condition is applied. The primes indicate differentiation with respect to η, and Pr is a parameter that depends on the fluid (see Problem 9.21). Solve this system of equations by converting it into five first-order equations and employing the fourth-order Runge–Kutta method, with shooting. Obtain the velocity and temperature distributions, $f'(\eta)$ and $\theta(\eta)$, respectively, for $Pr = 1.0$. Take $\eta = 8$ as being sufficiently large to apply the conditions at infinity.

9.35. In a radiating fin (see Problem 9.23), the temperature $T(x)$ is governed by

$$\frac{d^2T}{dx^2} - \frac{P}{kA}\left[h\left(T - T_\infty\right) - \varepsilon\sigma\left(T^4 - T_\infty^4\right)\right] = 0$$

with the following boundary conditions:

$$\text{At } x = 0: \quad T = T_0$$

$$\text{At } x = L: \quad \frac{dT}{dx} = 0$$

Here, ε is a property of the surface known as *emissivity*, and σ is a constant (see Problem 9.7). Employing the finite difference method, solve this problem for the parametric values given in Problem 9.23, with ε given as 0.5 and $\sigma = 5.67 \times 10^{-8}$ W/m^2 K^4.

9.36. Solve Problem 9.33 by taking two, three, or four subdivisions of the rod for generating the finite difference equations and obtaining the polynomial from its characteristic determinant. Compare the computed value of the smallest eigenvalue λ_{min} with the analytical result π^2/L^2.

9.37. Using the finite difference method, repeat Problem 9.30. Discuss the accuracy given by the two approaches. When would the finite difference approach be the preferred one for such problems?

9.38. Study the stability of the second-order Runge–Kutta method by considering the differential equation $dy/dx = -ay$, where a is a constant. Also determine the TE.

9.39. Outline a numerical scheme for solving the differential equation $dy/dx = Ax^{-1/2}$, where A is a constant and $y = 0$ at $x = 0$, without using the analytical solution.

9.40. Consider the differential equation $d^2y/dx^2 = ay$, which is to be solved by finite difference methods. Discuss the nature of the resulting algebraic equations and put them in matrix form. Outline the numerical scheme for solving this set of equations and give reasons for your choice.

9.41. Show that the modified Euler's method is second-order accurate.

9.42. Solve the problem of Example 9.2 by Milne's predictor–corrector method, and compare the results with those obtained earlier by the Runge–Kutta method. Comment on the difference between the computational effort involved in the two methods.

9.43. Consider a third-order ODE of the form $d^3y/dx^3 = F(x, y, dy/dx)$. Develop a finite difference scheme for solving this equation if $y = A$ at $x = a$; $y = B$ at $x = b$; and $dy/dx = C$ at $x = a$. Assume that F is a linear function in the dependent variable y.

9.44. Consider a nonlinear second-order ODE of the form $d^2y/dx^2 = F(x, y, dy/dx)$, where F is nonlinear in y. Using the finite difference approach, obtain a numerical scheme for solving this equation. Assume the problem to be a boundary-value one with $y = A$ at $x = a$ and $y = B$ at $x = b$.

10 Numerical Solution of Partial Differential Equations

10.1 INTRODUCTION

In the preceding chapter, the numerical solution of ODEs, which involve a single independent variable, was discussed. However, for a wide variety of problems in science and engineering, the dependent variables are functions of two or more independent variables, such as time and the spatial coordinate distances. Consequently, the differential equations that govern such problems involve partial derivatives and are known as partial differential equations (PDEs). These equations arise in almost all areas of engineering, for instance, in fluid mechanics, elasticity, heat transfer, energy systems, environmental flows, hydraulics, neutron diffusion in nuclear reactors, and structural analysis. The numerical solution of PDEs is generally more involved than that of ODEs because of the presence of several independent variables, each with its own initial and boundary conditions. Therefore, effort is often made, whenever possible, by the use of simplifying approximations and transformations, to reduce the governing PDE to an ODE. However, this simplification is possible in only a limited number of cases. Because of the complicated nature of PDEs, analytical solutions are rarely obtained, and numerical methods are necessary for most problems of practical interest.

10.1.1 CLASSIFICATION

Many of the classifications outlined in Chapter 9 for ODEs also apply for PDEs. Therefore, the equations may be linear or nonlinear, homogeneous or inhomogeneous, of first or higher order, and may involve a single equation or a system of equations. The initial and boundary conditions are specified in terms of the various independent variables, making it possible for the problem to be an initial-value problem in relation to one independent variable and a boundary value problem in relation to another variable. However, the suitable initial and boundary conditions that may be imposed for a given equation are determined by the type of the equation. Each type demands a particular set of initial and boundary conditions that must be specified to obtain a *well-posed* problem that is amenable to an analytical or a numerical solution.

Let us consider the general form of a second-order PDE in two independent variables, given as

$$A\frac{\partial^2\phi}{\partial x^2}+B\frac{\partial^2\phi}{\partial x\partial y}+C\frac{\partial^2\phi}{\partial y^2}+D\left(x, y, \phi, \frac{\partial\phi}{\partial x}, \frac{\partial\phi}{\partial y}\right)=0 \tag{10.1}$$

where A, B, and C may also be functions of the two independent variables x and y and of the dependent variable ϕ and its first-order derivatives. If ϕ appears in the first power throughout, the equation is said to be *linear*. This requires that A, B, and C be functions of only x and y and that D contain only linear functions of ϕ and its first-order derivatives. We shall concern ourselves mainly with linear PDEs in this chapter. Nonlinear equations are also frequently encountered, for example, in fluid flow. However, the solution of nonlinear equations is usually much more involved than linear equations, and only a brief outline of the applicable numerical techniques is given later in the chapter.

The classification of the above PDE is based on the sign of $B^2 - 4AC$. The equation is said to be *elliptic* when $B^2 - 4AC < 0$, *parabolic* when $B^2 - 4AC = 0$, and *hyperbolic* when $B^2 - 4AC > 0$. This classification is related to the nature of characteristics, which are lines or surfaces along with a disturbance or information can propagate. A PDE in two dimensions, as given by Equation 10.1, reduces to an ODE along a characteristic. A hyperbolic equation has two real and distinct characteristics, which are often employed to obtain the solution. In this case, a disturbance propagates at finite speed over a finite region, bounded by the two families of characteristics. For parabolic equations, the two families of characteristics merge, giving rise to an infinite propagation speed and information flow in one direction. Therefore, in a parabolic equation, the solution at a given point depends only on the results obtained along this direction up to the point and not on results in the region beyond it. For time as an independent variable, this implies that the solution is affected by the occurrences in the past, but not by those in the future. For elliptic equations, complex characteristics are obtained, and, therefore, no directional restrictions arise and a disturbance propagates in all directions. The solution at a given point is affected by disturbances at every other point in the region where the elliptic equation applies. There are no preferred directions, and the solution must be obtained over the entire region simultaneously. Therefore, the mathematical character of the equation is indicated by its classification, which also determines the suitable boundary conditions and the solution procedure.

10.1.2 Examples

A few examples of the three types of PDEs are considered here for illustration of the preceding discussion. A very common parabolic equation is the one-dimensional, unsteady, heat conduction equation, which is given as

$$\alpha\frac{\partial^2 T}{\partial x^2}=\frac{\partial T}{\partial t} \tag{10.2}$$

where $T(x, t)$ is the temperature, x is a spatial coordinate, t is the time, and α is a constant, known as the *thermal diffusivity* of the material. This equation requires the specification of an initial condition, in time, and two boundary conditions, in x. Another important parabolic equation is the transient convective–diffusive transport equation, which may be written for a one-dimensional problem as

$$\frac{\partial \phi}{\partial t} + C\frac{\partial \phi}{\partial x} = D\frac{\partial^2 \phi}{\partial x^2} \tag{10.3}$$

where $\phi(x, t)$ is the dependent variable, such as temperature, concentration, or density, and C and D are constants. Again, an initial condition in time and two boundary conditions in x are needed. This equation applies for practical problems such as material movement in chemical reactors, extrusion of plastics, and transport of discharged effluents in a water stream.

Two elliptic equations that are frequently encountered in engineering problems are Laplace's and Poisson's equations, written, respectively, as follows:

$$\frac{\partial^2 \phi}{\partial x^2} + \frac{\partial^2 \phi}{\partial y^2} = 0 \tag{10.4}$$

$$\frac{\partial^2 \phi}{\partial x^2} + \frac{\partial^2 \phi}{\partial y^2} = \beta(x, y) \tag{10.5}$$

where $\phi(x, y)$ is the dependent variable and β may be a constant or a function of x and y. These equations arise in several areas such as fluid mechanics, electrostatics, elasticity, and conduction heat transfer. In conduction, the dependent variable becomes the temperature T and β is a distributed heat source. Boundary conditions are needed at all the edges of the solution domain. Frequently, these are specified as the value of ϕ, of its derivative, or of a linear combination of the two, at the boundaries.

A common hyperbolic equation is the wave equation, written as

$$\frac{\partial^2 \phi}{\partial t^2} = c^2\frac{\partial^2 \phi}{\partial x^2} \tag{10.6}$$

where c is the propagation velocity of the wave and $\phi(x, t)$ is the physical dependent variable, such as the displacement of a string. Two initial conditions are needed, and the spatial domain may or may not be bounded. If bounded, boundary conditions are needed at the two boundaries of the region. This equation governs the vibration of a string as well as the behavior of waves in a given medium. Another simple hyperbolic equation is the first-order convection equation, given as

$$\frac{\partial \phi}{\partial t} + c\frac{\partial \phi}{\partial x} = 0 \tag{10.7}$$

where the physical quantity $\phi(x, t)$ is convected at constant velocity c. The characteristics are straight lines, given by $x - ct = \text{constant}$, for this equation, which may be differentiated with respect to x or t to obtain a second-order PDE of the form given by Equation 10.1.

10.1.3 Basic Considerations

The preceding discussion illustrates that one must determine the type of the given PDE before proceeding to its numerical solution. If more than two independent variables are to be considered, the equation retains the characteristics of the three types of equations discussed above, as determined by the highest derivatives in each of the independent variables. Unsteady, two-dimensional, mass diffusion, for instance, is governed by the equation

$$\frac{\partial C}{\partial t} = D\left(\frac{\partial^2 C}{\partial x^2} + \frac{\partial^2 C}{\partial y^2}\right) \tag{10.8}$$

where C is the concentration of a diffusing chemical species and D is known as the *mass diffusivity*. The problem retains the parabolic nature with respect to its time dependence and the elliptic behavior with respect to the spatial coordinates. Therefore, one marches in time, to obtain the concentration distribution at each time interval, using the distribution at the preceding interval. The concentration distribution at a specified time, which is generally taken as zero, is needed as the initial condition. Also, conditions involving the concentration and/or its derivatives must be specified at all the boundaries of the region.

The classification of PDEs is discussed here in terms of second-order equations, which are the most frequently encountered equations in engineering applications. However, higher-order equations are also of interest in many problems. Fourth-order equations arise, for instance, in fluid flow and in solid mechanics. These equations can usually be solved by the methods applicable to the second-order elliptic equations, although the finite difference equations are obviously more involved. In fact, a fourth-order PDE may often be broken down into two second-order equations, which are solved simultaneously. First-order equations also occur in a few cases. Very often, these equations have real characteristics and can be solved by the methods employed for the hyperbolic equation of second order.

The main approach to the solution of a PDE is based on the reduction of the equation to a system of algebraic equations, which are then solved by direct or iterative methods to yield the value of the dependent variable at a finite number of mesh points. Two techniques that are commonly employed for generating the governing system of algebraic equations are the *finite difference* and the *finite element* methods. Finite difference methods apply the approximations to the PDE at a finite number of grid points in the computational domain. Finite element methods divide the region into a finite number of subdivisions. The integral form of the PDE is then applied to each element, and the integrals are minimized or the integral statement satisfied, using interpolation functions that contain adjustable parameters, in order to satisfy the conservation principles. In both cases, a system of algebraic equations is obtained from the application of the method at the boundaries and in the interior region.

The finite element method has become very important in recent years because of its versatility in treating a wide variety of boundary conditions, material property variations, and complicated geometries. However, it is generally more involved than the finite difference method and is advantageous to use only when some of the complications mentioned above are present. We shall restrict our discussion in this chapter to relatively simple problems and largely to the finite difference methods of solving them. A brief discussion of the finite element approach is also given later in the chapter, including a comparison with the finite difference technique. Also, there are other methods available for solving PDEs, such as *boundary element*, *control volume*, and *spectral methods*. In most cases, the PDE is reduced to a system of algebraic equations, which are solved by the methods given in Chapter 6.

In this chapter, the three types of PDEs mentioned above are considered, and an introductory treatment of their solution is given. The subject is an extensive one, and various complexities arise in diverse engineering applications; see, for instance, the book by Jaluria and Torrance (2003) on the numerical solution of heat transfer problems. Here, we are interested mainly in a consideration of the basic approach to the numerical solution of these different types of equations. Therefore, a few simple equations are taken and their solution is discussed. The parabolic equations are treated first since the methods for solving them are similar to those used for ODEs and since they also form the basis for some of the methods used for elliptic equations.

10.2 PARABOLIC PDEs

The solution domain in parabolic equations stretches outward indefinitely in one coordinate direction, say, z, from the given initial values, as shown in Figure 10.1. The equation is solved for the dependent variable ϕ by advancing, or marching, in the direction of increasing or decreasing value of the independent variable z, depending

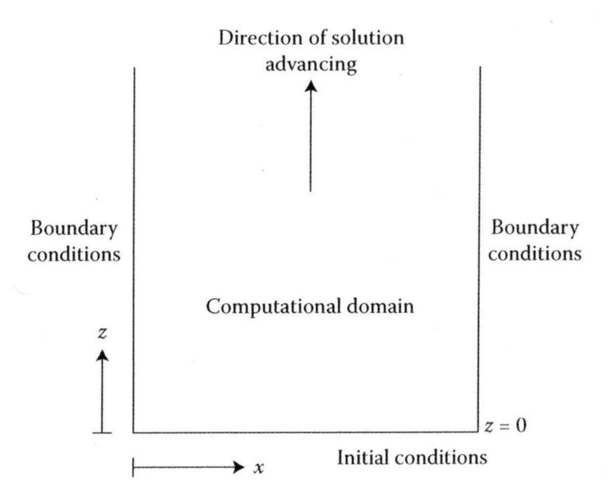

FIGURE 10.1 Solution domain for a parabolic PDE, along with the necessary boundary and initial conditions.

on the problem and starting with the initial conditions at $z = 0$. The solution must satisfy the prescribed boundary conditions, in the other independent variable x, as the solution advances in z. For results obtained with increasing z, the solution at any z depends on the values of ϕ at smaller z but is independent of those at larger z. This implies a definite direction for disturbance propagation. A physical analogy to this circumstance is a fast-flowing river in which a disturbance at a given location travels downstream but does not affect the flow upstream. The solution procedures for parabolic equations are, therefore, based on marching outward from the initial conditions in one of the independent variables, while satisfying the given boundary conditions in the other.

10.2.1 Numerical Solution with an Explicit Scheme

Let us consider the one-dimensional transient heat conduction problem, given by Equation 10.2, as an example of a parabolic PDE. Therefore, the dependent variable is the temperature $T(x, t)$, which is governed by the equation

$$\alpha \frac{\partial^2 T}{\partial x^2} = \frac{\partial T}{\partial t} \tag{10.2}$$

Since the equation contains the second derivative in x and only the first derivative in time t, two boundary conditions are needed in x and a single initial condition in t. These may be prescribed in terms of $T(x, t)$ as

$$\begin{aligned} T(a, t) &= A \\ T(b, t) &= B \\ T(x, 0) &= T_0 \end{aligned} \tag{10.9}$$

where $x = a$ and $x = b$ represents the boundaries of the region and T_0 is the initial condition. In general, T_0 may be a function of x, and A and B may be functions of time. However, for simplicity these are taken as constants here. The problem, as given above, is properly posed for obtaining the solution for $t > 0$.

The above parabolic PDE may be solved numerically by finite difference methods. A space-time grid, with Δx and Δt denoting the corresponding mesh sizes, is taken, as shown in Figure 10.2. Then, the finite difference approximations to the derivatives are applied to the given equation at each grid point. Several finite difference approximations can be obtained, depending on the representations used for the derivatives, as discussed in Chapter 4. For instance, a forward difference representation may be used for the partial derivative in time to obtain

$$\frac{\partial T}{\partial t} = \frac{T_{i+1,j} - T_{i,j}}{\Delta t} \tag{10.10}$$

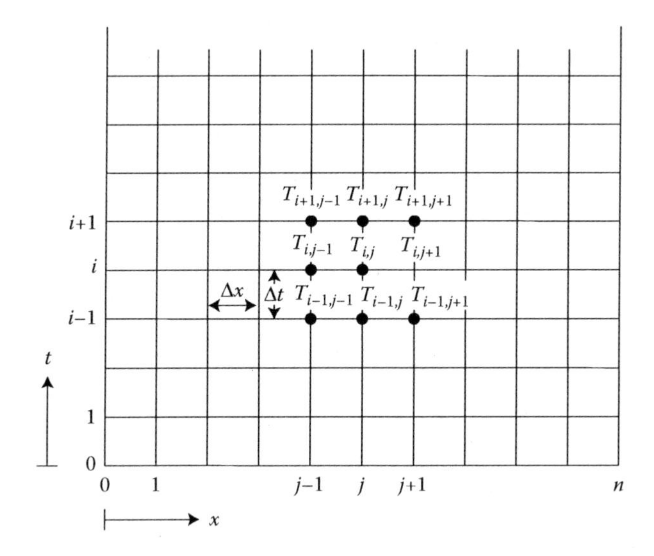

FIGURE 10.2 Space-time grid for the solution of a parabolic PDE by finite difference methods.

where the subscript $(i + 1)$ denotes the values at time $(t + \Delta t)$, and i those at time t. The spatial location is given by j. For the mesh shown in Figure 10.2, $x = j\Delta x$ and $t = i\Delta t$. The TE is $O(\Delta t)$, as obtained in Chapter 4.

A central difference approximation may be employed for the second derivative to obtain the approximation at time t and location x as

$$\left(\frac{\partial^2 T}{\partial x^2}\right)_{i,j} = \frac{T_{i,j+1} - 2T_{i,j} + T_{i,j-1}}{(\Delta x)^2} \tag{10.11}$$

with a TE of $O[(\Delta x)^2]$. In this expression, the second derivative in x is approximated at time t. If the above finite difference representations are substituted into Equation 10.2, the resulting equation may be written as

$$T_{i+1,j} = T_{i,j} + \frac{\alpha \Delta t}{(\Delta x)^2}(T_{i,j+1} - 2T_{i,j} + T_{i,j-1}) \tag{10.12}$$

where the TE is $O(\Delta t) + O[(\Delta x)^2]$. Therefore,

$$T_{i+1,j} = (1 - 2F)T_{i,j} + F(T_{i,j+1} + T_{i,j-1}) \tag{10.13}$$

where F is a constant known as the *grid Fourier number* and is given by

$$F = \frac{\alpha \Delta t}{(\Delta x)^2} \tag{10.14}$$

Equation 10.13 gives the temperature at time $(t + \Delta t)$ at the grid point whose spatial coordinate is $x = j\Delta x$, in terms of the temperatures at time t at the grid points with coordinates $(x - \Delta x)$, x, and $(x + \Delta x)$. The temperatures at $x = a$ and $x = b$ remain constant at A and B, respectively, because of the given boundary conditions. The initial condition gives the temperatures at the grid points at time $t = 0$ as T_0, where T_0 may be given as a constant or as a function of x. Using Equation 10.13, the temperature distribution at time $t = \Delta t$ may be determined from the given temperatures at $t = 0$. Similarly, the distribution at the second time step $t = 2\Delta t$ is obtained, employing the computed values at the first time step. Therefore, the solution is obtained at increasing values of time t. This time marching may be continued indefinitely. However, the process is generally terminated when a specified time t_{max} has been attained or when the steady state, as given by a negligible change in the solution with increasing time, is reached. In most cases, a steady-state circumstance is attained at large time, and a convergence criterion may be applied to the solution in order to terminate the computation when the solution is sufficiently close to the steady state. A convergence criterion is generally needed since the solution may approach the steady state asymptotically, thus taking infinite time, theoretically, to reach it exactly. The choice of the convergence criterion, therefore, affects the computational time taken to obtain the solution. Also, as discussed in Chapter 2, the convergence criterion must be varied to ensure a negligible dependence of the numerical results on the chosen parameter.

If F is chosen as 1/2, Equation 10.13 becomes

$$T_{i+1,j} = \frac{T_{i,j+1} + T_{i,j-1}}{2} \tag{10.15}$$

which implies that the new temperature at a grid point is the average of the old temperatures at the two adjacent grid points. A graphical method, known as the *Schmidt–Binder method*, has been developed on the basis of this equation. A uniform distribution of grid points is taken, and the temperature at a point, for the next time step, is simply given by the intersection of the normal to the x-axis at this point with the line joining the graphical points representing the present temperatures at the two adjacent grid points, as shown in Figure 10.3. Therefore, the solution to the differential equation, Equation 10.2, may be obtained graphically over a desired time interval, starting with the initial conditions.

The computational scheme of Equation 10.13 gives the temperatures at time $(t + \Delta t)$ in terms of known temperatures at time t. Therefore, the temperatures at a given time step may be determined explicitly from known values at the previous time step. This method is known as the *explicit Euler method* and is also referred to as the *forward time central space (FTCS) method*, because of the finite

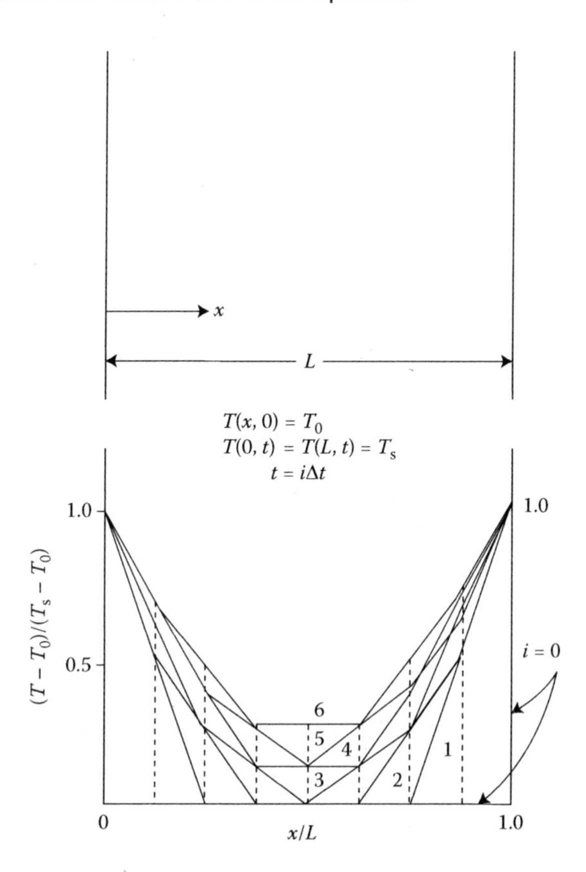

FIGURE 10.3 Graphical solution of transient heat conduction in a wall, with $F = 1/2$ (Schmidt–Binder method).

difference approximations used. It provides the simplest computational procedure for computing the time-dependent temperature distribution, starting with the initial conditions.

10.2.2 Stability of Euler's (FTCS) Method

The above method becomes unstable at large values of F, and stability is assured only if

$$F = \frac{\alpha \Delta t}{(\Delta x)^2} \leq \frac{1}{2} \tag{10.16}$$

Therefore, an amplification of the round-off and TEs arises if $F > 1/2$ and may lead to an unbounded growth in the solution as the computation advances in time, resulting in overflow. This condition for numerical stability is obtained by stability analysis, considering the growth of errors introduced into the solution; see, for instance, Roache (1976),

Ferziger (1998), and Jaluria and Torrance (2003). A physical explanation may also be given in terms of Equation 10.13. If $F > 1/2$, the coefficient of $T_{i,j}$ on the right-hand side of Equation 10.13 is negative. This implies that a larger value of the temperature $T_{i,j}$ at time t gives rise to a smaller value of the temperature $T_{i+1,j}$ at the same location at the next time step. Similarly, a smaller value of $T_{i,j}$ at time t results in a larger value of $T_{i+1,j}$ at $(t + \Delta t)$. The result is an oscillatory and unstable solution. Figure 10.4 shows the nature of this instability in terms of the computed results for transient conduction in a plate of thickness L at different values of F (from Jaluria and Torrance, 2003). The temperature distributions are obtained by solving Equation 10.2 with $A = B = T_s$, where T_s is a constant temperature. The results are shown in terms of dimensionless coordinate distance x/L and temperature $\theta = (T - T_s)/(T_0 - T_s)$. Only half the conduction region is shown because of symmetry. Clearly, instability arises as F increases beyond 0.5.

The major problem with the stability criterion given by Equation 10.16 is the constraint that it imposes on the allowable time step for a given grid spacing Δx, which is generally chosen as small to keep the TE small. For a given value of the thermal diffusivity α, the maximum permissible time step Δt is given by the stability constraint as

$$\Delta t \leq \frac{(\Delta x)^2}{2\alpha} \tag{10.17}$$

A small value of Δt is desirable for keeping the TE, which is of order Δt in this formulation, down to a desired level. However, the stability criterion generally limits the time step to a value that is much smaller than that needed for maintaining the accuracy of the solution. Therefore, the explicit method often severely constrains the time step and results in excessive computational time. Consequently, other methods have been developed which, although often more involved than the FTCS method, have better stability characteristics.

10.2.3 Implicit Methods

In the FTCS explicit method, the finite difference approximation for the second spatial derivative, $\partial^2 T/\partial x^2$, is written, in Equation 10.11, at time t. A family of implicit methods may be obtained by approximating this derivative at a different time, between t and $t + \Delta t$. The resulting finite difference equation is

$$\frac{T_{i+1,j} - T_{i,j}}{\Delta t} = \alpha\left[\gamma \frac{T_{i+1,j+1} - 2T_{i+1,j} + T_{i+1,j-1}}{(\Delta x)^2} + (\gamma - 1)\frac{T_{i,j+1} - 2T_{i,j} + T_{i,j-1}}{(\Delta x)^2}\right] \tag{10.18}$$

where γ is a parameter that lies between 0 and 1. Therefore, the second derivative is written as a weighted average of the finite difference approximations corresponding to time levels t and $t + \Delta t$.

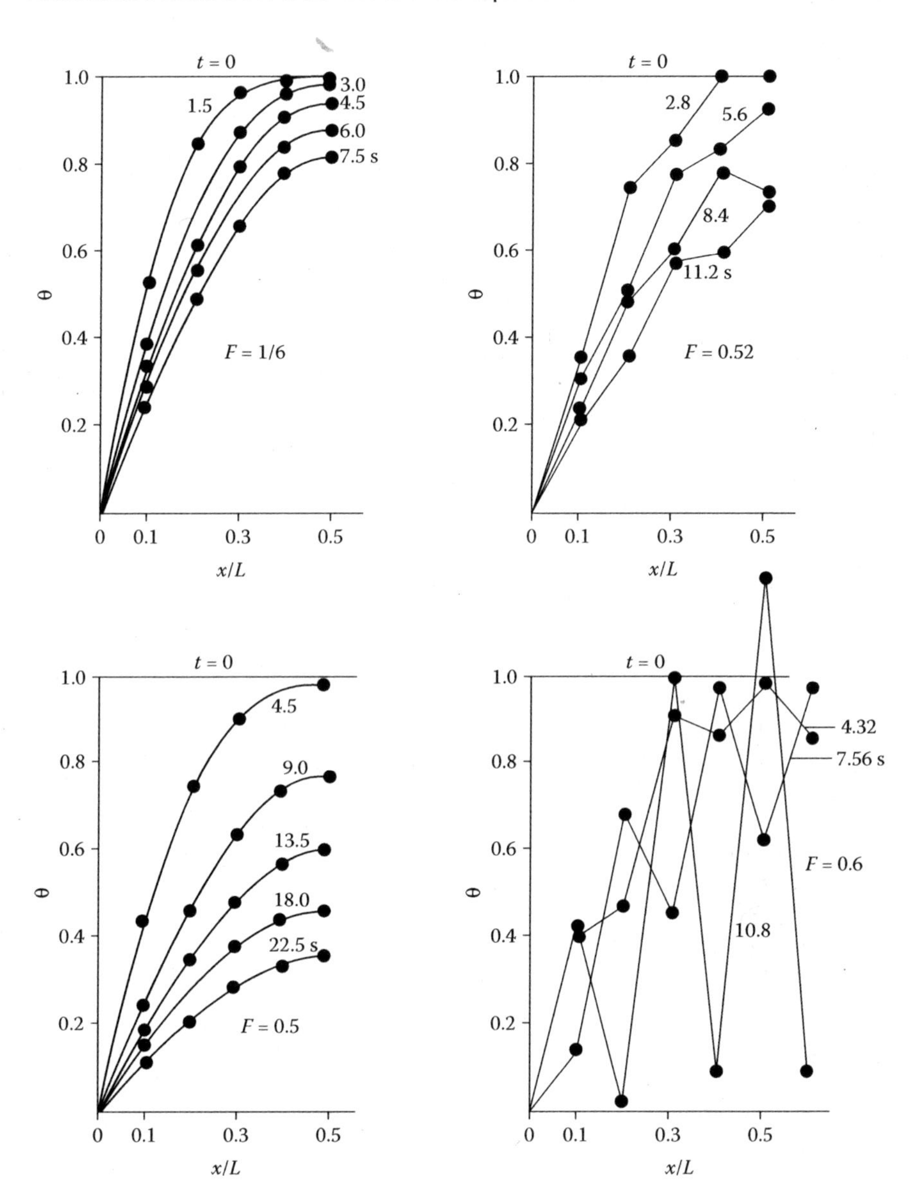

FIGURE 10.4 Time-dependent temperature distributions for one-dimensional conduction in a plate, governed by Equation 10.2, at various values of the grid Fourier number F, with time t in seconds. (Adapted from Jaluria, Y. and Torrance, K.E., *Computational Heat Transfer*, 2nd edn, Taylor & Francis, New York, NY, 2003.)

If $\gamma = 0$, the FTCS explicit method, given in Section 10.2.1, is obtained. For $\gamma = 1/2$, the second derivative is evaluated midway between the two time levels, and the TE can be shown to become $O[(\Delta t)^2] + O[(\Delta x)^2]$. This method, known as the *Crank–Nicolson method*, is very popular for the solution of parabolic equations. If $\gamma = 1$, the second derivative is evaluated at time $t + \Delta t$, and the formulation is known as the *fully implicit* or the *Laasonen method.* The TE is the same as that for the FTCS explicit method. From Equation 10.18, the finite difference equations for the Crank–Nicolson and the fully implicit methods are, respectively,

$$-FT_{i+1,j+1} + 2(1+F)T_{i+1,j} - FT_{i+1,j-1} = FT_{i,j+1} + 2(1-F)T_{i,j} + FT_{i,j-1} \quad (10.19)$$

$$-FT_{i+1,j+1} + (1+2F)T_{i+1,j} - FT_{i+1,j-1} = T_{i,j} \quad (10.20)$$

As seen from Equation 10.18, a set of simultaneous linear algebraic equations must be solved for implicit methods to obtain the temperature distribution at time $t + \Delta t$. A tridiagonal system arises which is conveniently solved at each time step by Gaussian elimination, as discussed in Chapter 6, to obtain the time-dependent temperature distribution. The solution marches in time, starting with the known initial values, until steady state or a specified time t_{max} is reached. The numerical procedure is more involved than the FTCS explicit method. However, the implicit methods generally have much better stability characteristics. The Crank–Nicolson implicit method is unconditionally stable, and much larger time steps can be taken as compared to the FTCS explicit method. The only constraint on the time step is generally because of accuracy considerations. However, oscillations that generally remain bounded do arise in the solution for certain problems at large values of F and may lead to a restriction on Δt, although at much larger values than that given by Equation 10.17.

The FTCS method, on the other hand, often restricts the time step to much smaller values than those demanded by the desired accuracy of the results. The Crank–Nicolson method also has a lower TE in time, $O[(\Delta t)^2]$, as mentioned above, allowing a larger time step for given accuracy. This arises because the finite difference approximation for the time derivative is in effect obtained midway between the two time levels t and $t + \Delta t$, making it a central difference approximation with TE of $O[(\Delta t)^2]$. Therefore, this method yields numerical results of greater accuracy than those from the FTCS method with a smaller computational cost. Even for nonlinear equations, which require iteration for solving the resulting set of algebraic equations, the Crank–Nicolson method is generally superior because only a few iterations are often needed for the linearized tridiagonal set. The numerical procedure may be graphically represented in terms of a computational molecule, which illustrates the grid points involved in the computation. Figure 10.5 shows the computational molecules for the FTCS, Crank–Nicolson, and fully implicit methods. These indicate the relationships between the values at the neighboring grid points. Examples 10.1 and 10.2 demonstrate the application of the FTCS and Crank–Nicolson methods, respectively.

10.2.4 Other Methods and Considerations

The major advantage of implicit methods over the explicit methods is the numerical stability, which allows much larger time steps in the computation. This consideration

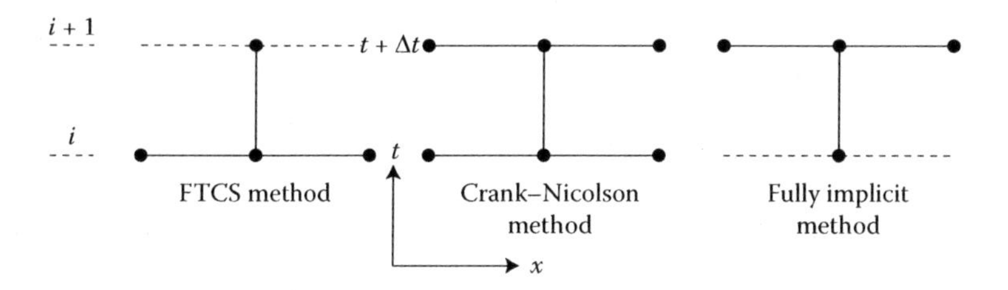

FIGURE 10.5 Computational molecules for the explicit Euler (FTCS), Crank–Nicolson, and fully implicit methods.

is particularly important in regions where the solution varies slowly with time, for example, near the steady state. Because the resulting algebraic equations are in the tridiagonal form, the number of arithmetic operations needed to solve the set for each time step is only $O(n)$, n being the number of grid points where the temperatures are to be computed. This is of the same order as the number of arithmetic operations necessary for taking one time step using the explicit method. Generally, the computer time taken by the implicit methods per time step is around twice that for the FTCS method. Since, in many cases, the time step Δt for the implicit methods may be taken as large as 10 to 100 times that allowed by the explicit method, due to stability considerations, a substantial reduction in computer time may be obtained by the use of implicit methods. The explicit method has the advantage of simpler programing.

Since the computation for the explicit method requires only the known values from the previous time step, the method can be used for solving nonlinear equations without much difficulty, whereas the solution of the simultaneous nonlinear equations that arise in the implicit methods requires iteration. Nonlinear equations arise in many physical problems, for instance, in the one-dimensional transient conduction problem if the material properties are not constant but vary with the temperature. Explicit methods are also advantageous to use if the boundary conditions are time-dependent.

Because of this advantage of an explicit procedure over implicit methods, for nonlinear problems, for time-dependent boundary conditions, and for other complexities in the problem, several other explicit methods with better stability characteristics than the FTCS method have been developed. Two explicit methods that are unconditionally stable for the problem under consideration are the Saul'yev and Dufort–Frankel methods. The corresponding finite difference equations for these methods, for the one-dimensional unsteady conduction problem, are

$$\frac{T_{i+1,j} - T_{i,j}}{\Delta t} = \alpha\left[\frac{T_{i+1,j-1} - T_{i+1,j} - T_{i,j} + T_{i,j+1}}{(\Delta x)^2}\right] \tag{10.21a}$$

$$\frac{T_{i+2,j} - T_{i+1,j}}{\Delta t} = \alpha\left[\frac{T_{i+1,j-1} - T_{i+1,j} - T_{i+2,j} + T_{i+2,j+1}}{(\Delta x)^2}\right] \tag{10.21b}$$

and

$$\frac{T_{i+1,j} - T_{i-1,j}}{2\Delta t} = \alpha\left[\frac{T_{i,j+1} - T_{i+1,j} - T_{i-1,j} + T_{i,j-1}}{(\Delta x)^2}\right] \tag{10.22}$$

In the first case, two equations are used, with the computation in the first one proceeding in the positive x direction for the $(n + 1)$th time interval and in the second one in the negative x direction for the $(n + 2)$th time interval. The two equations are used consecutively to advance by two time steps. In each equation, the right-hand side is explicitly known from previous calculations, one obtained with increasing j and the other with j decreasing. In the second method, a central difference approximation is used for the derivative in time, giving a TE of $O[(\Delta t)^2]$, and the temperature at the jth grid point is split into the values at two time steps t and $t + \Delta t$. Although this method is unconditionally stable, it is not self-starting because values at $t - \Delta t$ are needed for computing those at $t + \Delta t$. Thus, it requires another method to start the computation. Also, it can behave poorly under certain conditions and is not widely used. Various other explicit and implicit methods are given by Carnahan et al. (1969), Smith (1978), and Ferziger (1998).

10.2.5 Multidimensional Problems

The methods discussed here for solving the one-dimensional transient problem can easily be extended to multidimensional problems. Two-dimensional, unsteady diffusion processes are governed by Equation 10.8 for mass transfer and the following equation for heat conduction:

$$\frac{\partial T}{\partial t} = \alpha\left(\frac{\partial^2 T}{\partial x^2} + \frac{\partial^2 T}{\partial y^2}\right) \tag{10.23}$$

Using the FTCS explicit formulation, we obtain the finite difference equation as follows:

$$\frac{T_{i+1,j,k} - T_{i,j,k}}{\Delta t} = \alpha\left[\frac{T_{i,j+1,k} - 2T_{i,j,k} + T_{i,j-1,k}}{(\Delta x)^2} + \frac{T_{i,j,k+1} - 2T_{i,j,k} + T_{i,j,k-1}}{(\Delta y)^2}\right] \tag{10.24}$$

where the first subscript refers to time, the second subscript to the location in the x direction, and the third subscript to the location in the y direction. The corresponding step sizes are Δt, Δx, and Δy, respectively. Therefore, the temperature distribution at the next time step, $t + \Delta t$, may be obtained in terms of the known values at time t. Two grid Fourier numbers F_1 and F_2 arise, where

$$F_1 = \frac{\alpha \Delta t}{(\Delta x)^2} \quad \text{and} \quad F_2 = \frac{\alpha \Delta t}{(\Delta y)^2} \tag{10.25a}$$

Also, the finite difference equation is

$$T_{i+1,j,k} = (1-2F_1-2F_2)T_{i,j,k} + F_1(T_{i,j+1,k} + T_{i,j-1,k}) + F_2(T_{i,j,k+1} + T_{i,j,k-1}) \quad (10.25b)$$

Stability considerations, similar to those for the one-dimensional problem, arise, and the implicit methods may be employed advantageously. If $\Delta x = \Delta y$, the grid Fourier number F $(= 2F_1 = 2F_2) \leq 1/4$ for numerical stability in the explicit FTCS method. From Equation 10.25b, stability requires that $1 - 2F_1 - 2F_2 \geq 1$. As mentioned earlier, the equation retains the characteristics of a parabolic equation in time and those of an elliptic equation in the spatial coordinates. Therefore, the solution is obtained by marching in time, while satisfying the boundary conditions at the boundaries of the region. Similarly, the three-dimensional transient conduction problem may be solved. The constraint on F, for $\Delta x = \Delta y = \Delta z$, where z is the third coordinate, is obtained as $F \leq 1/6$ from stability considerations. For further details, see Smith (1978), Carnahan et al. (1969), and Jaluria and Torrance (2003).

The numerical solution of parabolic differential equations has been discussed here in terms of the transient heat conduction problem, since it is an important problem and also because several other parabolic equations of engineering interest are of similar form. For instance, the equation that governs the motion of fluid due to a plate being suddenly set into motion, from rest, in an infinite medium is

$$\frac{\partial u}{\partial t} = \nu \frac{\partial^2 u}{\partial x^2} \quad (10.26)$$

where u is the velocity in the direction of motion x, and ν is a property of the fluid known as *kinematic viscosity*. The treatment given here may easily be applied to this equation and also extended to other forms of parabolic equations. In several cases, there is no time dependence, and the equation is parabolic in one of the spatial coordinates. An example of such a circumstance is the boundary-layer flow over a surface. In this case, the variation with time is replaced by a variation with x, the direction in which the main flow occurs; see Figure 10.6. Information travels downstream to larger x from a given point, but is assumed not to travel upstream to smaller x, similar to the time-dependent problem. However, this problem is nonlinear, as are several problems of practical interest, and explicit methods are often easier to use in such cases.

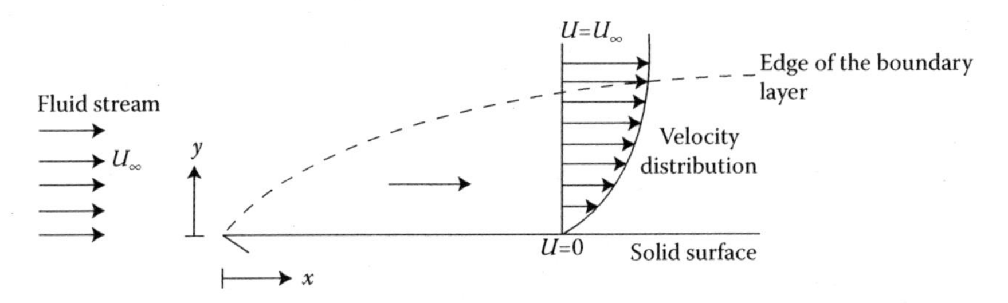

FIGURE 10.6 Sketch of the boundary-layer flow over a flat surface. This flow is governed by a parabolic PDE.

Several methods that employ the useful features of both implicit and explicit methods have also been developed and are among the most popular techniques for solving parabolic PDEs in two or more dimensions. One such method is the alternating direction implicit (ADI) method, discussed in the next section. This method employs the implicit formulation in one direction and treats the other direction explicitly, the two directions being interchanged from one step to the next. The result is a tridiagonal system at each step. The ADI method is the most important method in a class of methods known as *splitting methods*, several of which are available in the literature.

The imposition of the boundary conditions for parabolic PDEs has been considered here simply in terms of the value of the dependent variable, such as T, being specified at the boundaries. However, in practical problems, several other boundary conditions arise, particularly those related to the gradient of the dependent variable. Such boundary conditions and the relevant finite difference formulations are considered later and also in Example 10.2.

Example 10.1

In a chemical manufacturing system, a process involves the diffusion of salt into a layer of water. The layer is of thickness L and initially has a uniform salt concentration C_0. At time $t = 0$, the layer is brought into contact with saline solution at one surface, and the concentration at this surface is raised to C_s, while the other surface is maintained at concentration C_0. The mass diffusivity, also known as the *diffusion coefficient*, of salt in water is denoted by D. The governing PDE for the concentration $C(x, t)$ in the water layer, where x is the coordinate distance measured from the surface whose concentration is raised to C_s, is

$$D\frac{\partial^2 C}{\partial x^2} = \frac{\partial C}{\partial t} \tag{10.27}$$

Figure 10.7 illustrates this problem. The corresponding initial and boundary conditions are as follows:

$$\text{For } t \le 0: C = C_0 \text{ at all } x$$

$$\text{For } t > 0: \; C = C_s \;\text{ at } x = 0 \quad \text{and} \quad C = C_0 \;\text{ at } x = L \tag{10.28}$$

Using the FTCS explicit method, solve this one-dimensional transient diffusion problem to obtain the time-dependent concentration distribution in the material.

SOLUTION

In the problem, the numerical values of the physical quantities, such as concentration and the water layer thickness, are not given, so that the problem may be solved in generalized terms. Thus, the governing equation and the boundary conditions may be nondimensionalized to obtain a general solution that can be used for different sets of physical quantities.

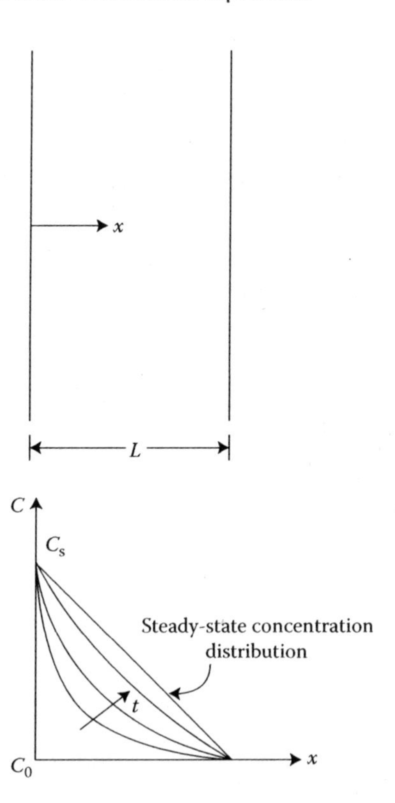

FIGURE 10.7 The physical problem, the coordinate system, and the expected transient behavior of the concentration distribution for Example 10.1.

We start by defining the following dimensionless quantities:

$$X = \frac{x}{L}, \quad \bar{t} = \frac{Dt}{L^2}, \quad \theta = \frac{C - C_0}{C_s - C_0} \tag{10.29}$$

where X, $\bar{t}$, and θ are the dimensionless distance, time, and concentration, respectively. We can formulate the given problem in terms of these quantities by using the above definitions to replace the physical variables by dimensionless ones. Then the governing equation is obtained as

$$\frac{\partial^2 \theta}{\partial X^2} = \frac{\partial \theta}{\partial \bar{t}} \tag{10.30}$$

The initial and boundary conditions become

$$\theta(X, 0) = 0, \quad \theta(0, \bar{t}) = 1, \quad \theta(1, \bar{t}) = 0 \tag{10.31}$$

Therefore, the above dimensionless, parabolic, PDE may be solved to obtain $\theta(X, \bar{t})$. If the concentrations at the boundaries, mass diffusivity, and thickness of the layer are given, the physical concentration distributions can be also determined from Equation 10.29. This implies that the problem must be solved only once in dimensionless terms, instead of separately for each set of physical parameters. For this reason, the equations are often nondimensionalized and results are obtained in generalized terms, as outlined here.

For the FTCS explicit method, the forward difference approximation is used for the first derivative in time, and central difference for the second derivative in the spatial coordinate x. This yields

$$\theta_{i+1,j} = (1 - 2F)\theta_{i,j} + F(\theta_{i,j+1} + \theta_{i,j-1}) \tag{10.32}$$

where

$$F = \frac{\Delta\bar{t}}{(\Delta X)^2} \tag{10.33}$$

Here, the subscript i represents the time step, and j the spatial grid point. Therefore, $\bar{t} = i\Delta\bar{t}$ and $X = j\Delta X$. Appendix B.31 gives the MATLAB® script file and Appendix C.17 gives the Fortran computer program for solving this problem by the explicit method. An interactive program is written so that the number of grid points, or grid size, and the initial concentration may be given as inputs. The time step is taken as the largest value from Equation 10.17 to avoid numerical instability. For the chosen total number of grid points, the mesh length is determined so as to obtain a total dimensionless distance of 1.0. The output is printed at specified time intervals, given by a chosen number of time steps, and the computation is carried out until a specified time is attained. A convergence criterion may also be employed to stop the computation when steady-state conditions are reached, as indicated by a concentration distribution that does not vary with time. The computed results may be stored, printed, or plotted. The variable names employed in the program are defined at the beginning of the program, and the various important steps in the computation are indicated.

Figures 10.8 and 10.9 present the numerical results obtained with $F = 0.5$, which gives the maximum time step for a stable numerical scheme. The initial dimensionless concentration θ is zero throughout the plate, and the steady-state distribution is a linear variation from 1.0 at one surface to 0.0 at the other. We can easily obtain the steady-state result by setting the transient term in Equation 10.30 equal to zero and solving the ODE $d^2\theta/dX^2 = 0$, to obtain $\theta = 1 - X$ as the steady-state distribution. Figure 10.8 shows the concentration distribution at various time intervals. Note that the steady-state distribution is attained by $\bar{t} = 0.5$. Figure 10.9 shows the variation of the dimensionless concentration θ with time $\bar{t}$ at several locations within the plate. Note that the temperatures increase sharply from the initial value of 0.0, as time $\bar{t}$ increases. The final approach to the steady-state value is a gradual one. Also, the concentration starts changing from zero at a later time for points which are farther away from the surface $x = 0$, where the step change in concentration occurs. This indicates a finite speed for the propagation of the mass diffusion effects in the FTCS method. At each time step, only the next grid point is affected, as seen from Equation 10.32 and Figure 10.3. For details on the physical aspects of this problem and other similar ones in heat and mass

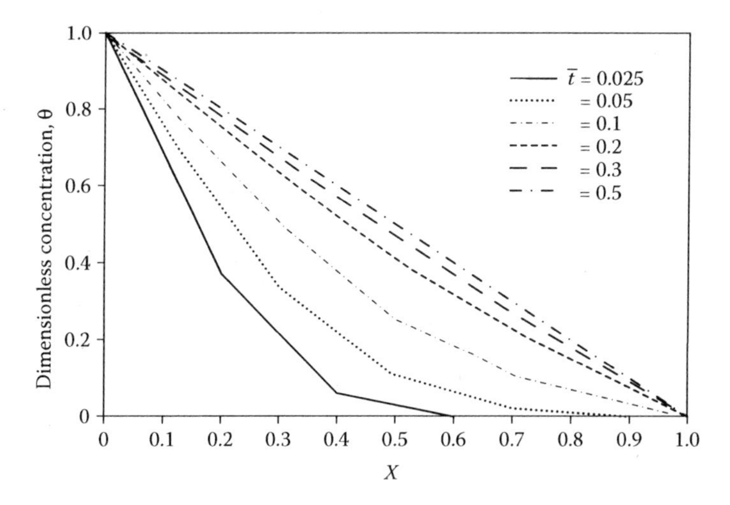

FIGURE 10.8 Computed concentration distribution at various time intervals for Example 10.1. The dimensionless time step $\Delta\bar{t}$ and the grid size ΔX are taken as 0.01 and 0.1, respectively.

transfer, standard textbooks in the area, such as Incropera et al. (2006), may be consulted.

Example 10.2

A flat plate of thickness L is initially at a uniform temperature T_0. At time $t = 0$, the temperature at one surface is raised to T_s, while the other surface is kept

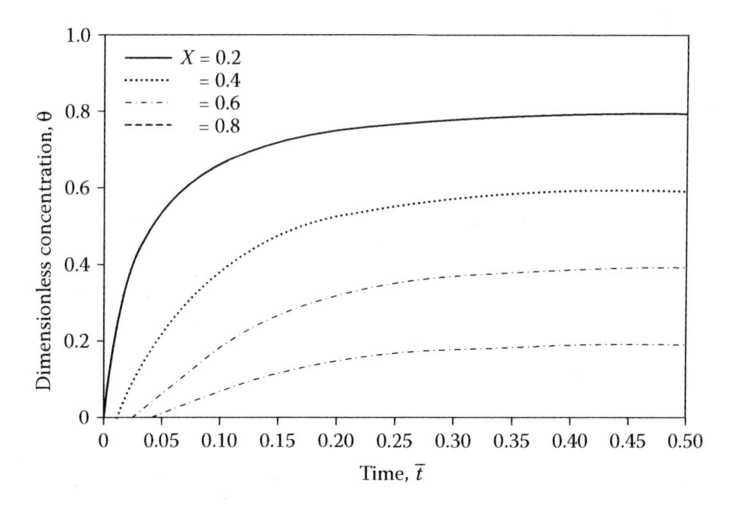

FIGURE 10.9 Computed variation of the concentration in Example 10.1, at various locations with time, indicating the approach to steady-state conditions at large time.

perfectly insulated. The thermal diffusivity of the material is denoted by α. Solve this problem by the Crank–Nicolson method.

SOLUTION

This problem is very similar to the one discussed in Example 10.1. The governing equation is Equation 10.2. The given initial and boundary conditions may be written for this problem as follows:

$$\text{For } t \le 0\text{: } T = T_0 \text{ at all } x$$

$$\text{For } t > 0\text{: } T = T_s \text{ at } x = 0 \text{ and } \frac{\partial T}{\partial x} = 0 \text{ at } x = L \tag{10.34}$$

The last condition implies a perfectly insulated surface. The heat transfer at the surface is proportional to $\partial T/\partial x$ and is zero if the temperature gradient is zero. Dimensionless quantities similar to those defined in Equation 10.29 may be employed to obtain the governing dimensionless equation. Thus, the nondimensionalization employed here is

$$X = \frac{x}{L}, \quad \bar{t} = \frac{\alpha t}{L^2}, \quad \theta = \frac{T - T_0}{T_s - T_0} \tag{10.35}$$

and the governing equation is

$$\frac{\partial^2 \theta}{\partial X^2} = \frac{\partial \theta}{\partial \bar{t}} \tag{10.36}$$

The initial and boundary conditions become

$$\theta(X,0) = 0, \quad \theta(0,\bar{t}) = 1, \quad \frac{\partial \theta}{\partial X}(1,\bar{t}) = 0 \tag{10.37}$$

The governing equation, with the above initial and boundary conditions, is solved by the Crank–Nicolson iterative scheme. The finite difference equation for this method is obtained for Equation 10.36 as

$$-F\theta_{i+1,j+1} + 2(1+F)\theta_{i+1,j} - F\theta_{i+1,j-1} = F\theta_{i,j+1} + 2(1-F)\theta_{i,j} + F\theta_{i,j-1} \tag{10.38}$$

where $F = \Delta\bar{t}/(\Delta X)^2$, i represents the time step, and j is the spatial grid location. This equation may be written more concisely as

$$A\theta_{j-1} + B\theta_j + C\theta_{j+1} = R \tag{10.39}$$

where the θ values are at the next time step, $i+1$, and R is the expression on the right-hand side of Equation 10.38. Therefore, R is a function of the θ values at the present time step i and is thus known. The constants A, B, and C are the coefficients on the left-hand side of Equation 10.38 and depend on the value of F chosen. No constraints arise in this problem due to stability considerations, and the time step and the grid size are chosen on the basis of desired accuracy. However, as mentioned earlier, bounded oscillations may arise in this method for some problems at large values of F. In most cases, accuracy is the main consideration in the choice of the grid size and the time step.

The MATLAB script file for this problem is given in Appendix B.32 and the corresponding Fortran computer program is given in Appendix C.18. An interactive program is written to allow the user to enter the input parameters, such as time step, initial conditions and the number of grid points. A tridiagonal matrix is generated from Equation 10.38, which is divided by 2 to simplify the computation. The form of the equation is given by Equation 10.39. The appropriate boundary conditions, given by Equation 10.37, are also incorporated to obtain the tridiagonal matrix, as discussed earlier in Example 6.2. The total number of grid points n is given, the left boundary being $i = 1$ and the right one being $i = n$. For the right boundary, $X = 1$, the second-order backward difference approximation, given in Figure 4.8, is used so that the error is $O[(\Delta x)^2]$. Thus,

$$\left(\frac{\partial \theta}{\partial X}\right)_{i,j} = \frac{\theta_{i,j-2} - 4\theta_{i,j-1} + 3\theta_{i,j}}{2\Delta X} = 0 \tag{10.40}$$

or

$$\theta_{i,j} = \frac{4}{3}\theta_{i,j-1} - \frac{1}{3}\theta_{i,j-2} \tag{10.41}$$

where j is replaced by n for the right boundary. A tridiagonal matrix, with rows from 2 to $n - 1$, corresponding to the interior points in the computational domain, is obtained. This tridiagonal matrix is solved to obtain the time-dependent temperature distribution, which then serves as the input for the computation of the distribution at the next time step. The boundary temperatures are obtained using the equations given above. This process is repeated until a specified time limit, or the steady-state circumstance, is attained. Results are obtained and plotted at specified time intervals.

Figure 10.10 shows the computed temperature distributions at different time intervals. The initial temperature is zero throughout the plate, and then at time $\bar{t} = 0$, the temperature θ at the left surface, $X = 0$, is raised to 1.0 and held at this value. The right surface, $X = 1$, is insulated. Steady-state conditions are obtained when $\theta = 1.0$ throughout the plate, within the chosen convergence criterion. This figure shows that the temperature distributions approach the steady-state distribution as time elapses. Steady state is attained when time $\bar{t}$ reaches a value of around 4.5. This is much larger than the time taken to reach steady state in Example 10.1; see Figure 10.9. However, in the previous example, one surface was maintained at $\theta = 0$, whereas in this example, the entire plate is heated or cooled. This implies a greater transfer of energy in the present case, as compared to the mass transfer in Example 10.1. Figure 10.11 shows the variation of the temperature at several locations in the plate with time. The temperatures are found to rise sharply from the initial value of 0 and to approach the steady-state value of 1.0 gradually as time increases.

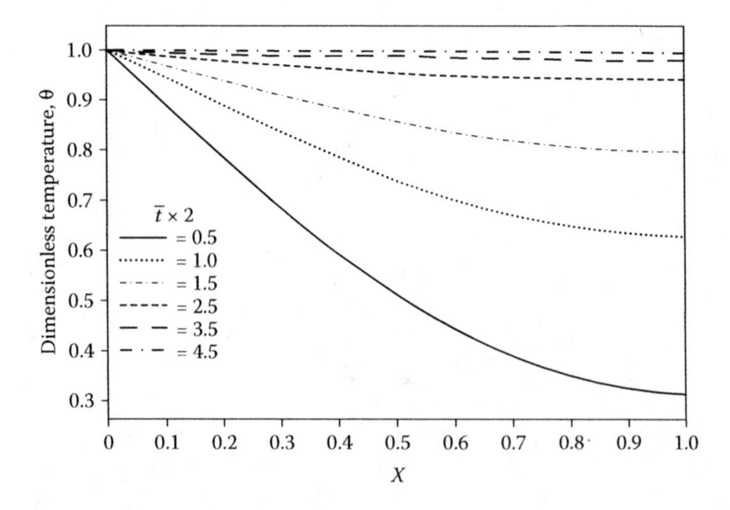

FIGURE 10.10 Computed temperature distribution at various time intervals for Example 10.2. Here, $\Delta\bar{t} = 0.05$ and $\Delta X = 0.1$.

In both Examples 10.1 and 10.2, we have considered one-dimensional transient mass and heat diffusion problems in order to relate the computational procedure to the physical aspects of such problems. However, the numerical schemes discussed here can easily be extended to other physical circumstances that are governed by parabolic PDEs. Such problems arise, for instance, in fluid flow as given by Equation 10.26, diffusion of moisture in porous media, neutron diffusion in nuclear reactors, and water seepage into the ground.

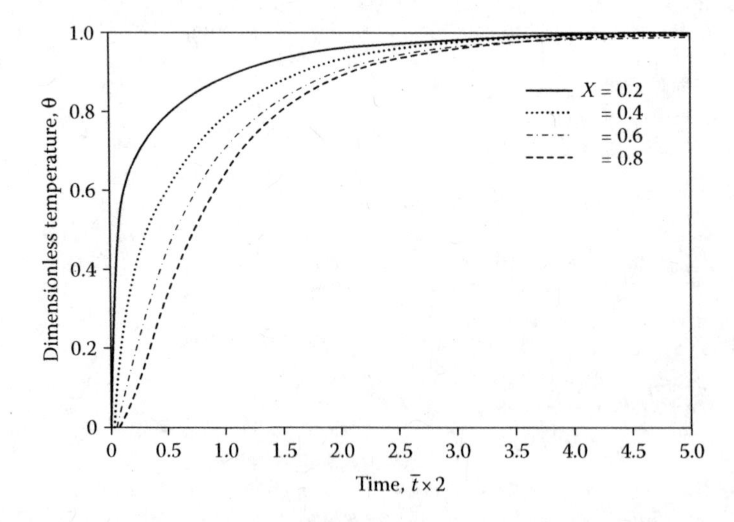

FIGURE 10.11 Variation of the temperature at several locations in the plate with dimensionless time $\bar{t}$ for Example 10.2. The approach to steady state is again seen at large time.

10.3 ELLIPTIC PDEs

In an elliptic PDE, a disturbance at a given point propagates in all directions, in contrast to a parabolic PDE in which there is a definite direction for the flow of information. Therefore, the solution domain in an elliptic PDE is an enclosed one, with boundary conditions specified everywhere along the edges of this domain, as shown in Figure 10.12. The solution at each point is influenced by the solution at every other point in the region where the elliptic PDE applies. Therefore, the numerical solution at the finite number of grid points taken in the region must be obtained simultaneously. This characteristic of elliptic PDEs generally makes the numerical solution more involved than that for parabolic PDEs, in which a marching procedure may be adopted to advance the solution in a particular direction, say, in the direction of increasing time, starting with the initial conditions. Because of the advantages of such a marching scheme, particularly in numerical stability and convergence characteristics, elliptic equations are often formulated as time-dependent parabolic equations, which are solved by time marching to yield the desired solution to the elliptic equations at steady state.

10.3.1 Finite Difference Approach

Several important physical processes are governed by elliptic PDEs. These include conductive and convective heat transfer, mass transfer, the diffusion of neutrons in a nuclear reactor, deflection of a membrane or a plate, interaction of electromagnetic fields, and fluid flow. In order to discuss the numerical techniques for solving elliptic PDEs, let us consider a specific physical problem, namely, that of the two-dimensional steady-state heat conduction in the rectangular region shown in Figure 10.13. In the absence of heat sources in the region, the temperature $T(x, y)$ is governed by Laplace's equation

$$\frac{\partial^2 T}{\partial x^2} + \frac{\partial^2 T}{\partial y^2} = 0 \tag{10.42}$$

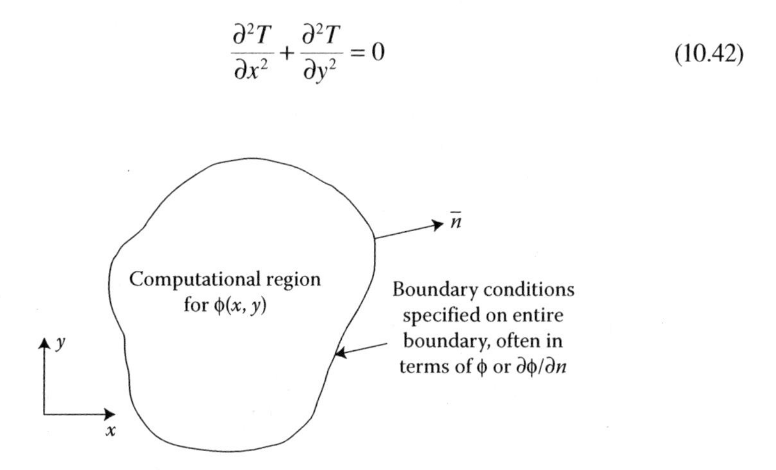

FIGURE 10.12 Solution domain for an elliptic PDE, along with the necessary boundary conditions.

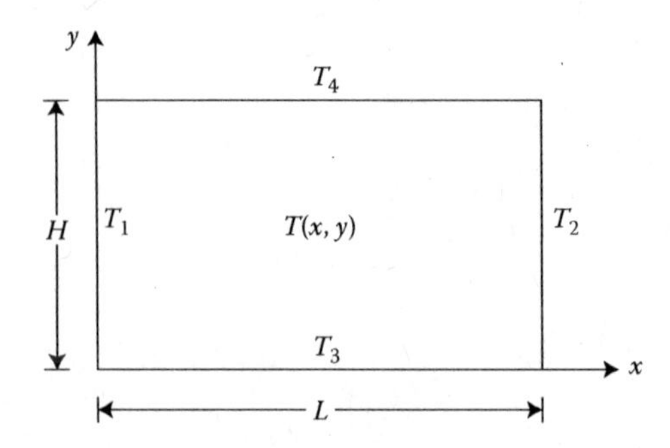

FIGURE 10.13 Coordinate system and boundary conditions for steady-state heat conduction in a rectangular region.

where x and y are the coordinate axis, as indicated in Figure 10.13. The boundary conditions are given in terms of specified values of the temperatures. Such a problem in which the value of the unknown variable, being temperature in this case, is given at the boundaries is known as a *Dirichlet problem*, and the conditions as *Dirichlet boundary conditions*. If the gradient of the variable is specified instead at the given boundary, the condition is known as *Neumann boundary condition*, considered in Example 10.2 and also later in this section. If a relationship between the gradient and the value of the variable is given at the boundary, the condition is known as mixed boundary condition. The following discussion of the numerical methods for the solution of elliptic PDEs is directed at the above elliptic equation with Dirichlet boundary conditions. However, most of the methods considered are applicable to other boundary conditions and other elliptic equations as well.

We wish to determine the temperature $T(x, y)$ in the interior of the region shown in Figure 10.13 by solving the governing elliptic PDE, Equation 10.42. The boundary conditions, shown in Figure 10.13, may be written as follows:

$$\begin{aligned}
&\text{At } x=0: T(x,y)=T_1 \quad \text{for } 0 \le y \le H \\
&\text{At } x=L: T(x,y)=T_2 \quad \text{for } 0 \le y \le H \\
&\text{At } y=0: T(x,y)=T_3 \quad \text{for } 0 < x < L \\
&\text{At } y=H: T(x,y)=T_4 \quad \text{for } 0 < x < L
\end{aligned} \tag{10.43}$$

Therefore, the value of the dependent variable $T(x, y)$ is completely specified on the boundaries of the region in which Equation 10.42 applies. To obtain a numerical solution of the given elliptic equation by finite difference methods, we impose a grid with a mesh size of Δx by Δy on the region, as shown in Figure 10.14. Then the numerical solution consists of determining the temperatures at the finite number of grid points in the solution domain. As done earlier for parabolic PDEs, the temperature $T(x, y)$ at a grid point (i, j) is denoted by $T_{i,j}$, where

$$x = i\Delta x \quad \text{and} \quad y = j\Delta y \tag{10.44}$$

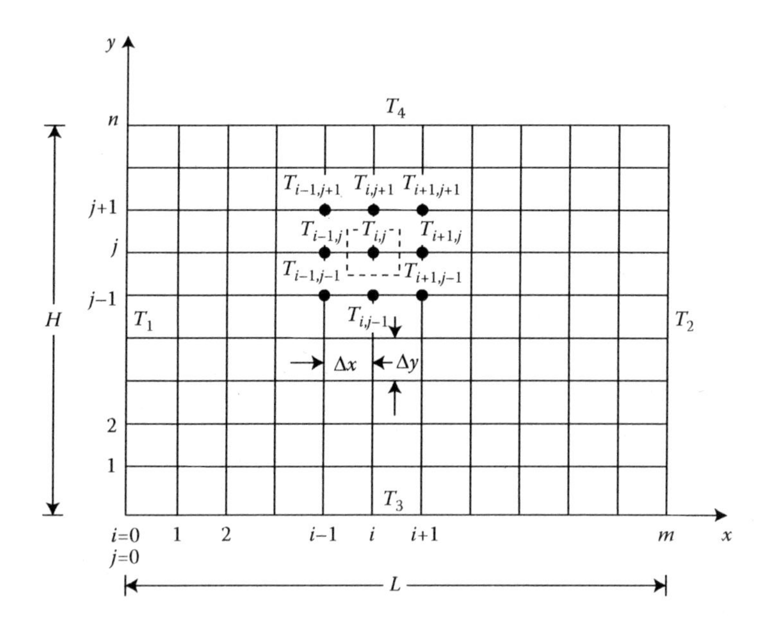

FIGURE 10.14 Subdivision of the computational region by means of a grid with a mesh size of Δx by Δy in the two directions x and y. The nomenclature for the labeling of the temperatures at the grid, or mesh, points is also indicated.

Similarly, the temperatures at other grid points are labeled, as shown in Figure 10.14. If the length L in the x direction is divided into m equal subdivisions, and the height H in the y direction into n equal subdivisions, then

$$\Delta x = \frac{L}{m} \quad \text{and} \quad \Delta y = \frac{H}{n} \tag{10.45}$$

Thus, i varies from 0 to m and j from 0 to n.

We may now proceed to obtain a finite difference approximation to the given elliptic PDE. The second partial derivatives at the grid point (i, j) may be approximated, in central difference form and in terms of the temperatures at the neighboring grid points, as follows:

$$\frac{\partial^2 T}{\partial x^2} = \frac{T_{i+1,j} - 2T_{i,j} + T_{i-1,j}}{(\Delta x)^2} \tag{10.46}$$

$$\frac{\partial^2 T}{\partial y^2} = \frac{T_{i,j+1} - 2T_{i,j} + T_{i,j-1}}{(\Delta y)^2} \tag{10.47}$$

where the TE in Equation 10.46 is $O[(\Delta x)^2]$ and that in Equation 10.47 is $O[(\Delta y)^2]$, as obtained in Chapter 4. Substituting these finite difference approximations into Equation 10.42, we obtain

$$\frac{T_{i+1,j} - 2T_{i,j} + T_{i-1,j}}{(\Delta x)^2} + \frac{T_{i,j+1} - 2T_{i,j} + T_{i,j-1}}{(\Delta y)^2} = 0 \quad (10.48)$$

The above finite difference equation can be written at each of the interior points in the computational domain. Therefore, a system of $[(m-1)\,(n-1)]$ simultaneous linear equations is obtained. These equations may be solved by the methods discussed in Chapter 6 to obtain the $[(m-1)\,(n-1)]$ unknown temperatures at the interior grid points. Frequently, a square mesh, with $\Delta x = \Delta y$, is employed. In this case, Equation 10.48 may be written as

$$T_{i,j} = \frac{T_{i+1,j} + T_{i-1,j} + T_{i,j+1} + T_{i,j-1}}{4} \quad (10.49)$$

which implies that the temperature at a given grid point is simply an average of the temperatures at the four adjacent grid points. The computational molecule, which indicates the effect of the values at the neighboring grid points on that at a given node, is shown in Figure 10.15a for this second-order approximation of Laplace's

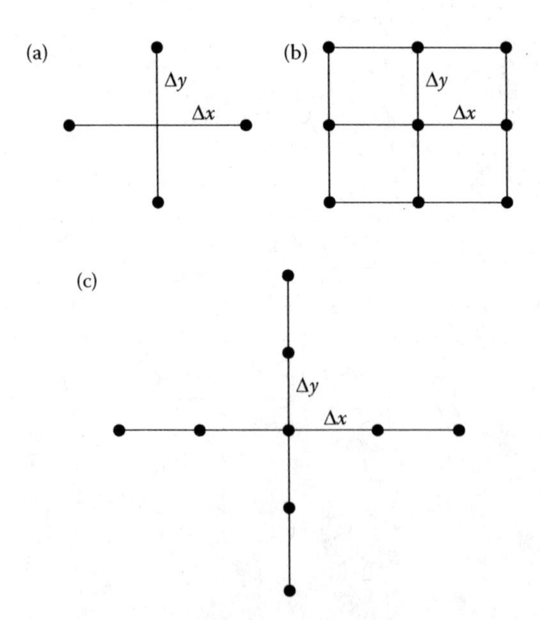

FIGURE 10.15 Computational molecules for various finite difference schemes for solving Laplace's equation: (a) second-order approximation; (b) and (c) two different fourth-order approximations.

equation. We can also obtain higher-order approximations by using a larger number of points in the neighborhood of the grid point being considered; see Figure 4.9. Figures 10.15b and c show, for instance, the computational molecules for fourth-order approximations of Laplace's equation. The accuracy of the numerical results can be improved by the use of a higher-order difference method or by a reduction of the mesh size. However, the first approach has problems near the boundaries because of the large number of neighboring points needed for the approximation at a given nodal point. Therefore, grid refinement, with the spacing between the grid points being reduced until the numerical results are essentially independent of the mesh size, is often preferred for improving the accuracy.

As shown by Equation 10.48, we are faced with the task of solving a large set of linear algebraic equations. If the number of points at which the numerical solution is to be obtained is M in the x direction and N in the y direction, where $M = m - 1$ and $N = n - 1$ for the problem considered above, the number of unknowns is MN. The set of equations to be solved for this problem may be written as

$$AT = B \tag{10.50}$$

where the coefficient matrix A is of size $MN \times MN$ and B is a vector whose elements are all zero except for those that arise from the boundary conditions. The unknown temperatures constitute a vector T whose elements are $T_{1,1}, T_{1,2}, \ldots, T_{1,N}, T_{2,1}, \ldots, T_{M,N}$. Then the coefficient matrix A from Equation 10.49 is of the following form:

$$A = \begin{bmatrix} -4 & 1 & & & 1 & & & \\ 1 & -4 & 1 & & & 1 & & \\ & 1 & -4 & 1 & & & \ddots & \\ 1 & & & & & & & 1 \\ & 1 & & & & & & \\ & & \ddots & & & & & \\ & & & 1 & & 1 & -4 & 1 \\ & & & & 1 & & 1 & -4 \end{bmatrix} \tag{10.51}$$

where only the elements at the diagonal, on either side of it, and in the two distant bands shown are nonzero.

Therefore, the coefficient matrix is not tridiagonal but has two additional bands, which are one element wide and are far removed from the main diagonal. In fact, the last nonzero element in the first row and the lowest nonzero element in the first column are both at the $(N + 1)$th position. Since the coefficient matrix is very sparse, although not tridiagonal, iterative methods can be employed advantageously as compared to direct methods for solving this system of equations. The number of equations is generally large, since even for a coarse grid with $M = N = 20$, we have 400 equations. We shall first consider iterative methods for solving the system of linear equations obtained from the finite difference formulation, followed by a discussion of some direct methods that have been developed in recent years.

10.3.2 Numerical Solution by Iterative and Direct Methods

Several iterative methods for solving simultaneous linear equations were discussed in Chapter 6. These included the Jacobi, the Gauss–Seidel, and the SOR or SUR methods. It was indicated that diagonal dominance is needed for the convergence of these methods. The finite difference equation, Equation 10.48, can be written for each grid point. Then the coefficient of $T_{i,j}$ is the largest one in magnitude and its absolute value is equal to the sum of the coefficients of the other terms. The system of equations can be arranged so that the dominant terms appear along the diagonal. As discussed in Section 6.6, the absolute value of the diagonal coefficient must be *larger* than the sum of the absolute values of the remaining coefficients in each row of the matrix for a diagonally dominant system that is guaranteed to converge. However, the present system of equations, where the absolute value of the diagonal coefficient is *equal to* the sum of the absolute values of the remaining coefficients in each row, has adequate diagonal dominance to converge in most cases. Therefore, for the application of iterative methods, Equation 10.48 is solved for $T_{i,j}$, which constitutes the diagonally dominant term, to give

$$T_{i,j} = \frac{T_{i+1,j} + T_{i-1,j} + (\Delta x/\Delta y)^2 (T_{i,j+1} + T_{i,j-1})}{2[1 + (\Delta x/\Delta y)^2]} \tag{10.52}$$

This equation yields Equation 10.49 if $\Delta x = \Delta y$.

In the Jacobi iteration method, we start with initial, assumed values of the dependent variable at all the grid points in the computational domain. Using this assumed initial distribution, we obtain the next approximation to the solution from Equation 10.52 and compare it with the starting solution. If a specified convergence criterion is not satisfied, the computed results are used in Equation 10.52 to obtain the next iteration. This process is repeated until the given convergence criterion is satisfied. Generally, the convergence criterion demands that the change in the value of the dependent variable from one iteration to the next be less than a prescribed small quantity ε, at each grid point. The computed results from two successive iterations are, therefore, stored, and the values are updated only after the completion of the computation for a given iteration. This numerical scheme is given by the recursive formula

$$T_{i,j}^{(l+1)} = \frac{T_{i+1,j}^{(l)} + T_{i-1,j}^{(l)} + (\Delta x/\Delta y)^2 \left(T_{i,j+1}^{(l)} + T_{i,j-1}^{(l)}\right)}{2[1 + (\Delta x/\Delta y)^2]} \quad \text{for } 1 \le i \le m-1, \quad 1 \le j \le n-1 \tag{10.53}$$

where the superscript l or $l + 1$ refers to the number of the iteration. The starting values are denoted by the superscript (0). As discussed in Chapter 6, this method is inefficient for conventional, or single-processor, computers, since the old values of the unknown are replaced by the new ones only after all the values for a given iteration have been computed and since both of the iterative solution vectors must be stored.

A considerable improvement in the computational procedure for single-processor machines, such as PCs and workstations, is obtained by the Gauss–Seidel method, which employs the most recent values of the unknowns in the computation. Generally, a systematic traverse is used, for instance, by increasing i at a given value of j, which is itself increased by 1 after each traverse in the x direction. Then $T_{i-1,j}$ and $T_{i,j-1}$ are calculated before $T_{i,j}$ for a given iteration. Therefore, the iterative scheme for the Gauss–Seidel method is given by

$$T_{i,j}^{(l+1)} = \frac{T_{i+1,j}^{(l)} + T_{i-1,j}^{(l+1)} + (\Delta x/\Delta y)^2 \left(T_{i,j+1}^{(l)} + T_{i,j-1}^{(l+1)}\right)}{2[1+(\Delta x/\Delta y)^2]} \quad \text{for } 1 \le i \le m-1 \quad 1 \le j \le n-1 \tag{10.54}$$

The previous value of an unknown is replaced by the new value as soon as it is obtained, and, therefore, only one value of each unknown needs to be stored. The programming is also simplified, since we must deal with only one iterative value of the temperature at a given grid point. The method converges if the system is diagonally dominant, which is adequately achieved in the problem being considered. Convergence is generally obtained with even weaker diagonal dominance. The systems of linear equations obtained from the finite difference approximation of the differential equations that arise in common engineering problems generally have sufficient diagonal dominance for the iterative methods to be employed satisfactorily.

The convergence of the iterative scheme is given in terms of the change in the computed values from one iteration to the next. If the magnitude of this change, at each grid point, is less than a specified small number ε, which is known as the *convergence parameter*, the scheme is assumed to have converged. This convergence criterion may be given in terms of the absolute or the normalized value of the change in the temperatures. Therefore, the iterative process is terminated if

$$\left|T_{i,j}^{(l+1)} - T_{i,j}^{(l)}\right| \le \varepsilon \quad \text{for } 1 \le i \le m-1 \quad 1 \le j \le n-1 \tag{10.55a}$$

or

$$\left|\frac{T_{i,j}^{(l+1)} - T_{i,j}^{(l)}}{T_{i,j}^{(l)}}\right| \le \varepsilon \quad \text{for } 1 \le i \le m-1 \quad 1 \le j \le n-1 \tag{10.55b}$$

The value of ε is taken as small, say, 10^{-4} for the second convergence criterion, and is varied over a few orders of magnitude to ensure that the computed results are independent of the value chosen.

10.3.2.1 Point Relaxation

The Gauss–Seidel method converges about twice as fast as the Jacobi method, on conventional computing machines with a single CPU, for a given convergence criterion. The rate of convergence can be improved considerably by the use of the SOR method, discussed in Chapter 6. This method is given by the formula

$$T_{i,j}^{(l+1)} = \omega[T_{i,j}^{(l+1)}]_{GS} + (1-\omega)T_{i,j}^{(l)} \tag{10.56}$$

where ω is a constant, known as the *relaxation factor*, and $[T_{i,j}^{(l+1)}]_{GS}$ is the value obtained from the Gauss–Seidel iteration formula, such as Equation 10.54. For SOR, ω lies between 1 and 2. The method diverges for $\omega > 2$; the Gauss–Seidel scheme is obtained for $\omega = 1$; and SUR is obtained if $0 < \omega < 1$. Substituting Equation 10.54 into Equation 10.56, we can write the recursion formula for the SOR method for the problem under consideration as

$$T_{i,j}^{(l+1)} = \omega\left[\frac{T_{i+1,j}^{(l)} + T_{i-1,j}^{(l+1)} + (\Delta x/\Delta y)^2\left(T_{i,j+1}^{(l)} + T_{i,j-1}^{(l+1)}\right)}{2(1+(\Delta x/\Delta y)^2)}\right] + (1-\omega)T_{i,j}^{(l)} \quad \text{for } 1 \le i \le m-1, \quad 1 \le j \le n-1 \tag{10.57}$$

There is an optimum value of the relaxation factor, ω_{opt}, at which convergence is the fastest. For a square region, with $n = m$, the Gauss–Seidel method converges about twice as fast as the Jacobi method and the SOR method, at the optimum value of the relaxation factor, six and 19 times faster, respectively, than the Gauss–Seidel method for $n = 10$ and $n = 30$ (Jaluria and Torrance, 2003). Therefore, if ω_{opt} is known, the SOR method is very efficient. However, ω_{opt} varies with the PDE, the boundary conditions, the grid spacing, the geometry of the computational domain, and so on. It is not known in most cases, and the analytical determination of its value is quite involved. Therefore, one generally determines it by solving the problem at different values of ω to obtain the optimum or by employing the information available on other similar problems. If several problems of a particular type are to be solved, it would be worthwhile to spend the effort and time to determine ω_{opt}.

The rate of convergence is quite sensitive to the value of ω and, for a value far from the optimum value, the convergence rate is close to that for the Gauss–Seidel method. For some simple cases, ω_{opt} may be obtained analytically. For Laplace's equation in a rectangular region with Dirichlet conditions (see Figure 10.13), the optimum value is given by

$$\omega_{opt} = \frac{2}{1+(1-a^2)^{1/2}} \tag{10.58a}$$

where

$$a = \frac{1}{1+(\Delta x/\Delta y)^2}\left[\cos\frac{\pi}{m} + \left(\frac{\Delta x}{\Delta y}\right)^2 \cos\frac{\pi}{n}\right] \tag{10.58b}$$

Therefore, for $m = n = 20$, $\omega_{opt} = 1.7295$; and for $m = n = 30$, it is 1.8107, with $\Delta x = \Delta y$ in these cases. Figure 10.16 shows the dependence of the number of iterations, for convergence, on ω for a rectangular region. The need to employ a value close to the optimum is clear. It may also be mentioned here that SUR is generally used to improve the convergence characteristics of the iterative process, particularly for nonlinear equations which may diverge when Gauss–Seidel iteration is applied. Though we have discussed only point relaxation here, several other similar relaxation

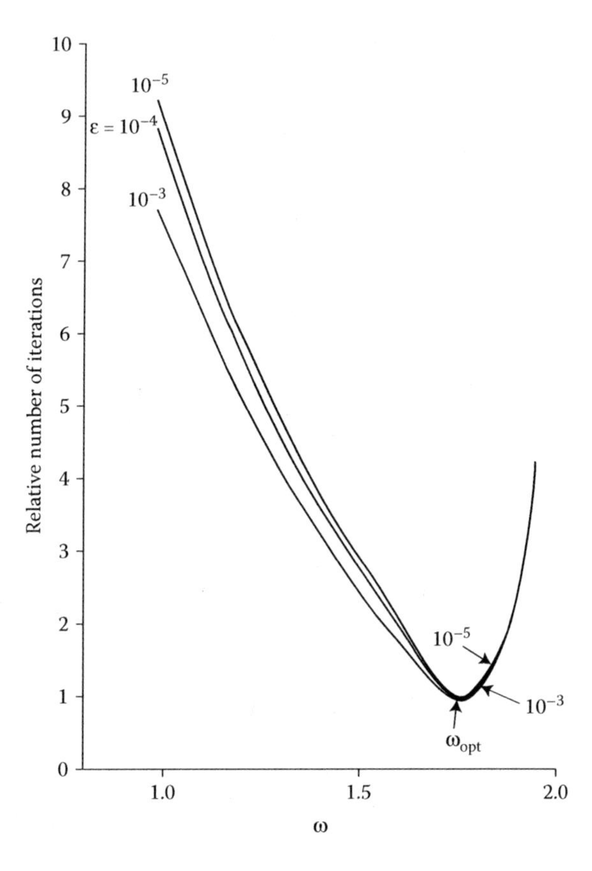

FIGURE 10.16 Variation of the number of iterations, normalized by the number at the optimum, for convergence of the second-order finite difference scheme for Laplace's equation in a square region, with the relaxation factor ω, for the SOR method. Note the strong dependence on ω and the considerable reduction in number of iterations as ω varies from 1.0 (Gauss–Seidel) to the optimum value ω_{opt}. Here, the number of subdivisions in either direction is 20.

methods have been developed to employ blocks of unknowns, rather than a single unknown, in order to increase the efficiency of the method. Such methods, known as *block relaxation* are commonly used in the solution of elliptic PDEs, as discussed in greater detail by Jaluria and Torrance (2003).

10.3.2.2 Direct Methods

Several direct methods, based on elimination, were discussed in Chapter 6. Among the most important of these are the Gaussian elimination and the matrix decomposition methods. Many other methods, such as Gauss–Jordan and matrix inversion methods, are based on Gaussian elimination. For a tridiagonal matrix system, Gaussian elimination may be used very effectively, as demonstrated in Example 6.2. In this case, the number of arithmetic operations required are of order n, instead of n^3 for a general system of n equations. In the triangular decomposition method, such as Crout's method, the coefficient matrix A is factored into lower and upper triangular matrices, each of which may be solved by forward and backward substitution. However, except for tridiagonal systems, these direct methods are often not as efficient as the iterative methods, such as the optimized SOR method, and also give rise to larger round-off errors. Therefore, iterative methods are frequently used for solving the large systems of algebraic equations obtained from the finite difference approximation of elliptic PDEs. Nonlinear algebraic equations are obtained if the elliptic PDE is nonlinear. In such cases, iteration is generally necessary for the solution of the equations, and iterative methods, such as the Gauss–Seidel and relaxation methods, are particularly appropriate.

Recently, specialized direct methods for solving finite difference approximations of the Poisson and Laplace equations in simple geometries have been developed. These methods include the cyclic reduction and the fast Fourier transform methods, which are among the most efficient means for solving these equations in simple, two-dimensional regions for Dirichlet or Neumann boundary conditions. A discussion of these methods is beyond the scope of this book. Further details and references may be obtained from Ferziger (1998) and Jaluria and Torrance (2003). Generally, available computer software is used for the application of these methods, since the algorithms tend to be very involved.

10.3.3 Other Methods

An efficient iterative method for solving elliptic PDEs is the ADI method, which gives rise to a tridiagonal set in each iterative step. The method employs the unknown values of the dependent variable from the current iteration along one direction and known values from the previous iteration along the other direction. In the next step, these directions are reversed. An acceleration parameter $\tilde{\omega}$, similar to that in the SOR method, is used to improve the rate of convergence. The recursion formulas for two iterative steps are

$$\frac{T_{i+1,j}^{(l+1)} - (2+\tilde{\omega})T_{i,j}^{(l+1)} + T_{i-1,j}^{(l+1)}}{(\Delta x)^2} + \frac{T_{i,j+1}^{(l)} - (2-\tilde{\omega})T_{i,j}^{(l)} + T_{i,j-1}^{(l)}}{(\Delta y)^2} = 0 \qquad (10.59a)$$

$$\frac{T_{i+1,j}^{(l+1)} - (2-\tilde{\omega})T_{i,j}^{(l+1)} + T_{i-1,j}^{(l+1)}}{(\Delta x)^2} + \frac{T_{i,j+1}^{(l+2)} - (2+\tilde{\omega})T_{i,j}^{(l+2)} + T_{i,j-1}^{(l+2)}}{(\Delta y)^2} = 0 \quad (10.59b)$$

These two steps are considered together to constitute one complete iteration. The tridiagonal sets obtained are solved by Gaussian elimination, and the iteration is repeated until convergence is attained. This method, developed by Peaceman and Rachford (1955), is used extensively for two-dimensional steady-state diffusion problems, governed by elliptic equations, and transient problems, as outlined below.

In several cases, particularly for nonlinear problems, the elliptic PDE is solved by considering an equivalent time-dependent problem, which is parabolic in time. Laplace's equation may, for instance, be solved by obtaining the transient solution of the equation

$$\frac{\partial T}{\partial t} = \frac{\partial^2 T}{\partial x^2} + \frac{\partial^2 T}{\partial y^2} \quad (10.60)$$

where the steady-state solution at large time is the required solution of the elliptic PDE. One could use time marching to solve this problem, using the various techniques outlined in the preceding section. The ADI method may be employed without iteration for this problem and is one of the most efficient methods for such two-dimensional transient problems. The corresponding finite difference equations, with the superscripts denoting the time step, are as follows:

$$\frac{T_{i,j}^{(l+1)} - T_{i,j}^{(l)}}{\Delta t} = \frac{T_{i+1,j}^{(l+1)} - 2T_{i,j}^{(l+1)} + T_{i-1,j}^{(l+1)}}{(\Delta x)^2} + \frac{T_{i,j+1}^{(l)} - 2T_{i,j}^{(l)} + T_{i,j-1}^{(l)}}{(\Delta y)^2} \quad (10.61a)$$

$$\frac{T_{i,j}^{(l+2)} - T_{i,j}^{(l+1)}}{\Delta t} = \frac{T_{i+1,j}^{(l+1)} - 2T_{i,j}^{(l+1)} + T_{i-1,j}^{(l+1)}}{(\Delta x)^2} + \frac{T_{i,j+1}^{(l+2)} - 2T_{i,j}^{(l+2)} + T_{i,j-1}^{(l+2)}}{(\Delta y)^2} \quad (10.61b)$$

The two equations are employed together consecutively to advance by two time steps. Tridiagonal systems are obtained in both cases. This approach for solving an elliptic PDE is frequently employed for nonlinear equations, such as those encountered in fluid mechanics and in heat transfer. The main advantage is that time marching generally yields better stability and convergence characteristics.

10.3.4 Other Geometries and Boundary Conditions

We have considered simple rectangular regions and Dirichlet boundary conditions in the above discussion. However, there are many complexities that arise due to irregularly shaped regions and more involved boundary conditions. Since these considerations are particularly important in real physical problems, a brief discussion of these aspects is included here.

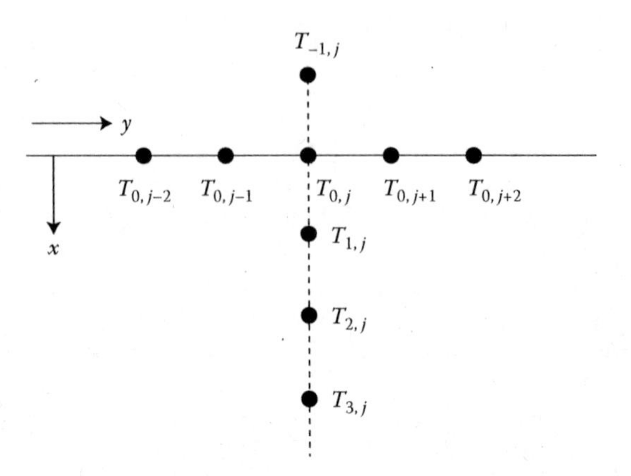

FIGURE 10.17 Distribution of grid points at a boundary, showing the fictitious point $T_{-1,j}$ outside the region.

Consider a boundary at $x = 0$, as shown in Figure 10.17 with a distribution of grid points. If the boundary condition is of Neumann type, that is $\partial T/\partial x = B$, where B is a constant, the value $T_{0,j}$ at the boundary is not known and must be obtained from a finite difference approximation of the derivative in terms of the neighboring grid points. The simplest formulation is

$$\frac{\partial T}{\partial x} \cong \frac{T_{1,j} - T_{0,j}}{\Delta x} = B \tag{10.62}$$

which employs a forward difference in x and is accurate only to order Δx. We can derive a more accurate approximation of the derivative by employing the Taylor-series expansions for the three points adjacent to the boundary, as discussed in Chapter 4. This gives

$$\frac{\partial T}{\partial x} \cong \frac{-T_{2,j} + 4T_{1,j} - 3T_{0,j}}{2\Delta x} = B \tag{10.63}$$

This formulation has a TE of order $(\Delta x)^2$ and may be written for all the points on the boundary. Therefore, an equation for $T_{0,j}$, in terms of the values at the neighboring grid points, is obtained. Similar equations may be written at other boundaries.

Another approach is to employ a fictitious point $T_{-1,j}$ outside the boundary, as shown in Figure 10.17, and write $\partial T/\partial x$ in the central difference approximation; that is,

$$\frac{\partial T}{\partial x} \cong \frac{T_{1,j} - T_{-1,j}}{2\Delta x} = B \tag{10.64}$$

Then $T_{-1,j}$ is eliminated between this equation and the finite difference equation of the given PDE, written for the grid point $(0, j)$ at the boundary. This also gives an error of $O[(\Delta x)^2]$. However, a row of unknowns at points outside the boundary is introduced, increasing the computational effort. Similarly, other, more involved, boundary conditions may be treated. The resulting equations for the surface grid points are used along with the equations for the interior region to obtain the solution.

An approach frequently employed in fluid flow and in heat and mass transfer is based on the mass, momentum, and energy balance equations for the finite regions represented by the surface grid point. For instance, a rectangular region of dimensions $\Delta y \times (\Delta x/2)$ may be placed symmetrically surrounding the grid point with temperature $T_{0,j}$. Then finite difference equations are written to balance the mass, momentum, and energy transported across the boundaries of this finite region against those stored in the region. This approach gives an accurate and physically representative equation for the dependent variable at the surface node. The equation will also be consistent with basic physical or chemical laws governing the transport processes under consideration. This approach is the preferred one for many transport phenomena of interest in engineering applications. For further details, see Jaluria and Torrance (2003).

Frequently, we are faced with an irregular region, and it becomes necessary to obtain an equation applicable to an interior grid point that lies near such a boundary. Consider a point C in a square mesh, as shown in Figure 10.18, with points A and B at the boundary. Since these points A and B are at distances $\beta_1 \Delta x$ and $\beta_2 \Delta y$, respectively, away from C, where β_1 and β_2 are constants that are both less than 1.0, the finite difference equation, such as Equation 10.49, derived for the interior region does not apply at C. One method of determining the value at C is to use interpolation between the points A and 3, or B and 4. An average of these interpolations may also be employed to represent the value at C.

Another method is to employ Taylor series expansions to derive the finite difference approximations for the derivatives at the point C in terms of the values at the points A, B, C, 3, and 4. These are then substituted in the given PDE to yield the

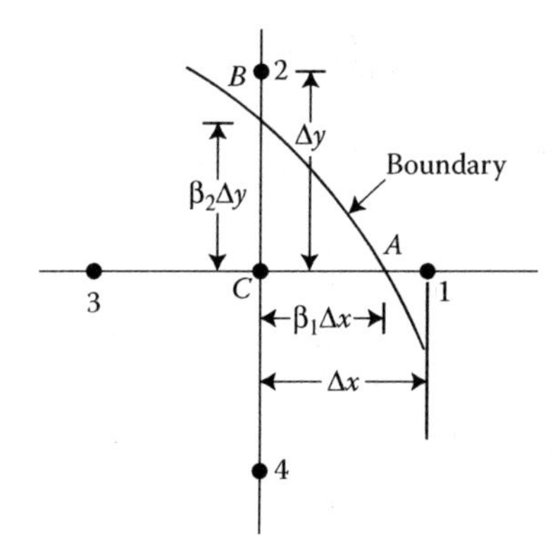

FIGURE 10.18 Grid points of a rectangular mesh near an irregular boundary.

equation that applies at C. For instance, the finite difference approximation for Laplace's equation at point C is obtained by this method as follows:

$$\frac{2T_A}{\beta_1(1+\beta_1)}+\frac{2T_B}{\beta_2(1+\beta_2)}+\frac{2T_3}{1+\beta_1}+\frac{2T_4}{1+\beta_2}-\left(\frac{2}{\beta_1}+\frac{2}{\beta_2}\right)T_C=0 \tag{10.65}$$

Therefore, T_C may be obtained in terms of the values at the boundary and the interior points. This applies for Dirichlet conditions. For further details and for other boundary conditions, see Forsythe and Wasow (1960) and Smith (1978).

We conclude this discussion on the finite difference solution of elliptic PDEs by repeating that a finite difference approximation is obtained for the given PDE, employing a chosen grid in the computational region, to yield a system of algebraic equations. For Dirichlet conditions, the values at the boundary grid points are known. For other boundary conditions, algebraic equations are obtained that relate the values at the boundaries with those at the interior grid points. The resulting system of equations may be solved by direct or iterative methods to yield the desired solution. Iterative methods are more frequently used because of the large number of equations that arise and the simplicity in programming. Iteration is usually necessary for nonlinear equations. Time marching may be employed in some cases, and specialized direct methods are also available for some simple problems. Examples 10.3 and 10.4 discuss the Gauss–Seidel and the SOR methods, respectively, for solving elliptic PDEs.

10.3.5 Finite Element and Other Solution Methods

In the preceding sections, we considered the solution of parabolic and elliptic PDEs by means of finite difference approximations, which are applied to the governing differential equations. However, in recent years, the *finite element method* has gained in popularity for practical problems in engineering. Finite difference methods are simpler to comprehend, and it is easier to develop computer programs for them. They are still widely used for engineering problems because of this ease in programming, and, therefore, we discussed them in detail here. However, practical circumstances often involve complexities, such as complicated geometries, boundary conditions, and material property variations. In such cases, the finite element approach provides a very versatile method that can be employed for a wide range of engineering problems. Frequently, available software is used, since the development of the computer program is generally involved and time-consuming.

The finite element method is based on the integral formulation of the conservation principles. The computational region is divided into a number of finite elements, several forms and types of which are available for different geometries and governing equations. Triangular elements for two-dimensional problems and tetrahedral elements for three-dimensional problems are commonly employed, as shown in Figure 10.19. The variation of the dependent variable is generally taken in terms of simple polynomials and frequently as linear within the elements. Integral equations that apply for each element are derived, and the conservation postulates are satisfied by minimization of the integrals or by reducing their weighted residuals to zero. The latter gives rise to a

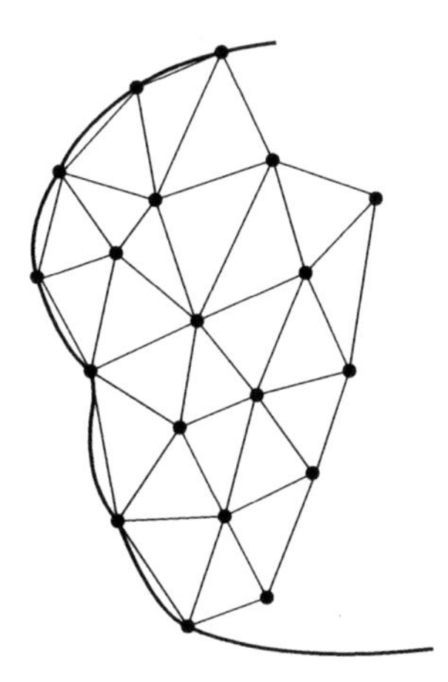

FIGURE 10.19 Finite element discretization, employing triangular elements.

commonly used method known as *Galerkin's method*. Thus, the distribution of the dependent variable within the elements, and then in the entire region, is obtained. As mentioned above, the method is particularly suitable for irregular boundaries and complicated boundary conditions. Consequently, it is widely used for practical problems in engineering. For details on finite element methods, see the books by Mitchell and Wait (1977), Huebner et al. (1995), and Reddy and Gartling (2010).

Two other approaches have gained in importance in recent years. These are the boundary element method and the control volume approach. The former is similar to the finite element method, except that the integral formulation for the computational domain is transformed to one that applies for the bounding surface. Although somewhat limited in its applicability, this method is finding much interest for many problems of practical interest, particularly for those where the phenomena at the surface are of main concern. The method has the advantage, over finite element methods, of a smaller number of elements and unknowns. See the books by Brebbia (1977), Banerjee and Butterfield (1981), and Beer et al. (2010) on the background and application of this method.

The control volume approach is also based on the integral formulation. The physical region is divided into a set of nonoverlapping control volumes, such as those obtained by drawing lines parallel to the coordinate axes midway between the nodes; see the dashed lines in Figure 10.14. The integral conservation statement is applied to each control volume, using interpolation between the node points to approximate the integrands. Thus, the volume and surface integrals are approximated, using values at the nodes. The resulting algebraic equations are similar to those obtained from the finite difference approach, which is based on the differential equations. However, the finite volume method satisfies the conservation principles more accurately and is particularly valuable for the numerical formulation of the boundary conditions. Greater flexibility and versatility is obtained as compared to the finite difference methods and the programming is generally much simpler than that for the finite and boundary element methods. See Patankar (1980) and Jaluria and Torrance (2003) for details on this method.

Example 10.3

The transverse deflection ϕ of a flexible membrane, which cannot resist any bending, is governed by the Poisson equation

$$\frac{\partial^2\phi}{\partial x^2} + \frac{\partial^2\phi}{\partial y^2} = -\frac{p}{T} \tag{10.66}$$

where p is the pressure on the membrane and T is the tension per unit length at the edges. For small deflections, T may be assumed to be constant. A square membrane, of 1.0 m side, is fixed at its boundaries and is subjected to a pressure of 4×10^7 N/m^2; see Figure 10.20. The tension T is 10^8 N/m. Employing the Gauss–Seidel method, compute the variation of the deflection ϕ across the membrane. Take $\Delta x = \Delta y = 0.1$ m.

SOLUTION

The elliptic PDE to be solved numerically is

$$\frac{\partial^2\phi}{\partial x^2} + \frac{\partial^2\phi}{\partial y^2} = -0.4 \tag{10.67}$$

with the boundary conditions

$$\begin{aligned} &\text{At } x = 0 \quad \text{and} \quad x = 1.0 \text{ m: } \phi = 0 \quad \text{for } 0 \le y \le 1.0 \text{ m} \\ &\text{At } y = 0 \quad \text{and} \quad y = 1.0 \text{ m: } \phi = 0 \quad \text{for } 0 < x < 1.0 \text{ m} \end{aligned} \tag{10.68}$$

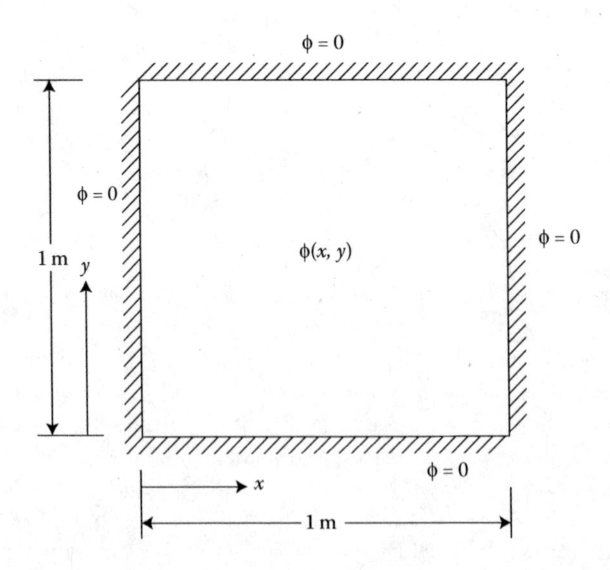

FIGURE 10.20 Coordinate system for computing the deflection $\phi(x, y)$ of a square membrane, as considered in Example 10.3.

where ϕ is also in meters. The origin is taken at the lower left corner of the membrane and x and y are along two sides of the square, as shown in Figure 10.20. Taking the grid spacing to be equal in both directions, that is, $\Delta x = \Delta y$, we obtain the finite difference approximation of the governing equation as follows:

$$\phi_{i,j} = \frac{\phi_{i+1,j} + \phi_{i-1,j} + \phi_{i,j+1} + \phi_{i,j-1}}{4} + 0.4\frac{(\Delta x)^2}{4} \tag{10.69}$$

where $x = i\Delta x$ and $y = i\Delta y$, as shown in Figure 10.14. Also, for $\Delta x = 0.1$, the last term becomes 0.001.

The iterative scheme for the Gauss–Seidel method is obtained from Equation 10.54 for this problem as follows:

$$\phi_{i,j}^{(l+1)} = \frac{[\phi_{i+1,j}^{(l)} + \phi_{i,j+1}^{(l)} + \phi_{i-1,j}^{(l+1)} + \phi_{i,j-1}^{(l+1)}]}{4} + 0.001 \quad \text{for} \quad \begin{matrix} 1 \le i \le n-1 \\ 1 \le j \le n-1 \end{matrix} \tag{10.70}$$

where n is the number of subdivisions in each of the two directions. Appendix B.33 gives the MATLAB script file for this problem. An initially uniform ϕ distribution is assumed in the computational domain, and Equation 10.70 is employed to compute the values for the next iteration. Only the most recent value of ϕ at any given grid point is stored, so that the values are updated as soon as they are computed. The program allows an interactive input of parameters such as number of grid points in each direction, initial uniform value of ϕ in the domain, and convergence parameter. The grid size and the number of grid points in the x and y directions may be taken as different, if the computational domain is not a square or if the boundary conditions are nonsymmetric. Note that the constant 0.001 in Equation 10.70 must be replaced, in the program, by $0.4/[2/(\Delta x)^2 + 2/(\Delta y)^2]$ for arbitrary Δx and Δy.

The computation is terminated if the number of iterations exceeds a specified limit or if the following convergence criterion is satisfied:

$$|\phi_{i,j}^{(l+1)} - \phi_{i,j}^{(l)}| \le \varepsilon \quad \text{for } 1 \le i \le n-1 \quad \text{and} \quad 1 \le j \le n-1 \tag{10.71}$$

where ε is the convergence parameter, denoted by *ep* in the program. Figure 10.21 shows the variation of ϕ with x at different values of y. Because of the symmetry of the given problem, a similar plot of ϕ versus y is obtained at the corresponding values of x. Note in this figure that, as imposed by the boundary conditions, ϕ is zero at the boundaries and is maximum midway between the boundaries. Thus, the maximum deflection is at the center of the square region and is around 2.9 cm for the given values of the physical variables. The deflection ϕ increases as y increases from 0 at the boundary to 0.5 m at the midway point and then decreases toward the far boundary at $y = 1.0$ m. Because of symmetry, one could also consider only one-fourth of the membrane, employing the zero slope conditions of $\partial\phi/\partial x = 0$ at $x = 0.5$ m and $\partial\phi/\partial y = 0$ at $y = 0.5$ m.

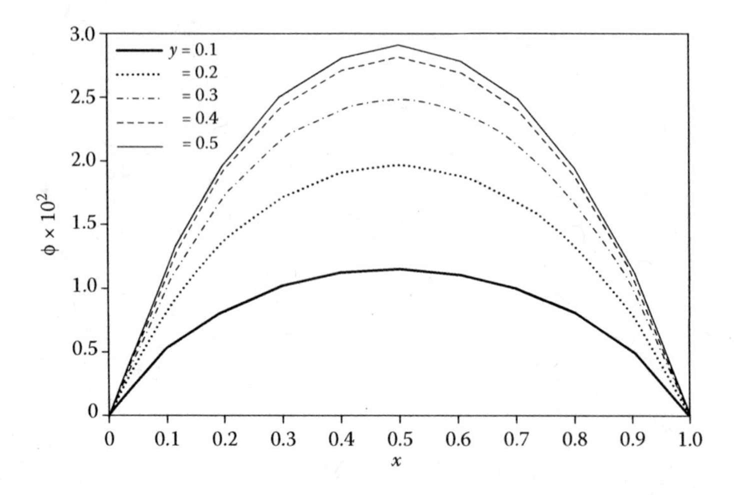

FIGURE 10.21 Computed distributions of the deflection $\phi(x, y)$ at various values of the coordinate distance y, for Example 10.3. The grid spacing $\Delta x = \Delta y = 0.1$ m.

Example 10.4

a. The temperature in a long bar of square cross section is governed by the Laplace equation. The temperature at one surface is T_1, while the other three surfaces are maintained at temperature T_2, see Figure 10.22. Using the SOR method, compute the temperature distribution in the bar. Also determine the optimum value of the relaxation factor ω.
b. Using the program developed in Part (a), solve the equation $\nabla^2\psi = 0$, with the boundary conditions shown in Figure 10.23. This equation governs the

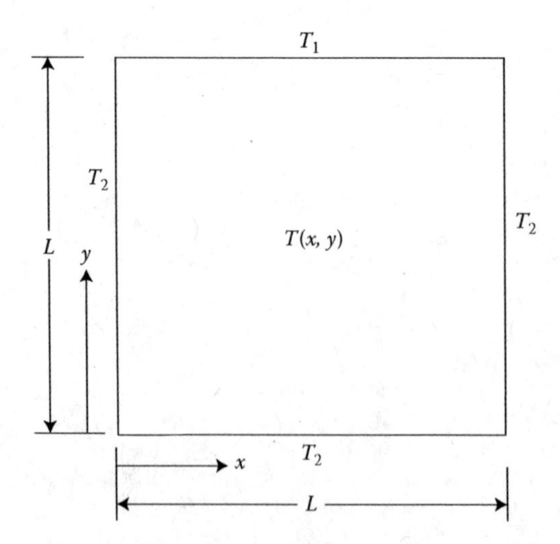

FIGURE 10.22 Physical problem considered in Example 10.4(a), along with the coordinate system.

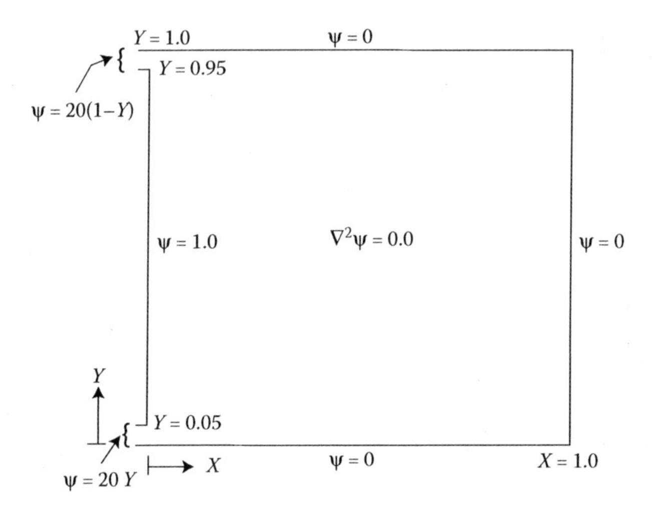

FIGURE 10.23 Flow problem governed by Laplace's equation, as considered in Example 10.4(b).

flow of a fluid in the absence of viscous, or frictional, and rotational effects. Here, ψ is known as the *dimensionless stream function*. It is related to the flow rate and, hence, to the flow field. Compute the ψ distribution in the flow region due to the inflow and outflow as shown in Figure 10.23, and obtain streamlines, or contours of constant ψ, corresponding to $\psi = 0$, 0.05, 0.1, 0.25, 0.5, 0.75, and 1.0.

SOLUTION

a. We can formulate the given problem in dimensionless terms by defining the nondimensional temperature ϕ and coordinate distances X and Y as

$$\phi = \frac{T - T_2}{T_1 - T_2}, \quad X = \frac{x}{L}, \quad Y = \frac{y}{L} \tag{10.72}$$

where $T(x, y)$ is the temperature at an arbitrary location, given by the coordinates x and y in the computational domain, and L is the length or width of the region, as shown in Figure 10.22. The governing equation is obtained as

$$\frac{\partial^2 \phi}{\partial X^2} + \frac{\partial^2 \phi}{\partial Y^2} = 0 \tag{10.73}$$

with the following boundary conditions:

$$\begin{aligned} &\text{At } X = 0 \text{ and } X = 1.0: && \phi = 0 && \text{for } 0 \le Y \le 1.0 \\ &\text{At } Y = 0: && \phi = 0 && \text{for } 0 < X < 1.0 \\ &\text{At } Y = 1.0: && \phi = 1.0 && \text{for } 0 < X < 1.0 \end{aligned} \tag{10.74}$$

If a rectangular region of length L and width W is considered instead, the above nondimensionalization may again be used, so that the governing equation remains unchanged. However, L/W will appear as a parameter in the boundary conditions in that case.

For the SOR method, the iterative scheme is obtained from Equation 10.57 as follows:

$$\phi_{i,j}^{(l+1)} = \omega\left[\frac{\phi_{i+1,j}^{(l)} + \phi_{i-1,j}^{(l+1)} + \phi_{i,j+1}^{(l)} + \phi_{i,j-1}^{(l+1)}}{4}\right] + (1-\omega)\phi_{i,j}^{(l)} \qquad \begin{array}{l}\text{for } 1 \le i \le n-1 \\ \quad\; 1 \le j \le n-1\end{array} \tag{10.75}$$

where the grid spacings Δx and Δy are taken as equal and n is the number of subdivisions in each direction, as shown in Figure 10.14 for a rectangular region.

For SOR, ω lies between 1 and 2. We need to solve the problem for several values of ω to determine the optimum value. The Gauss–Seidel method is obtained for $\omega = 1$. Appendix B.34 gives the MATLAB script file and Appendix C.19 the computer program in Fortran for the SOR method. The program allows one to choose the number of grid points in the two directions, from which the grid sizes are determined. The initial, guessed distribution of ϕ is taken as uniform throughout the computational domain. The value of this initial guess may be specified. A convergence criterion on $\phi_{i,j}$ is employed to check for convergence, see Equation 10.71. We can also specify the maximum number of iterations in order to terminate the computation if convergence is not attained. The convergence parameter and the grid size, or number of grid points, are varied to ensure that the results obtained are not significantly dependent on the values chosen. A subroutine is used in the Fortran program to specify the boundary conditions.

Figure 10.24 shows the variation of the number of iterations for convergence with the relaxation factor ω. The optimum value is found to be

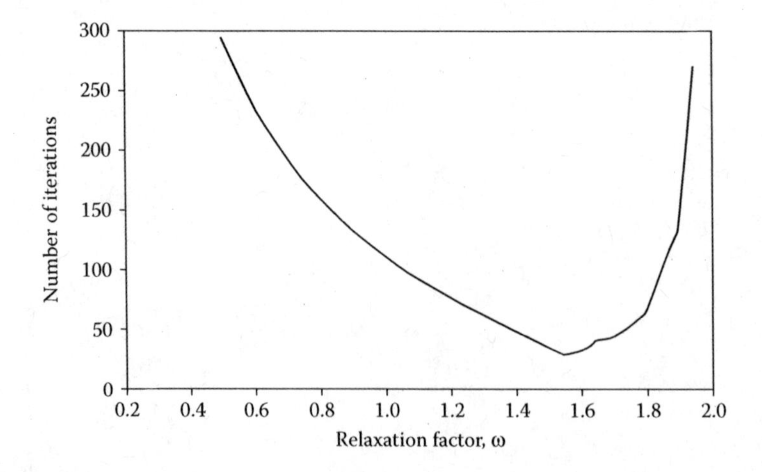

FIGURE 10.24 Variation of the number of iterations, for convergence in Example 10.4(a), with the relaxation factor ω, employing $\Delta X = \Delta Y = 0.1$.

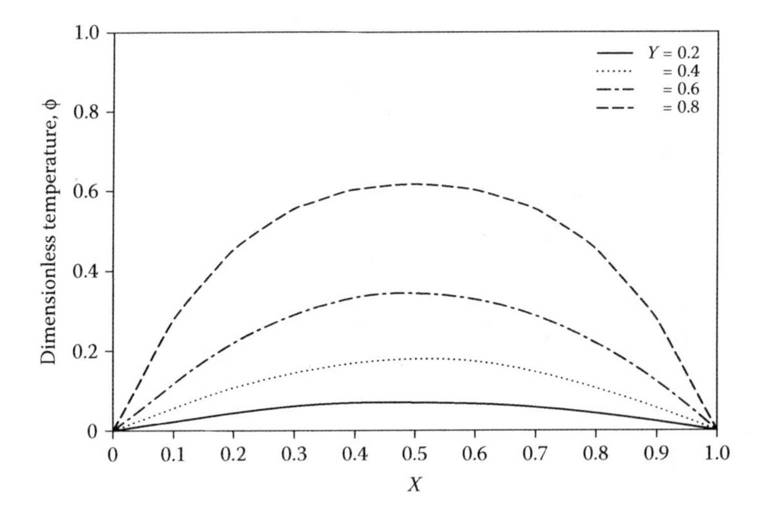

FIGURE 10.25 Computed temperature distributions in Example 10.4(a), with $\omega = 1.6$.

around 1.55. Note that the number of iterations increases sharply as ω is varied away from the optimum value. Therefore, the SOR method is advantageous to use if the value of ω is close to the optimum. Also, see Figure 6.12, which illustrates the application of the SOR method to a system of linear equations. Figures 10.25 and 10.26 show the temperature distributions in the region. Three surfaces are at temperature $\phi = 0$, and the fourth one is at $\phi = 1.0$. The temperatures decrease as one moves away from the hot surface. The distributions are symmetric about $X = 0.5$, as expected.

b. The $\psi(X, Y)$ distribution in the flow region is governed by Equation 10.73 with ϕ replaced by ψ. The problem, as presented in Figure 10.23, is given

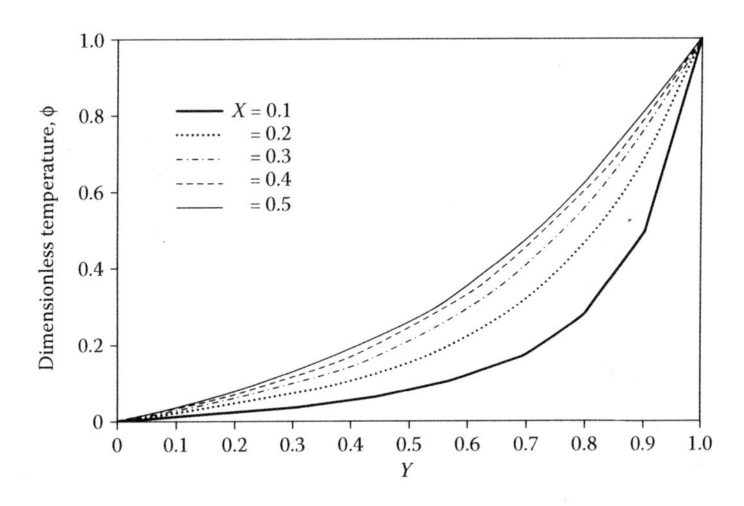

FIGURE 10.26 Computed temperature variation with Y at various values of X in Example 10.4(a), with $\omega = 1.6$.

in nondimensional terms, so that ψ varies from 0 to 1, as do the coordinate distances X and Y. The boundary conditions are written as follows:

$$\begin{aligned}
&\text{At } Y = 0 \quad \text{and} \quad Y = 1.0: && \psi = 0 && \text{for } 0 \le X \le 1.0 \\
&\text{At } X = 1.0: && \psi = 0 && \text{for } 0 < Y < 1.0 \\
&\text{At } X = 0: && \psi = 1.0 && \text{for } 0.05 \le Y \le 0.95 \\
&\text{At } X = 0: && \psi = 20Y && \text{for } 0 < Y < 0.05 \\
&\text{At } X = 0: && \psi = 20(1-Y) && \text{for } 0.95 < Y < 1.0
\end{aligned} \tag{10.76}$$

These equations imply that $\psi = 0$ on three sides of the enclosure and that it varies linearly to 1.0 at the inflow/outflow channels. Although more involved than Equation 10.74, these conditions can easily be incorporated into the program by a suitable modification of the subroutine BCOND in the Fortran program given earlier, or by modifying the conditions in the MATLAB program. One can then employ the main program to obtain the results for chosen values of the grid spacing ΔX and ΔY.

Figure 10.27 shows the contours of constant ψ, or streamlines, at ψ values of 0, 0.05, 0.1, 0.25, 0.5, 0.75, and 1.0. The computed ψ values at the nodal points are used with a simple linear interpolation scheme to determine the X, Y locations where these ψ values are attained. Using graphical procedures, similar to those for Part (a), we draw the contours by joining the various locations on the $X - Y$ plane where a given value of ψ is obtained. The grid spacing $\Delta X = \Delta Y$ was taken as 0.025 at the start and reduced to 0.01 to confirm that the results were not significantly dependent on the value chosen. A convergence parameter ε of 10^{-5} was employed in Equation 10.71, with ϕ replaced by ψ. Again, ε was varied to ensure a negligible effect of the value chosen on the numerical results.

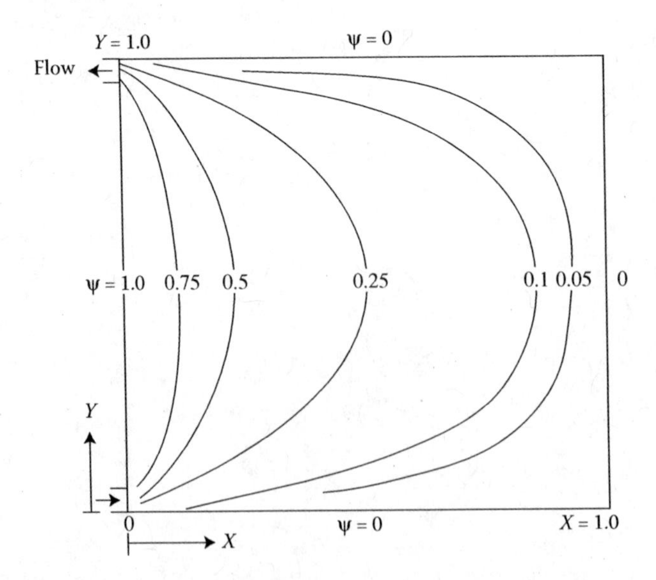

FIGURE 10.27 Computed contours of constant stream function ψ for Example 10.4(b).

This problem represents an important circumstance encountered in several engineering applications, particularly in mechanical and civil engineering. The problem concerns the flow field generated in an enclosed region due to the inflow and outflow of a fluid, assuming the viscous and rotational effects to be absent. If rotational effects are present, with viscous effects still negligible, the Poisson equation is obtained and may be solved in a similar way.

10.4 HYPERBOLIC PDEs

10.4.1 BASIC ASPECTS

Hyperbolic PDEs arise in several problems of engineering interest, such as vibration of rods and strings, transmission of sound in air, and supersonic flow. As discussed earlier, hyperbolic PDEs have two real and distinct characteristics. Information travels at finite speed in regions defined by these characteristics, as shown in Figure 10.28. An observer at point 0 in Figure 10.28a is affected by disturbances only in the region of dependence of the point 0, and a disturbance at 0 can be felt only in the region of influence, where both of these regions are marked by the two families of characteristics. The movement of an object in a stationary fluid or flow of the fluid past a stationary object at speeds greater than the speed of sound in that fluid is known as *supersonic flow* and is also governed by a hyperbolic PDE. Figure 10.28b shows the supersonic flow of air over an airplane, and the region of influence is given by the two characteristic lines shown. A disturbance at 0 is, therefore, felt only in the region of influence, and an observer outside this region is not affected by the presence of the airplane. For this reason, the sound of a supersonic airplane is heard only after it has passed overhead. In this figure, the angle θ is given by $\sin\theta = a/V$, where a is the speed of sound in air and V is the speed of the air flow, with $V > a$ for supersonic flow. If $V/a = 2$, for instance, the angle θ, which is known as the *angle of the Mach cone*, is 30°. This flow circumstance is analogous to that of an airplane moving at speed V in quiescent air.

10.4.2 METHOD OF CHARACTERISTICS

An important numerical technique for solving hyperbolic PDEs is the method of characteristics. In this method, the computational domain is divided into finite regions by the two families of characteristics, taking the grid points at the intersections of these lines, as shown in Figure 10.28a. The properties of characteristics are used to reduce the problem to a system of ODEs, which are solved by methods similar to those discussed in Chapter 9 for ODEs. The main advantage of this method is that the important properties of the exact solution are preserved in the numerical solution. Discontinuities, such as shock waves in supersonic flow, can easily be treated, since discontinuities can occur only along characteristics. However, this method is difficult to use in complicated geometries, because of the problem of keeping track of the characteristics, and in problems where an elliptic or parabolic PDE may apply in one portion of the computational domain and a hyperbolic PDE in the

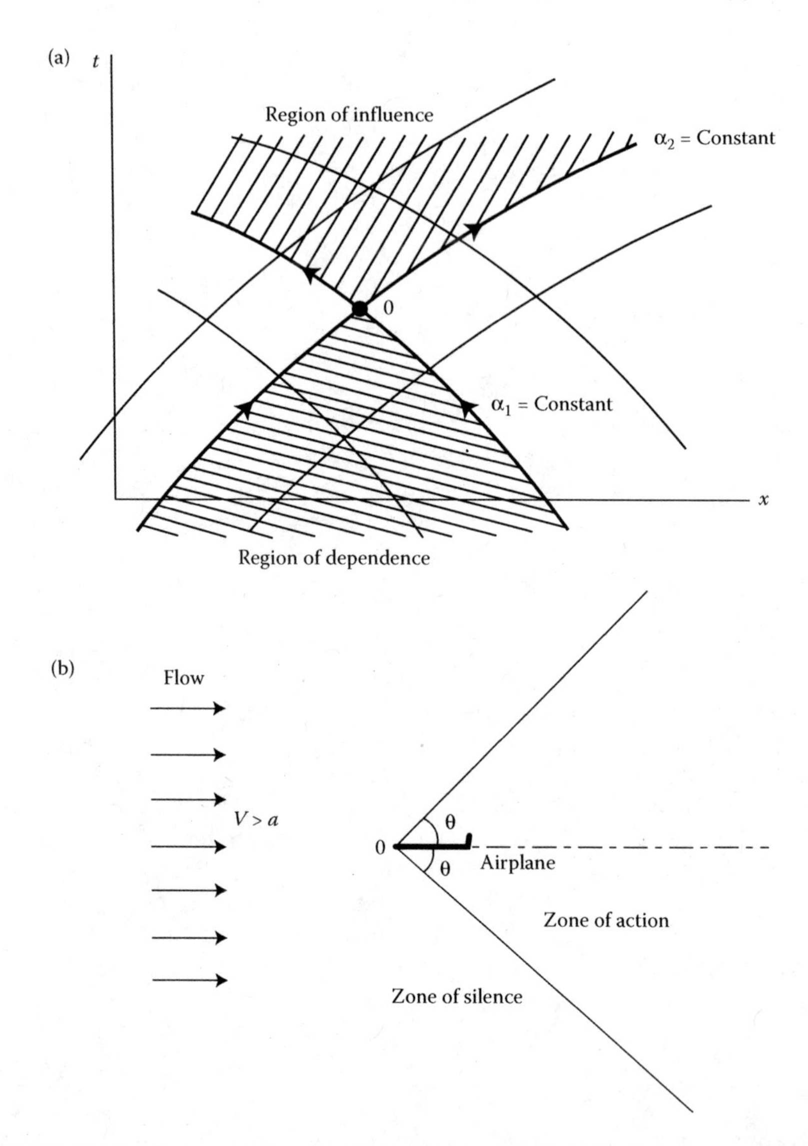

FIGURE 10.28 Characteristics associated with a hyperbolic PDE. (a) Information travel as limited by characteristics in the computational region; (b) supersonic flow of air over an airplane, indicating the zones of action and silence.

other. For further details on the methods based on characteristics for solving hyperbolic PDEs, see Smith (1978) and Ferziger (1998).

10.4.3 Finite Difference Methods

Much of the recent work on hyperbolic PDEs has been based on finite difference methods, which are quite similar to those discussed earlier for parabolic and elliptic equations. Some of these methods are outlined here. Two important hyperbolic equations that we considered earlier are the first-order convection equation

$$\frac{\partial \phi}{\partial t} + c\frac{\partial \phi}{\partial x} = 0 \tag{10.7}$$

and the second-order wave equation

$$\frac{\partial^2 \phi}{\partial t^2} = c^2 \frac{\partial^2 \phi}{\partial x^2} \tag{10.6}$$

where c is the convection velocity in the former case and the propagation velocity of the wave in the latter. Also, ϕ is a dependent variable such as temperature in the first case and deflection of a string in the second.

The initial and boundary conditions for the wave equation may be written as follows:

$$\begin{aligned} &\text{At } t = 0\text{:} \quad \phi = \alpha_1(x) \quad \frac{\partial \phi}{\partial t} = \alpha_2(x) \\ &\text{At } x = 0\text{:} \quad \phi = \beta_1 \quad \text{for } t > 0 \\ &\text{At } x = \text{L:} \quad \phi = \beta_2 \quad \text{for } t > 0 \end{aligned} \tag{10.77}$$

where α_1 and α_2 are given constants, or functions of x, and β_1 and β_2 are constants. It may be pointed out that Equation 10.7 yields Equation 10.6, on differentiation, as follows:

$$\frac{\partial^2 \phi}{\partial t^2} = \frac{\partial}{\partial t}\left(-c\frac{\partial \phi}{\partial x}\right) = -c\frac{\partial}{\partial x}\left(\frac{\partial \phi}{\partial t}\right) = -c\frac{\partial}{\partial x}\left(-c\frac{\partial \phi}{\partial x}\right) = -c^2\frac{\partial^2 \phi}{\partial x^2} \tag{10.78}$$

Therefore, the methods for solving the first-order equation may also be used for solving the second-order equation.

The solution domain for the wave equation is shown in Figure 10.29. The boundary conditions are specified at two values of the spatial coordinate x, and the initial conditions are given at $t = 0$. The dependent variable ϕ is to be computed at all of the grid points over a given time interval. Starting with the known values at $t = 0$, the numerical solution progresses in the direction of increasing time, or increasing i, with the boundary conditions being satisfied at each time step. If central difference approximations for the second derivatives are substituted into the wave equation, we obtain the finite difference equation

$$\frac{\phi_{i+1,j} - 2\phi_{i,j} + \phi_{i-1,j}}{(\Delta t)^2} = c^2 \frac{\phi_{i,j+1} - 2\phi_{i,j} + \phi_{i,j-1}}{(\Delta x)^2} \tag{10.79}$$

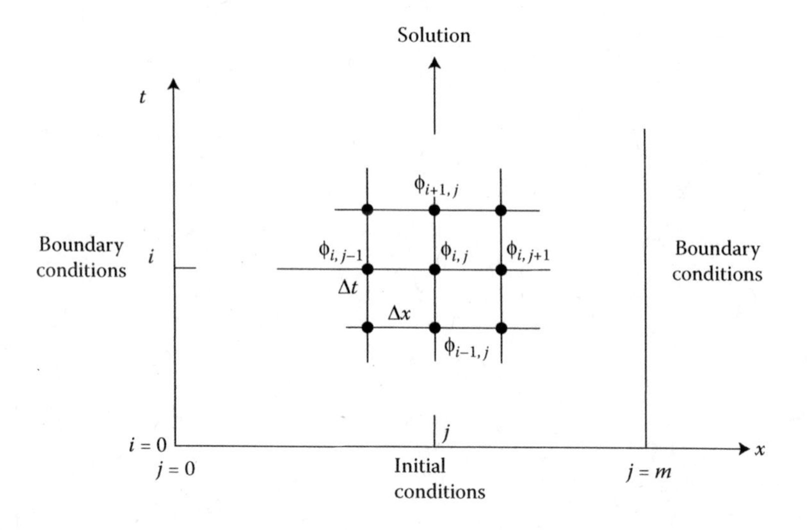

FIGURE 10.29 Solution domain for the wave equation, indicating the initial and boundary conditions needed and the mesh employed.

which gives

$$\phi_{i+1,j} = -\phi_{i-1,j} + c^2 \frac{(\Delta t)^2}{(\Delta x)^2}(\phi_{i,j+1} + \phi_{i,j-1}) + 2\left[1 - c^2 \frac{(\Delta t)^2}{(\Delta x)^2}\right]\phi_{i,j} \tag{10.80}$$

This equation is an explicit representation since the values to be computed at a given time step are obtained directly from known values at earlier time steps. The initial conditions are given in terms of the function ϕ and its derivative $\partial\phi/\partial t$. The condition on the derivative provides a relation between $\phi_{1,j}$ and $\phi_{0,j}$, and, therefore, the values of $\phi_{1,j}$ for starting the computational scheme may be determined. The stability of the above representation may be considered in terms of our earlier discussion on the stability of the explicit schemes for parabolic equations. Then we would expect the numerical scheme to become unstable when the coefficient of $\phi_{i,j}$ becomes negative. Therefore, the method would be expected to be stable if

$$\frac{c\Delta t}{\Delta x} \leq 1 \tag{10.81}$$

This stability condition is frequently known as the *Courant condition*, and the dimensionless parameter $c\Delta t/\Delta x$ as the *Courant number.* A more detailed stability analysis of the method also yields the above stability constraint. Example 10.5 demonstrates the solution of the wave equation by this explicit method.

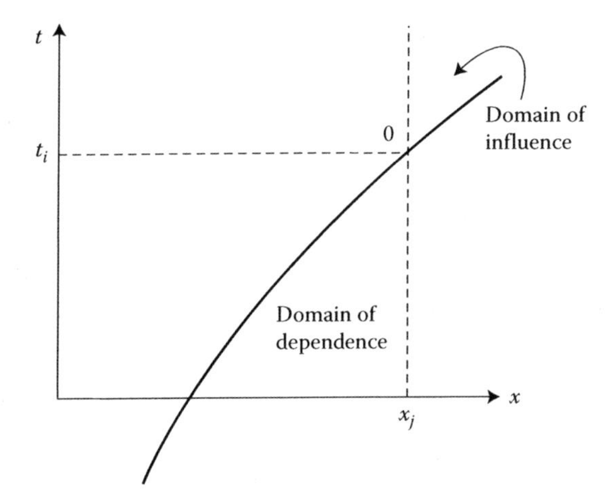

FIGURE 10.30 Domains of dependence and influence for the first-order convection equation, which is a hyperbolic PDE.

Similarly, the finite difference approximation for Equation 10.7 may be written as

$$\frac{\phi_{i+1,j} - \phi_{i,j}}{\Delta t} = -c\frac{\phi_{i,j} - \phi_{i,j-1}}{\Delta x} \tag{10.82}$$

Backward differences are used so that the solution depends only on the domain of dependence, as shown in Figure 10.30. This method is known as the *backward* or *upwind difference method*. The formulation is, therefore, of accuracy $[O(\Delta t), O(\Delta x)]$. Again, the necessary condition for stability is Equation 10.81. This condition basically ensures that the solution at a given location at a given time is affected only by the values at the grid points in its domain of dependence. Example 10.6 discusses the numerical solution of this equation, along with the stability considerations.

A more accurate explicit method is the Lax–Wendroff method, which is second-order accurate in both time and space. It is also stable for Courant numbers less than unity and is frequently employed for linear hyperbolic equations. When applied to the first-order convection equation, the finite difference form of this method is obtained as follows:

$$\phi_{i+1,j} = \phi_{i,j} - \frac{c\Delta t}{2\Delta x}(\phi_{i,j+1} - \phi_{i,j-1}) + \frac{c^2(\Delta t)^2}{2(\Delta x)^2}(\phi_{i,j+1} - 2\phi_{i,j} + \phi_{i,j-1}) \tag{10.83}$$

We can also apply the method to the second-order wave equation by breaking the equation down into two coupled first-order equations as

$$\frac{\partial \phi}{\partial t} = c\frac{\partial u}{\partial x} \tag{10.84a}$$

$$\frac{\partial u}{\partial t} = c\frac{\partial \phi}{\partial x} \tag{10.84b}$$

The stability constraint is given by Equation 10.81. It can be shown that the results are theoretically exact when the Courant number is 1.0 and that the accuracy of the solution decreases as the Courant number decreases below 1.0. Therefore, the value of this parameter is generally chosen to be around unity.

Several other explicit and implicit methods, similar to those for parabolic PDEs, have been developed for hyperbolic PDEs. These methods include the unconditionally stable Crank–Nicolson method, which yields a tridiagonal system of equations for the one-dimensional problem. For two or three space dimensions, splitting methods similar to the ADI method, outlined earlier, are employed. These methods are also unconditionally stable and give rise to tridiagonal systems, which are easily solved by Gaussian elimination. Many such methods have been developed and applied to aerodynamic applications. For further details, see Ferziger (1998).

Example 10.5

Consider the vibration of a string, 1 m in length, stretched between two supports with an initial tension of 40 N. The mass of the string is 0.04 kg/m. The string is displaced from its equilibrium position, as shown in Figure 10.31, held at rest in this configuration, and then released. Compute the variation of the displacement at various points along the string with time. For approximately one period of vibration, following the release of the string from rest, determine the configuration of the string at various intermediate time intervals. Also consider the case when the string is plucked in the middle, instead of at the one-fourth point, and obtain the string configuration as a function of time.

SOLUTION

The governing equation for this problem is the wave equation, written as

$$\frac{\partial^2 u}{\partial t^2} = c^2\frac{\partial^2 u}{\partial x^2} \tag{10.85}$$

where u is the vertical displacement at a point on the string, indicated by coordinate distance x, t is the time following the release of the string from rest, and c^2 is a constant. For a vibrating string, it can be shown from the derivation of the

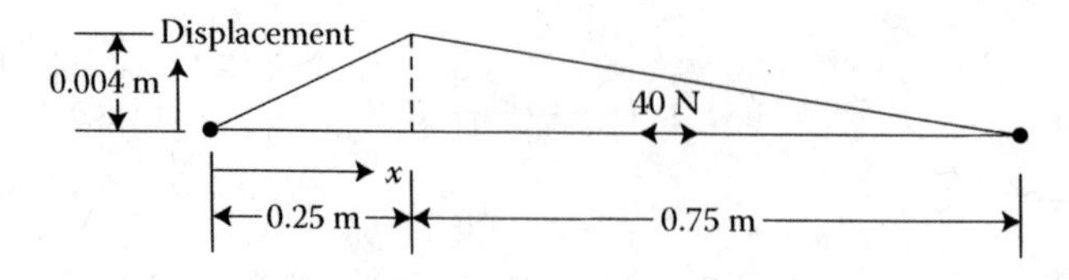

FIGURE 10.31 Physical circumstance of a vibrating string, considered in Example 10.5.

governing equation that $c^2 = T/m$, where T is the tension and m is the mass per unit length of the string. Therefore, in the given problem,

$$c^2 = \frac{T}{m} = \frac{40 \text{ N}}{0.04 \text{ kg/m}} = 1000 \text{ m}^2/\text{s}^2 \tag{10.86}$$

We now select the value of Δx as 0.05 m, giving 21 grid points along the string. We must consider the Courant number $C = c\Delta t/\Delta x$ to select a suitable time step. As discussed in Section 10.4, $C \leq 1$ for numerical stability, and a greater accuracy is obtained if C is close to 1.0, that is,

$$\frac{T}{m}\frac{(\Delta t)^2}{(\Delta x)^2} = 1 \tag{10.87}$$

This gives the value of Δt as 1.58×10^{-3} s. Therefore, Δt is chosen as 0.0015 s for convenience.

The initial conditions, on $u(x, t)$ in meters, for the string plucked at the one-fourth point, are as follows:

$$u(x,0) = \begin{bmatrix} 0.016x & \text{for } 0 \leq x \leq 0.25 \\ \dfrac{0.016}{3}(1-x) & \text{for } 0.25 < x \leq 1.0 \end{bmatrix} \tag{10.88}$$

$$\frac{\partial u}{\partial t}(x,0) = 0 \tag{10.89}$$

The boundary conditions are

$$u(0,t) = u(1,t) = 0 \tag{10.90}$$

Similarly, the initial conditions for plucking the string in the middle may be written.

The given problem may be solved by finite difference methods. If central differences are used, the finite difference equation is Equation 10.80, which may be written for the present case as

$$u_{i+1,j} = -u_{i-1,j} + C^2(u_{i,j+1} + u_{i,j-1}) + 2(1 - C^2)u_{i,j} \tag{10.91}$$

where $C^2 = (T/m)(\Delta t)^2/(\Delta x)^2$. The initial condition given by Equation 10.89 is written, using central differencing, as

$$\frac{u_{2,j} - u_{0,j}}{2\Delta t} = 0 \tag{10.92}$$

where $u_{0,j}$ is the value at a fictitious point one time step before the initial condition, $i = 1$. Thus,

$$u_{2,j} = u_{0,j} \tag{10.93}$$

Substituting this relationship into Equation 10.91 for $i = 1$, we obtain

$$u_{2,j} = \frac{C^2}{2}(u_{1,j+1} + u_{1,j-1}) + (1 - C^2)u_{1,j} \tag{10.94}$$

Equation 10.94 is used for advancing from the initial configuration to the first time interval, $t = \Delta t$. Beyond that, Equation 10.91 is employed. A first-order approximation may also be used for Equation 10.89, giving $u_{2,j} = u_{1,j}$. The boundary conditions give the displacement at grid points 1 and 21 as zero. For the remaining points, we compute the displacement for the next time step, using the values at the adjacent points corresponding to the previous time steps. Since the values of the displacements for only the last two time steps are needed to advance the solution, only three arrays corresponding to time intervals t, $t + \Delta t$, and $t + 2\Delta t$ need to be considered, where the values at $t + 2\Delta t$ are obtained in terms of the other two arrays.

Appendix B.35 shows the MATLAB script file and Appendix C.20 the Fortran computer program for solving this problem. The various symbols employed are defined in the programs. Arrays u, u_1, and u_2 contain the displacements at time t, $t + \Delta t$, and $t + 2\Delta t$. The input values are entered and the maximum number of time steps, as well as the time intervals after which the output is obtained, may be specified. As indicated above, $\Delta x = 0.05$ m, $\Delta t = 0.0015$ s, and $n = 21$, where n is the total number of grid points. The initial distribution is entered and the boundary conditions are given to proceed with the computations.

The displacements at various locations along the string are computed as time elapses. Figure 10.32 shows the variation of the displacement at four locations with time. A periodic behavior is clearly seen, as expected. Also, the time period is found to be about 0.2 s. Figure 10.33 shows the configuration of the string at various time intervals. The string starts at its initial distribution, given in Figure 10.31, and, as time elapses, the displacement at each point undergoes a periodic process. At different time intervals, different displacements exist at the various grid points,

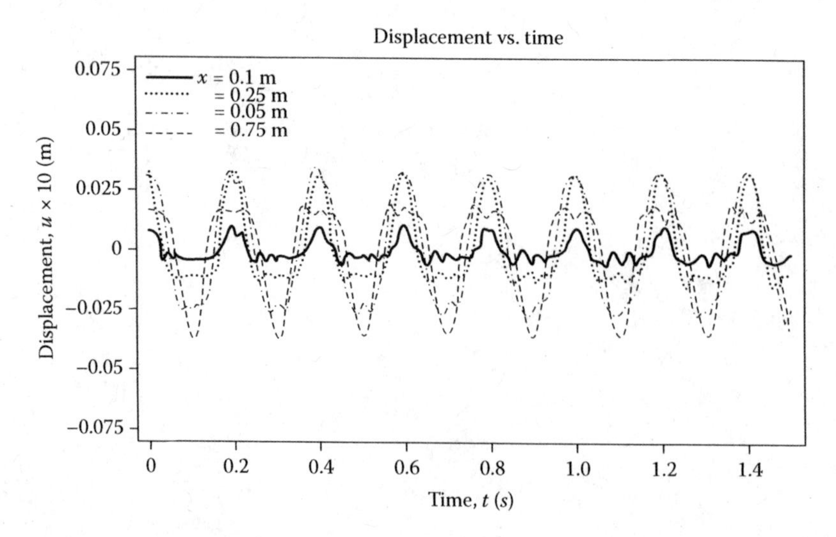

FIGURE 10.32 Computed displacements at four locations along the string as functions of time in Example 10.5.

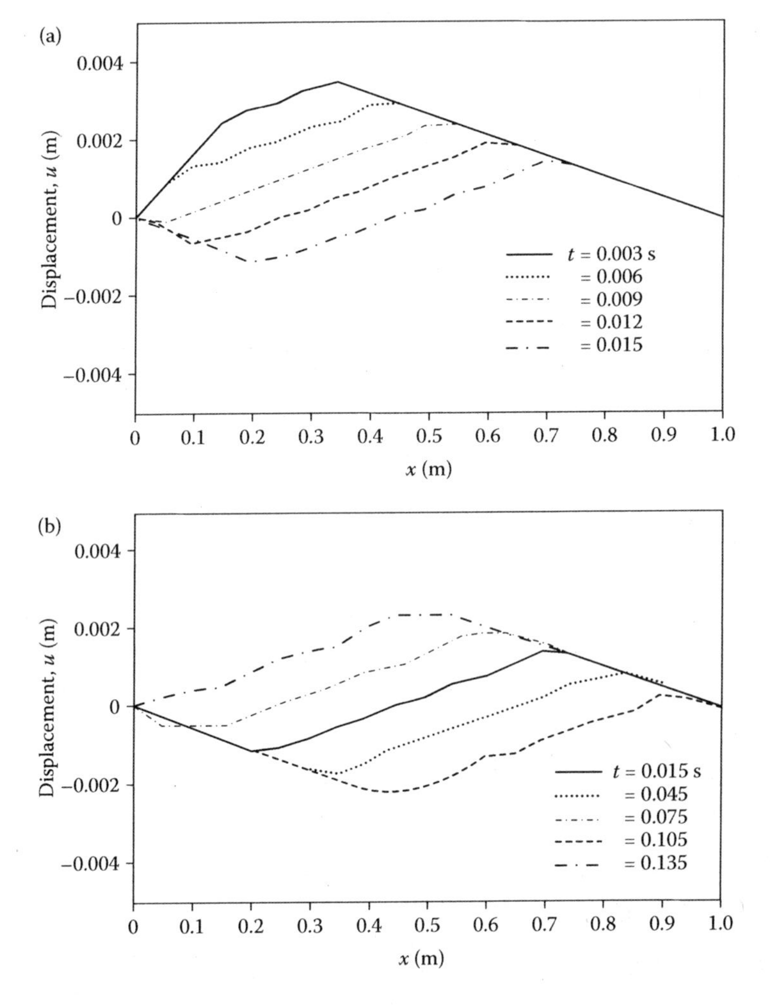

FIGURE 10.33 Calculated configuration of the string in Example 10.5 at various time intervals, when it is plucked at the one-fourth point, as shown in Figure 10.31.

giving rise to different configurations of the string. Initially, the displacements are all positive, that is, on one side of the equilibrium position. Then, with time, the displacements become negative over portions of the string. Figure 10.34 shows the corresponding results for the case when the string is plucked in the middle. An expected symmetry arises in the displacement.

Example 10.6

Consider the first-order convection equation

$$\frac{\partial P}{\partial t} + c\frac{\partial P}{\partial x} = 0 \tag{10.95}$$

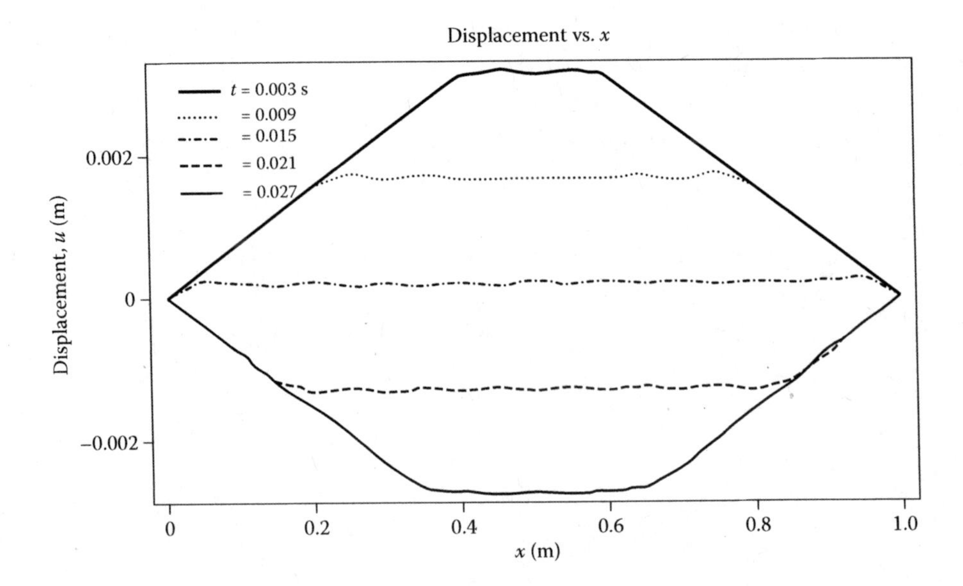

FIGURE 10.34 Calculated configuration of the string at various time intervals, when it is plucked in the middle.

which governs the transport of a physical or chemical quantity $P(x, t)$ by convection. Here, x is the spatial coordinate distance, t is time, and P represents a convected quantity such as concentration or temperature. Using Euler's method, the backward or upwind differencing method, and the Lax–Wendroff method, solve this hyperbolic equation. The initial and boundary conditions are given as follows:

$$\begin{aligned} &\text{At } t \leq 0: \quad P = 0 \quad \text{for } x \geq 0 \\ &\text{At } t > 0: \quad P = 1 \quad \text{for } x = 0 \end{aligned} \tag{10.96}$$

Take the convection velocity c as 2.5 m/s. Solve for x up to 5.0 m, taking the grid size Δx as 0.5 m. Compute the results up to time $t = 2.0$ s, taking the step size Δt as 0.05, 0.1, and 0.2 s.

SOLUTION

Several important transport problems in fluid flow and heat transfer are governed by first-order hyperbolic equations such as the one given here. For instance, wave propagation in a shallow water body is governed by a nonlinear first-order hyperbolic equation. The solution to the problem given here is simply the movement of the step change, at $x = 0$ and $t = 0$, downstream, with no change in amplitude and at the convection velocity c. Thus, we can use this analytical result to evaluate the solution of the given equation by the three methods considered.

The finite difference equation for the solution of Equation 10.95 by Euler's method is

$$\frac{P_{i+1,j} - P_{i,j}}{\Delta t} = -c\frac{P_{i,j+1} - P_{i,j-1}}{2\Delta x} \tag{10.97}$$

where the first subscript refers to the time step and the second to the spatial location. Thus, this method uses forward difference for the derivative with respect to time and second-order central difference for the spatial derivative. The finite difference equations for the backward difference method and the Lax–Wendroff method are obtained from Equations 10.82 and 10.83 as

$$\frac{P_{i+1,j} - P_{i,j}}{\Delta t} = -c\frac{P_{i,j} - P_{i,j-1}}{\Delta x} \tag{10.98}$$

$$P_{i+1,j} = P_{i,j} - \frac{c\Delta t}{2\Delta x}(P_{i,j+1} - P_{i,j-1}) + \frac{c^2(\Delta t)^2}{2(\Delta x)^2}(P_{i,j+1} - 2P_{i,j} + P_{i,j-1}) \tag{10.99}$$

Thus, all three methods are explicit, and the only parameter that arises is the Courant number C, given by

$$C = \frac{c\Delta t}{\Delta x} \tag{10.100}$$

For the values given in the problem, $C = 0.25$, 0.5, and 1.0. As pointed out earlier, the last two methods are unstable for $C > 1.0$.

A computer program in MATLAB is written for solving the given problem, as shown in Appendix B.36. The input quantities and the initial and boundary conditions are entered. The numerical method for solving the problem is chosen interactively by the user. Computations for the three values of the time step Δt are then carried out up to time $t = 2.0$ s at each step size. The computed results are printed at five locations in x, corresponding to $x = 1$, 2, 3, 4, and 5 m. The various symbols employed are defined in the program. Here, p and pn are used to denote values at the previous and present time steps, respectively, and i denotes the spatial location. Calculations are needed for the ten grid points corresponding to $i = 2$ to $i = 11$. However, both Euler's method and the Lax–Wendroff method need the value at $(i + 1)$ to calculate the value at i. One approach to avoiding the problem that arises at $i = 11$ is to compute the value at $i = 11$ using the backward difference method which does not require the value at $i = 12$. This is done in the program, and the computed results are obtained for all three methods.

The results are shown in Figure 10.35 in terms of the computed variation of P at a few locations as a function of time. Interestingly, the numerical solution is exact for both the upwind differencing and the Lax–Wendroff methods at Courant number $C = 1.0$. Generally, accuracy is expected to increase as Δt is decreased, at a given value of Δx, because of smaller TE in time. However, for this particular problem, $C = 1.0$ yields the exact solution. In fact, as mentioned earlier, C is generally taken as close to 1.0 for greater accuracy. The results from both of these methods at $C = 0.5$ and $C = 0.25$ are unable to capture the expected step variation in P. However, that is not surprising since any numerical method will introduce computational errors.

The Lax–Wendroff method is expected to be more accurate due to second-order accuracy in space. Euler's method is found to be unstable. The solution oscillates at low values of C, and these oscillations grow without bound as time increases. This instability was found to be worse at large C, as expected. As shown

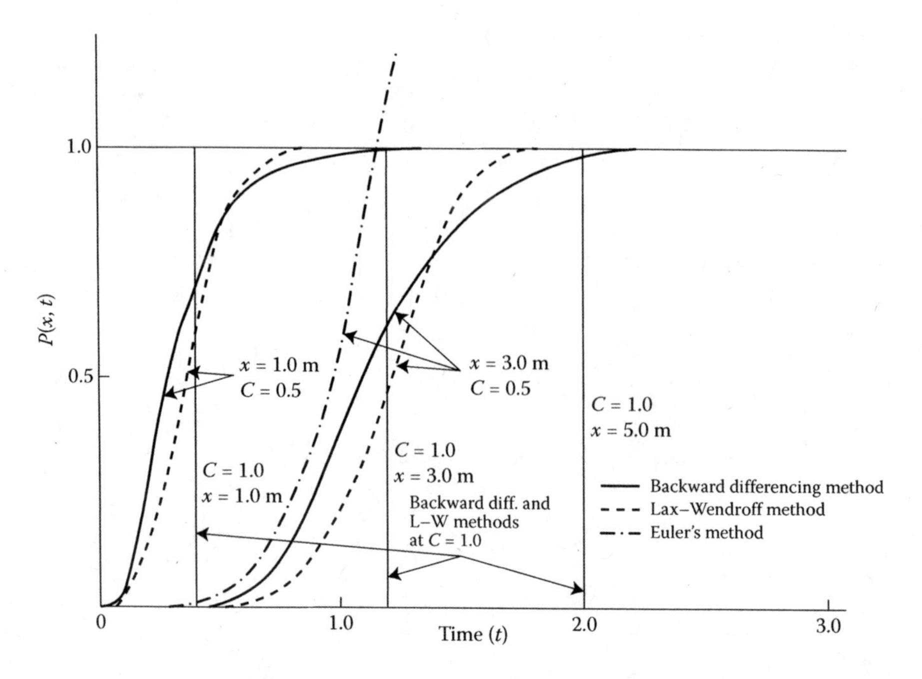

FIGURE 10.35 Numerical results in terms of the computed variation of the dependent variable P with time t for Example 10.6.

by Ferziger (1998), this method is unconditionally unstable for this problem, and, therefore, instability arises in all cases as time elapses. For $C > 1.0$, the other two methods were also found to indicate numerical instability.

This problem is a fairly simple example of a hyperbolic equation. However, it has been chosen to demonstrate the use of three numerical techniques for solving hyperbolic equations and the constraints imposed by numerical instability.

10.5 SUMMARY

This chapter gives a brief discussion on the numerical solution of PDEs. The solution procedure is dependent on the type of the PDE: parabolic, elliptic, or hyperbolic. Employing simple examples of these three types of PDEs, various important numerical methods for solving them are outlined. Only linear equations are considered to illustrate the methods, since nonlinear PDEs, although important in many engineering applications, are beyond the scope of this book. However, in several cases, a nonlinear PDE may be linearized and then solved by the methods discussed in this chapter. Still, the nonlinear problem is generally much more involved than the linear one. Frequently, the stability and the convergence characteristics of the numerical scheme are not known for nonlinear equations, and numerical experimentation is needed to ensure the accuracy and correctness of the solution.

The main approach to the solution of PDEs, considered in this chapter, is by means of finite difference methods, which give rise to a system of algebraic equations.

A solution of this system of equations yields the value of the dependent variable at a finite number of grid points in the computational domain. Parabolic equations are solved by marching in one coordinate direction. Explicit methods allow the computation of values at a given step from the known values at earlier steps. However, the step size is generally constrained in explicit methods due to considerations of numerical stability. Implicit methods have better stability characteristics, but they require the solution of a system of simultaneous algebraic equations. Direct methods are usually employed if the system is tridiagonal. Otherwise, iterative methods, such as the Gauss–Seidel and the SOR methods, are used. A tridiagonal system is obtained in one-dimensional problems, and Gaussian elimination is used for these. For multidimensional problems, splitting methods, which treat one direction as implicit and alternate between the various directions, are frequently employed, since these methods also give rise to tridiagonal systems.

Elliptic PDEs are often solved by iterative methods. Direct methods are applicable in a few special cases. Splitting methods, such as the ADI method, can also be used, with an acceleration parameter to obtain a faster convergence. In some cases, a pseudotransient term is added to the elliptic PDE. Then the resulting equation is parabolic in time and may be solved by time-marching techniques, giving the required solution at large time.

Hyperbolic PDEs are solved by methods similar to those for parabolic PDEs. A specialized method, known as the *method of characteristics*, is also an important method for hyperbolic equations, since it allows the treatment of discontinuities which frequently arise in these equations. In recent years, finite difference methods have become very popular for the solution of hyperbolic PDEs, and several very efficient schemes have been developed.

Another approach to the solution of PDEs is the finite element method. In this method, the solution domain is subdivided into finite regions, and the PDE is integrated over each region, employing weight functions with the equation. The solution and the weight functions are taken as polynomials, and the integrals or the weighted residuals are minimized, or reduced to zero, to yield a system of algebraic equations that is solved by the usual methods. Although more complicated in implementation, the finite element methods have become very popular in recent years because of their advantages in the treatment of irregular boundaries and complex boundary conditions. A brief outline of the finite element, the boundary element, and the control volume methods for solving PDEs is included in this chapter.

Several programs are given here in both MATLAB and in Fortran to solve linear PDEs, such as the common two-dimensional elliptic equation and the well-known parabolic and hyperbolic equations. A function *pdepe* is also available to solve initial-BVP for parabolic–elliptic PDEs in 1-D for small systems of parabolic and elliptic PDEs in one space variable x and time t to modest accuracy. The function *pdeval* is then used to evaluate/interpolate the solution obtained. However, for more complicated problems, particularly nonlinear equations, the use of the PDE Toolbox in MATLAB is probably the best approach. Different algorithms are available and complicated geometries and boundary conditions can be handled by employing the commands available in this toolbox. Various plotting routines are also available to obtain the computed results in appropriate graphical forms, such as contour plots.

PROBLEMS

10.1. Consider the governing PDE for the dependent variable $\phi(x, y)$, given as

$$\frac{\partial \phi}{\partial x} + A\frac{\partial \phi}{\partial y} = B\frac{\partial^2 \phi}{\partial y^2}$$

where A and B are constants. Determine the nature of this equation, and give a set of boundary conditions that may be applied to it.

10.2. The temperature $T(x, y)$ in a steady, two-dimensional flow with heat transfer is governed by the equation

$$u\frac{\partial T}{\partial x} + v\frac{\partial T}{\partial y} = \alpha\left(\frac{\partial^2 T}{\partial x^2} + \frac{\partial^2 T}{\partial y^2}\right) + Q$$

where $u(x, y)$ and $v(x, y)$ are the two velocity components, α is a constant known as *thermal diffusivity*, and $Q(x, y)$ is a function that gives the energy generation per unit volume in the fluid. Determine the nature of this equation, and specify suitable boundary conditions. Also, for each of the three special circumstances of (a) $u = v = 0$, (b) $\alpha = 0$, and (c) $\partial^2 T/\partial x^2 = 0$, classify the resulting reduced equations, and give the relevant boundary conditions.

10.3. In a one-dimensional diffusion problem, the governing equation is $\partial\phi/\partial t = C\, \partial^2\phi/\partial x^2$, where $\phi(x, t)$ is the dependent variable, C is a constant, x is the spatial coordinate, and t is time. The boundary condition at $x = 0$ is given as $\partial\phi/\partial x = B$, where B is a constant. The condition at the other boundary, at $x = L$, is given as $\phi = 0$. Obtain the finite difference equation for solving this problem by the Crank–Nicolson method. Write the gradient boundary condition in terms of forward differences, using both the first-order and the second-order approximations. Is the resulting system of equations tridiagonal? If not, can it be obtained in tridiagonal form by simple elimination?

10.4. For the numerical solution of a one-dimensional transient diffusion problem, governed by

$$\frac{\partial \phi}{\partial t} = A\frac{\partial^2 \phi}{\partial x^2}$$

the explicit Euler method is to be used. If the grid size Δx is taken as 0.1 m, find the maximum time step that may be employed for a stable numerical scheme, if $A = 10^{-6}$ m^2/s. Also find the limitation on the time step if the grid size is reduced to 0.01 m.

10.5. A long bar of rectangular cross section is initially at a uniform temperature T_0. At time $t = 0$, the temperature at the outer surface is raised to T_s and held at this value. Write down the governing PDE and obtain the finite difference equations for solving this problem by the explicit FTCS and the Crank–Nicolson methods. Also give the equations for the relevant boundary and initial conditions. Indicate the constraints, if any, on the time step, due to stability considerations, for chosen values of Δx and Δy. What method would you employ for solving the system of algebraic equations obtained in the Crank–Nicolson method? Justify your choice.

10.6. Consider the one-dimensional conduction heat transfer in a plate of thickness 3 cm. The plate is initially at 1000°C. At time $t = 0$, the temperature at two surfaces is dropped to 0°C and maintained at this value. The thermal diffusivity is given as 5×10^{-6} m²/s. Employing $\Delta x = 0.3$ cm, solve this problem numerically to obtain the time-dependent temperature distributions for $F = 1/6$, 0.5, 0.52, and 0.6. Does numerical instability arise for $F > 0.5$? Discuss.

10.7. Solve the preceding problem graphically, by the Schmidt–Binder method, for $F = 0.5$.

10.8. If a plate in a stationary fluid is suddenly set into motion, at a velocity U, the governing equation is $\partial u/\partial t = \nu\, \partial^2 u/\partial x^2$, where u is the local velocity, ν is a constant known as *kinematic viscosity of the fluid*, t is time, and x is the distance out from the plate, which is at $x = 0$, as shown in the figure. The boundary conditions are, therefore, as follows:

$$\text{At } x = 0: \quad u = U \quad \text{for } t > 0$$

$$\text{as } x \to \infty: \quad u \to 0$$

The initial condition is the following:

$$\text{For} \quad t \le 0: u = 0 \quad \text{for } x \ge 0$$

This problem is to be solved by the explicit FTCS method. The values of u are computed, at each time step, outward from the plate until u is zero. Taking $U = 1$ m/s, $\nu = 10^{-5}$ m²/s, and $\Delta x = 0.01$ m, find the maximum time step that may be employed if numerical instability is to be avoided. Using this maximum time step, solve this problem.

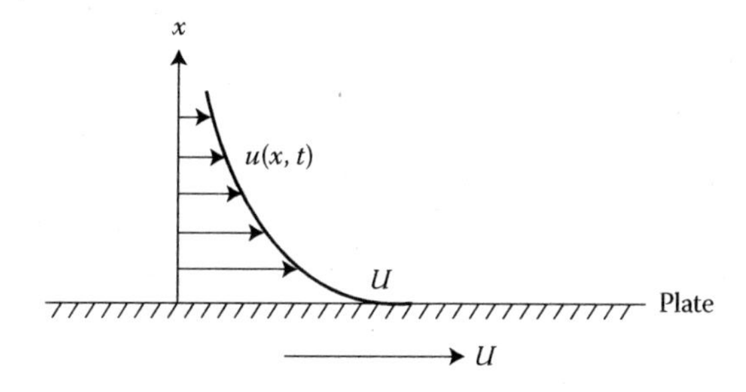

10.9. For specifying the boundary condition $\partial\phi/\partial x = B$ in Problem 10.3, a fictitious grid point is taken outside the computational domain, as shown. The boundary condition is then written in central difference form, using this point. The finite difference equation for the PDE is also written for a point at the boundary, again employing this fictitious point. The unknown value of ϕ at this grid point outside the region is eliminated by using the two equations thus obtained. The resulting equation gives the finite difference equation for the boundary condition. Compare this result with that given in Equation 10.63.

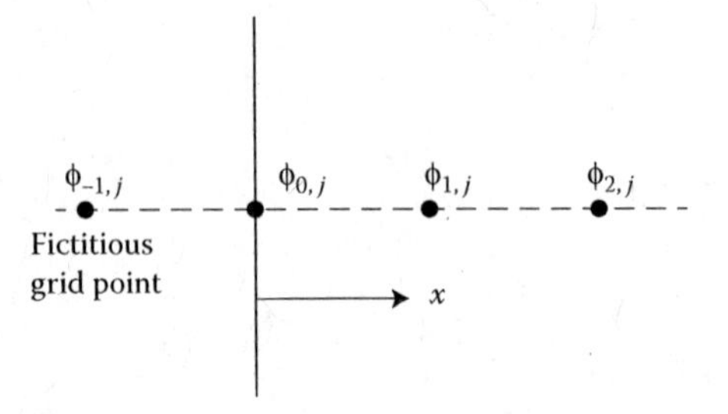

10.10. We wish to solve the following equation for $\phi(x, y)$

$$\frac{\partial^2\phi}{\partial y^2} = f(\phi)\frac{\partial\phi}{\partial x}$$

This equation is nonlinear because of the presence of the function $f(\phi)$. Formulate a simple numerical scheme, based on the discussion in the text, for solving this problem.

10.11. One-dimensional conduction in a rod of length L is governed by the equation

$$\frac{1}{\alpha}\frac{\partial T}{\partial t} = \frac{\partial^2 T}{\partial x^2} - H(T - T_\alpha)$$

where x is the distance from one end, as shown, t is time, T_a is the ambient temperature, and H is a heat loss parameter. For time $t < 0$, the temperature throughout the rod is T_a. At $t = 0$, the temperatures at the two ends, at $x = 0$ and $x = L$, are raised to 100°C and held at this value for $t > 0$. Using any suitable numerical method, solve this problem. Take $T_a = 15$°C, $\alpha = 10^{-6}$ m²/s, $L = 0.4$ m, $\Delta x = 0.04$ m, and $H = 100$ m⁻².

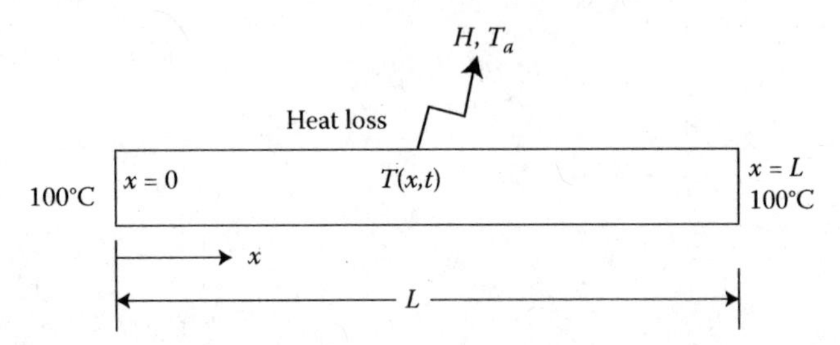

10.12. The one-dimensional diffusion of water in a porous medium is governed by the equation $D\,\partial^2C/\partial x^2 = \partial C/\partial t$, where C is the concentration of water in kg/m³ and D is the mass diffusivity in m²/s. A long, hollow cylinder of outer diameter 0.2 m and inner diameter 0.1 m is initially dry; that is, water concentration is zero. Then, at time $t = 0$, the outer surface is brought in contact with water, raising the concentration there to 1000 kg/m³, while the concentration at the inner surface is held at zero. If $D = 10^{-4}$ m²/s, obtain the time-dependent concentration profiles in the cylinder, taking ten grid points across the annular region. Neglect the effect of curvature in the problem and treat the region as a slab.

10.13. Consider the problem discussed in Example 10.1. Study the effect of varying the grid size on the numerical results, by solving the problem with Δx half and also twice the value taken in the example. Take $\Delta x = \Delta y$. Compare the results obtained with the earlier results presented in Example 10.1, and discuss the dependence on grid size.

10.14. Steady-state mass diffusion in an enclosed region is governed by Laplace's equation for the concentration C. Consider the diffusion in a rectangular region of length 0.3 m and width 0.1 m. The third dimension is given as large. For this two-dimensional mass diffusion problem, the concentration, in nondimensional terms, is given as 1.0 at one surface and as 0.0 at the remaining three surfaces. Compute the concentration distribution in the region by the SOR method, and determine the optimum value of the relaxation factor. Compare this value with that obtained from the analytical expression, Equation 10.58, given in the text.

10.15. A rectangular trampoline may be considered as a rubber membrane, of length L and width W, fastened securely at the boundary. As given in Example 10.3, the vertical deflection f is governed by the equation

$$\frac{\partial^2 f}{\partial x^2} + \frac{\partial^2 f}{\partial y^2} = -p/T$$

where p is the pressure and T the tension. Solve this problem by the Gauss–Seidel method, taking $p/T = 1.0\ \text{m}^{-1}$, $L = 2.0$ m, and $W = 1.0$ m.

10.16. If fluid friction, or viscosity, is taken as negligible in a flow, the flow is termed *inviscid*. In the absence of rotational effects, the flow is then governed by the equation $\nabla^2\psi = 0$, where ψ is the stream function. A line of constant ψ is known as a *streamline*, and the velocity field may be obtained from a given ψ distribution. Consider the flow in a channel whose cross-sectional area varies as shown. The boundary conditions on ψ are also given, as linear distributions at the inflow and outflow. Compute the ψ distribution in the channel, and obtain the streamlines, or contours of constant ψ, corresponding to $\psi = 0$, 0.2, 0.4, 0.6, 0.8, and 1.0. Use the Gauss–Seidel iterative scheme. See also Example 10.4(b).

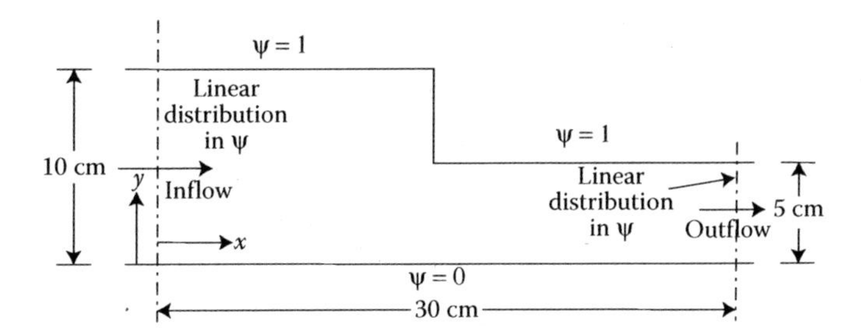

10.17. A solid cylinder of diameter D and length L has its two ends at temperature T_r, and the curved, lateral surface at temperature T_s. The temperature distribution may be assumed to be independent of the angular position. Thus, the problem becomes two-dimensional, or axisymmetric. Write the governing PDE and obtain the relevant finite difference equation for solving this problem by the SOR method. Use polar coordinates.

10.18. The flow of a very viscous fluid in a circular tube, as shown, is governed by the equation

$$\nabla^2 u = \frac{1}{\mu}\frac{dp}{dz}$$

where the vector operator ∇^2 may be written in polar coordinates, $u(x, y)$ is the velocity in the axial direction z, μ is the coefficient of viscosity, and dp/dz is the constant pressure drop along the flow. The velocity is zero at the boundary. Formulate this problem for a numerical solution by the Gauss–Seidel method. Give the governing equation and the relevant boundary conditions in finite difference form, and outline the numerical procedure.

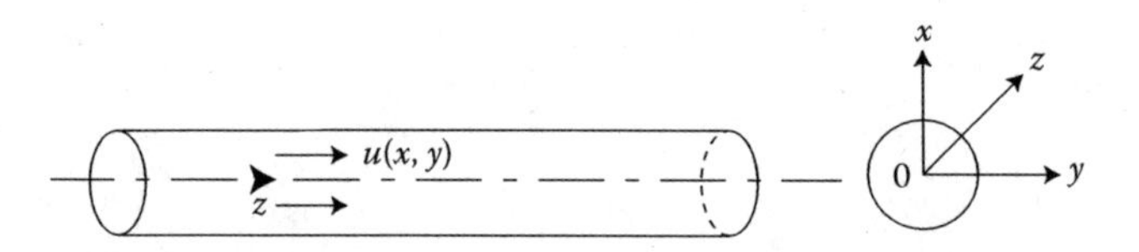

10.19. We are interested in the steady-state temperature distribution in a hollow cylinder of length L and inner and outer diameters D_i and D_o, respectively. The inner and outer surfaces are at temperature T_s, and the ends at T_0. Formulate this problem as a two-dimensional transient heat conduction problem whose solution yields the steady-state results at large time. Also give the finite difference equations for solution by the ADI method. What is the order of the TE in this formulation?

10.20. A rectangular finite difference mesh is used for solving Laplace's equation in a circular region. Consider the grid points near the circular boundary, and derive the applicable finite difference equation, as done in the text, taking a uniform grid distribution along both directions.

10.21. A long rod of rectangular, 10 cm × 5 cm, cross section has all the surfaces maintained at 100°C. Due to nuclear reaction, energy is generated within the material at a uniform rate $\dot{Q}$ of 5×10^7 W/m^3. The thermal conductivity k of the material is 50 W/m K. The temperature distribution in steady-state conduction with energy generation is governed by the Poisson equation

$$\frac{\partial^2 T}{\partial x^2} + \frac{\partial^2 T}{\partial y^2} + \frac{\dot{Q}}{k} = 0$$

Using the Gauss–Seidel iterative method, solve this problem to obtain the temperature distribution. Plot the temperature variation along the two axes of the rectangular region.

10.22. Consider the one-dimensional convection equation

$$\frac{\partial \phi}{\partial t} + c\frac{\partial \phi}{\partial x} = 0$$

Determine the nature of this equation and give a set of relevant boundary conditions. Also obtain the finite difference equation for solving it by the Crank–Nicolson method.

10.23. In Example 10.5, if the initial deflection of the string results from being plucked at the one-third point instead of at the one-fourth point, compute the time-dependent displacements at various locations on the string after the string has been released.

10.24. The longitudinal vibration of a beam is governed by the equation

$$\frac{\partial^2 u}{\partial x^2} = \frac{\rho}{E}\frac{\partial^2 u}{\partial t^2}$$

where x is the coordinate along the axis, t is time, u is the longitudinal displacement, ρ is the material density, and E is a constant known as the *elastic modulus* for the material. The two ends of the beam, at $x = 0$ and $x = L$, are fixed. A deflection u_0 is given at the midpoint of the beam and then released from rest. Using the explicit method, formulate this problem for a numerical solution. Give the relevant finite difference equations and outline the numerical procedure.

10.25. A string is fixed at its two ends. The initial deflection u is given as

$u = 2(x - 0.2)$ for $0.2 < x \leq 0.6$

$u = 2(0.8 - x)$ for $0.6 < x \leq 0.8$

$u = 0$ at all other values of x in the range 0 and 1.0

Also, the time derivative of u is zero, that is,

$$\frac{\partial u}{\partial t}(x,0) = 0$$

Using the explicit method, solve this equation with the constant c in the governing wave equation given as 1.0. Take $\Delta x = 0.1$ and Courant number $= 0.5$ and 1.0. Compute the results up to $t = 2.0$. Discuss the observed trends in terms of the nature of hyperbolic equations.

10.26. Determine the nature of the following equation which governs the propagation of waves in a nonuniform medium:

$$\frac{\partial^2 u}{\partial t^2} = \frac{\partial}{\partial x}\left[c^2(x)\frac{\partial u}{\partial x}\right]$$

where u represents the displacement and c^2 varies with location. Outline a numerical method for solving this problem.

10.27. If the problem discussed in Example 10.5 is to be solved by the Crank–Nicolson method, give the resulting finite difference equations. Also outline the numerical method that may be adopted for solving these equations.

10.28. If in Example 10.5, the initial rate of change of displacement $\partial u/\partial t(x, 0)$ is given as 0.1 m/s for $0 < x < 1$, compute the resulting displacement as a function of time at four points on the string.

10.29. If in Example 10.6, $P = 0$ for $t \leq 0$ and e^{-t} at $x = 0$ for $t > 0$, solve the given hyperbolic PDE.

10.30. In a rectangular region of length L and width W, with $L/W = 2$, the Laplace equation governs the electric field ϕ. The value of ϕ is zero on three sides and is given as $\sin(\pi x/L)$ on the fourth side $y = W$. Compute the ϕ distribution in the region.

10.31. The dimensionless concentration C of a diffusing species in a square region is governed by

$$\frac{\partial^2 C}{\partial X^2} + \frac{\partial^2 C}{\partial Y^2} = \frac{\partial C}{\partial t}$$

Starting with an initial value of C as zero in the entire region, calculate the transient and steady-state distributions, using the FTCS method. C is given as zero on two opposite sides of the region and as 1.0 on the other two. Both X and Y vary from 0 to 1.0. Determine if the initial conditions affect the steady-state distribution.

Appendix A: Some Common Commands in MATLAB®

FOR MATRICES a AND b

`a.*b  a./b  a.\b`	Element by element arithmetic; a and b must have identical rows and columns
`a*b  a/b  a\b`	Matrix algebra; a and b must have appropriate rows and columns to perform these operations
`rand(n)`	Generates random numbers between 0 and 1 for a n × n matrix
`b=26*rand(3)-10`	Generates 3 × 3 matrix of random numbers between −10 and 16
`max(a)`	Gives maximum element in one-dimensional array a
`min(a)`	Gives minimum element in array a
`max(max(a))`	Gives maximum element in matrix a
`min(min(a))`	Gives minimum element in matrix a
`[i,j]=find(a==max(max(a)))`	Gives row and column where maximum element is located

FOR SYSTEM OF EQUATIONS ax = b

`inv(a)`	Gives inverse a^{-1} of the matrix a
$aa^{-1} = I$	Identity matrix I
$x = a^{-1}b$	Yields the solution x; b is column vector
`x=inv(a)*b`	Yields the solution vector x
`x=a\b`	Backslash operator; also gives the solution x
`[l,u,p]=lu(a)`	Decomposition of matrix a into upper and lower triangular matrices; p is permutation matrix
`y=l\(p*b); x=u\y`	Yields the solution x

OUTPUT

```
>> a=2.0;
>> b=4.5;
>> s=['The number that is obtained is', num2str(a)]
```

yields

The number that is obtained is 2.0

```
>> s=sprintf('The number %.5g is modified to %.5g.',a,b)
```

yields

```
The number 2.0 is modified to 4.5.
```

Another command is fprintf, which is similar to sprintf and displays data to the formatted string in the command window, whereas sprintf creates a character string which can be displayed or modified like a character array.

Similarly, try other formats: %.0g gives integers; %.3f is used for fixed-point numbers, with three places after the decimal; %8.4f gives the width of characters as 8 and 4 places after the decimal; g gives the best of fixed or floating point formats; bank gives fixed format for dollars and cents, that is, 2 decimal digits; e is the exponent format.

Use of a semi-colon at the end of a statement suppresses printing of the given or calculated number. Similarly, disp(s) suppresses the printing of s = and simply gives the specified or calculated value.

POLYNOMIALS

Specification of Polynomials

```
>> p = [1 -4 7 -6 2]
```

represents

$x^4 - 4x^3 + 7x^2 - 6x + 2$

Coefficients are arranged in descending powers of the independent variable.

ROOTS

```
>> r = roots(p)
```

gives the roots of the polynomial as

```
1.00 + 1.00 i
1.00 - 1.00 i
1.00
1.00
```

```
>> pp = poly(r)
pp =
1.00 -4.00 7.00 -6.00 2.00
```

gives the polynomial with the array r as the roots.

ALGEBRA OF POLYNOMIALS

```
>> a = [1 2 3 4];
>> b = [1 4 9 16];
```

```
>> c = conv(a,b)    % convolution (multiplication) of polynomials
c =
1 6 20 50 75 84 64
```

```
>> d=a+b                    % addition
d =
2 6 12 20

>> d=b-a                    % subtraction
d =
0 2 6 12

>> [q,r]=deconv(c,b)        % division
q =
1 2 3 4                     % quotient polynomial
r =
0 0 0 0                     % remainder polynomial

>> g=[1 6 20 48 69 72 44];
>> h=polyder(g)             % differentiation of a polynomial
                              with coefficients given by vector g
h =
6 30 80 144 138 72
```

CURVE FITTING, PLOTTING

For the following data:

```
>> x=[0 .1 .2 .3 .4 .5 .6 .7 .8 .9 1.0];
>> y=[-.45 1.98 3.28 6.16 7.08 7.34 7.66 9.56 9.48 9.30 11.2];
```

Curve fitting is obtained by

```
>> n=2;                      % specify order of the polynomial
                               for best fit
>> p=polyfit(x,y,n)          % gives best fit with nth order
                               polynomial
p =                          % Coefficients arranged in
-9.8147 20.1338 -0.0327        descending powers of x

>> xi=linspace(0, 1, 100);  % 100 evenly spaced points between
                               0 and 1
>> yi=polyval(p, xi);        % values of given polynomial at x
                               values of xi
>> plot (x,y,'g*',xi,yi,'b-')% plotting with given symbols,
                               color, labels and title
>> xlabel('x'), ylabel('y=f(x)')
>> title('Second Order Curve Fitting')
```

DEFINITION OF A FUNCTION

A function may be defined inline or as part of a script file, as, for example,

```
f=inline('x.^3+3*x.^2-4*x+2')
g=inline('x.*exp(x)-x.^0.5+3')
```

where .* and .^ are used to allow x to be a vector, or array.

The function may also be defined by a function file f.m and g.m as

```
function z=f(x)
z=x.^3+3*x.^2-4*x+2;
end
```

and

```
function z=g(x)
z=x.*exp(x)-x.^0.5+3;
end
```

Then, the value of the function at given x, where x may be scalar or vector, is obtained by

```
feval(f,x) or feval ('f',x)
```

where the former is used for the function defined inline and the latter for the function defined by a function file.

Appendix B: Computer Programs in MATLAB®

B.1

```
%       SEARCH METHOD FOR FINDING THE ROOTS OF AN ALGEBRAIC
        EQUATION
%
%       This program finds the real roots of the equation
%       f(x)=0 by the incremental search method
%
%       Here, eps is the convergence parameter, dx the
%       increment in x, dx1 the increment at sign change of
%       f(x), and f1 and f2 the values of the function f(x) at
%       two consecutive x values
%
%       Define function f(x)
%
f=inline('0.8*5.67*10^(-8)*(1000^4-x^4)-50*(x-500)- ...
   (25/0.15)*(x-300)');
%
%       Enter starting values
%
eps=10.0;
for i=1:6
x=300;dx=50;
dx1=dx;
a=f(x)*f(x);
fprintf('EPS=%.5f\n',eps)
%
%       Check for convergence to the root
%
    while dx1>eps
%
%       Check for sign change in f(x)
%
while a>0
f1=f(x);
x=x+dx;
f2=f(x);
a=f1*f2;
end
fprintf('X=%.5f        F1=%10.4f        F2=%10.4f\n',x,f1,f2)
%
```

```
%       Reduce increment size
%
dx1=dx;
     x=x-dx;
     dx=dx/10;
f1=f(x);
x=x+dx;
f2=f(x);
a=f1*f2;
end
%
%       Print numerical results
%
     fprintf('TEMPERATURE=%.5f        F(X)=%.4f\n\n',x,f1)
%
%       Vary convergence parameter
%
     eps=eps/10;
end
```

B.2

```
%       BISECTION METHOD FOR FINDING THE ROOTS OF AN ALGEBRAIC
%       EQUATION
%
%       This program finds the real roots of the equation
%       f(x)=0 by the Bisection method
%
%       eps is the convergence parameter; fa, fb and fc are
%       values of the function f(x) at the two ends, a and b,
%       of the domain containing the root and at the mid point,
%       respectively.
%
format short
eps=0.02;
%
%       Enter limits of the domain
%
a=input('Enter lowest value of interval, a=');
b=input('Enter highest value of interval, b=');
%
%       Apply Bisection method
%
for i=1:40
fa=f(a);
fb=f(b);
c(i)=(a+b)/2;
fc=f(c(i));
%
```

```
%       Check for convergence
%
if(abs(fc) <= eps)
        disp(sprintf('Iteration converged'))
break
end
%
%       Next iteration
%
if(fa*fc<0)
        b=c(i);
        else
a=c(i);
end
end
c=c';
%
%       Print results
%
disp(c)
```

B.3

```
%       SECANT METHOD FOR ROOT SOLVING
%
function [p1,err,k]=secant(f,p0,p1,delta,max1)
%
%       f is the function in the equation f(x)=0 entered as a
%       string, p0 and p1 are the two ends of the domain, given
%       as inputs in the function call, delta is the convergence
%       parameter and max1 is the specified maximum number of
%       iterations
%
%       Apply Secant method
%
for k=1:max1
p2=p1 - feval(f,p1)*(p1-p0)/(feval(f,p1)-feval(f,p0));
fprintf('Approximation to the root=%.4f\n',p2);
%
%       Calculate error
%
err=abs(p2-p1);
%
%       Update values
%
p0=p1;
p1=p2;
%
%       Apply convergence condition
```

```
%
if (k>2)&(err<delta)
fprintf('The root is=%8.4f\n',p1);
break
end
end
%
%      Stop if convergence not achieved
%
if(k==max1)
          disp('Max number of iterations reached')
end
```

B.4

```
% NEWTON-RAPHSON METHOD FOR ROOT SOLVING
%
% Given equation: f(x)=0
%
%      eps is the convergence parameter, fd is the derivative
%      of the function at the present approximation x(i), and
%      the next approximation to the root is x(i+1)
%
f=inline('294*w*(1 - exp(-1000/(21*(5+20*w)))) - 250');
%
%      Enter convergence parameter
%
eps=input('Enter the convergence parameter, eps=');
fprintf('EPS= %.4f\n',eps);
%
%      Enter starting value of the root
%
x(1)=input('Enter the initial guess, x(1)=');
%
%      Apply Newton-Raphson method
%
for i=1:20
fprintf('X= %.4f      FUNCTION F(X)= %0.6f\n',x(i),f(x(i)));
fd=(f(x(i)+0.001)-f(x(i)))/0.001;
x(i+1)=x(i) - f(x(i))/fd;
%
%      Check for convergence and print results
%
if(abs(x(i+1)-x(i)) <= eps)
         fprintf('FLOW RATE X= %.4f      FUNCTION F(X)= ...
%.6f\n',x(i+1),f(x(i+1)));
break
end
end
```

B.5

```
%       SUCCESSIVE SUBSTITUTION METHOD FOR ROOT SOLVING
%
%       Given equation: f(x) = 0, rewritten as x = z = g(x)
%
%       conv is the convergence parameter, x is the present
%       approximation to the root and z the next approximation
%
%       Enter initial guess for the root and convergence parameter
%
x = input('Enter the value of x, x = ');
conv = input('Enter the value of Convergence Parameter, conv = ');
fprintf('X= %.2f       CONV= %.4f\n',x,conv);
%
%       Apply successive substitution
%
for i = 1:20
z = (((((15-x)/(7.5*10^-5))^ .5)-80)/10.5)^ .6;
fprintf('X = %.4f        Z = %.4f\n',x,z);
%
%       Check for divergence of scheme
%
if abs(z-x) > 1/conv
disp('Convergence not achieved');
break
end
%
%       Check for convergence
%
if abs(z-x) < conv
%
%       Print results
%
fprintf('THE REQUIRED ROOT IS X = %.4f\n',x);
break
elseif abs(z-x) >=conv
x = z;
end
end
```

B.6

(a)

```
%       GAUSSIAN ELIMINATION METHOD FOR SOLVING
%       A SYSTEM OF LINEAR EQUATIONS
%
%       a is the coefficient matrix, b the constant vector, x
%       the vector of unknowns and tr the transformed upper
%       triangular matrix
%
```

```
%      Input data
%
function [x,tr] = gauss(a,b)
[n n] = size(a);
x = zeros(n,1);
c = zeros(1,n + 1);
%
%      Form the augmented matrix
%
aug = [a b];
%
%      Partial pivoting
%
for p = 1:n-1
  [y,j] = max(abs(aug(p:n,p)));
  c = aug(p,:);
  aug(p,:) = aug(j + p-1,:);
  aug(j + p-1,:) = c;
%
%      Check if matrix is singular
%
if aug(p,p) ==0
  'a was singular. No unique solution'
  break
end
%
%      Obtain upper triangular matrix
%
  for k = p + 1:n
      m = aug(k,p)/aug(p,p);
      aug(k,p:n + 1) = aug(k,p:n + 1)-m*aug(p,p:n + 1);
  end
end
%
%      Apply back-substitution
%
tr = aug(1:n,1:n);
x = backsub(aug(1:n,1:n),aug(1:n,n + 1));
%
```

(b)

```
%      Back Substitution
%
function x = backsub(a,b)
n = length(b);
x = zeros(n,1);
x(n) = b(n)/a(n,n);
for k = n-1:-1:1
 x(k) = (b(k)-a(k,k + 1:n)*x(k + 1:n))/a(k,k);
end
end
```

B.7

(a)

```
%       GAUSSIAN ELIMINATION METHOD FOR A TRIDIAGONAL
%       COEFFICIENT MATRIX
%
%       n is the number of unknowns, s is a parameter from the
%       problem being solved, a, b and c are coefficients in
%       the tridiagonal matrix, f is the constant vector and tp
%       is the physical temperature
%
%       Enter input data
%
s=0.071^2;
n=29;
f(1)=100;f(29)=100;
f(2:28)=0;
a(2:n)=-1;b(1:n)=2+s;c(1:n-1)=-1;
%
%       Apply tridiagonal matrix algorithm
%
for i=2:n;
d=a(i)./b(i-1);
b(i)=b(i)-c(i-1).*d;
f(i)=f(i)-f(i-1).*d;
end
%
%       Apply back-substitution
%
t(n)=f(n)./b(n);
for i=1:n-1;
j=n-i;
t(j)=(f(j)-c(j).*t(j+1))./b(j);
end
%
%       Plot the results obtained
%
tp(2:30)=t(1:29)+20;
tp(1)=120;tp(31)=120;
x=linspace(0,30,31);
plot(x,tp,'k')
xlabel('Distance x (cm)', 'Fontsize', 14)
ylabel( 'Physical Temperature Tp (Degrees C)', 'Fontsize', 14)
```

(b) TRIDIAGONAL MATRIX ALGORITHM

```
function t=tdma( a,b,c,f,n)
for i=2:n;
d=a(i)./b(i-1);
b(i)=b(i)-c(i-1).*d;
```

```
f(i)=f(i)-f(i-1).*d;
end
%
%       Apply back-substitution
%
t(n)=f(n)./b(n);
for i=1:n-1;
j=n-i;
t(j)=(f(j)-c(j).*t(j+1))./b(j);
end
```

B.8

```
%       GAUSS-JORDAN METHOD FOR SOLVING
%       A SYSTEM OF LINEAR EQUATIONS
%
%       a is the coefficient matrix, b is the constant vector,
%       x is the vector of the n unknowns and tr is the
%       transformed matrix which should be an identity matrix
%
function [x,tr]=jordan(a,b)
%
%       Enter input data
%
[n n]=size(a);
x=zeros(n,1);
c=zeros(1,n+1);
aug=[a b];
%
%       Partial pivoting
%
for p=1:n
  [y,j]=max(abs(aug(p:n,p)));
  c=aug(p,:);
  aug(p,:)=aug(j+p-1,:);
  aug(j+p-1,:)=c;
%
%       Check if coefficient matrix is singular
%
  if aug(p,p)==0
     'a was singular. No unique solution'
     break
  end
%
%       Apply Gauss-Jordan method
%
  for k=p+1:n+1
      aug(p,k)=aug(p,k)/aug(p,p);
  end
  aug(p,p)=1;
```

```
  for i=1:n
      if i~=p
         for j=p+1:n+1
         aug(i,j)=aug(i,j)-aug(i,p)*aug(p,j);
         end
         aug(i,p)=0;
      end
  end
end
%
%      Output solution
%
tr=aug(1:n,1:n);
x=aug(:,n+1)
```

B.9

```
%      SOLVING A SYSTEM OF LINEAR EQUATIONS
%      BY MATRIX INVERSION
%
%      a is the coefficient matrix, b is the constant vector,
%      x is the vector of the n unknowns, d is the calculated
%      inverse of the matrix a and tr is the transformed
%      matrix which should be an identity matrix
%
function [x,tr]=matinv(a,b)
%
%      Enter input data
%
[n n]=size(a);
x=zeros(n,1);
c=zeros(1,n+1);
d=eye(n);
%
%      Form augmented matrix
%
aug=[a d];
%
%      Partial pivoting
%
for p=1:n
  [y,j]=max(abs(aug(p:n,p)));
  c=aug(p,:);
  aug(p,:)=aug(j+p-1,:);
  aug(j+p-1,:)=c;
%
%      Check if coefficient matrix is singular
%
  if aug(p,p)==0
      'a was singular. No unique solution'
```

```
      break
   end
%
%       Apply Gauss-Jordan method
%
   for k=p+1:2*n
        aug(p,k) = aug(p,k)/aug(p,p);
   end
   aug(p,p) = 1;
   for i=1:n
        if i ~=p
           for j=p+1:2*n
           aug(i,j) = aug(i,j)-aug(i,p)*aug(p,j);
           end
           aug(i,p) = 0;
        end
   end
end
%
%       Output results
%
tr=aug(1:n,1:n);
d=aug(1:n,n+1:2*n);
disp(d);
x=d*b;
disp(x);
```

B.10

```
%       SOLVING THE SYSTEM OF LINEAR EQUATIONS
%       IN EXAMPLE 6.5 BY THE GAUSS-SEIDEL METHOD
%
%       s is a parameter in the problem, eps is the convergence
%       parameter, tp is the physical temperature
%
%       Input given data
%
s=0.071^2;
eps=0.001;
%
%       Enter initial guess
%
x=zeros(1,31)
x(1)=100;
x(31)=100;
%
%       Gauss-Seidel iteration
%
```

```
for k=1:1000
%
%      Store old values
%
xold=x;
%
%      Calculate new values
%
for i=2:30
x(i)=(x(i+1)+x(i-1))/(2+s);
end
%
%      Check for convergence
%
if abs(x-xold)<=eps
fprintf('No. of iterations=%g\n',k);
fprintf('The Solution is:\n');
tp=x+20;
disp(tp');
break
end
end
for j=1:31;
    y(j)=(j-1)*1.0;
end
plot(y,tp)
```

B.11

```
%      GAUSS-SEIDEL METHOD FOR SOLVING
%      A SYSTEM OF LINEAR EQUATIONS
%
function x=gseid(a,b,p,ep,max1)
%
%      a is the coefficient matrix, b is the constant vector,
%      x is the vector of the n unknowns, p is the initial
%      guess for the vector of unknowns, ep is the convergence
%      parameter, and max1 is the specified maximum number of
%      iterations
%
%      Determine number of unknowns
%
n=length(b);
%
%      Apply Gauss-Seidel iteration
%
for k=1:max1
```

```
for j=1:n
if j==1
x(1)=(b(1)-a(1,2:n)*p(2:n))/a(1,1);
elseif j==n
       x(n)=(b(n)-a(n,1:n-1)*(x(1:n-1))')/a(n,n);
else
x(j)=(b(j)-a(j,1:j-1)*(x(1:j-1))'- ...
  a(j,j+1:n)*p(j+1:n))/a(j,j);
end
end
%
%       Calculate error and apply convergence criterion
%
err=abs(norm(x'-p));
p=x';
if(err<ep)
break
end
end
x=x';
```

B.12

```
%       POWER METHOD FOR SOLVING AN EIGENVALUE PROBLEM
%
%       x is the eigenvector, xo is the eigenvector at the
%       previous iteration, a is the coefficient matrix, eps is
%       the convergence parameter, and c is the largest
%       eigenvalue
%
%       Enter initial guess
%
x=input('Initial guess of unknown eigenvector=');
a=input('Coefficient matrix a=');
eps=0.0001;
%
%       Apply Power method
%
for i=1:30
xo=x;
x=a*x;
c=max(x);
x=x/c;
%
%       Check for convergence
%
if (abs(x-xo))<eps
%
%       Print results
%
```

```
c,x
break
end
end
```

B.13

```
%       SUCCESSIVE SUBSTITUTION METHOD FOR SOLVING
%       A SYSTEM OF NONLINEAR ALGEBRAIC EQUATIONS
%
%       ep is the convergence parameter, b, p, f1, f2 are
%       parameters in the problem, and c is the total flow rate
%       of the mixture entering the plant
%
ep=0.0000001;
b=0.1;
c=180.0;
bo=b;
      disp('ARGON               TOTAL FLOW         AMMONIA')
for i=1:50
    f1=0.9/(1.0-b);
    p=1.0-0.57*exp(-0.0155*f1);
    f2=90.0/(1.0-b*p);
    b=1.0-23.5/(4.0*f2*p+f1);
    c=f1+4.0*f2;
    d=0.57*exp(-0.0155*f1)*2.0*f2;
      fprintf('%.4f              %.4f                 %.4f\n',f1,c,d)
    if (abs(b-bo))<ep
      disp('Iteration has converged')
      disp('Converged results are')
      fprintf('ARGON=%.4f    TOTAL FLOW= %.4f ...
        AMMONIA= %.4f\n',f1,c,d)
      break
    end
    bo=b;
end
```

B.14

```
%       NEWTON'S METHOD FOR SOLVING A SYSTEM
%       OF NONLINEAR ALGEBRAIC EQUATIONS
%
%       r and p are parameters in the problem, ep is the
%       convergence parameter and dr, dp are the increments in
%       r and p, respectively
%
%       Enter starting values
%
r=input('Enter the value of parameter r, r =');
p=input('Enter the value of parameter p, p =');
ep=input('Enter the value of convergence parameter ...
  ep, ep =');
```

```
for i=1:10
    r1=((p-80)/10.5)^0.6-r;
    p1=((15-r)*(10^6)/75)^0.5-p;
    b=r1^2+p1^2;
%
%      Check for convergence
%
          if b<ep
             disp('THE REQUIRED SOLUTION IS:')
    fprintf('The flow rate R=%.4f The pressure ...
      P=%.4f\n',r,p)
    break
          end
%
%      Calculate partial derivatives
%
    rr=-1;
    rp=3/(5*(10.5^0.6)*((p-80)^0.4));
    pr=-1/(2*((7.5*10^-5)^0.5)*((15-r)^0.5));
    pp=-1;
    d=rr*pp-rp*pr;
%
%      Determine increments for the next iteration
%
    dr=(-r1*pp+p1*rp)/d;
    dp=(-p1*rr+r1*pr)/d;
%
%      Calculate values of r and p for the next iteration
%
    r=r+dr;
    p=p+dp;
%
%      Print results
%
    fprintf('R =%.4f          P =%.4f\n',r,p)
end
```

B.15

```
%      INTERPOLATION WITH AN EXACT FIT
%
%      Fifth order polynomial, y=f(x)
%
%      c is the coefficient matrix, a is the vector
%      representing the constants of the polynomial in
%      ascending powers of x, and p is the vector of constants
%      of the polynomial in descending powers of x
%
%      Enter given data
x=[1 2 3 4 5 6];
y=[106.4 57.79 32.9 19.52 12.03 7.67]';
%
```

```
%       Form Matrix
%
c=[x.^0;x;x.^2;x.^3;x.^4;x.^5]';
%
%       Find coefficients of polynomial
%
disp('Coefficients of the polynomial are:')
a=c\y
plot(x,y,'*')
hold
%
%       Find value at x=3.4
%
p=a(6:-1:1);
x1=3.4;
y1=polyval(p,x1);
fprintf('Interpolated value from exact fit y=%.4f\n',y1)
%
%       Use of Matlab functions
%
y2=interp1(x,y',x1,'linear');
y3=interp1(x,y',x1,'spline');
fprintf('Value from linear interpolation y=%.4f\n',y2)
fprintf('Value from spline interpolation y=%.4f\n',y3)
x=linspace(1,6,20);
y=polyval(p,x);
plot(x,y,'-g')
xlabel('x','Fontsize',14);ylabel('y','Fontsize',14)
```

B.16

```
%       LAGRANGE INTERPOLATION
%
%       w is the number of data points and c is the vector of
%       constants of the polynomial in descending powers of x
%
%       Enter given data
x=[0.5 1.0 1.5 2.0 2.5];
y=[3.0 3.9 5.2 7.3 10.5];
%
%       Get number of data points or unknowns
%
w=length(x);
n=w-1;
l=zeros(w,w);
%
%       Calculate coefficients of the general polynomial
%
for k=1:n+1
v=1;
```

```
for j=1:n+1
if k~=j
v=conv(v,poly(x(j)))/(x(k)-x(j));
end
end
l(k,:)=v;
end
c=y*l;
%
%       Print coefficients of the polynomial
%
disp('Coefficients of the polynomial in descending powers of x ...
  are:')
disp(c')
%
%       Check accuracy of polynomial
%
xp=[0 0.5 0.75 1.0 1.25 1.5 1.8 2.0 2.2 2.5 3.0];
yp=polyval(c,xp);
disp('Interpolated values:')
for k = 1:11
fprintf('xp=%.4f        yp=%.4f\n',xp(k),yp(k))
end
```

B.17

```
%       NEWTON'S DIVIDED DIFFERENCE METHOD FOR INTERPOLATION
%
%       c is the vector of the coefficients of the polynomial
%       for Newton's divided differences
%
%       Enter input data
%
n=input('Enter the number of data points, n=');
x=input('Enter values of the independent variable, x=');
y=input('Enter corresponding values of the dependent ...
   variable, y=');
%
f(1:5,1)= y;
%
%       Apply Newton's Divided Difference method
%
for k=1:(n-1)
l=k+1;
for m=1:(n-k)
f(m,l) = (f(m+1,k)-f(m,k))/(x(m+k)-x(m));
end
end
disp('Coefficients of the polynomial c0, c1, c2 ... are:')
c=f(1,1:n)'
```

```
%
%      Enter value of independent variable for interpolation
%
for i=1:6
   xp=input ('\nEnter x where interpolation is desired, ...
     xp=');
   fprintf('xp =%.3f\n',xp)
%
%      Calculate interpolated results and remainder
%
b=1;z=0;
for i=1:n
z=z+f(1,i)*b;
fprintf('Interpolated value of y =%.3f\n',z)
b=b*(xp-x(i));
if i<n
r=b*f(1,i+1);
fprintf('Remainder term =%.3f\n',r)
end
end
end
```

B.18

```
%      POLYNOMIAL REGRESSION
%
%      n is the number of data points, c is the vector
%      representing the constants of the polynomial in
%      ascending powers of x, and p is the vector of constants
%      of the polynomial in descending powers of x
%
%      Input given data
%
x=input('Enter values of the independent variable, x=');
y=input('Enter values of the dependent variable, y=');
np=input('Enter order of polynomial for best fit, np=');
%
n=length(x);
m=np+1;
%
%      Initialize matrices
%
a=zeros(m,m);
b=zeros(m,1);
%
%      Apply polynomial regression
%
for i=1:m
for j=1:m
n1=i+j-2;
for k=1:n
```

```
    a(i,j)=a(i,j)+x(k)^n1;
end
end
for k=1:n
    b(i)=b(i)+y(k)*x(k)^(i-1);
end
end
%
%      Print polynomial constants and values calculated from
%      best fit
%
disp('The constants of the polynomial are:')
c=a\b
p=c(m:-1:1);
disp('The values calculated from the best fit are:')
s=polyval(p,x);
y=s'
```

B.19

(a)

```
%      NUMERICAL INTEGRATION BY TRAPEZOIDAL RULE
%
function s=trap(f,a,b,m)
%
%      f is the function, entered as a string, a and b are the
%      limits of integration, m is the number of subintervals,
%      and s the sum or quadrature
%
%      Calculate step or segment size
%
h=(b-a)/m;
fprintf('Step Size=%.4f\n',h)
for i=1:10
%
%      Apply Trapezoidal rule
%
s=0;
for k=1:m-1
x=a+h*k;
s=s+feval(f,x);
end
s=h*(feval(f,a)+feval(f,b))/2+h*s;
%
%      Print results
%
fprintf('Time=%2g      Charge=%.4f      Voltage= ...
%.4f\n',b,s,s/0.025);
%
```

```
%      Vary upper limit for integration
%
b=2+b;m=(b-a)/h;
end
```

(b)

```
%      ALTERNATIVE IMPLEMENTATION OF ALGORITHM
%
%      Apply Trapezoidal rule
%
x=a:h:b;
f=feval('f81',x);
s=h*(0.5*f(1)+sum(f(2:n-1))+0.5*f(n));
%
```

B.20

(a)

```
%      NUMERICAL INTEGRATION BY SIMPSONS'S RULE
%
function s=simp2(f,a,b,n)
%
%      f is the function, entered as a string, a and b are the
%      limits of integration, n is the number of subintervals,
%      m is the number of two-segment intervals and s the sum
%      or quadrature
%
for i=1:10
%
%      Calculate segment size h and number of two-segment
%      sections m
%
h=(b-a)/n;
m=n/2;
%
%      Apply Simpson's rule
%
s1=0;
s2=0;
for k=1:m
x=a+h*(2*k-1);
s1=s1+feval(f,x);
end
for k=1:(m-1)
x=a+h*2*k;
s2=s2+feval(f,x);
end
s=h*(feval(f,a)+feval(f,b)+4*s1+2*s2)/3;
%
```

```
%      Print results
%
fprintf('n=%4g    Flow Rate=%.4f    Avg. Vel.=%.4f\n',n,s,...
 s/(pi*0.01));
%
%      Vary number of segments n
%
n=2*n;
end
```

(b)

```
%      ALTERNATIVE IMPLEMENTATION OF ALGORITHM
%
%      Apply Simpson's rule
%
  x=a:h:b;
  f=feval('f82',x);
  s=(h/3)*(f(1)+4*sum(f(2:2:n))+ ...
     2*sum(f(3:2:n-1))+f(n+1));
%
```

B.21

```
%      ROMBERG INTEGRATION
%
%      This program obtains the integral of a given function
%      f(x) over specified lower limit xmin and upper limit
%      xmax by using Romberg integration with a convergence
%      parameter ep; h is the step size
%
%      Define the function to be integrated
%
f=inline('(2.0/sqrt(pi))*exp(-x^2)');
%
%      Enter input values
%
ep=0.00001;
dif=1.0;
xmin=0.0;
xmax=input('Enter the value of z =');
h=xmax-xmin;
%
%      Carry out first order (Trapezoidal rule) calculations
%
n=1;
y(1,1)=0.5*h*(f(xmin)+f(xmax));
fprintf('No. of iterations=%2g    Erf(z)=%.6f\n',n,y(1,1));
%
%      Apply convergence criterion
%
```

```
while dif>ep
%
%       Calculate higher order extrapolations
%
m=2^(n-1);
h=h/2.0;
n=n+1;
y(1,n)=0.5*y(1,n-1);
for k=1:m
    x=xmin+(2*k-1)*h;
    y(1,n)=y(1,n)+h*f(x);
end
for k=2:n
       y(k,n)=(4^(k-1)*y(k-1,n)-y(k-1,n-1))/(4^(k-1)-1);
end
dif=abs(y(n,n)-y(n-1,n));
%
%       Print results
%
fprintf('No. of iterations=%2g     Erf(z)=%.6f\n',n,y(n,n));
end
```

B.22

```
%       INTEGRATION WITH SEGMENTS OF UNEQUAL WIDTH
%
%       This program calculates the integral from experimental
%       data on the dependent variable v given at unevenly
%       distributed values of the independent variable t; eps is
%       a specified small number and s is the sum or
%       quadrature
%
%       Enter given data
%
t=[0 .1 .2 .3 .5 .7 .8 1 1.1 1.3 1.5 1.6 1.7 1.8 2.0];
v=[9.5 10 10.57 11.24 12.97 15.38 16.93 20.9 23.41...
   29.74 38.17 43.33 49.21 55.88 71.9];
%
%       Specify small quantity eps to compare segment widths
%
eps=1.0e-6;
%
%       Starting values
%
m=length(t);
i=1;
s=0;
%
%       Compare adjacent segment widths dt1, dt2 and dt3
%
while i<m
```

```
    dt=t(i+1)-t(i);
    if i==m-1
        dt1=dt;dt2=0;dt3=0;
    elseif i==m-2
        dt1=dt;dt2=t(i+2)-t(i+1);dt3=0;
    else
        dt1=dt;dt2=t(i+2)-t(i+1);dt3=t(i+3)-t(i+2);
    end
%
%       Apply Trapezoidal rule
%
if abs(dt2-dt1)>eps
    s=s+(v(i+1)+v(i))*dt/2;
    i=i+1;
    disp('Trapezoidal rule')
%
%       Apply Simpson's one-third rule
%
elseif abs(dt3-dt2)>eps
        s=s+(v(i)+4.0*v(i+1)+v(i+2))*dt/3.0;
        i=i+2;
        disp('Simpson one-third rule')
%
%       Apply Simpson's three-eighths rule
%
    else
        s=s+(v(i)+3.0*v(i+1)+3.0*v(i+2)...
           +v(i+3))*dt*3.0/8.0;
        i=i+3;
        disp('Simpson three-eighths rule')
    end
%
%       Print results
%
        fprintf('I=%2g    Time=%.4f    Velocity=%.4f ...
Distance=%.4f\n',i,t(i),v(i),s)
end
```

B.23

```
%       NUMERICAL INTEGRATION OF AN IMPROPER
%       INTEGRAL BY SIMPSONS'S ONE-THIRD RULE
%
function s=simpimp(f,a,b,h)
%
%       f is the function, entered as a string, a and b are the
%       limits of integration, h is the width of each
%       subinterval, and s the sum or quadrature
%
%       Define starting parameters
%
```

```
for i=1:4
    xmin=a;
    se=0.0;
    fprintf('xmin=%.4f\n',xmin);
    for j=1:8
        xmax=b;
%
%      Calculate number of sub-intervals n and number of
%      two-segment sections m
%
n=(b-a)/h;
m=n/2;
%
%      Apply Simpson's rule
%
s1=0;
s2=0;
for k=1:m
x=a+h*(2*k-1);
s1=s1+feval(f,x);
end
for k=1:(m-1)
x=a+h*2*k;
s2=s2+feval(f,x);
end
s=h*(feval(f,a)+feval(f,b)+4*s1+2*s2)/3;
%
%      Print results
%
fprintf('Integral=%4g            xmax=%.4f\n',s,b);
%
%      Vary number of segments n
%
if abs(s-se)>0.00001
b=b+5.0;
se=s;
else
      break
end
      end
a=a+5.0;
b=a+10.0;
end
```

B.24

(a)

```
%      EULER'S METHOD FOR SOLVING A FIRST-ORDER ODE
%
function e=euler(f,a,b,y0,n)
```

```
%
%      f is the function entered as a string 'f'
%      a and b are the starting and end points
%      y0 is the initial condition y(1)
%      n is the number of steps
%      e[t' y'] is the output where t is the vector of
%      independent variable and y is the vector of dependent
%      variable
%
h=(b-a)/n;
t=zeros(1,n+1);
y=zeros(1,n+1);
t=a:h:b;
y(1)=y0;
for j=1:n
    y(j+1)=y(j)+h*feval(f,t(j),y(j));
end
e=[y' t'];
```

(b)

```
%      SOLUTION OF ODES IN EXAMPLE 8.1 BY EULER'S METHOD
%
%
%      Given ODE: dy/dt=f(t,y)
%
%      dt is step size, tn is total range of t, y0 is initial
%      value of y, and n is total number of t values
%
%      Enter given ODE
%
dydt=inline('2-0.5*y','t','y');
%
%      Choose step size and total time
%
dt=0.01;
tn=800*dt;
%
%      Enter initial conditions and starting values
%
y0=0;
n1=51;
t=(0:dt:tn)';
n=length(t);
y=y0*ones(n,1);
%
%      Apply Euler's Method
%
for j=2:n1;
    y(j)=y(j-1)+dt*dydt(t(j-1),y(j-1));
```

```
end
%
%       Second ODE
%
dydt=inline('-0.5*y','t','y');
for j=n1+1:n;
    y(j)=y(j-1)+dt*dydt(t(j-1),y(j-1));
end
%
%       Plot results
%
  plot(t,y,'-g')
```

B.25

(a)

```
%       HEUN'S METHOD FOR SOLVING A FIRST-ORDER ODE
%
function s=heun(f,a,b,ya,h)
%
%       f is the function entered as a string 'f'
%       a and b are the starting and end points
%       ya is the initial condition y(1)
%       h is the step size
%       s[t' y'] is the output where t is the vector of
%       independent variable and y is the vector of dependent
%       variable
%
m=(b-a)/h;
t=zeros(1,m+1);
y=zeros(1,m+1);
t=a:h:b;
y(1)=ya;
for j=1:m
k1=feval(f,t(j),y(j));
k2=feval(f,t(j+1),y(j)+h*k1);
y(j+1)=y(j)+(h/2)*(k1+k2);
end
s=[t' y']
```

(b)

```
function z=fe1( x,y )
z=4-2*y;
end
```

(c)

```
function z=fe2( x,y )
z=-2*y;
end
```

```
(d)
s1=heun('fe1',0,0.5,0,0.01);
s2=heun('fe2',0.5,8,s1(51,2),0.01);
plot(s1(:,1),s1(:,2))
hold
plot(s2(:,1),s2(:,2))
```

B.26

```
%       FOURTH-ORDER RUNGE-KUTTA METHOD
%
%       Enter the function f for the ODE dv/dt=f(t,v)
%
f=inline('-9.8-(0.01*v+0.001*v^2)');
%
%       Choose time step and enter initial conditions
%
dt=input('Step size dt =');
t=0;
x=0;
v=100.0;
i=1;
%
while v>=0
%
%       Initialize variables
%
        q=x;z=v;
        tp(i)=t;xp(i)=x;vp(i)=v;
%
%       Apply 4th order Runge-Kutta formulas
%
rk1x=dt*z;
rk1v=dt*f(z);
rk2x=dt*(z+rk1v/2);
rk2v=dt*f(z+rk1v/2);
rk3x=dt*(z+rk2v/2);
rk3v=dt*f(z+rk2v/2);
rk4x=dt*(z+rk3v);
rk4v=dt*f(z+rk3v);
x=q+(rk1x+2*rk2x+2*rk3x+rk4x)/6;
v=z+(rk1v+2*rk2v+2*rk3v+rk4v)/6;
%
%       Advance to next time step
%
t=t+dt;
i=i+1;
end
%
```

```
%       Plot results
%
plot(tp,xp,'-',tp,vp,'--')
```

B.27

```
%       ADAM'S PREDICTOR-CORRECTOR METHOD
%
%       dt is time step, ep is the convergence parameter for
%       steady state and ep1 is the convergence parameter for
%       the corrector
%
%       Enter function f(y) in ODE dy/dt=f(y)
%
f=inline('10-0.05*y');
%
%       Enter initial conditions
%
t(1)=0;
y(1)=100;
%
ep=0.0001;
ep1=0.00001;
%
%       Choose time step
%
dt=input('Time step dt=');
for i=1:3
%
%       Apply Runge-Kutta for first 3 steps
%
    rk1=dt*f(y(i));
    rk2=dt*f(y(i)+rk1/2);
    rk3=dt*f(y(i)+rk2/2);
    rk4=dt*f(y(i)+rk3);
    y(i+1)=y(i)+(rk1+2*rk2+2*rk3+rk4)/6;
    t(i+1)=t(i)+dt;
end
s=abs((y(i+1)-y(i))/(y(i)*dt));
%
%       Apply convergence criterion
%
while s>=ep
    i=i+1;
%
%       Apply predictor
%
    y(i+1)=y(i)+dt*(55*f(y(i))-59*f(y(i-1))...
            +37*f(y(i-2))-9*f(y(i-3)))/24;
    yp(i+1)=y(i+1);
```

```
    dy=abs(y(i+1)-y(i));
%
%      Apply corrector with iteration
%
    while dy>=ep1
    y(i+1)=y(i)+dt*(9*f(y(i+1))+19*f(y(i))...
            -5*f(y(i-1))+f(y(i-2)))/24;
    dy=abs(y(i+1)-yp(i+1));
    end
    t(i+1)=t(i)+dt;
    s=abs((y(i+1)-y(i))/(y(i)*dt));
end
%
%      Plot results
%
plot(t,y,'-')
```

B.28

```
%      HAMMING'S PREDICTOR-CORRECTOR METHOD
%
%      Define function f(x,y) in dy/dx=f(x,y)
%
%      dt is time step and ep is convergence parameter for
%      steady state
%
f=inline('9.8-(2*y+0.1*y^2)');
%
%      Enter initial conditions
%
t(1)=0;
y(1)=0;
%
ep=0.0001;
%
%      Choose step size dt
%
dt=input('Time step dt=');
%
%      Apply Runge-Kutta method for the first three steps
%
for i=1:3
    rk1=dt*f(y(i));
    rk2=dt*f(y(i)+rk1/2);
    rk3=dt*f(y(i)+rk2/2);
    rk4=dt*f(y(i)+rk3);
    y(i+1)=y(i)+(rk1+2*rk2+2*rk3+rk4)/6;
    t(i+1)=t(i)+dt;
end
yp(i+1)=y(i+1);ym(i+1)=yp(i+1);yc(i+1)=yp(i+1);
```

```
s=abs((y(i+1)-y(i))/(y(i)*dt));
%
%      Apply convergence criterion
%
while s>=ep
      i=i+1;
%
%      Apply Predictor
%
      yp(i+1)=y(i-3)+dt*(4/3)*(2*f(y(i))-f(y(i-1))...
              +2*f(y(i-2)));
%
%      Apply Modifier
%
      ym(i+1)=yp(i+1)-(112/121)*(yp(i)-y(i));
%
%      Apply Corrector
%
      yc(i+1)=(1/8)*(9*y(i)-y(i-2))+(3/8)*dt*...
              (f(ym(i+1))+2*f(y(i))-f(y(i-1)));
%
%      Update results
%
      y(i+1)=yc(i+1);
      t(i+1)=t(i)+dt;
      s=abs((y(i+1)-y(i))/(y(i)*dt));
end
%
%      Plot results
%
plot(t,y,'-')
```

B.29

(a)

```
%      SOLUTION OF A THIRD-ORDER BOUNDARY-VALUE PROBLEM BY
%      RUNGE-KUTTA METHOD WITH SHOOTING TECHNIQUE
%
%      t is the independent variable, v represents the three
%      dependent variables, ep is the convergence parameter,
%      er is the difference between two values of variable
%      v(2), or velocity, at the second boundary, y0
%      represents the initial conditions, s is the unknown
%      variable v(3) to be determined at t=0
%
%      Enter convergence parameter and starting values
%
ep=0.001;
s=0.5;
edge=6.0;
```

```
e1=0;
%
for i=1:20;
%
%       Apply convergence criterion
%
   if abs(e1-1)<ep
      break
   end
%
%       Apply ode45 to solve the ODEs
%
y0=[0;0;s];
[t,v]=ode45('rhs1',edge,y0);
e1=v(length(v),2);
%
%       Compute the derivative for Newton-Raphson
%
y0=[0;0;s+0.001];
[t,v]=ode45('rhs1',edge,y0);
e2=v(length(v),2);
er=e2-e1;
der=(er)/0.001;
%
%       Apply Newton-Raphson method
%
s=s-(e1-1)/der;
end
%
%       Plot the results
%
plot(t,v(:,1),'-',t,v(:,2),'--',t,v(:,3),'-.')
```

(b)

```
%       Defining the three ODEs in Example 8.5
%
function dydt=rhs1(t,y)
b=0.5;
dydt=[y(2);y(3);-y(1)*y(3)-b*(1-y(2)^2)];
end
```

B.30

```
%       FINITE DIFFERENCE METHOD FOR SOLVING SECOND-ORDER ODE
%
%
%       s, p are parameters in the problem, nt is the total
%       number of grid points, a, b and c are coefficients in
%       the tridiagonal matrix, f is the constant vector, t is
```

```
%       the dimensionless temperature and tp is the physical
%       temperature
%
%       Enter input data
%
p=input('Parameter P=');
nt=input('Total number of grid points=');
n=nt-2;
s=2+(p^2)*((1.0/(nt-1))^2);
%
%       Enter boundary conditions and form tridiagonal matrix
%
f(1)=1;f(n)=0.5;
f(2:n-1)=0;
a(2:n)=-1;b(1:n)=s;c(1:n-1)=-1;
%
%       Apply tridiagonal matrix algorithm
%
for i=2:n;
d=a(i)./b(i-1);
b(i)=b(i)-c(i-1).*d;
f(i)=f(i)-f(i-1).*d;
end
%
%       Apply back-substitution
%
t(n)=f(n)./b(n);
for i=1:n-1;
j=n-i;
t(j)=(f(j)-c(j).*t(j+1))./b(j);
end
%
%       Calculate resulting temperature distribution
%
tp(2:nt-1)=t(1:n);
tp(1)=1;tp(nt)=0.5;
%
%       Plot the results obtained
%
x=linspace(0,1,51);
plot(x,tp,'k')
xlabel('Distance X', 'Fontsize', 14)
ylabel( 'Temperature T', 'Fontsize', 14)
```

B.31

```
%       FORWARD TIME CENTRAL SPACE (FTCS) METHOD
%
%       th is the unknown theta, or dimensionless
```

```
%       concentration, tint is the initial value of th taken as
%       uniform, kmax is the maximum number of time steps,
%       kprint the steps after which results are printed or
%       plotted,dx is the grid size, dt the time step, n the
%       number of grid points, k represents the time step and i
%       the spatial grid point
%
%       Enter starting values
%
tint=input('Enter the initial condition tint=');
n=input('Enter number of grid points n=');
kmax=input('Enter maximum number of time steps kmax=');
kprint=input('Time steps after which results are plotted ...
  kprint=');
%
%       Specify boundary conditions
%
th(1,2:n)=tint;
th(1:kmax,1)=1.0;
dx=1/(n-1);
%
%       Calculate maximum time step to avoid numerical instability
%
dt=(dx^2)/2;
     for k=2:kmax;
%
%       Apply FTCS method
%
           for i=2:n-1;
           th(k,i)=th(k-1,i)+dt*(th(k-1,i+1)-2*th ...
                    (k-1,i)+th(k-1,i-1))/(dx^2);
           end
        end
%
%       Store results for plotting
%
    for j=1:10;
         m=kprint*j+1;
         time=(m-1)*dt;
         fprintf('Time=%.4f\n',time)
         tp(j,1:n)=th(m,1:n);
    end
%
%       Plot results
%
x=linspace(0,1,n);
plot(x,tp)
xlabel('X');ylabel('Dimensionless concentration, \theta');
title('Concentration Versus Distance at Different Times')
```

B.32

```
%       CRANK-NICOLSON METHOD
%
%       t is the unknown dimensionless temperature, tint is the
%       initial value of t taken as uniform, kmax is the maximum
%       number of time steps, kprint the steps after which
%       results are printed or plotted, dx is the grid size, dt
%       the time step, n the number of grid points, k
%       represents the time step and i the spatial grid point,
%       and a, b, c and f are the parameters of the tridiagonal
%       system
%
%       Enter starting values
%
tint=input('Enter the initial condition tint=');
n=input('Enter number of grid points n=');
dt=input('Enter the time step dt=');
kmax=input('Enter maximum number of time steps kmax=');
kprint=input('Time steps after which results are plotted ...
   kprint=');
%
%       Specify boundary conditions
%
t(1,2:n)=tint;
t(1:kmax,1)=1.0;
dx=1/(n-1);
%
%       Calculate the parameters of the tridiagonal system
%
for k=2:kmax;
     a(1:n-2)=-dt/(2*dx^2);
     b(1:n-2)=1+dt/(dx^2);
     c(1:n-2)=-dt/(2*dx^2);
     for i=2:n-1;
          f(i-1)=t(k-1,i)+dt*(t(k-1,i+1)-2*t ...
                    (k-1,i)+t(k-1,i-1))/(2*dx^2);
     end
     f(1)=f(1)-a(1)*t(k,1);
     a(n-2)=a(n-2)-c(n-2)/3;
     b(n-2)=b(n-2)+4*c(n-2)/3;
%
%       Use the TDMA function file to obtain temperatures at
%       the next time step
%
     t(k,2:n-1)=tdma(a,b,c,f,n-2);
%
%       Apply boundary condition at the right boundary
%
     t(k,n)=4*t(k,n-1)/3-t(k,n-2)/3;
end
```

```
%
%      Store results for plotting
%
   for j=1:10;
        m=kprint*j+1;
        time=(m-1)*dt;
        fprintf('Time=%.4f\n',time)
        tp(j,1:n)=t(m,1:n);
   end
%
%      Plot results
%
x=linspace(0,1,n);
plot(x,tp)
xlabel('X');ylabel('Dimensionless Temperature, \theta');
title('Temperature Versus Distance at Different Times')
```

B.33

```
%      GAUSS-SEIDEL METHOD FOR AN ELLIPTIC PDE
%
%      m and n are grid points in x and y directions, imax is
%      maximum number of iterations, phi is the unknown
%      dependent variable, phiol the value of phi at the
%      previous iteration, and ep the convergence parameter
%
%      Input given data
%
m=input('Enter number of grid points in x direction m=');
n=input('Enter number of grid points in y direction n=');
phint=input('Enter initial guess for phi taken as uniform ...
  phint=');
imax=input('Enter maximum number of iterations imax=');
ep=input('Enter convergence parameter ep=');
%
%      Calculate grid or mesh lengths
%
dx=1/m;
dy=1/n;
%
%      Apply boundary conditions
%
phi(2:m-1,2:n-1)=phint;
phi(1,1:n)=0;
phi(m,1:n)=0;
phi(1:m,1)=0;
phi(1:m,n)=0;
%
%      Apply Gauss-Seidel iterative scheme
%
```

```
for i=1:imax;
    phiol(1:m,1:n)=phi(1:m,1:n);
    for j=2:m-1;
        for k=2:n-1;
            phi(j,k)=((phi(j+1,k)+phi(j-1,k))/...
            (dx^2)+(phi(j,k+1)+phi(j,k-1))/(dy^2))/(2/...
            (dx^2)+2/(dy^2))+0.001;
        end
    end
%
%      Check for convergence
%
    if abs(phi-phiol)<ep
       break
    end
end
%
%      Plot results
%
xp=linspace(0,1,m);
nn=(n+1)/2;
plot(xp,phi(1:m,nn-4),xp,phi(1:m,nn-3),xp,phi(1:m,nn-2),...
xp,phi(1:m,nn-1),xp,phi(1:m,nn))
```

B.34

```
%      SOR METHOD FOR AN ELLIPTIC PDE
%
%      m and n are grid points in x and y directions,
%      respectively, imax is maximum number of iterations, phi
%      is the unknown dependent variable,phiol the value of
%      phi at the previous iteration, and ep the convergence
%      parameter
%
%      Input given data
%
m=input('Enter number of grid points in x direction m=');
n=input('Enter number of grid points in y direction n=');
phint=input('Enter initial guess for phi taken as uniform ...
  phint=');
imax=input('Enter maximum number of iterations imax=');
ep=input('Enter convergence parameter ep=');
%
%      Calculate grid or mesh lengths
%
dx=1/m;
dy=1/n;
%
%      Specify relaxation parameter w
```

```
%
w=0.5;
for ni=1:14;
%
%       Apply boundary conditions
%
phi(2:m-1,2:n-1)=phint;
phi(1,1:n)=0;
phi(m,1:n)=0;
phi(1:m,1)=0;
phi(1:m,n)=1;
%
%       Apply SOR iterative scheme
%
for i=1:imax;
    phiol(1:m,1:n)=phi(1:m,1:n);
    for j=2:m-1;
        for k=2:n-1;
            phi(j,k)=w*((phi(j+1,k)+phi(j-1,k))/...
                     (dx^2)+(phi(j,k+1)+phi(j,k-1))/(dy^2))/...
                     (2/(dx^2)+2/(dy^2))+(1-w)*phiol(j,k);
        end
    end
%
%       Check for convergence
%
    if abs(phi-phiol)<ep
       break
    end
end
%
%       Plot results
%
s(ni)=i;
w=w+0.1;
end
rf=linspace(0.5,1.9,14);
plot(rf,s)
```

B.35

```
%       SOLUTION OF THE WAVE EQUATION
%
%       This script file solves a second-order hyperbolic
%       partial differential equation by the finite difference
%       method. dx is grid size, dt is time step, n is number
%       of grid points, c is the Courant number, being taken close
%       to 1.0, u is the dependent variable, with u, u1 and u2
%       representing values at time t, t+dt and t+2dt, nlim is
%       the maximum number of time steps, and nprint is number
```

```
%       of steps after which results are displayed
%
%       Enter input quantities
%
dx=input('Grid size dx=');
dt=input('Time step=');
nlim=input('Maximum number of time steps nlim=');
nprint=input('Time steps after which results are displayed ...
  nprint=');
c=1000*(dt^2)/(dx^2);
m=1;
j=0;
%
%       Set the boundary conditions
%
n=(1/dx)+1;
nn=(0.25/dx)+1;
u1(1:nn)=linspace(0,0.004,nn);
u1(nn:n)=linspace(0.004,0,(n+1-nn));
u2(1:n)=0;
%
%       Initialize the variables
%
for i=2:n-1;
     u2(i)=u1(i)+c*(u1(i+1)-2*u1(i)+u1(i-1))/2;
end
time=0;
t=dt;
time=t;
while t<nlim*dt
                    m=m+1;
%
%       Save previous values
%
      u(1:n)=u1(1:n);
      u1(1:n)=u2(1:n);
%
%       Obtain results for next time step
%
      t=t+dt;
      time=t;
          for i=2:n-1;
               u2(i)=2*u1(i)-u(i)+c*(u1(i+1)-...
                       2*u1(i)+u1(i-1));
          end
%
%       Store results for display
%
              if m==nprint;
                  j=j+1;
                  up(j,1:n)=u2(1:n);
```

```
            m=0;
            fprintf('Time=%.4f\n',time)
        else
        end
end
%
%      Plot results
%
x=linspace(0,1,n);
plot(x,up)
```

B.36

```
%      SOLUTION OF FIRST-ORDER CONVECTION (HYPERBOLIC) PDE
%
%      x is the coordinate distance, t is the time, p(i)
%      represents the values of the dependent variable p at
%      previous time and pn(i) those at the present time step,
%      n is number of spatial grid points, c is convection
%      velocity, co is the Courant number, dx is the step
%      size, dt is the time step, tmax is the maximum time for
%      the computation and t is the time
%
%      Enter input quantities
%
dx=input('Enter grid size dx=');
dt=input('Enter time step dt=');
c=2.5;
n=5/dx+1;
tmax=2.0;
%
%      Choose solution method: 1 for backward difference, 2
%      for Euler's and 3 for Lax-Wendroff method
%
m=input('Choose solution method m=');
%
%      Input initial and boundary conditions
%
p(2:n)=0;
pn(2:n)=0;
p(1)=1.0;
pn(1)=1;
t=0;
j=1;
p1(1)=0;p2(1)=0;p3(1)=0;p4(1)=0;
co=c*dt/dx;
%
%      Compute results for next time step
%
while t<tmax-eps;
    t=t+dt;
```

```
    j=j+1;
for i=2:n;
    if (i-n+1)|m==1
%
%     Backward or upwind difference method
%
      pn(i) = (1-co)*p(i) +co*p(i-1);
    elseif m==2
%
%     Euler's method
%
      pn(i)=p(i)-co*(p(i+1)-p(i-1))/2;
    else m==3
%
%   Lax-Wendroff method
%
     pn(i)=p(i)-co*(p(i+1)-p(i-1))/2+(co^2)*...
           (p(i+1)-2*p(i)+p(i-1))/2;
    end
end
%
%       Store results for display and update previous values
%
p1(j)=pn(3);p2(j)=pn(5);p3(j)=pn(7);p4(j)=pn(9);
p(2:n)=pn(2:n);
end
%
%       Plot results
%
ts=linspace(0,2,(2/dt+1));
plot(ts,p1,ts,p2,ts,p3,ts,p4)
```

Appendix C: Computer Programs in FORTRAN

C.1: Search method for finding the roots of an algebraic equation
C.2: Root solving with the bisection method
C.3: Root solving with the secant method
C.4: Newton–Raphson method for finding the roots of an algebraic equation
C.5: Gaussian elimination method for a system of linear equations
C.6: Tridiagonal matrix algorithm
C.7: The Gauss–Jordan elimination method
C.8: SOR method for solving a system of linear equations
C.9: Lagrange interpolation
C.10: Spline interpolation
C.11: Least-squares method for polynomial regression
C.12: Trapezoidal rule for numerical integration
C.13: Romberg integration
C.14: Euler's method for solving an ODE
C.15: Runge–Kutta method for solving a second-order ODE
C.16: Finite difference method for a second order-ODE
C.17: Forward time central space (FTCS) method for a parabolic PDE
C.18: Crank–Nicolson method for a parabolic PDE
C.19: SOR method for an elliptic PDE
C.20: Solution of the wave equation

C.1

```
C       SEARCH METHOD FOR FINDING THE ROOTS OF AN ALGEBRAIC
C       EQUATION
C
C       THIS PROGRAM FINDS THE REAL ROOTS OF THE EQUATION
C       F(X)=0
C       BY THE SEARCH METHOD
C
C       HERE X IS THE UNKNOWN, XMAX IS THE MAXIMUM VALUE OF X,
C       F1,F2 ARE THE VALUES OF THE FUNCTION F(X) AT TWO
C       CONSECUTIVE X VALUES, DX IS THE INCREMENT IN X, AND
C       EPS IS THE CONVERGENCE CRITERION ON X
C
C        DEFINE GIVEN FUNCTION F(X)
C
        F(X)=0.8*5.67E-8*(1000.0**4.0-X**4.0)-50.0*(X-500.0)
C     $ -(25.0/0.15)*(X-300.0)
```

```
C
C      SPECIFY INITIAL PARAMETERS
C
        EPS=10.0
        DO 14 I=1,6
        XMAX=1000.0
        X=300.0
        DX=50.0
        WRITE(6,16)EPS
   16   FORMAT(2X,'EPS=',F10.5/)
        F1=F(X)
    4   X=X+DX
        F2=F(X)
        A=F1*F2
C
C      CHECK FOR CHANGE IN SIGN OF F(X)
C
        IF(A .LT. 0.0)THEN
        WRITE(6,15)X,F1,F2
C
C      CHECK FOR CONVERGENCE TO THE ROOT
C
        IF(DX .LT. EPS) GO TO 7
        X=X-DX
        DX=DX/10.0
        GO TO 4
        ELSE IF(A .EQ. 0.0) THEN
    7   WRITE(6,10)X,F1
   15   FORMAT(2X,'X=',F10.4,4X,'F1=',F10.4,4X,'F2=',F10.4)
   10   FORMAT(/2X,'TEMPERATURE =',F10.4,4X,'F(X) = ',F10.4//)
        ELSE
        IF(X .GT. XMAX) STOP
        F1=F2
        GO TO 4
        END IF
C
C      VARY CONVERGENGE CRITERION
C
   14   EPS=EPS/10.0
   12   STOP
        END
```

C.2

```
C      ROOT SOLVING WITH THE BISECTION METHOD
C
C      X IS THE INDEPENDENT VARIABLE, FUN(X) IS THE GIVEN
C      FUNCTION, X1 AND X2 ARE THE TWO EXTREME VALUES OF X
C      BOUNDING THE REGION WHICH CONTAINS THE ROOT AT A
```

```
C       GIVEN ITERATION, X3 IS THE APPROXIMATION TO THE ROOT,
C       F1, F2 AND F3 ARE THE CORRESPONDING VALUES OF THE
C       FUNCTION, AND EPS IS THE CONVERGENCE CRITERION
C
C
C       DEFINE THE GIVEN FUNCTION
C
        FUN(X)=ALOG10(X)+X*X-6.0
        EPS=1.0
        DO 4 I=1,5
        X1=2.0
        X2=5.0
    1   F1=FUN(X1)
        F2=FUN(X2)
C
C       COMPUTE APPROXIMATION TO THE ROOT
C
        X3=(X1+X2)/2.0
        F3=FUN(X3)
C
C       CHECK FOR CONVERGENCE
C
        IF (ABS(F3) .LE. EPS) GO TO 2
        IF ((F1*F3) .GE. 0.0) THEN
        X1=X3
        GO TO 1
        ELSE
        X2=X3
        GO TO 1
        END IF
    2   WRITE(6,3)EPS,X3,F3
    3   FORMAT(2X,'EPS=',F8.5,4X,'TERMINAL VELOCITY=',F10.4,4X,
       $ 'FUN(X)=',F8.4)
C
C       VARY CONVERGENCE CRITERION
C
    4   EPS=EPS/10
        STOP
        END
```

C.3

```
C       ROOT SOLVING WITH THE SECANT METHOD
C
C       X IS THE INDEPENDENT VARIABLE, FUN(X) IS THE GIVEN
C       FUNCTION, X1 AND X2 ARE THE X VALUES FROM THE TWO
C       PREVIOUS ITERATIONS, STARTING WITH THE TWO POINTS
C       BOUNDING THE REGION, X3 IS THE APPROXIMATION TO THE
C       ROOT, F1, F2 AND F3 ARE THE CORRESPONDING VALUES OF
```

```
C       THE FUNCTION, AND EPS IS THE CONVERGENCE CRITERION
C
C
C
C       DEFINE FUNCTION
C
        FUN(X) = 0.2275*X*X/(465.9+ALOG(X)**2.58)-0.017*X - 9.8
        X1 = 150.0
        X2 = 200.0
        WRITE(6,12)X1,X2
   12   FORMAT(/10X,'INITIAL X1=',F7.2,10X,'INITIAL X2=',F7.2//)
        EPS = 1.0
        DO 2 I=1,5
    1   F1 = FUN(X1)
        F2 = FUN(X2)
C
C       COMPUTE THE APPROXIMATION TO THE ROOT
C
        X3 = (X1*F2 - X2*F1)/(F2 - F1)
        F3 = FUN(X3)
C
C       CHECK FOR CONVERGENCE
C
        IF (ABS(F3) .GT. EPS) THEN
        X1 = X2
        X2 = X3
        GO TO 1
        ELSE
   11   WRITE(6,13)EPS,X3,F3
   13   FORMAT(2X,'EPS=',F8.5,4X,'TERMINAL VELOCITY=',F10.4,4X,
     $ 'FUN(X) = ',F8.4)
        END IF
C
C       VARY CONVERGENCE CRITERION
C
    2   EPS = EPS/10
        STOP
        END
```

C.4

```
C       NEWTON-RAPHSON METHOD FOR FINDING THE ROOTS OF AN
C       ALGEBRAIC EQUATION
C
C       THIS PROGRAM FINDS THE REAL ROOTS OF AN EQUATION
C       F(X) = 0 BY THE NEWTON-RAPHSON METHOD
```

```
C
C
C
C      HERE X IS THE INDEPENDENT VARIABLE, Y1 THE VALUE OF THE
C      FUNCTION AT X, Y2 THE FUNCTION AT X+0.001, YD THE
C      DERIVATIVE, DX THE INCREMENT IN X FOR THE NEXT
C      ITERATION, EPS THE CONVERGENCE CRITERION ON THE
C      FUNCTION AND XMAX THE MAXIMUM VALUE OF X
C
C
C      DEFINE FUNCTION AND SPECIFY INPUT PARAMETERS
C
       Y(X)=294.0*X*(1.0-EXP(-1000.0/(21.0*(5.0+20.0*X))))-250.0
       EPS=0.001
       WRITE(6,15)EPS
  15   FORMAT(2X,'EPS=',F8.4/)
       X=0.1
       XMAX=5.0
   1   Y1=Y(X)
       WRITE(6,10) X,Y1
C
C      CHECK FOR CONVERGENCE
C
       IF(ABS(Y1) .GT. EPS) THEN
       XN=X+0.001
       Y2=Y(XN)
       YD=(Y2-Y1)/0.001
C
C      CHECK IF RESULTS DIVERGE
C
       IF(YD .GE. (1.0/EPS))GO TO 20
C
C      COMPUTE NEW APPROXIMATION TO THE ROOT
C
       DX=-Y1/YD
       X=X+DX
       IF(X .GE. XMAX)GO TO 20
       GO TO 1
       ELSE
   5   WRITE(6,12) X,Y1
  12   FORMAT(/2X,'FLOW RATE X=',F8.4,4X,'FUNCTION
      $ F(X)=',F12.6)
  10   FORMAT(2X,'X=',F8.4,4X,'FUNCTION F(X)=',F12.6)
       END IF
  20   STOP
       END
```

C.5

```
C       GAUSSIAN ELIMINATION METHOD FOR A SYSTEM OF LINEAR
C       EQUATIONS
C
C       A(I,J) REPRESENTS THE ELEMENTS OF THE AUGMENTED MATRIX
C       BEING REDUCED BY THE GAUSSIAN ELIMINATION METHOD,
C       A1(I,J) ARE THE ELEMENTS OF THE ORIGINAL AUGMENTED
C       MATRIX, X(I) ARE THE UNKNOWN VARIABLES, N IS THE
C       NUMBER OF EQUATIONS, M IS THE NUMBER OF COLUMNS IN
C       THE AUGMENTED MATRIX, K REPRESENTS THE NUMBER OF THE
C       PIVOT ROW AND B(I) REPRESENTS THE CONSTANTS ON THE
C       RIGHT-HAND SIDE OF THE GIVEN SYSTEM OF EQUATIONS
C
C
          PARAMETER (IN=10)
          DIMENSION A(IN,IN+1),A1(IN,IN+1),X(IN)
          PRINT *, 'NUMBER OF EQUATIONS ARE :'
          READ(5,*)N
          M=N+1
C
C       ENTER THE COEFFICIENT MATRIX
C
          PRINT *, 'THE ELEMENTS OF THE ORIGINAL AUGMENTED MATRIX
       $    ARE :'
          READ(5,*)((A(I,J),J=1,M),I=1,N)
          DO 101 J=1,N
               DO 1 I=1,M
                     A1(I,J)=A(I,J)
    1          CONTINUE
  101     CONTINUE
C
C       CALL SUBROUTINE TO SOLVE THE SYSTEM OF EQUATIONS
C
          CALL GAUSS(N,A,X)
          WRITE(6,9)
    9     FORMAT(2X,'THE SOLUTION TO THE EQUATIONS IS:'//)
          DO 10 I=1,N
               WRITE(6,11)I,X(I)
   10     CONTINUE
   11     FORMAT(2X,'X(',I1,') = ',F12.5)
          WRITE(6,12)
   12     FORMAT(//2X,'THE CONSTANT VECTOR OF THE EQUATIONS
       $    IS:'//)
C
C       CALCULATE THE CONSTANT VECTOR B USING THE SOLUTION
C       OBTAINED TO CHECK THE ACCURACY OF THE RESULTS
C
          DO 13 I=1,N
               Y=0.0
               DO 14 J=1,N
```

```
                    Y=Y+X(J)*A1(I,J)
 14               CONTINUE
                  WRITE(6,15)I,Y
 15               FORMAT(2X,'B(',I1,')=',F12.5)
 13         CONTINUE
            STOP
            END
C
C
            SUBROUTINE GAUSS(N,A,X)
            DIMENSION A(10,11),X(10)
            N1=N-1
            M=N+1
C
C     FIND THE ROW WITH THE LARGEST PIVOT ELEMENT
C
            DO 2 K=1,N1
                 K1=K+1
                 K2=K
                 B0=ABS(A(K,K))
                 DO 3 I=K1,N
                      B1=ABS(A(I,K))
                      IF((B0-B1) .LT. 0.0) THEN
                              B0=B1
                              K2=I
                      END IF
 3               CONTINUE
                 IF((K2-K) .NE. 0) THEN
C
C     INTERCHANGE ROWS TO OBTAIN THE LARGEST PIVOT ELEMENT
C
                DO 5 J=K,M
                      C=A(K2,J)
                      A(K2,J)=A(K,J)
 5              A(K,J)=C
               END IF
               DO 2 I=K1,N
C
C     APPLY THE GAUSSIAN ELIMINATION ALGORITHM
C
                DO 6 J=K1,M
                      A(I,J)=A(I,J)-A(I,K)*A(K,J)/A(K,K)
 6              CONTINUE
                A(I,K)=0.0
 2          CONTINUE
C
C     APPLY BACK SUBSTITUTION
C
           X(N)=A(N,M)/A(N,N)
           DO 7 I1=1,N1
                I=N-I1
```

```
          S=0.0
          J1=I+1
          DO 8 J=J1,N
 8              S=S+A(I,J)*X(J)
 7        X(I) = (A(I,M)-S)/A(I,I)
      RETURN
      END
```

C.6

```
C     TRIDIAGONAL MATRIX ALGORITHM
C
      SUBROUTINE TDMA(A,B,C,F,N,T)
C
C     A, B AND C ARE THE THREE ELEMENTS IN EACH ROW, WITH B
C     AT THE DIAGONAL, F IS THE CONSTANT ON THE RIGHT-HAND
C     SIDE OF EACH EQUATION, N IS THE NUMBER OF EQUATIONS
C     AND T IS THE VARIABLE TO BE COMPUTED
C
          DIMENSION A(N),B(N),C(N),F(N),T(N)
C
C     REDUCE THE A'S TO ZERO BY GAUSSIAN ELIMINATION AND
C     DETERMINE THE NEW COEFFICIENTS
C
      NN=N-1
      DO 5 I=2,N
          D=A(I)/B(I-1)
          B(I)=B(I)-C(I-1)*D
 5        F(I)=F(I)-F(I-1)*D
C
C     APPLY BACK SUBSTITUTION
C
      T(N)=F(N)/B(N)
      DO 6 I=1,NN
          J=N-I
 6        T(J) = (F(J)-C(J)*T(J+1))/B(J)
      RETURN
      END
```

C.7

```
C     THE GAUSS-JORDAN ELIMINATION METHOD
C
C
C     A(I,J) REPRESENTS THE ELEMENTS OF THE AUGMENTED MATRIX,
C     X(I) DENOTES THE UNKNOWN VARIABLES, K IS THE NUMBER OF
C     THE PIVOT ROW, N IS THE NUMBER OF EQUATIONS, AND M IS
```

```
C       N+1.
C
          DIMENSION A(10,11),X(10)
          PRINT *, 'NUMBER OF EQUATIONS N IS: '
          READ(5,*)N
          M=N+1
C
C       READ COEFFICIENTS OF THE AUGMENTED MATRIX
C
          PRINT *, 'THE ELEMENTS OF THE AUGMENTED MATRIX ARE: '
          READ(5,*)((A(I,J),J=1,M),I=1,N)
          N1=N-1
          DO 6 K=1,N
          K1=K+1
          K2=K
C
C       SEARCH FOR ROW WITH LARGEST PIVOT ELEMENT
C
          B0=ABS(A(K,K))
          DO 1 I=K,N
          B1=ABS(A(I,K))
          IF((B0-B1) .LT. 0.0)THEN
          B0=B1
          K2=I
          END IF
    1     CONTINUE
C
C       DECIDE IF ROW INTERCHANGE IS NEEDED FOR MAXIMUM PIVOT
C       ELEMENT
C
          IF((K2-K) .NE. 0)THEN
C
C       INTERCHANGE ROW FOR OBTAINING LARGEST PIVOT ELEMENT
C
          DO 2 J=K,M
          C=A(K2,J)
          A(K2,J)=A(K,J)
    2     A(K,J)=C
          END IF
C
C       APPLY GAUSS JORDAN ELIMINATION
C
    3     DO 4 J=K1,M
    4     A(K,J)=A(K,J)/A(K,K)
          A(K,K)=1.0
          DO 6 I=1,N
          IF (I .NE. K) THEN
          DO 5 J=K1,M
    5     A(I,J)=A(I,J)-A(I,K)*A(K,J)
          A(I,K)-0.0
```

```
          END IF
   6      CONTINUE
C
C      DETERMINE THE UNKNOWNS
C
          DO 7 I=1,N
   7      X(I)=A(I,M)
          WRITE(6,8)
   8      FORMAT(2X,'THE SOLUTION TO THE EQUATIONS IS:'//)
          DO 9 I=1,N
   9      WRITE(6,10)I,X(I)
  10      FORMAT(2X,'X(',I1,')=',F12.5)
          WRITE(6,11)
  11      FORMAT(//2X,'THE REDUCED MATRIX IS'//)
          DO 13 I=1,N
          WRITE(6,12)(A(I,J),J=1,N)
  12      FORMAT(10F10.3)
  13      CONTINUE
          STOP
          END
```

C.8

```
C      SOR METHOD FOR SOLVING A SYSTEM OF LINEAR EQUATIONS
C
C
C      T(I) REPRESENTS THE TEMPERATURE DIFFERENCES FROM THE
C      AMBIENT TEMPERATURE, TO(I) DENOTES THE TEMPERATURE
C      DIFFERENCES AFTER THE PREVIOUS ITERATION, TP IS THE
C      ACTUAL TEMPERATURE,S IS A CONSTANT DEFINED IN THE
C      PROBLEM AND N IS THE NUMBER OF EQUATIONS
C
C
C      ENTER VALUES OF RELEVANT PARAMETERS
C
          DIMENSION T(31),TO(31)
          S=(0.071**2)*(1.0**2)+2.0
          W=1.8
          N=29
          NN=N-1
          EPS=0.0001
          T(0)=100.0
          T(30)=100.0
C
C      INPUT STARTING VALUES
C
          J=0
          DO 1 I=1,N
   1      T(I)=0.0
```

```
C
C     STORE COMPUTED VALUES AFTER EACH ITERATION
C
    2   DO 3 I=1,N
    3   TO(I)=T(I)
C
C     COMPUTE THE END VALUES T(1) AND T(N)
C
        T(1)=(T(2)+100.0)/S
        T(N)=(100.0+T(N-1))/S
C
C     COMPUTE INTERMEDIATE VALUES
C
        DO 4 I=2,NN
    4   T(I)=W*(T(I+1)+T(I-1))/S+(1.0-W)*T(I)
C
C     CHECK FOR CONVERGENCE
C
        J=J+1
        DO 5 I=1,N
        IF(ABS(TO(I)-T(I)) .GT. EPS) GO TO 2
    5   CONTINUE
        WRITE(6,6)EPS
    6   FORMAT(//2X,'EPS=',F10.5)
        WRITE(6,7)J
    7   FORMAT(/2X,'NUMBER OF ITERATIONS=',I4/)
C
C     COMPUTE ACTUAL TEMPERATURES
C
        DO 8 I=0,N+1
        TP=T(I)+20.0
    8   WRITE(6,9)I,TP
    9   FORMAT(2X,'TP(',I2,')=',F12.4)
        STOP
        END
```

C.9

```
C     LAGRANGE INTERPOLATION
C
C
C     X IS THE INDEPENDENT VARIABLE AND Y THE DEPENDENT
C     VARIABLE, WITH X(I) AND Y(I) REPRESENTING THE GIVEN
C     DATA POINTS. N IS THE NUMBER OF DATA POINTS, XL THE
C     VALUE OF X AT WHICH INTERPOLATION IS DESIRED AND YL THE
C     CORRESPONDING COMPUTED VALUE OF Y AT X=XL. A(I)
C     REPRESENTS THE COEFFICIENTS OF THE LAGRANGE POLYNOMIAL
C     AND M IS THE NUMBER OF POINTS
C     AT WHICH INTERPOLATED VALUES ARE NEEDED.
```

```
C
C
      DIMENSION X(10),Y(10),A(10)
C
C     ENTER THE GIVEN DATA
C
      READ(5,*)N
      READ(5,*)M
      READ(5,*)(X(I),I=1,N)
      READ(5,*)(Y(I),I=1,N)
      WRITE(6,10)
 10   FORMAT(2X,'THE VALUES FROM LAGRANGE INTERPOLATION
     $ ARE:'//)
      DO 6 K=1,M
      READ(5,*)XL
C
C     COMPUTE THE COEFFICIENTS OF THE LAGRANGE POLYNOMIAL
C
      DO 2 J=1,N
      A(J)=Y(J)
      DO 1 I=1,N
      IF(I .NE. J) THEN
      A(J)=A(J)/(X(J)-X(I))
      END IF
 1    CONTINUE
 2    CONTINUE
C
C     CALCULATE THE INTERPOLATED VALUE OF THE DEPENDENT
C     VARIABLE
C
      YL=0.0
      DO 4 J=1,N
      S=1.0
      DO 3 I=1,N
      IF(I .NE. J) THEN
      S=S*(XL-X(I))
      END IF
 3    CONTINUE
 4    YL=YL+S*A(J)
C
C     PRINT THE CALCULATED RESULTS
C
      WRITE(6,5)XL,YL
 5    FORMAT(2X,'XL=',F9.4,4X,'YL=',F9.4)
 6    CONTINUE
      WRITE(6,7)
 7    FORMAT(//2X,'COEFFICIENTS OF THE LAGRANGE POLYNOMIAL
     $ ARE:')
      DO 9 I=1,N
      WRITE(6,8)I,A(I)
 8    FORMAT(/4X,'A(',I1,')=',F9.4)
```

```
    9     CONTINUE
          STOP
          END
```

C.10

```
C      SPLINE INTERPOLATION
C
C      V IS THE INDEPENDENT VARIABLE, T THE DEPENDENT
C      VARIABLE, M THE NUMBER OF DATA POINTS, T2 THE SECOND
C      DERIVATIVE OF THE DEPENDENT VARIABLE, VP THE VALUE OF V
C      AT WHICH THE INTERPOLATED VALUE TP IS DESIRED AND
C      V(I),T(I) REPRESENT THE VALUES AT THE DATA POINTS.
C
C
          DIMENSION V(15),T(15),T2(15)
C
C      ENTER INPUT VARIABLES AND DATA
C
          PRINT *,'ENTER THE NUMBER OF DATA POINTS'
          READ *,M
          OPEN(UNIT=11,FILE='V.DAT')
          OPEN(UNIT=12,FILE='T.DAT')
          READ (11,*) (V(I),I=1,M)
          READ (12,*) (T(I),I=1,M)
          CLOSE(UNIT=11)
          CLOSE(UNIT=12)
C
C      CALL SUBROUTINE TO COMPUTE THE SECOND DERIVATIVE T2
C
          CALL DERIVATIVE(M,V,T,T2)
C
C      SPECIFY VALUE OF V FOR INTERPOLATION
C
    2     PRINT *,'ENTER THE VALUE OF V FOR INTERPOLATION'
          READ *,VP
C
C      CALL SUBROUTINE TO USE SPLINE INTERPOLATION
C
          CALL SPLINE(M,V,T,T2,VP,TP)
C
C      OUTPUT RESULTS
C
          WRITE(6,4)VP,TP
    4     FORMAT(2X,'VOLTAGE V=',F9.5,4X,'TEMPERATURE
     $      T=',F9.5//)
          PRINT *,'IF YOU WANT ADDITIONAL INTERPOLATION, TYPE 1'
          READ *,MORE
          IF (MORE .EQ. 1) GO TO 2
          STOP
          END
```

```
C
C
C                               SUBROUTINE DERIVATIVE
C
C      THIS SUBROUTINE CALCULATES THE SECOND DERIVATIVE VALUES
C      T2 NEEDED FOR A CUBIC SPLINE INTERPOLATION. A,B AND C
C      ARE THE ELEMENTS IN EACH ROW OF THE TRIDIAGONAL MATRIX
C      AND D REPRESENTS THE CONSTANTS ON THE RIGHT-HAND SIDE
C      OF THE EQUATIONS THAT YIELD THE T2 VALUES.
C
C
        SUBROUTINE DERIVATIVE(M,V,T,T2)
        DIMENSION V(15),T(15),T2(15),A(15),B(15),C(15),D(15)
C
C      COMPUTE THE ELEMENTS OF THE TRIDIAGONAL MATRIX
C
        C(1)=V(2)-V(1)
        DO 1 I=2,M-1
        A(I)=V(I)-V(I-1)
        B(I)=2.0*(V(I+1)-V(I-1))
        C(I)=V(I+1)-V(I)
    1   D(I)=6.0*((T(I+1)-T(I))/C(I)-(T(I)-T(I-1))/A(I))
C
C      SOLVE THE TRIDIAGONAL SYSTEM FOR THE SECOND DERIVATIVE
C
        DO 2 I=3,M-1
        B(I)=B(I)-A(I)*C(I-1)/B(I-1)
    2   D(I)=D(I)-A(I)*D(I-1)/B(I-1)
        T2(1)=0.0
        T2(M)=0.0
        T2(M-1)=D(M-1)/B(M-1)
        DO 3 I=2,M-2
        IN=M-I
    3   T2(IN)=(D(IN)-C(IN)*T2(IN+1))/B(IN)
        RETURN
        END
C
C
C                               SUBROUTINE SPLINE
C
C      THIS SUBROUTINE OBTAINS THE RELEVANT CUBIC SPLINE AND
C      COMPUTES THE DESIRED INTERPOLATED VALUE OF THE
C      DEPENDENT VARIABLE
C
        SUBROUTINE SPLINE(M,V,T,T2,VP,TP)
        DIMENSION V(15),T(15),T2(15)
C
C      DETERMINE THE INTERVAL IN WHICH VP LIES
C
```

```
      DO 1 I=1,M-1
      IF (VP .LE. V(I+1)) THEN
      S1=V(I+1)-V(I)
      S2=VP-V(I)
      S3=V(I+1)-VP
C
C     COMPUTE THE INTERPOLATED VALUE FROM THE CUBIC SPLINE
C
      TP=T2(I)*S3*(S3**2/S1-S1)/6.0+T2(I+1)*S2
     $            *(S2**2/S1-S1)/6.0+T(I)*S3/S1+T(I+1)*S2/S1
      GO TO 2
      END IF
    1 CONTINUE
    2 RETURN
      END
```

C.11

```
C     LEAST-SQUARES METHOD FOR POLYNOMIAL REGRESSION
C
C
      DIMENSION A(10,11),C(10),X(25),Y(25)
C
C     ENTER THE INPUT DATA
C
      OPEN(UNIT=15,FILE='REGRES.DAT')
      READ(15,*)MP
      READ(15,*)ND
      READ(15,*)(X(I),I=1,ND)
      READ(15,*)(Y(I),I=1,ND)
      N=MP+1
      M=N+1
C
C     INITIALIZE THE COEFFICIENT MATRIX
C
      DO 1 I=1,N
      DO 1 J=1,M
    1 A(I,J)=0.0
C
C     COMPUTE ELEMENTS OF THE AUGMENTED MATRIX
C
      DO 5 I=1,N
      DO 3 J=1,N
      L=I+J-2
      DO 2 K=1,ND
    2 A(I,J)=A(I,J)+X(K)**L
    3 CONTINUE
      DO 4 K=1,ND
    4 A(I,M)=A(I,M)+Y(K)*X(K)**(I-1)
```

```
    5     CONTINUE
C
C     CALL SUBROUTINE TO SOLVE THE SYSTEM OF EQUATIONS
C
          CALL GAUSS(N,A,C)
          WRITE(6,12)MP
   12     FORMAT(2X,'THE ORDER OF THE POLYNOMIAL=',I2/)
          WRITE(6,9)
    9     FORMAT(2X,'THE CONSTANTS OF THE POLYNOMIAL ARE:'/)
          DO 10 I=1,N
   10     WRITE(6,11)I,C(I)
   11     FORMAT(2X,'C(',I1,')=',F12.5)
C
C     CALCULATE THE VALUES OBTAINED FROM THE POLYNOMIAL IN
C     ORDER TO CHECK THE ACCURACY OF THE RESULTING BEST FIT
C
          WRITE(6,13)
   13     FORMAT(/2X,'THE VALUES CALCULATED FROM THE BEST FIT
     $     ARE:'/)
          DO 7 I=1,ND
          Y(I)=0.0
          DO 6 J=1,N
    6     Y(I)=Y(I)+C(J)*X(I)**(J-1)
    7     WRITE(6,8)I,X(I),I,Y(I)
    8     FORMAT(2X,'X(',I2,')=',F10.4,5X,'Y(',I1,')=',
     $     F10.4)
          CLOSE(UNIT=15)
          STOP
          END
C
C
          SUBROUTINE GAUSS(N,A,C)
          DIMENSION A(10,11),C(10)
          N1=N-1
          M=N+1
C
C     FIND THE ROW WITH THE LARGEST PIVOT ELEMENT
C
          DO 2 K=1,N1
          K1=K+1
          K2=K
          B0=ABS(A(K,K))
          DO 3 I=K1,N
          B1=ABS(A(I,K))
          IF((B0-B1) .LT. 0.0) THEN
          B0=B1
          K2=I
          END IF
    3     CONTINUE
          IF((K2-K) .NE. 0) THEN
C
```

```
C       INTERCHANGE ROWS TO OBTAIN THE LARGEST PIVOT ELEMENT
C
           DO 5 J=K,M
           D=A(K2,J)
           A(K2,J)=A(K,J)
    5      A(K,J)=D
           END IF
           DO 2 I=K1,N
C
C       APPLY THE GAUSSIAN ELIMINATION ALGORITHM
C
           DO 6 J=K1,M
    6      A(I,J)=A(I,J)-A(I,K)*A(K,J)/A(K,K)
    2      A(I,K)=0.0
C
C       APPLY BACK SUBSTITUTION
C
           C(N)=A(N,M)/A(N,N)
           DO 7 I1=1,N1
           I=N-I1
           S=0.0
           J1=I+1
           DO 8 J=J1,N
    8      S=S+A(I,J)*C(J)
    7      C(I)=(A(I,M)-S)/A(I,I)
           RETURN
           END
```

C.12

```
C       TRAPEZOIDAL RULE FOR NUMERICAL INTEGRATION
C
C       F(X) IS THE FUNCTION TO BE INTEGRATED AND REPRESENTS
C       THE ELECTRIC CURRENT AS A FUNCTION OF TIME T IN
C       SECONDS, V IS THE VOLTAGE, Q IS THE ELECTRICAL CHARGE
C       IN COULOMBS, C IS THE CAPACITANCE IN FARADS, DT IS THE
C       TIME STEP, N IS THE NUMBER OF SUBDIVISIONS, AND TMIN
C       AND TMAX ARE THE MINIMUM AND MAXIMUM VALUES OF T.
C
           IMPLICIT REAL (A-H,O-Z)
C
C       DEFINE FUNCTION TO BE INTEGRATED
C
           F(X)=4.0*(1.0-EXP(-0.5))*(EXP(-0.5* (X-1.0)))
        $          *(1.0-EXP(-X))
C
C       ENTER INPUT VALUES
C
           TMIN=1.0
           C=0.025
```

```
      PRINT *,'ENTER THE STEP SIZE DT'
      READ *,DT
      DO 6 J=1,6
      WRITE(6,7)DT
7     FORMAT(//5X,'STEP SIZE DT=',F7.5)
      WRITE(6,1)
1     FORMAT(/6X,'TIME T',19X,'CHARGE Q',17X,'VOLTAGE V')
      WRITE(6,2)
2     FORMAT(5X,8('-'),17X,10('-'),15X,11('-')/)
C
C    VARY TIME AT WHICH CHARGE IS TO BE COMPUTED
C
      DO 3 TMAX=2.0,20.0,2.0
      N=(TMAX-TMIN)/DT
C
C    COMPUTE SUM OF INTERIOR ORDINATES FOR TRAPEZOIDAL RULE
C
      SUM=0.0
      T=TMIN+DT
      DO 4 I=1,N-1
      SUM=SUM+F(T)
      T=T+DT
4     CONTINUE
C
C    APPLY TRAPEZOIDAL RULE
C
      Q=(DT/2.0)*(F(TMIN)+2.0*SUM+F(TMAX))
      V=Q/C
      WRITE(6,5)TMAX,Q,V
5     FORMAT(5X,F5.2,21X,E9.4,16X,E9.4)
3     CONTINUE
6     DT=DT/2
      STOP
      END
```

C.13

```
C     ROMBERG INTEGRATION
C
C     F(X) IS THE FUNCTION TO BE INTEGRATED, X THE
C     INDEPENDENT VARIABLE, XMIN AND XMAX THE MINIMUM AND
C     MAXIMUM VALUES OF X, DX THE SEGMENT WIDTH, ERF(Z) THE
C     ERROR FUNCTION AT Z, EPS THE CONVERGENCE CRITERION, Y
C     THE VALUE OF THE INTEGRAL CORRESPONDING TO AN
C     EXTRAPOLATION, M THE NUMBER OF SEGMENTS,
C     AND DIF THE DIFFERENCE BETWEEN THE RESULTS FOR THE TWO
C     HIGHEST ORDERS OF EXTRAPOLATION AT A GIVEN NUMBER OF
C     SEGMENTS.
C
      DIMENSION Y(8,8)
```

```
C       DEFINE FUNCTION TO BE INTEGRATED
C
          F(X) = (2.0/SQRT(3.14159))*EXP(-X**2)
C
C       ENTER INPUT VALUES
C
          EPS = 0.00001
          XMIN = 0.0
          DO 7 J = 1,4
              PRINT *,'ENTER THE VALUE OF Z'
              READ *,XMAX
              DX = XMAX - XMIN
C
C       FIRST ORDER (TRAPEZOIDAL RULE) CALCULATION
C
              N = 1
              Y(1,1) = 0.5*DX*(F(XMIN) + F(XMAX))
              WRITE(6,1)N,Y(1,1)
    1         FORMAT(2X, 'NO. OF ITERATIONS = ', I2, 5X, 'ERF(Z)
     $=', F9. 6)
    2     M = 2**(N-1)
          DX = DX/2.0
          N = N+1
          Y(1,N) = 0.5*Y(1,N-1)
          DO 3 K = 1,M
              X = XMIN+ (2*K-1)*DX
    3     Y(1,N) = Y(1,N) +DX*F(X)
C
C       COMPUTE HIGHER ORDER EXTRAPOLATIONS
C
          DO 4 K = 2,N
    4     Y(K,N) = (4.0**(K-1)*Y(K-1,N) - Y(K-1,N-1))
     $                  /(4.0**(K-1) - 1.0)
          DIF = ABS(Y(N,N) - Y(N-1,N))
          WRITE(6,5)N,Y(N,N)
    5     FORMAT(2X,'NO. OF ITERATIONS = ',I2,5X,
     $            'ERF(Z) = ',F9.6)
C
C       APPLY CONVERGENCE CRITERION
C
          IF(DIF .GT. EPS)THEN
              IF(N .LT. 8)THEN
                  GO TO 2
              ELSE
                  PRINT *,'MORE THAN 8 ITERATIONS'
              END IF
          ELSE
```

```
        WRITE(6,6)XMAX,Y(N,N)
 6      FORMAT(/4X,'VALUE OF Z=',F5.3,5X,'ERF(Z)=',
     $    F9.6//)
        END IF
 7      CONTINUE
        STOP
        END
```

C.14

```
C       EULER'S METHOD FOR SOLVING AN ODE
C
C       THIS PROGRAM NUMERICALLY SOLVES A FIRST ORDER
C       DIFFERENTIAL EQUATION USING EULER'S METHOD
C
C
C
C
C       IN THE FOLLOWING PROGRAM
C
C       T STANDS FOR TIME T     EE STANDS FOR E.M.F. OF THE
     $  BATTERY
C
C       EI STANDS FOR CURRENT   ER STANDS FOR RESISTANCE
C
C       EPS IS THE CONVERGENCE CRITERION
C
C       DT IS STEP SIZE IN T    EL STANDS FOR INDUCTANCE
C
C
        IMPLICIT REAL (A-H,O-Z)
        OPEN(UNIT=10,FILE='ET')
        OPEN(UNIT=11,FILE='EI')
C
C       FILE 'ET' CONTAINS VALUES OF TIME T
C       FILE 'EI' CONTAINS VALUES OF EI AT CORRESPONDING T
C
C         INPUT PARAMETERS
          PRINT*,'INPUT PARAMETERS'
          PRINT*,'EE=','ER=','EL ='
          READ*,EE,ER,EL
          PRINT*,'STEP SIZE DT ='
          READ*,DT
          PRINT*,'CONVERGENCE CRITERION EPS ='
          READ*,EPS
C
C       SET INITIAL CONDITIONS:
C
          T=0.0
```

```
         EI = 0.
         WRITE(10,*)T
         WRITE(11,*)EI
C
C     CALCULATIONS FOR THE NEXT STEP USING EULER'S METHOD
C
  11     T = T + DT
         EI = EI + DT*(EE/EL - ER*EI/EL)
C
C     AT T = 0.5 SEC. THE E.M.F. IS REMOVED FROM THE CIRCUIT
C
         IF(T.GT.0.5)GO TO 99
C
         WRITE(10,*)T
         WRITE(11,*)EI
         GO TO 11
  99     PRINT*,'T = ',T,'EI = ',EI
         PRINT*,'AT THIS STAGE E.M.F. IS REMOVED FROM THE
     $   CIRCUIT'
 100     EI = EI + DT*(-ER*EI/EL)
         IF(EI.LE.EPS) GO TO 199
         WRITE(10,*)T
         WRITE(11,*)EI
         T = T + DT
         GO TO 100
 199     PRINT*,'TOTAL TIME FOR EI TO BECOME LESS THAN
     $   EPS= ',T,' SEC.'
         PRINT*,'THE VALUES OF TIME T ARE IN THE FILE ET'
         PRINT*,'THE VALUES OF CURRENT EI ARE IN THE FILE EI'
         PRINT*,'*************'
         CLOSE(10)
         CLOSE(11)
         STOP
         END
```

C.15

```
C     RUNGE-KUTTA METHOD FOR A SECOND-ORDER ODE
C
C     THIS PROGRAM NUMERICALLY SOLVES A SECOND ORDER
C     DIFFERENTIAL EQUATION USING THE 4TH ORDER RUNGE-KUTTA
C     METHOD
C
C     IN THE FOLLOWING PROGRAM
C
C       T STANDS FOR TIME          DT STANDS FOR STEP SIZE IN T
C
C       X STANDS FOR DISPLACEMENT  V STANDS FOR VELOCITY
C
```

```
C         A AND B ARE THE CONSTANTS APPEARING IN THE
C         DIFFERENTIAL EQUATION
C
C         G IS THE ACCELERATION DUE TO GRAVITY=9.8 M/(SEC**2)
C
          IMPLICIT REAL (A-H,O-Z)
          OPEN(UNIT=14,FILE='RT')
          OPEN(UNIT=15,FILE='RX')
          OPEN(UNIT=16,FILE='RV')
C
C      VALUES OF T ARE WRITTEN IN FILE RT
C      VALUES OF X ARE WRITTEN IN FILE RX
C      VALUES OF V ARE WRITTEN IN FILE RV
C
C
C      INPUT PARAMETERS
          PRINT*,'INPUT PARAMETERS'
          PRINT*,'A= ','B= '
          READ*,A,B
          PRINT*,'DT= '
          READ*,DT
          G=9.8
C
C      SET INITIAL CONDITIONS
C
          T=0.
          X=0.
          V=100.
          WRITE(14,*)T
          WRITE(15,*)X
          WRITE(16,*)V
C
C      NEXT TIME STEP
C
   11     Q=X
          Z=V
C
C      Q AND Z ARE VALUES OF X AND V RESPECTIVELY, AT PREVIOUS
C      TIME STEP
C
C      CALCULATIONS FOR THE NEXT STEP USING 4TH ORDER
C      RUNGE-KUTTA METHOD
C
          RK1X=DT*Z
          RK1V=DT*(-G -A*Z -B*(Z**2))
          RK2X=DT*(Z+RK1V/2.)
          RK2V=DT*(-G -A*(Z+RK1V/2.) -B*(Z+RK1V/2.)**2)
          RK3X=DT*(Z+RK2V/2.)
          RK3V=DT*(-G -A*(Z+RK2V/2.) -B*(Z+RK2V/2.)**2)
          RK4X=DT*(Z+RK3V)
          RK4V=DT*(-G -A*(Z+RK3V) -B*(Z +RK3V)**2)
```

```
      X=Q +(RK1X +2.*RK2X+2.*RK3X+RK4X)/6.
      V=Z +(RK1V+2.*RK2V+2.*RK3V+RK4V)/6.
      T=T+DT
C
C     CALCULATIONS ARE STOPPED WHEN V BECOMES ZERO.
C
      IF(V.GT.0.) THEN
      WRITE(14,*)T
      WRITE(15,*)X
      WRITE(16,*)V
      GO TO 11
      END IF
C
C     OUTPUT RESULTS
C
      PRINT*,'THE VELOCITY HAS BECOME ZERO OR NEGATIVE'
      PRINT*,'TOTAL TIME TAKEN TO REACH MAXIMUM HEIGHT =',
     $ T,'SEC'
      PRINT*,'TOTAL HEIGHT REACHED BY THE
        PROJECTILE=',X,'METERS'
      CLOSE(UNIT=14)
      CLOSE(UNIT=15)
      CLOSE(UNIT=16)
      STOP
      END
```

C.16

```
C     FINITE DIFFERENCE METHOD FOR A SECOND-ORDER ODE
C
C     THIS PROGRAM SOLVES A SECOND ORDER ORDINARY
C     DIFFERENTIAL EQUATION
C
C     USING THE FINITE DIFFERENCE METHOD
C
C     NOMENCLATURE:
C
C     X: DIMENSIONLESS X COORDINATE
C     TH : ARRAY FOR TEMPERATURE (DIMENSIONLESS)
C     P : CONSTANT APPEARING IN THE DIFFERENTIAL EQUATION
C     A : LOWER DIAGONAL OF THE TRIDIAGONAL MATRIX
C     B : MAIN DAIGONAL OF THE TRIDIAGONAL MATRIX
C     C : UPPER DIAGONAL OF THE TRIDIAGONAL MATRIX
C     D : ARRAY FOR RIGHT HAND SIDE COLUMN MATRIX
C     T : ARRAY CONTAINING SOLUTIONS OF THE TRIDIAGONAL
C         SYSTEM
C     N : NUMBER OF GRID POINTS
C     DX : GRID SIZE
C
```

```
C      EQUATION SOLVED: (TH)'' = (P*P)*(TH) FIN PROBLEM
C
         IMPLICIT REAL (A-H,O-Z)
         PARAMETER(N=51)
         DIMENSION A(N-2),B(N-2),C(N-2),D(N-2),T(N-2),TH(N)
         OPEN(UNIT=50,FILE='IX')
         OPEN(UNIT=51,FILE='ITH')
C
C      FILE IX CONTAINS VALUES OF X
C      FILE ITH CONTAINS VALUES OF TEMPERATURE, TH
C
C      INPUT PARAMETERS
C
         PRINT*,'P='
         READ*,P
         DX=1./(N-1)
         PRINT*,'DX=',DX
C
C      VALUE OF N CAN BE CHANGED. THE DIMENSION STATEMENT
C      SHOULD BE MODIFIED ACCORDINGLY.
C
C      SUBROUTINE 'BC' PROVIDES THE BOUNDARY CONDITIONS
C
         CALL BC(TH,N)
C
C      'FMTDM' FORMS THE TRIDIAGONAL MATRIX AND THE RIGHT HAND
C      SIDE COLUMN MATRIX
C
         CALL FMTDM(DX,P,N,TH,A,B,C,D)
C
C      THE TRIAGINAL MATRIX THUS GENERATED IS OF DIMENSION N-2
C
C      THE SUBROUTINE 'TRIDIAG' SOLVES THE TRIDIAGONAL SYSTEM.
C      THE SOLUTIONS ARE RETRIEVED IN THE ARRAY 'T'.
C
         CALL TRIDIAG(A,B,C,D,T,N-2)
C
C      BACK SUBSTITUTION FROM MATRIX 'T' TO 'TH'
C
         DO 5 I=2,N-1
            TH(I)=T(I-1)
  5      CONTINUE
C
C      CALCULATIONS OVER
C
         X=-DX
         DO 6 I=1,N
            X=X+DX
            WRITE(50,*)X
            WRITE(51,*)TH(I)
  6      CONTINUE
```

```
      PRINT *,'THE VALUES OF X ARE STORED IN FILE IX'
      PRINT *, 'THE VALUES OF TEMPERATURE ARE STORED IN FILE
     $ ITH'
      STOP
      END
C
C     THE FOLLOWING SUBROUTINE FORMS THE TRIDIAGONAL MATRIX
C     OF THE FORM
C
C     A*T(I-1)+B*T(I)+C*T(I+1)=R
C
C
      SUBROUTINE FMTDM(P,DX,N,T,A,B,C,R)
      DIMENSION T(N),A(N),B(N),C(N),R(N)
      A(1)=0.0
      DO 1 I=2,N-2
         A(I)=1.0
 1    CONTINUE
      DO 2 I=1,N-3
         C(I)=1.0
 2    CONTINUE
      C(N-2)=0.0
      DO 3 I=1,N-2
         B(I)= -(2.0 +(P**2)*(DX**2))
 3    CONTINUE
      R(1)= -T(1)
      R(N-2)= -T(N)
      DO 4 I=2,N-3
         R(I)=0.0
 4    CONTINUE
      RETURN
      END
C
C     THE FOLLOWING SUBROUTINE SOLVES THE TRIDIAGONAL SYSTEM
C     USING THE THOMAS ALGORITHM
C
C     A,B,C ARE THE DIAGONALS AS MENTIONED IN THE MAIN
C     PROGRAM.
C     F CONTAINS THE RIGHT HAND SIDE. T CONTAINS SOLUTIONS.
C
      SUBROUTINE TRIDIAG(A,B,C,F,T,M)
      DIMENSION A(M),B(M),C(M),F(M),T(M)
      PRINT*,'SOLVING TRIDIAG'
C
      DO 2 I=2,M
         D=A(I)/B(I-1)
         B(I)= B(I) -C(I-1)*D
         F(I)=F(I)-F(I-1)*D
 2    CONTINUE
      T(M)=F(M)/B(M)
      DO 3 I=1,M-1
```

```
         J=M-I
         T(J)=(F(J) - C(J)*T(J+1))/B(J)
    3  CONTINUE
       RETURN
       END
C
C     SET THE BOUNDARY CONDITIONS
C
       SUBROUTINE BC(T,IL)
       DIMENSION T(IL)
       T(1)=1.0
       T(IL)=0.5
       RETURN
       END
```

C.17

```
C     FORWARD TIME CENTRAL SPACE (FTCS) METHOD FOR A
C     PARABOLIC PDE
C
C     THIS PROGRAM SOLVES A PARABOLIC EQUATION BY THE FTCS
C     METHOD
C
C     WHEN THE PROGRAM IS RUN, IT PROMPTS FOR THE INPUT
C     VALUES REQUIRED. TYPE IN THE INPUT VALUES AND THE
C     OUTPUT WILL BE STORED IN A FILE CALLED 'FTCS.DAT'.
C
C
C     DESCRIPTION OF THE INPUT PARAMETERS:
C
C     IL    IS THE NUMBER OF GRID POINTS.
C     DX    IS THE GRID SIZE.
C     TINIT IS THE INITIAL VALUE OF THE SOLUTION VECTOR,
C           THETA, TAKEN AS UNIFORM OVER THE WHOLE DOMAIN.
C     NLIM IS THE MAXIMUM NUMBER OF TIME STEPS BEFORE
C           STOPPING.
C     NSTEP IS THE NUMBER OF TIME STEPS AFTER WHICH PRINTOUT
C           OCCURS.
C
C
C     DESCRIPTION OF OTHER VARIABLES USED:
C
C     T     IS THE SOLUTION, THETA, AT THE NTH TIME STEP.
C     TOL   IS THE SOLUTION, THETA, AT THE (N-1)TH TIME STEP.
C     DT    IS THE TIME STEP USED. THE PROGRAM USES THE
C           MAXIMUM TIME STEP ALLOWED FROM STABILITY
C           CONSIDERATIONS.
C
```

```
C      ENTER INPUT PARAMETERS
C
          IMPLICIT REAL*8(A-H,O-Z)
          DIMENSION T(50),TOL(50)
          PRINT*,'ENTER NO. OF GRID POINTS, IL='
          READ(5,*)IL
          PRINT*,'ENTER GRID SIZE, DX='
          READ(5,*)DX
          PRINT*,'ENTER INITIAL VALUE OF CONCENTRATION TAKEN AS'
          PRINT*,'UNIFORM OVER THE WHOLE DOMAIN'
          READ(5,*)TINT
          PRINT*,'ENTER MAXIMUM NO. OF TIME STEPS BEFORE
     $    STOPPING'
          READ(5,*)NLIM
          PRINT*,'ENTER NO. OF TIME STEPS AFTER WHICH PRINTOUT
     $    OCCURS'
          READ(5,*)NSTEP
          ISTEP1=0
          ISTEP2=0
          TIME=0.
C
C      OPEN FILES FOR STORING NUMERICAL RESULTS
C
          OPEN(UNIT=10,FILE='FTCS.DAT')
C
C      SET THE INITIAL CONDITIONS
C
          DO 10 I=1,IL
          T(I)=TINT
          TOL(I)=TINT
   10     CONTINUE
C
C      CALCULATE THE MAXIMUM POSSIBLE TIME STEP TO AVOID
C      INSTABILITY
C
          DT=DX**2/2.
          PRINT*,'TIME STEP=',DT
          WRITE(10,120)DX,DT
          WRITE(10,130)IL
          WRITE(10,140)TIME
          WRITE(10,150)(T(I),I=1,IL)
C
C      INCREMENT THE ITERATION COUNTER AND CHECK IF THE CHOSEN
C      MAXIMUM NUMBER OF ITERATIONS IS EXCEEDED.
C
   15     ISTEP1=ISTEP1+1
          ISTEP2=ISTEP2+1
          TIME=TIME+DT
          IF(ISTEP1.GT.NLIM)GO TO 50
C
C      SAVE THE SOLUTION AT THE PREVIOUS TIME STEP
```

```
C
         DO 20 I=1,IL
         TOL(I)=T(I)
  20     CONTINUE
C
C      APPLY FTCS SCHEME AT INTERIOR POINTS
C
         DO 30 I=2,IL-1
         T(I)=TOL(I)+DT*(TOL(I+1)-2.*TOL(I)+TOL(I-1))/DX**2
  30     CONTINUE
C
C      APPLY BOUNDARY CONDITIONS
C
         T(1)=1.
         T(IL)=0.
C
C      OUTPUT THE RESULTS
C
         IF(ISTEP2.EQ.NSTEP)THEN
         WRITE(10,140)TIME
         WRITE(10,150)(T(I),I=1,IL)
         ISTEP2=0
         GO TO 15
         END IF
         GO TO 15
 120     FORMAT(/,4X,'DX=',F4.2,4X,'DT=',F4.2)
 130     FORMAT(//,4X,'IL=',I3)
 140     FORMAT(/,1X,'AT T=',F7.3,1X,'CONCENTRATION FIELD IS:')
 150     FORMAT(1X,20(F8.4,2X))
  50     CLOSE(UNIT=10)
         STOP
         END
```

C.18

```
C      CRANK-NICOLSON METHOD FOR A PARABOLIC PDE
C
C
C      THIS PROGRAM SOLVES 1D, UNSTEADY HEAT EQUATION BY
C      EMPLOYING IMPLICIT CRANK-NICOLSON SCHEME. EQUATION
C      SOLVED IS THE ONE IN EXAMPLE 10.2.
C
C      THE OUTPUT WILL BE IN CN.DAT
C
C      SUBROUTINE 'FMTDIG' FORMS THE TRIDIAGONAL MATRIX.
C      'TDIG'INVERTS THE MATRIX AND SOLVES FOR TEMPERATURE.
C
C
C      DESCRIPTION OF INPUT PARAMETERS:
```

```
C
C       IL    NUMBER OF GRID POINTS.
C       DX    DIMENSIONLESS GRID SIZE.
C       DT    DIMENSIONLESS TIME STEP.
C       TINT  THE INITIAL CONDITIONS TAKEN AS UNIFORM OVER THE
C             WHOLE DOMAIN.
C       NLIM  THE MAXIMUM NUMBER OF TIME STEPS TAKEN BEFORE
              STOPPING.
C       NSTEP THE NUMBER OF TIME STEPS AFTER WHICH PRINTOUT
              OCCURS.
C
C
C       DESCRIPTION OF OTHER VARIABLES USED:
C
C       T     THE DIMENSIONLESS SOLUTION AT NTH TIME STEP.
C       TOL   THE DIMENSIONLESS SOLUTION AT (N-1)TH TIME STEP.
C
C
         PARAMETER (IN=50)
         DIMENSION T(IN),TOL(IN)
         DIMENSION A(IN),B(IN),C(IN),R(IN),SOLN(IN)
         PRINT*,'ENTER NUMBER OF GRID POINTS, IL='
         READ(5,*)IL
         PRINT*,'ENTER GRID SIZE, DX='
         READ(5,*)DX
         PRINT*,'ENTER TIME STEP,DT='
         READ(5,*)DT
         PRINT*,'ENTER INITIAL CONDITIONS, TAKEN AS UNIFORM'
         PRINT*,'OVER THE WHOLE DOMAIN'
         READ(5,*)TINT
         PRINT*,'ENTER MAXIMUM NO. OF TIME STEPS BEFORE
     $   STOPPING'
         READ(5,*)NLIM
         PRINT*,'ENTER NO. OF TIME STEPS AFTER WHICH PRINTOUT
     $   OCCURS'
         READ(5,*)NSTEP
C
C       OPEN THE OUTPUT FILE
C
         OPEN(UNIT=10,FILE='CN.DAT')
         WRITE(10,100)DX,DT
  100    FORMAT(/,4X,'DX=',F4.2,2X,'DT=',F4.2)
         WRITE(10,110)IL
  110    FORMAT(/,4X,'IL=',I3,//)
         ISTEP1=0
         ISTEP2=0
         TIME=0.
C
C       SET THE INITIAL CONDITION
C
```

```
      DO 10 I=1,IL
        T(I)=TINT
        TOL(I)=TINT
10    CONTINUE
C
C     SET THE BOUNDARY CONDITIONS
C
      CALL BCOND(T,DX,DT,IL)
      WRITE(10,120)TIME
      WRITE(10,130)(T(I),I=1,IL)
C
C     SOLVE FOR T ON INTERIOR POINTS AT NTH TIME STEP
C
C     INCREMENT THE ITERATION COUNTERS AND CHECK FOR THE
C     MAXIMUM LIMIT OF ITERATIONS
C
20    ISTEP1=ISTEP1+1
      ISTEP2=ISTEP2+1
      TIME=TIME+DT
      IF(ISTEP1.GT.NLIM)GO TO 40
C
C     FORM THE TRIDIAGONAL SYSTEM OF EQUATIONS
C
      CALL FMTDIG(DX,DT,IL,T,TOL,A,B,C,R)
      N=IL-1
C
C     SOLVE THE TRIDIAGONAL SYSTEM OF EQUATIONS
C
      CALL TDIG(A,B,C,R,SOLN,N)
C
C     OBTAIN DESIRED SOLUTION
C
      DO 26 I=2,IL
        T(I)=SOLN(I)
26        CONTINUE
C
C     IMPOSE THE BOUNDARY CONDITIONS
C
      CALL BCOND(T,DX,DT,IL)
C
C     SAVE SOLUTION FOR NEXT TIME STEP
      DO 25 I=1,IL
        TOL(I)=T(I)
25        CONTINUE
C
C     OUTPUT THE RESULTS
C
      IF(ISTEP2.EQ.NSTEP)THEN
        WRITE(10,120)TIME
120     FORMAT(/,1X,'AT T=',F7.3,1X,'TEMPERATURE FIELD IS:')
        WRITE(10,130)(T(I),I=1,IL)
```

```
 130         FORMAT(1X,20(F8.4,2X))
             ISTEP2=0
             GO TO 20
           END IF
           GO TO 20
  40       CLOSE(UNIT=10)
           STOP
           END
C*************************************************************
           SUBROUTINE FMTDIG(DX,DT,IL,T,TOL,A,B,C,R)
C
C      THIS SUBROUTINE FORMS THE TRIDIAGONAL MATRIX FOR THE
C      CRANK-NICOLSON METHOD. THE GENERIC FORM OF THE EQUATION
C      IS:
C
C      A*T(I-1)+B*T(I)+C*T(I+1)=R
C
C
           DIMENSION T(IL),TOL(IL),A(IL),B(IL),C(IL),R(IL)
           DO 10 I=2,IL-1
             A(I-1)=-DT/(2.*DX**2)
             C(I-1)=-DT/(2.*DX**2)
  10       CONTINUE
           DO 20 I=2,IL-1
             B(I-1)=1.+DT/DX**2
             R(I-1)=TOL(I)+DT*(TOL(I+1)-2.*TOL(I)+TOL(I-1))
      $                /(2.*DX**2)
C
C      INCORPORATE THE APPROPRIATE BOUNDARY CONDITIONS:
C
C      LEFT BOUNDARY:
C
           IF(I.EQ.2) R(I-1)=R(I-1)-A(I-1)*T(I-1)
           IF(I.EQ.IL-1)THEN
           A(I)=-DT/DX**2
           B(I)=(1.+DT/DX**2)
           R(I)=(1.-DT/DX**2)*TOL(IL)+(DT/DX**2)*TOL(IL-1)
           END IF
C
C
  20       CONTINUE
           RETURN
           END
C*************************************************************
           SUBROUTINE TDIG(A,B,C,R,SOLN,N)
C
C      THIS SUBROUTINE INVERTS A TRIDIAGONAL MATRIX BY THOMAS
C      ALGORITHM.
C
C      SOLUTION IS RETURNED IN THE ARRAY CALLED 'SOLN'.
C
```

```
      DIMENSION A(N),B(N),C(N),R(N),SOLN(N)
      DO 20 I=2,N
        D=A(I)/B(I-1)
        B(I)=B(I)-C(I-1)*D
        R(I)=R(I)-R(I-1)*D
20    CONTINUE
      SOLN(N+1)=R(N)/B(N)
      DO 30 I=1,N-1
        J=N-I
        SOLN(J+1)=(R(J)-C(J)*SOLN(J+2))/B(J)
30    CONTINUE
      RETURN
      END
C***********************************************************
      SUBROUTINE BCOND(T,DX,DT,IL)
C
C     THIS SUBROUTINE IMPLEMENTS THE APPROPRIATE BOUNDARY
C     CONDITIONS
C
        DIMENSION T(IL)
C
C     LEFT BOUNDARY:
C     ISOTHERMAL; DIMENSIONLESS TEMPERATURE FIXED AT 1.0
C
         T(1)=1.
C
C     RIGHT BOUNDARY:
C     ADIABATIC
C
C     T(IL)=4.*T(IL-1)/3.-T(IL-2)/3.
C
        RETURN
        END
```

C.19

```
C     SUCCESSIVE OVER RELAXATION (SOR) METHOD FOR AN ELLIPTIC
C     PDE
C
C     THIS PROGRAM SOLVES THE LAPLACE EQUATION BY EMPLOYING
C     THE SUCCESSIVE OVER RELAXATION (SOR) ITERATION METHOD.
C
C     WHEN THE PROGRAM IS RUN IT PROMPTS FOR THE INPUT VALUES
C     REQUIRED.
C     ENTER THE INPUT VALUES AND THE OUTPUT WILL BE IN A FILE
C     CALLED 'SOR.DAT'
C
C
C     DESCRIPTION OF INPUT PARAMETERS:
```

```
C
C       IL     IS THE NUMBER OF GRID POINTS IN THE X DIRECTION.
C       JL     IS THE NUMBER OF GRID POINTS IN THE Y DIRECTION.
C       DX     IS THE GRID SIZE IN X DIRECTION.
C       DY     IS THE GRID SIZE IN Y DIRECTION.
C       OMEGA  IS THE RELAXATION PARAMETER
C       PHIINT IS THE INITIAL GUESS FOR PHI TAKEN UNIFORM OVER
               THE WHOLE DOMAIN.
C       ITMAX  IS THE NUMBER OF MAXIMUM ITERATIONS BEFORE
               STOPPING.
C       EPSI   IS THE CONVERGENCE CRITERION.
C
C
C       DESCRIPTION OF OTHER VARIABLES:
C
C       PHI    IS THE SOLUTION VARIABLE AT NTH TIME STEP.
C       PHIOL  IS THE SOLUTION VARIABLE AT N-1TH TIME STEP.
C
C
          CHARACTER*2 XFILE(5)
          CHARACTER*2 YFILE(5)
          DIMENSION PHI(21,21),PHIOL(21,21)
          PRINT*,'ENTER INITIAL GUESS FOR PHI TAKEN UNIFORM'
          PRINT*,'OVER THE WHOLE DOMAIN'
          READ(5,*)PHIINT
          PRINT*,'ENTER GRID SIZE DX=, DY='
          READ(5,*)DX,DY
          PRINT *,'ENTER NO. OF GRID POINTS IL=, JL= '
          PRINT*,' MAXIMUM POSSIBLE IS 21 FOR BOTH IL AND JL,'
          PRINT*,'UNLESS DIMENSION STATEMENTS ARE CHANGED.'
          READ(5,*)IL,JL
          PRINT *,'ENTER THE RELAXATION PARAMETER'
          READ(5,*)OMEGA
          PRINT*,'ENTER MAXIMUM NO. OF ITERATIONS ALLOWED BEFORE
       $  STOPPING'
          READ(5,*)ITMAX
          PRINT *,'ENTER CONVERGENCE CRITERION'
          READ(5,*)EPSI
          PRINT*,'THE INPUT VALUES ARE:'
          PRINT*,'INITIAL GUESS FOR PHI=',PHIINT
          PRINT*,'DX=',DX,'DY=',DY
          PRINT*,'IL=',IL,'JL=',JL
          PRINT*,'MAX NO. OF ITERATIONS=',ITMAX
          PRINT*,'CONVERGENCE CRITERION=',EPSI
          ITERATION=0
C
C       SET INITIAL DISTRIBUTION OF PHI
C
          DO 51 I=1,IL
            DO 5 J=1,JL
              PHI(I,J)=PHIINT
```

```
   5       CONTINUE
  51      CONTINUE
C
C      START SOLVING FOR PHI.
C
  15      ITERATION = ITERATION + 1
          IF(ITERATION.GE.ITMAX)GO TO 40
C
C      SAVE THE FIELD AT PREVIOUS TIME STEP.
C
          DO 101 I = 1,IL
            DO 10 J = 1,JL
              PHIOL(I,J) = PHI(I,J)
  10        CONTINUE
 101      CONTINUE
C
C      EMPLOY SOR ITERATIVE METHOD FOR PHI AT INTERIOR POINTS.
C
          DO 20 J = 2,JL - 1
            DO 20 I = 2,IL - 1
              PHIGS = (PHI(I + 1,J) + PHI(I - 1,J))/DX**2 +
     $             (PHI(I,J + 1) + PHI(I,J - 1))/DY**2
              PHIGS = PHIGS/(2./DX**2 + 2./DY**2)
              PHI(I,J) = OMEGA*PHIGS + (1.-OMEGA)*PHIOL(I,J)
  20        CONTINUE
C
C      IMPOSE THE BOUNDARY CONDITIONS
C
          CALL BCOND(PHI,IL,JL)
C
C      CHECK FOR CONVERGENCE
C
          DO 35 I = 1,IL
            DO 35 J = 1,JL
              IF(ABS(PHI(I,J) - PHIOL(I,J)).GE.EPSI)GO TO 15
  35        CONTINUE
          GO TO 50
  40      PRINT*,'SOLN. DOES NOT CONVERGE
     $    IN',ITMAX,'ITERATIONS'
  50      OPEN(UNIT = 10,FILE = 'SOR.DAT')
          WRITE(10,110)EPSI
 110      FORMAT(1X,'CONVERGENCE CRITERION ='1X,E9.1)
          WRITE(10,115)OMEGA
 115      FORMAT(//,1X,'RELAXATION PARAMETER =',F5.2)
          WRITE(10,120)ITERATION
 120      FORMAT(//,1X,'NO. OF ITERATIONS TO
     $    CONVERGE = ',1X,I4,//)
          WRITE(10,130)
 130      FORMAT(1X,'PHI DISTRIBUTION IS:',//)
          WRITE(10,140)(I,I = 1,IL)
 140      FORMAT(1X,'I = ',8X,11(I2,8X))
          DO 60 J = 1,JL
```

```
          WRITE(10,100)J,(PHI(I,J),I=1,IL)
   60   CONTINUE
  100   FORMAT(1X,'J=',I2,3X,11(F8.5,2X))
        STOP
        END
C*******************************************************
        SUBROUTINE BCOND(PHI,IL,JL)
C
C     THIS SUBROUTINE IMPLEMENTS APPROPRIATE BOUNDARY
C     CONDITIONS.
C
        DIMENSION PHI(IL,JL)
C
C     SET THE CONDITIONS ON I=1 AND I=IL SURFACES.
C
        DO 25 J=1,JL
          PHI(1,J)=0.
          PHI(IL,J)=0.
   25   CONTINUE
C
C     SET THE CONDITIONS ON J=1 AND J=JL SURFACES
C
        DO 30 I=1,IL
          PHI(I,1)=0.
          PHI(I,JL)=1.
   30   CONTINUE
        RETURN
        END
```

C.20

```
C     SOLUTION OF THE WAVE EQUATION
C
C     THIS PROGRAM SOLVES A SECOND-ORDER HYPERBOLIC PARTIAL
C     DIFFERENTIAL EQUATION BY THE FINITE DIFFERENCE METHOD.
C
C     SUBROUTINE INPUT PROVIDES THE INPUT DATA NECESSARY TO
C     RUN THE PROGRAM
C
C     DESCRIPTION OF VARIABLES:
C
C     DX IS THE GRID SIZE.
C     IL IS THE NUMBER OF GRID POINTS.
C     DT IS THE TIME STEP.
C     C IS THE COURANT NUMBER. CHOOSE DX AND DT SUCH THAT C
C     IS APPROXIMATELY 1.0.
C     U, U1, U2 CONTAIN THE U FIELD AT THE THREE TIME STEPS T,
C     T+DT AND T+2DT, WHERE U IS THE DEPENDENT VARIABLE.
```

```
C      NLIM=MAXIMUM NUMBER OF TIME STEPS BEFORE TERMINATION
C      OF THE CALCULATION.
C      NSTEP=NUMBER OF TIME STEPS AFTER WHICH PRINTOUT
       OCCURS.
C
C
        DIMENSION U(25),U1(25),U2(25)
C
C      ENTER THE INPUT VALUES
C
        CALL INPUT(DX,DT,IL,U1,NLIM,NSTEP,ASQR)
C
        OPEN(UNIT=10,FILE='HPB.DAT')
        C=ASQR*DT**2/DX**2
C
C      SET THE BOUNDARY CONDITIONS
C
        CALL BCOND(U2,IL)
C
C      INITIALIZE THE VARIABLES
C
        DO 10 I=2,IL-1
        U2(I)=U1(I)+C*(U1(I+1)-2.*U1(I)+U1(I-1))/2.
   10   CONTINUE
C
        ISTEP1=1
        ISTEP2=1
        T=DT
C
C      SOLVE FOR U AT SUCCESSIVE TIME STEPS
C      AND SAVE THE PREVIOUS VALUES
C
   30   DO 20 I=1,IL
        U(I)=U1(I)
        U1(I)=U2(I)
   20   CONTINUE
C
C      INCREMENT THE TIME
C
        T=T+DT
        ISTEP1=ISTEP1+1
        IF(ISTEP1.GT.NLIM)GO TO 50
        ISTEP2=ISTEP2+1
C
C      CALCULATE NEW 'U2'
C
        DO 40 I=2,IL-1
        U2(I)=2.*U1(I)-U(I)+C*(U1(I+1)-2.*U1(I)+U1(I-1))
   40   CONTINUE
C
C      OUTPUT THE RESULTS
```

```
C
         IF(ISTEP2.EQ.NSTEP)THEN
         WRITE(10,100)T
 100     FORMAT(//,1X,'AT TIME = ',F8.4,1X,'U FIELD IS')
         WRITE(10,110)(U2(I),I=1,IL)
 110     FORMAT(/,1X,20(F8.4,2X))
         ISTEP2 = 0
         GO TO 30
         END IF
         GO TO 30
  50     STOP
         END
C*************************************************************
         SUBROUTINE INPUT(DX,DT,IL,U1,NLIM,NSTEP,ASQR)
C
C      THIS SUBROUTINE PROVIDES THE INPUT VALUES TO THE MAIN
C      PROGRAM.
C
C      DESCRIPTION OF THE VARIABLES:
C
C      DX = GRID SIZE.
C      DT = TIME STEP.
C      IL = NUMBER OF GRID POINTS.
C      NLIM = MAXIMUM NUMBER OF TIME STEPS TO BE COMPUTED.
C      NSTEP = NUMBER OF TIME STEPS AFTER WHICH PRINTOUT
       OCCURS.
C      ASQR = CONSTANT IN THE DIFFERENTIAL EQUATION.
C      U1 CONTAINS THE INITIAL DISTRIBUTION OF THE DEPENDENT
C      VARIABLE.
C
         DIMENSION U1(25)
         DX = 0.05
         DT = 0.0015
         ASQR = 1000.
         IL = 21
         PRINT*,'ENTER MAXIMUM NUMBER OF TIME STEPS ALLOWED'
         READ(1,*)NLIM
         PRINT*,'ENTER NO. OF TIME STEPS AFTER WHICH OUTPUT
      $  OCCURS'
         READ(1,*)NSTEP
C
C      INITIAL DISTRIBUTION OF U1
C
         DO 10 I = 1,IL
         IF(I.LE.6)U1(I) = FLOAT(I - 1)*DX*0.0016
         IF(I.GT.6)U1(I) =-0.004*(FLOAT(I - 1)*DX - 1.)/0.75.
         PRINT*,U1(I)
  10     CONTINUE
         RETURN
         END
C**********************************************************
```

```
      SUBROUTINE BCOND(U2,IL)
C
C     THIS SUBROUTINE IMPOSES THE BOUNDARY CONDITIONS
C
      DIMENSION U2(25)
      U2(1)=0.
      U2(IL)=0.
      RETURN
      END
```

References

Abramowitz, M. and Stegun, I.A., Eds., *Handbook of Mathematical Functions with Formulas, Graphs and Mathematical Tables, National Bureau of Standards, Applied Mathematical Series*, US Govt. Printing Office, Washington, DC, Vol. 55, 1964.

Ahlberg, H.J., Nilson, E.N., and Walsh, J.L., *The Theory of Splines and Their Applications*, Academic Press, New York, NY, 1967.

Amazigo, J.C. and Ruhenfeld, L., *Advanced Calculus and Its Applications to the Engineering and Physical Sciences*, Wiley, New York, NY, 1980.

Anton, H., *Elementary Linear Algebra*, 10th edn, Wiley, New York, NY, 2010.

Atkinson, K., *An Introduction to Numerical Analysis*, 2nd edn, Wiley, New York, NY, 1989.

Banerjee, P.K. and Butterfield, R., *Boundary Element Method in Engineering Science*, McGraw-Hill, London, 1981.

Beer, G., Smith, I., and Duenser, C., *The Boundary Element Method with Programming: For Engineers and Scientists*, Springer, Heidelberg, 2010.

Brebbia, C.A., *The Boundary Element Method for Engineers*, 3rd edn, McGraw-Hill, London, 1977.

Brent, R., Some efficient algorithms for solving systems of nonlinear equations, *SIAM J. Num. Anal.*, 10, 327–344, 1973.

Bronson, R. and Costa, G.B., *Matrix Methods: Applied Linear Algebra*, 3rd edn, Academic Press, New York, NY, 2008.

Butcher, J.C., On Runge–Kutta processes of high order, *J. Austr. Math. Soc.*, 4, 179–194, 1964.

Carnahan, B.H., Luther, H.A., and Wilkes, J.O., *Applied Numerical Methods*, Wiley, New York, NY, 1969.

Chapra, S.C., *Applied Numerical Methods with MATLAB for Engineers and Scientists*, McGraw-Hill, New York, NY, 2005.

Chapra, S.C. and Canale, R.P., *Numerical Methods for Engineers*, 4th edn, McGraw-Hill, New York, NY, 2002.

Clocksin, W.F., *Prolog Programming for the Working Programmer*, Springer-Verlag, New York, NY, 2003.

Clocksin, W.F. and Mellish, C.S., *Programming in PROLOG: Using the ISO Standard*, 5th edn, Springer-Verlag, New York, NY, 2004.

Collatz, L., *The Numerical Treatment of Differential Equations*, 3rd edn, Springer-Verlag, Berlin, 1966.

Davis, P.J. and Rabinowitz, P., *Numerical Integration*, Ginn-Blaisdell, Waltham, MA, 1967.

Draper, N.R. and Smith, H., *Applied Regression Analysis*, 3rd edn, Wiley-Interscience, New York, NY, 1998.

Ferziger, J., *Numerical Methods for Engineering Applications*, 2nd edn, Wiley-Interscience, New York, NY, 1998.

Forsythe, G. and Wasow, W., *Finite Difference Methods for Partial Differential Equations*, Wiley, New York, NY, 1960.

Forsythe, G.E., Malcolm, M.A., and Moler, C.B., *Computer Methods for Mathematical Computations*, Prentice-Hall, Englewood Cliffs, NJ, 1977.

Fox, L., *Numerical Solution of Ordinary and Partial Differential Equations*, Pergamon Press, Oxford, 1962.

Francis, J.G.F., The QR transformation, *Comput. J.*, 4, 265–271, 1961; and 4, 332–345, 1962.

Gear, C.W., *Numerical Initial Value Problems in Ordinary Differential Equations*, Prentice-Hall, Englewood Cliffs, NJ, 1971.

Gebhart, B., *Heat Transfer*, 2nd edn, McGraw-Hill, New York, NY, 1971.

Gerald, C.F. and Wheatley, P.O., *Applied Numerical Analysis*, 7th edn, Addison-Wesley, Reading, MA, 2003.

Gilat, A., *MATLAB: An Introduction with Applications*, 3rd edn, Wiley, New York, NY, 2008.

Gill, S., A process for the step-by-step integration of differential equations in an automatic computing machine, *Proc. Camb. Philos. Soc.*, 47, 96–108, 1951.

Grama, A., Karypis, G., Kumar, V., and Gupta, A., *Introduction to Parallel Computing*, 2nd edn, Addison-Wesley, Wesley, MA, 2003.

Hall, G. and Watt, J.M., *Modern Numerical Methods for Ordinary Differential Equations*, Clarendon Press, Oxford, 1976.

Hall, T.E., Enright, W.N., Fellen, B.M., and Sedgewick, A.E., Comparing numerical methods for ordinary differential equations, *SIAM J. Num. Anal.*, 9, 603–637, 1972.

Halliday, D., Resnick, R., and Walker, J., *Fundamentals of Physics Extended*, 9th edn, Wiley, New York, NY, 2010.

Hamming, R.W., Stable predictor–corrector methods for ordinary differential equations, *J. Assoc. Comput. Mach.*, 6, 37–47, 1959.

Hornbeck, R.W., *Numerical Methods*, Prentice-Hall, Englewood Cliffs, NJ, 1982.

Householder, A.S., *The Numerical Treatment of a Single Nonlinear Equation*, McGraw-Hill, New York, NY, 1970.

Huebner, K.H., Thornton, E.A., and Byrom, T.G., *The Finite Element Method for Engineers*, 3rd edn, Wiley, New York, NY, 1995.

Incropera, F.P., Dewitt, D.P., Bergman, T.L., and Lavine, A.S., *Introduction to Heat Transfer*, 5th edn, Wiley, New York, NY, 2006.

Jaluria, Y., *Computer Methods for Engineering*, Taylor & Francis, Washington, DC, 1996.

Jaluria, Y., *Design and Optimization of Thermal Systems*, 2nd edn, CRC Press, Boca Raton, FL, 2008.

Jaluria, Y. and Torrance, K.E., *Computational Heat Transfer*, 2nd edn, Taylor & Francis, New York, NY, 2003.

James, M.L., Smith, G.M., and Wolford, J.C., *Applied Numerical Methods for Digital Computation*, 3rd edn, Harper & Row, New York, NY, 1985.

Keisler, H.J., *Elementary Calculus: An Infinitesimal Approach*, 2nd edn, Prindle, Weber & Schmidt, Boston, MA, 1986.

Keller, H.B., *Numerical Methods for Two-Point Boundary-Value Problems*, Ginn-Blaisdell, Waltham, MA, 1968.

Kernighan, B.W. and Ritchie, D.M., *C Programming Language*, 2nd edn, Prentice-Hall, Englewood Cliffs, NJ, 1988.

King, K.N., *C Programming: A Modern Approach*, 2nd edn, Norton & Co., New York, NY, 2008.

Kochan, S.G., *Programming in C*, 3rd edn, Sams, Indianapolis, IN, 2004.

Lalonde, W., *Discovering Smalltalk*, Pearson Tech., Indianapolis, IN, 2008.

Lambert, J.D., *Computational Methods in Ordinary Differential Equations*, Wiley, New York, NY, 1973.

Lancaster, P. and Tismenetsky, M., *Theory of Matrices: With Applications*, 2nd edn, Academic Press, New York, NY, 1985.

Larson, R. and Edwards, B.H., *Calculus*, 9th edn, Brooks Cole, Pacific Grove, CA, 2009.

Larson, R., Hostetler, R.P., and Edwards, B.H., *Calculus (with Analytical Geometry)*, 8th edn, Brooks Cole, Pacific Grove, CA, 2005.

Littlefield, B.L. and Hanselman, D.C., *Mastering MATLAB 7*, Prentice-Hall, Englewood Cliffs, NJ, 2005.

Matthews, J.H. and Fink, K.D., *Numerical Methods Using MATLAB*, 4th edn, Prentice-Hall, Englewood Cliffs, NJ, 2004.

Mitchell, A.R. and Wait, R., *The Finite Element Method in Partial Differential Equations*, Wiley, New York, NY, 1977.

Moore, H., *MATLAB for Engineers*, Prentice-Hall, Englewood Cliffs, NJ, 2006.

Ogata, K., *System Dynamics*, 4th edn, Prentice-Hall, Englewood Cliffs, NJ, 2003.

Ostrowsky, A.M., *Solution of Equations and Systems of Equations*, Academic Press, New York, NY, 1966.

Palm, W.J., III, *Introduction to MATLAB 7 for Engineers*, McGraw-Hill, New York, NY, 2005.

Patankar, S.V., *Numerical Heat Transfer and Fluid Flow*, Taylor & Francis, New York, NY, 1980.

Peaceman, D.W. and Rachford, H.H., The running solution of parabolic and elliptic differential equations, *J. Soc. Indust. Appl. Math.*, 3, 28–41, 1955.

Prata, S., *C++ Primer Plus*, 5th edn, Sams, Indianapolis, IN, 2005.

Ralston, A., *A First Course in Numerical Analysis*, McGraw-Hill, New York, NY, 1965.

Ralston, A. and Rabinowitz, P., *A First Course in Numerical Analysis*, 2nd edn, McGraw-Hill, New York, NY, 1978.

Rectenwald, G., *Numerical Methods with MATLAB*, Prentice-Hall, Englewood Cliffs, NJ, 2000.

Reddy, J.N. and Gartling, D.K., *The Finite Element Method in Heat Transfer and Fluid Dynamics*, 3rd edn, CRC Press, Boca Raton, FL, 2010.

Reiner, I., *Introduction to Matrix Theory and Linear Algebra*, Holt, Rinehart and Winston, New York, NY, 1971.

Reynolds, W.C. and Perkins, H.C., *Engineering Thermodynamics*, 2nd edn, McGraw-Hill, New York, NY, 1977.

Rice, J.R., *Numerical Methods, Software and Analysis*, McGraw-Hill, New York, NY, 1983.

Roache, P.J., *Computational Fluid Dynamics*, Revised Printing, Hermosa Pub., Albuquerque, NM, 1976.

Roache, P.J., *Fundamentals of Verification and Validation*, Hermosa Pub., Albuquerque, NM, 2010.

Rutishauser, H., *Solution of Eigenvalue Problems with the LR Transformation*, National Bureau of Standards, Applied Mathematical Series, US Govt. Printing Office, Washigton, DC, Vol. 49, pp. 47–81, 1958.

Salvadori, M.G. and Baron, M.L., *Numerical Methods in Engineering*, 2nd edn, Prentice-Hall, Englewood Cliffs, NJ, 1961.

Scott, L.R., Clark, T., and Bagheri, B., *Scientific Parallel Computing*, Princeton University Press, Princeton, NJ, 2005.

Shampine, L.P. and Gordon, M.K., *Computer Solution of Ordinary Differential Equations*, Freeman, San Francisco, CA, 1975.

Shanks, E.B., Solutions of differential equations by evaluations of functions, *Math. Comput.*, 20, 21–38, 1966.

Smith, G.D., *Numerical Solution of Partial Differential Equations*, 2nd edn, Oxford University Press, Oxford, 1978.

Stewart, J., *Calculus*, 6th edn, Brooks Cole, Pacific Grove, CA, 2007.

Stoecker, W.F., *Design of Thermal Systems*. 3rd edn, McGraw-Hill, New York, NY, 1989.

Stroud, A.H. and Secrest, D., *Gaussian Quadrature Formulas*, Prentice-Hall, Englewood Cliffs, NJ, 1966.

Stroustrup, B., *The C++ Programming Language: Special Edition*, Addison-Wesley, Reading, MA, 2000.

Stroustrup, B., *Programming: Principles and Practice Using C++*, Addison-Wesley, Reading, MA, 2009.

Thomas, G.B. and Finney, R.L., *Calculus and Analytic Geometry*, 9th edn, Addison-Wesley, Reading, MA, 1999.
Traub, J.F., *Iterative Methods for the Solution of Equations*, Prentice-Hall, Englewood Cliffs, NJ, 1964.
Wilkinson, J.H., *The Algebraic Eigenvalue Problem*, Oxford University Press, Oxford, 1988.
Williams, G., *Linear Algebra*, 5th edn, Jones and Bartlett Publishers, Sudbury, MA, 2004.
Winston, P.H. and Horn, B.K.P., *LISP*, 3rd edn, Addison-Wesley, Reading, MA, 1989.
Young, H.D., Freedman, R.A., Sandin, T.R., and Ford, A.L., *Sears and Zemansky's University Physics*, 10th edn, Addison-Wesley, Reading, MA, 2000.

Index

G

H

N

Q

R

U

V

W

Z